HANDBOOK OF NEUROCHEMISTRY

VOLUME I

CHEMICAL ARCHITECTURE
OF THE NERVOUS SYSTEM

HANDBOOK OF NEUROCHEMISTRY
Edited by Abel Lajtha

HANDBOOK OF NEUROCHEMISTRY

Edited by Abel Lajtha

New York State Research Institute
for Neurochemistry and Drug Addiction
Ward's Island
New York, New York

VOLUME I

CHEMICAL ARCHITECTURE
OF THE NERVOUS SYSTEM

PLENUM PRESS · NEW YORK · 1969

First Printing – May 1969

Second Printing – December 1971

Library of Congress Catalog Card Number 68-28097

ISBN 978-1-4615-7156-8   ISBN 978-1-4615-7154-4 (eBook)
DOI 10.1007/978-1-4615-7154-4

©1969 Plenum Press
Softcover reprint of the hardcover 1st edition 1969
A Division of Plenum Publishing Corporation
227 West 17th Street, New York, New York 10011

Contributors to this volume:

Harish C. Agrawal — Charing Cross Hospital Medical School, London, England (page 33)

H. S. Bachelard — Institute of Psychiatry, Department of Biochemistry, De Crespigny Park, London, England (page 25)

Samuel Bogoch — Foundation for Research on the Nervous System and Boston School of Medicine, Boston, Massachusetts (page 75)

Eric G. Brunngraber — Illinois State Psychiatric Institute, Chicago, Illinois (page 223)

Jørgen Clausen — The Neurochemical Institute, Copenhagen, Denmark (page 273)

R. R. Fritz — Department of Pediatrics, The University of Texas Medical Branch, Galveston, Texas (page 101)

R. Fumagalli — Institute of Pharmacology, University of Milan, Milan, Italy (page 195)

E. Grossi-Paoletti — Institute of Pharmacology, University of Milan, Milan, Italy (page 195)

Williamina A. Himwich — Thudichum Psychiatric Research Laboratory, Galesburg State Research Hospital, Galesburg, Illinois (page 33)

Bent Kofod — Psychochemical Laboratory, University Clinic of Psychiatry, Rigshospitalet, Copenhagen, Denmark (page 261)

Richard U. Margolis — Department of Pharmacology, New York University School of Medicine, New York, New York (page 245)

Lewis C. Mokrasch — Research Laboratory, McLean Hospital, Belmont, Massachusetts (page 171)

Blake W. Moore — Department of Psychiatry, Washington University School of Medicine, St. Louis, Missouri (page 93)

J. L. Myers — Department of Pediatrics, The University of Texas Medical Branch, Galveston, Texas (page 101)

R. Paoletti — Institute of Pharmacology, University of Milan, Milan, Italy (page 195)

John J. Pisano — Laboratory of Clinical Biochemistry, National Heart Institute, National Institutes of Health, Bethesda, Maryland (page 53)

Ole J. Rafaelsen — Psychochemical Laboratory, University Clinic of Psychiatry, Rigshospitalet, Copenhagen, Denmark (page 261)

D. A. Rappoport — Department of Pediatrics, The University of Texas Medical Branch, Galveston, Texas (page 101)

George Rouser — Section of Lipid Research, Division of Neurosciences, City of Hope Medical Center, Duarte, California (page 121)

William Sacks — Research Center, Rockland State Hospital, Orangeburg, New York (page 301)

N. Seiler — Max Planck Institut für Hirnforschung, Frankfurt am Main, Germany (page 325)

Donald B. Tower — Laboratory of Neurochemistry, National Institute of Neurological Diseases and Blindness, Bethesda, Maryland (page 1)

Akira Yamamoto — Section of Lipid Research, Division of Neurosciences, City of Hope Medical Center, Duarte, California (page 121)

PREFACE

Life, either as we think of it in the abstract in its highest sense, or life, as we think of it in terms of a compact living organism, is obviously the result of complex interaction of all of the components of the organism. One could therefore question the advisability of separating out the nervous system for a special detailed study in our age of overspecialization. The main purpose of the present *Handbook* is not to fragment further our approach or understanding of living phenomena, but, on the contrary, to try to summarize and integrate as much of the available information and thinking on the nervous system as is possible in a limited space. It is difficult to think of an area of modern biology that is more exciting to study and that has greater importance for mankind, from any point of view, than the study of the brain and of the nervous system. The influence that understanding of brain function in biological terms can exert on our future is not generally understood in its full impact.

Although our ignorance about even the most basic mechanisms in the nervous system is enormous, in recent years our knowledge has made most important advances, and as a consequence great masses of data have been accumulated. A gathering and evaluation of such data can put an insurmountable burden on the individual researcher trying to study the nervous system. The purpose of the *Handbook* is to be of service to those who are interested in the nervous system, by very briefly and critically summarizing the available information that can serve as a starting point for acquiring information on the function of the nervous system. Because of the size of the body of available knowledge, it is very difficult to summarize adequately all areas and still stay within any reasonable limitation of size, which is obviously necessary if such a book is to be easily available to the individual researcher. Even though the discussion was kept as much as possible to the neurochemical approaches, and the editor's pencil was sometimes very cruelly used, what was planned originally to be a tree, at its final stage, has become more a forest. Although similar chapters were grouped, if possible, in the same volume, many, if not most, subjects reach across artificial limits of strict classification just as any aspect of brain research reaches across many disciplines. The first volume deals more with the basic composition of the nervous system, but it is just as impossible to discuss composition without discussing metabolism as it is, in a later volume, to discuss metabolism without mentioning composition. Here, as in other fields, separations are more often than not arbitrary.

Today, perhaps more than ever, services and the time of the expert scientist are taken up by his numerous duties, and the editor must also here express his thanks to the authors who, with considerable personal sacrifice, contributed so many excellent chapters. It happened only rarely that a chapter had to be omitted because its author was unable to complete it in time, or an article could not be further improved because of time limitations. While the excellence of the many articles is attributable entirely to the authors, some of the faults may well be due to the limitations in space and approach imposed upon them by the editor. Since the main purpose of the *Handbook* is to be a help in the laboratory, in the clinic, and in the classroom, it was felt that it should be readily available and its contents should be easily found. These requirements necessitated that the book's total weight be not too great for one person to lift up, and the Procrustean scissors had to be employed often. Not only can the arrangement of chapters into volumes be rather arbitrary, but also the decision of what to include or exclude is at times capricious. There is of course no purely neurochemical approach to brain function that can ignore the exciting results or methods of pharmacology, morphology, physiology, and behavior, to name only a few, and employing scissors or other limitations was more often than not a painful decision.

Recent findings clearly established that although the brain is an incredibly complex dynamic organism it is not beyond our understanding in biological terms. Most importantly, we now begin to see connections between biology and brain function, and the connections do not seem to be beyond our understanding. If with its various chapters the *Handbook* offers help in designing new approaches to the understanding of the function of the nervous system in biological terms, the book served its purpose well.

Abel Lajtha

New York, New York
November 1968

CONTENTS

Chapter 5

Proteins 75
by Samuel Bogoch

Chapter 6

Acidic Proteins 93
by Blake W. Moore

Chapter 7

Chapter 8

Chapter 11

Chapter 12

CONTENTS OF VOLUME II: Structural Neurochemistry

CONTENTS OF VOLUME III: Metabolic Reactions in the Nervous System

FURTHER VOLUMES

Chapter 1

INORGANIC CONSTITUENTS

Donald B. Tower

Laboratory of Neurochemistry
National Institute of Neurological Diseases and Blindness
Bethesda, Maryland

I. GROSS COMPOSITION

A. Ash

The proportion of the total mass of neural tissues contributed by the inorganic constituents can hardly be taken as an indication of their functional importance. This is characteristic of a number of functionally important constituents of brain and nerve, including oxygen and glucose. Upon incineration, whole brain is reduced to approximately 1.1 % of its original wet weight.[1] This residual ash represents the nonvolatile inorganic constituents of brain, and in terms of brain solids, or dry weight, accounts for only about 5 % of the total (Table I). The principal components of the ash are the four commonest alkali metals, Na, K, Ca, and Mg. The latter two are known to exist in tissues primarily bound or complexed to lipids or proteins, but some of the Na and K may also be similarly sequestered.[2–7] Total proteins isolated from mammalian cerebral cortex (and representing some 8–10 % of the initial fresh tissue weight) exhibit an ash content amounting to 0.3 % of the isolated, dry protein.[7] Lipid-bound monovalent cations in cerebral cortex constitute

TABLE I

Brain Ash[a]

Source	Dry wt. (%)	Percentage of	
		Dry wt.	Wet wt.
Whole brain	23.0	4.8	1.10
Gray matter	16.0	6.1	0.98
White matter	30.0	4.2	1.26

[a] Data are derived from Tower.[1]

only 1–2 % of the total content of Na and K but are present at much higher levels in lipids of subcortical white matter; for example, in corpus callosum 14 % of the total Na and 5 % of the total K may be lipid-bound.[6]

B. Major Constituents

In terms of inorganic constituents, some basic data on the composition of neural tissues are summarized in Tables II (major constituents) and III (trace metals). These data have been selected for purposes of illustration and are reasonably representative. For more extensive consideration of major electrolytes, reviews by McIlwain,[8] Manery,[16] and Van Harreveld[17] should be consulted. Trace metals are covered in more detail by Cumings,[13,14] especially with respect to alterations associated with pathological conditions.

A number of studies have established that neural gray matter of most mammalian species contains approximately 100 μmoles of K and 40–60 μmoles of Na per gram of fresh tissue. Information on subcortical white matter is too incomplete to warrant any comparable generalization; values given in Table II must be considered tentative. Reports for white matter based on analyses of brains of small mammalian species are subject to some uncertainties since it is difficult to obtain pure samples of white matter in sufficient yield and free of manipulative artifacts for satisfactory analysis. Corpus callosum from brains of larger species is a suitable source[6] but may not necessarily be representative of all white matter.

There are few reliable data for the Ca and Mg contents of neural tissues because of the lack, until recently, of suitable analytical methods. With the advent of atomic absorption flame spectrophotometric analysis,[18–20] this situation will change, and reports based on use of this technique are now beginning to appear.[10,11] However, such analyses (particularly for Ca and Fe) are subject to many technical problems, so that some discrepancies between various studies are already apparent.[10,11,21] The values for Ca and Mg given in Table II are considered correct but may be subject to modification as further work is reported. On the basis of available data, it can be stated

TABLE II

Major Inorganic Constituents of Brain[a]

Type of sample	(μmoles/g fresh tissue)					
	Na	K	Ca	Mg	Cl	HCO$_3$
Cerebral cortex (cat)	57	95	2.2	5.1	45 ⎫	12
Corpus callosum (cat)	49	73	2.2	5.1	36 ⎭	
CSF (cat)	141	3	1.2	1.2	130	21
Neonatal cerebral cortex (kitten)	68	68	2.3	4.4	44	–

[a] Data are derived from Bourke and Tower,[5] Tower and Bourke,[6] McIlwain,[8,9] and Tower.[1,10]

TABLE III

Trace Metals in Brain[a]

Source of sample	μmoles/g fresh tissue			
	Cu	Mn	Zn	Fe
Cerebral cortex (rabbit)	0.050	0.023	0.25	0.45
Cerebral cortex (human)	0.080 [0.46][b]	–	–	0.75 [6.95][b]
Cerebral white (human)	0.057 [0.17][b]	–	–	0.75 [5.95][b]
Substantia nigra (human)	– [0.94][b]	–	–	3.3 [20.2][b,c]
Locus caeruleus (human)	– [3.16][b]	–	–	–
Cerebrospinal fluid (human)	0.004	–	–	–

[a] Data are derived from Hanig and Aprison,[11] Thompson,[12] Cumings,[13,14] and Hallgren and Sourander.[15]
[b] Values in brackets are in terms of μmoles/g dry wt.
[c] Values are for globus pallidus.

that the total cation content of neural gray matter amounts to 165–170 μeq (microequivalents) per gram of fresh tissue.

It is well recognized that the total anionic equivalent provided by chloride, bicarbonate, sulfates, phosphates, and organic acids (especially amino acids such as N-acetylaspartic acid) can, at most, account for only about one-half the necessary total. The deficit is presumed to be covered by the "fixed" anions of cellular lipids and proteins.[2,8,9,16] Since the majority of the chloride appears to be extracellular[5,6,23,24] (as discussed below), the major intracellular anions must then be organic and macromolecular in nature. Phylogenetically[22] there are some variations in this respect. In nerves from a number of invertebrate species, the principal intracellular anions are either aspartic and glutamic acids (crustacean nerve)[25] or isethionic acid (2-hydroxyethanesulfonate; in squid axon),[26] whereas the mammalian nervous system relies primarily on cellular lipids and proteins to provide the necessary additional anions.

With respect to chloride, there is a significant correlation of the content of Cl in cerebral cortex with species brain weight.[24] For example, the cortical Cl contents for representative species are the following (in micromoles per gram of fresh tissue): rat (brain wt. 1.6 g), 32.4; cat (brain wt. 31 g), 44.7; and beef (brain wt. 380 g), 55.2. The full significance of this correlation remains to be elucidated, but it appears to relate in part to cortical extracellular space (as discussed below). Most methods for determining Cl involve titration with Ag^+ and can yield falsely high values for Cl in tissues unless tissue sulfhydryl groups are oxidized prior to titration.[24,27] Failure to appreciate this factor can obscure the small but significant differences among species cited above.

Ontogenetic or developmental aspects of these subjects have not been sufficiently studied. For species such as the cat, where the central nervous system at birth is highly immature, some very striking changes in the contents of Na and K occur during cortical (and probably neuronal) maturation (Table II).[6] Concomitant changes of chloride content also occur.[6,28–31] Some of the potentialities for correlating such observations with morphological and functional parameters are discussed for kitten cerebral cortex in a paper by Tower and Bourke.[6]

C. Trace Metals

Only fragmentary data are available for levels and distribution of trace metals in neural tissues. The values cited in Table III provide an idea of the orders of magnitude demonstrable and of certain striking regional differences for the four most prevalent trace metals. As the atomic absorption method[20] is exploited more fully, more data on more elements and more reliable values can be expected. The functional significance of the relatively very high level of Cu in the locus cæruleus or of Fe in some of the basal ganglia is still unknown. Most, if not all, of the cerebral contents of these trace metals are in association with proteins or lipids. Approximately one-third of the cerebral Fe occurs in a protein-bound form similar to ferritin,[15] and only a minor percentage of the tissue total is contributed by the functionally important porphyrins such as the cytochromes which are engaged in oxidative phosphorylation.[9] Normally 40–60% of the cerebral Cu exists in the form of cerebrocuprein, specific copper-containing proteins[32,33] which differ from the ceruloplasmin of blood plasma and from the Cu-complexes responsible for elevated cerebral levels associated with hepatolenticular degeneration (Wilson's disease).[13,33]

No attempt is made here to cover pathological situations with any degree of completeness. For detailed consideration of certain types of heavy-metal poisoning, the reviews by Cumings[13,14] should be consulted. There are undoubtedly many aspects of the cerebral and nerve pathology of trace inorganic constituents still to be investigated. A few selected examples indicate some of the interesting facets of the general subject. An important factor may be the form in which the element is exhibited to neural tissue. Cumings[13] has discussed the differences in symptomatology and sequelae between inorganic Pb poisoning and poisoning from tetraethyl-Pb. Quastel and colleagues[34] pointed out the interesting relationship between penetrance into the central nervous system (or toxicity therein) and the valence of As in arsenical drugs administered systemically. An explanation for the sedative and toxic effects of Br on the nervous system is still awaited after more than a century of clinical use. The mechanisms by which heavy metals exert their toxicity and the relevance of therapies with such chelating agents as versene, BAL, and penicillamine are by no means clear.[13] Finally there is the interesting paradox posed by Li, which has recently come into vogue as a therapy for manic-depressive psychoses,[35,36] but which, if given in excess (especially in conjunction with a low Na intake) proves exceedingly toxic to the nervous system.[36–38] The toxicity is susceptible of explanation on the basis of tissue[39] and intracellular[5,40] accumulation of Li at the expense of Na and

K, since neural and muscle cells are unable to exclude passively accumulated Li by means of the Na pump.[5,44] The therapeutic benefits conferred by lesser concentrations of Li still require explanation.

II. FLUID COMPARTMENTATION AND ELECTROLYTE DISTRIBUTION *IN VIVO*

A. Interstitial Spaces

From the functional standpoint, data expressed as content per unit mass are of relatively little value. A comparison of the gross composition of neural tissue with that of the cerebrospinal fluid bathing it (Table II) illustrates this point. The electrolyte composition of CSF is very similar to that of blood plasma except for the modifications imposed by the much lower protein content and by a Cl content significantly higher than would be expected on the basis of a simple Gibbs–Donnan equilibrium with blood.[24,41] Recent studies by Cserr,[42] Katzman *et al.*[43] Pappenheimer *et al.*,[44] and Bradbury and Davson[45] demonstrate identity of composition of CSF and extracellular fluid of brain tissue for most of the major electrolytes. Therefore, if one could determine the size of the extracellular compartment of the neural tissue in question, a first approximation of the electrolyte composition of its intracellular compartments could be calculated.

Unlike skeletal muscle, for which there is a consensus that the fluid space of the tissue measured by Cl, sucrose, and inulin provides a close approximation of the extracellular compartment of muscle,[46–49] there is still a wide divergence of opinion about the analogous situation in brain. A detailed examination of this subject is beyond the scope of this chapter but may be found in discussions published elsewhere.[17,23,24,50] Suffice it to say here that the controversy has centered on the apparent discrepancy between biochemical or physiological estimates (based upon solute indicators like inulin or on impedance measurements) and estimates derived from electron micrographs of brain tissue. Although seemingly ignored or minimized by most electron microscopists, there are artifacts of fixation which can significantly influence the preservation of the morphology of interstitial spaces in electron micrographs of brain.[50–55] Recent electron microscope studies by Van Harreveld and co-workers[54,55] of cerebellar cortex, which were done by freeze-substitution techniques, have shown much larger interstitial spaces than are visualized by usual techniques. In addition, Brightman[56] by electron microscopy techniques has demonstrated the extracellular penetration of the large macromolecule, ferritin (approximate mol. wt. 750,000 and diameter about 100 Å), deep into the parenchyma of cerebral tissue.[56] Finally, the apparent demonstration of minimal ($<5\%$) extracellular spaces in brain by the use of systemically administered sulfate[57] and thiocyanate[58] has now been invalidated by studies which clearly indicate outward transport of these solutes from CSF to blood,[59–64] so that tissue space calculations referred to plasma levels of these solutes[57,58] will grossly underestimate the size of the tissue space.[24,60,61;65,66]

Thus, the preponderance of evidence is now shifting to favor the presence in significant proportions of an extracellular compartment in brain tissue. In view of the extensive experiences with inulin as a solute indicator of extracellular spaces in a variety of other tissues, values for brain tissue obtained under appropriate conditions with inulin are probably most nearly representative of the true situation *in situ* and *in vivo*. Following ventriculocisternal perfusion with constant concentrations of inulin-carboxyl-^{14}C, Rall *et al.*[67] obtained values for inulin spaces in periventricular structures of dog brain averaging 10–13% of the fresh weight of tissue. Inulin spaces of the same order of magnitude have been reported for rat brain by Woodbury *et al.*[65,66] More recently the studies by Bourke *et al.*[24] indicate that for cerebral cortex *in vivo* the size of the extracellular compartment varies as a function of species brain size. For species in which direct measurements could be made, this relationship is illustrated by the following examples of inulin or sucrose spaces or both; guinea pig (brain wt. 4.75 g), 19.4%; cat (brain wt. 31 g), 28.0%; and chimpanzee (brain wt. 380 g), 39.5%, respectively, of fresh tissue weights. By extrapolation, values of 8.5% for mouse and 14.5% for rat cerebral cortex could be predicted.[24]

Since Cl is characteristically an extracellular ion, it is not surprising that Bourke *et al.*[24] also found that Cl or thiocyanate spaces, or both, of cerebral cortex vary as a function of species brain size, thus providing a logical explanation for the differences among species in cortical Cl content, as already noted. The tissue spaces accessible to Cl or thiocyanate were consistently found to be slightly but significantly larger than those accessible to inulin or sucrose in the same species and in the same experiments (Table IV). Presumably this means that a fraction of the total tissue content of Cl is intracellular, as might be expected, if Cl is in electrochemical equilibrium with the membrane potential of neural cells.[68] The significance of these interspecies variations and of Cl distribution in neural tissues is not at all clear. Further investigation of these aspects is needed.

One important principle has emerged from the numerous studies devoted to these general problems. It is now clear that some of the phenomena attributed to the blood–brain "barrier" reflect for a variety of solutes transport systems oriented in the direction of CSF to blood and hence effecting, for practical purposes, exclusion of the solute from CNS compartments. Evidence is now available for such mechanisms of outward transport of sulfate, [59,60] thiocyanate,[61–64] iodide and bromide,[62,63,69,70a] and certain organic acids.[63,71] There has been a suggestion that Cl may be transported from blood to CSF,[72] and unpublished studies by D. M. Woodbury confirm this suggestion. Using 5 mM perchlorate ion as a competitive inhibitor of the Cl transport system, he observed significant decreases of CSF and brain tissue Cl levels in the face of unchanged serum levels. This transport system may represent the reverse of the outward transport system for Br, thiocyanate, and perchlorate. The constancy of K concentrations in CSF also represents

TABLE IV

Fluid Spaces in Cerebral Cortex Accessible *in Vivo* to Various Solute Indicators as Functions of Species Brain Size[a]

Species		μmoles/g or % of fresh tissue weight		
		Chloride content	Chloride or thiocyanate space	Inulin or sucrose space
Mouse	[0.4 g][b]	26.5	(17.0)[c]	(8.5)[c]
Rat	[1.6 g]	32.4	(23.0)	(14.5)
Guinea pig	[4.75 g]	36.9	27.5	19.4
Rabbit	[9.3 g]	39.7	30.1	22.5
Cat	[31 g]	44.7	34.9	28.0
Monkey	[88.5 g]	49.1	39.1	32.8
Beef	[380 g]	55.2	(44.8)	(39.5)
Chimpanzee	[380 g]	(55.2)	(44.8)	39.5

[a] Data are taken from Bourke *et al.*[24] The original report should be consulted for details, including evidence supporting significance of the differences in values between species. In these studies the space values for Cl and thiocyanate were identical, and those for sucrose and inulin were identical in a given species, but the differences between the two sets of values in a given species are statistically significant.
[b] Average species brain wt. given in brackets.
[c] Extrapolated values are indicated by parentheses.

control by such transport mechanisms.[42–45] In addition to frank transport systems, the distributions of solutes (and hence spaces calculated from them) can be influenced by diffusional factors, as illustrated by recent studies by Oldendorf and Davson[70b] on the "sink" phenomenon. In order to minimize distortions introduced by such factors, Davson believes that the solute in question should be held at constant levels in plasma and CSF (by perfusion) so that true tissue levels can be achieved and, hence, the true extent of solute distribution in the tissue can be calculated.[70b] In view of the various factors discussed briefly here, additional studies on solute distributions and fluid spaces of neural tissues would seem to be essential for verification of the values currently proposed for extracellular space in brain (Table IV).

B. Intracellular Compartments

As suggested above, fluid space values obtained under appropriate conditions with inulin as indicator solute are probably most nearly representative of brain extracellular spaces. On the basis of such measurements, the concentrations *in vivo* of the principal tissue electrolytes in intracellular compartments can be calculated for cerebral cortex (Table V). Comparable data for white matter cannot be derived with any degree of confidence. There is no

TABLE V

Electrolyte Concentrations in Intracellular Compartments of Cat Cerebral Cortex *in Vivo*

Measurements	Observed (tissue content)	Calculated[a] (intracellular conc.)
Solids (dry wt. %)	16.0	–
Inulin space (%)	27.6	0
Chloride (μmoles/g)[b]	44.9	9.0
Potassium (μmoles/g)[b]	94.5	166.0
Sodium (μmoles/g)[b]	57.4	18.0
Calcium (μmoles/g)[b]	2.2	[3.35][c]
Magnesium (μmoles/g)[b]	5.1	[8.5][c]

[a] For details, consult Bourke and Tower.[5] Concentrations of electrolytes in the extracellular compartment have been assumed to be the same as those in CSF (Table II).

[b] Intracellular concentrations are expressed as μmoles/ml of intracellular space.

[c] Most of the tissue Ca and Mg is bound, so that concentrations in intracellular fluids are likely to be fractions of these theoretical maxima.

direct information available for the size of the inulin space of white matter *in vivo*. Furthermore, the likelihood that significant fractions of the K and Na in white matter are lipid-bound would make calculations analogous to those used for cerebral cortex subject to considerable uncertainty.

The calculated intracellular concentrations of cortical electrolytes take no account of the heterogeneity of the intracellular "space." Waelsch[73] has discussed this problem in terms of compartmentation in the nervous system, considering it at the organ, tissue, cellular, and subcellular levels. Despite familiarity with this organization in morphological and physiological terms, the factors of compartmentation (or tissue and cellular heterogeneity) tend to be overlooked when dealing with electrolytes and fluid spaces. Differences between cerebral cortex and subcortical white matter might be anticipated, but very little attention has been devoted to such investigations or to the possibility of differences between functional areas of the brain or within a given subdivision, such as layers of cerebral cortex. An example illustrative of the latter is shown in Table VI, taken from studies by Lowry *et al.*[74] on Ammon's horn of rabbit cerebral cortex.

Six major cell types presumably contribute to the measured intracellular compartment of cerebral cortex, and no valid estimates are available to indicate the relative contributions from each. Kuffler *et al.*[75,76] have strong evidence indicating that glial cells in the nervous system of the leech and amphibia (*Necturus* and frog) are high K cells and, hence, do not contribute to the extracellular fluid space of the nervous system in these species.

TABLE VI

Distribution of Chloride in Layers of Cerebral Cortex (Ammon's Horn of Rabbit)[a]

Cortical layer	Thickness (μ)	Dry wt. (%)	Chloride (μmoles/g)
Moleculare	200	20.6	39.5
Lacunosum	150	27.1	49.7
Radiatum	400	23.3	50.3
Pyramidale	50	17.0	40.4
Oriens	250	20.4	38.8
Alveus	200	30.3	59.1

[a] Data are derived from Lowry et al.[74] The layers M and R consist mostly of dendrites; L and A are myelinated; P contains all the neuronal cell bodies; and O consists of axons and dendrites.

Whether or not these observations apply to neural tissues of mammalian species remains to be determined. The intracellular compartment of each cell is subdivided into several major subdivisions: nuclei, mitochondria, endoplasmic reticulum, nonparticulate cytoplasm, etc., and it is only reasonable to expect that there may be differences in electrolyte composition among these various intracellular subdivisions. For example, there is evidence to suggest that 10–15% of the cortical K comprises a "slowly exchangeable" fraction,[5,77–79] which may be mitochondrial in location.[79,80] In common with mitochondria from most tissues, brain mitochondria can concentrate Ca under appropriate conditions.[10,81–84] Despite these many uncertainties, resting transmembrane potentials calculated by the Nernst equation[68,85] and data in Table V yield a value, -69 mV,[5] which is of the same order of magnitude as reported for direct measurements *in vivo* by microelectrodes inserted into neurons of cat cerebral cortex *in situ*.[86] Much more information is needed before the true characteristics of the various components of the gross intracellular "space" can be defined. The most immediate need is for methods which will effect separation and isolation of the individual cortical cell types in good, pure yield and with reasonably intact compositional and metabolic characteristics. With regard to electrolytes, the currently available separation methods[87] do not appear to satisfy such stipulations.

III. ELECTROLYTE METABOLISM *IN VITRO*

A. Characteristics of Incubated Brain Slices

Many of the problems associated with brain electrolytes can be studied with incubated slices *in vitro*, where experimental conditions can be rigidly controlled and the effects of variables can be evaluated individually. However, most of the problems already discussed for studies *in vivo* also apply *in vitro* and, in addition, several other important problems peculiar to the *in vitro* situation are encountered. A detailed consideration of these problems is beyond the scope of this chapter; extensive discussions are available in recent papers by Varon and McIlwain,[88] Bachelard *et al.*,[89] Keesey *et al.*,[90] and Bourke and Tower.[5,6,23]

The principal differences or complications associated with studies *in vitro* are: (1) the occurrence of swelling (fluid imbibition) upon immersion and upon incubation of tissue slices; (2) initial outward leakage of K and inward leakage of Na and Cl in slices immersed in incubation media and incomplete or no reversal of this situation under optimal incubation conditions; and (3) markedly increased size of tissue spaces accessible to Cl. These and other characteristics of incubated slices of neural tissues are illustrated by selected examples in Table VII. Under optimal conditions, the swelling of incubated slices can be minimized[23,88,90] and at least some of the abnormalities associated with Cl may be obviated.[5,23] But the most serious drawback to use of incubated slices for studies of electrolyte metabolism *in vitro* is the inability to devise conditions which will permit the slices to achieve and maintain levels and behavior of electrolytes characteristic of the situation *in vivo* (compare Tables II and VII). A solution to these problems is badly

TABLE VII

Electrolytes in Incubated Slices of Cat Cerebral Cortex and Corpus Callosum[a]

Corpus callosum incubated 1 hr	Observations	Cerebral cortex			
			Incubated 1 hr		
		Immersed 5 min, 22°C	5 mM K medium	Anaerobic	Aerobic + ouabain[b]
28.3	Dry weight (%)	–	15.7	–	–
6.7	Swelling (%)	15.5	16.8	44.4	30.1
24.4	Inulin space (%)	–	47.5	46.6	51.2
47.3	Chloride space (%)	62.0	62.2	95.7	104.3
38.9	Slice K (μmoles/g)	69.3	74.9	27.1	28.4
77.4	Slice Na (μmoles/g)	103.7	95.4	166.3	132.8
58.0	Slice Cl (μmoles/g)	77.5	78.5	121.5	120.0
4.6	Slice Mg (μmoles/g)	–	4.8	–	4.85
3.0	Slice Ca (μmoles/g)	–	3.9	–	4.8

[a] Data are taken from studies reported by Bourke and Tower[5,6,10,23] and are all referred to initial fresh weight of slices. All incubations were carried out aerobically (unless otherwise noted) for 1 hr at 37°C in bicarbonate–saline–glucose media containing 5 mM K^+. For comparison with biopsy data, consult Table II.

[b] Ouabain concentration in the incubation medium was 10^{-5} M.

needed. In one approach, Ames and Nesbett[91] have utilized whole retina, but it does not appear to be entirely comparable to brain.

Nevertheless, slices of cerebral cortex, incubated *in vitro*, under optimal conditions, exhibit the most important characteristics of electrolyte metabolism. Li and McIlwain[92] were the first to demonstrate with microelectrodes the maintenance *in vitro* (in cells of incubating slices of guinea pig cerebral cortex) of resting membrane potentials of the order of -55 to -60 mV, values very close to those observed *in situ in vivo*.[86,92] In addition, injury potentials were occasionally observed, indicating that for such cells the membrane was still excitable. These studies have been continued so that now Gibson and McIlwain[68] have been able to correlate potential differences across cortical cell membranes as calculated with the Nernst equations from electrolyte analyses with potential differences measured directly in the same incubated slices by microelectrodes. These experiments demonstrated the effects of increased external K or of applied electrical pulses in depolarizing the cortical cells, as well as the ability, after removal of such stimuli, to repolarize—all *in vitro*. Most recently, Yamamoto and McIlwain[93] have succeeded in obtaining action potentials and conducted impulses under the same experimental conditions while incubating preparations of olfactory (pyriform) cortex *in vitro*. With the further development and refinement of these techniques, applications to studies of brain electrolyte metabolism *in vitro* should prove to be extraordinarily fruitful.

B. Cation Transport

The two most important characteristics of electrolyte metabolism in neural tissues, dependence upon continuing generation of metabolically derived energy and dependence upon operation of the transmembrane cation transport (or "pump") systems, are indicated in summary form in Table VII. The effects of incubation under anaerobic conditions exemplify the consequences of failure or deprivation of cellular energy supplies, viz., pronounced slice swelling (presumably intracellular), marked depletion of slice K, and marked increase of slice Na and Cl, so that such tissue samples exhibit an electrolyte composition essentially similar to that of the surrounding incubation media. Such experiments clearly show that brain slices incubated under more optimal conditions are indeed capable of maintaining electrolyte compositions against steep concentration gradients, albeit not as effectively as they are maintained *in vivo*. As discussed elsewhere in this *Handbook*,[94] it is now generally thought that cellular energy in the form of ATP, necessary to maintain high intracellular K and low intracellular Na against the concentration gradients across the nerve cell membrane, is utilized in a membrane transport or "pump" mechanism (usually called the Na pump) by means of the Mg-dependent, Na–K-activated ATPase. Part of the argument in favor of this viewpoint derives from the effects of the cardiac glycoside, ouabain (G-strophanthin), which inhibits both the ATPase activity and the transport of Na and K at similar concentrations (as discussed below). Hence, it is not surprising to find that inhibitory concentrations of ouabain affect electrolyte metabolism in incubated cortical slices in a manner very similar to anaerobiosis (Table VII)—the one interfering with energy utilization, the other with its production.

Further ramifications of these aspects are provided by studies in which other ions have been substituted for one of the principal ions of the incubation media (Table VIII). In the context of these experiments, Rb appears to be an effective substitute for K (in conformity with studies on other tissues), whereas choline and especially Li are poor substitutes since each accumulates within the incubated slices at the expense of both Na and K and the permeability characteristics of the slices toward Cl are markedly altered. The effects of the latter are in part related to the inability of neural cells to extrude or "pump out" the choline or Li which has entered the cells passively.[5] These observations illustrate the difficulty, if not impossibility, of finding an effective substitute for Na. Little or no effect on electrolyte metabolism of incubated slices occurs when Ca or Mg is omitted from the incubation medium, although it is well known that the omission of Ca is associated with stimulation of slice oxygen consumption (as discussed below).

Substitutions of other anions for chloride or bicarbonate, or both, in incubation media for brain slices are more common and generally quite innocuous. A variety of buffers can be used in place of bicarbonate, the commonest substitutes being phosphate or tris. Thomas and McIlwain[95] have shown that virtually all the Cl in incubated slices of cerebral cortex is diffusible and could be replaced by nitrate or acetate (the latter exhibiting some toxic effects). Bourke and Tower[5] confirmed this observation utilizing isethionate (2-hydroxyethanesulfonate) as the replacement anion. In this latter case, the lack of untoward effects on the slices is interesting in view of the fact that the distribution of the isethionate ion in the slices is apparently

TABLE VIII

Effects of Substitutions in the Incubation Medium on Electrolytes of Slices of Cat Cerebral Cortex in Vitro[a]

Control slices (27 mM K medium)	Observations	Substitutions			
		Rb/K	Li/Na	Choline/Na	Isethionate/Cl
32.7	Swelling (%)	33.3	36.5	24.1	14.4
99.9	Slice K (μmoles/g)	17.2	59.3	83.3	105.5
75.3	Slice Na (μmoles/g)	79.4	19.3	20.1	61.2
80.1	Slice Cl (μmoles/g)	73.2	104.9	90.1	15.5
—	[Substitute] (μmoles/g)	[86.4]	[117.4]	[~75]	[~35]

[a] Data are taken from Bourke and Tower[5] and are all referred to initial fresh weights of slices. All incubations were carried out aerobically for 1 hr at 37°C in bicarbonate–saline–glucose media containing 27 mM K^+.

restricted to the inulin or true extracellular space.[5] Thus, under otherwise identical conditions, the inulin space of incubated slices of cat cerebral cortex is 40.5%, isethionate space 43.3%, and chloride space(s) 64.1%. Furthermore, the swelling of incubated slices is markedly reduced when incubated in isethionate media (Table VIII).

The potentialities of the *in vitro* incubated slice technique for studies with radio-Na, K, Cl, etc., have only recently been demonstrated.[5,8,90] The most extensive data have been reported by McIlwain *et al.*[90] in experiments on slices of guinea pig cerebral cortex incubated *in vitro* with radio-Na. On the basis of these studies, the average rate of Na turnover in unstimulated slices was 175–275 μmoles/g/hr with an increase to 1050–1180 μmoles/g/hr during stimulation by electrical pulses. The stimulated rate represented a turnover of one-third of the Na in the non-inulin spaces per minute, and calculations suggested that about 25% of the energy available from slice respiration was required for Na and K transport at maximal rates in both unstimulated and stimulated preparations.

IV. EFFECTS OF OUABAIN

A. Cation Transport and Energy Metabolism

The use of relatively specific inhibitors of electrolyte transport, such as ouabain (G-strophanthin), serves to uncover many unsuspected facets of electrolyte metabolism. Reports of newly characterized inhibitors like tetrodotoxin, which appears to block Na conductance selectively,[96–98] or valinomycin, which appears to enhance K permeability selectively,[99–100] indicate that an increasing number of such tools are becoming available. Hence, it is useful to consider briefly the prototype represented by ouabain.

There are three well-recognized effects of ouabain on neural tissues (Table IX).[5] At concentrations of approximately 10^{-6} M, ouabain produces a 50% inhibition of maintenance of K and extrusion of Na in incubated slices of cerebral cortex, and higher concentrations (10^{-5} M or above) are completely inhibitory.[5,101–106] These concentration ranges are similarly effective in inhibiting in brain and cerebral cortex the Mg-dependent, Na–K-activated ATPase,[103,107–109] the enzyme generally presumed to be responsible for conversion of available cellular energy (as ATP) into transport of monovalent cations across neural membranes.[94] Both these effects were correlated with cellular energy metabolism by Whittam,[102,103] who proposed that 30–50% of cellular energy metabolism is controlled by the rate of monovalent cation transport, the control being mediated by means of cellular ADP levels, governed in turn by the activity of the Mg-dependent, Na–K-activated ATPase. Evidence adduced by Whittam and colleagues, particularly from studies on incubated erythrocytes and slices of renal cortex, made an impressive case in support of this proposal. A number of studies on cerebral cortex appeared to be at variance with this view, since at concentrations of ouabain

TABLE IX

Effects of Ouabain on Electrolytes and
Respiratory Metabolism of Slices of
Cat Cerebral Cortex Incubated *in Vitro*[a]

Observations	Control slices	+ Ouabain $(10^{-5}$ M)
Slice swelling (%)	32.7	45.1
Slice inulin (%)	42.3	45.9
Slice K (μmoles/g)	99.9	46.1
Slice Na (μmoles/g)	75.3	127.2
Slice Cl (μmoles/g)	80.1	131.0
Slice O_2 consumption (μmoles/g/hr)	64.3	89.4
Slice O_2 consumption (μmoles/g/hr) (Ca-free medium)	120.5	97.0
Cortical Mg–Na–K–ATPase (%)	100.0	5.0

[a] Data are taken from Bourke and Tower[5] and Bonting *et al.*[108] and are all referred to initial fresh weights of slices. All incubations were carried out aerobically for 1 hr at 37°C in bicarbonate–saline–glucose media containing 27 mM K$^+$. Except for the changes shown for oxygen consumption, incubation of slices in Ca-free media did not result in values significantly different from those shown here for slices incubated in the presence of 1.3 mM Ca^{2+}.

which markedly inhibited cation transport and Na–K-activated ATPase in incubated slices of cerebral cortex, oxygen consumption of the slices was either unaffected or stimulated rather than depressed,[5,101,104–106] a discrepancy which Whittam *et al.*[110] also eventually recognized.

Schwartz[104] had already suggested that the discrepancy could be explicable on the basis of whether or not Ca was present in the incubation medium. This explanation has now been amply confirmed[5,105,106,110] and is documented in Table IX. Recent studies[106,110] indicate that the stimulatory effect of ouabain on oxygen consumption by cortical slices in Ca-containing incubation media declines to eventual depression of oxygen consumption after 40–60 min of incubation.

B. Interactions with Calcium

As already pointed out,[5] these observations suggested to us the possibility of an underlying direct interaction between ouabain and tissue Ca. Two sets of observations seemed relevant. Firstly, although actions of ouabain are restricted to the outer (extracellular) face of cell membranes,[111,112] addition of Ca to mitochondrial preparations stimulates mitochondrial respiration in a manner simulating effects of added ADP or uncoupling agents[82,83] and coincides with uptake of Ca by mitochondria to result, in the presence of inorganic P, in its accumulation therein as insoluble Ca phosphates.[84] Secondly, it has been reported that the exposure of cardiac muscle to 10^{-6} M (or higher) ouabain mobilizes tissue Ca and, depending upon external Ca concentrations, permits a drop or rise of tissue Ca content.[113–117] On the basis of these analogies, it was suggested[5] that the effects of ouabain on brain Ca might be similar to those on cardiac muscle.

This hypothesis has been tested and shown to be correct (Table X).[10] The studies confirmed previous observations by Lolley[21] and Keesey et al.[90] that normally the Ca content of incubated slices of cerebral cortex was primarily a function of the Ca concentration of the incubation medium. Hence in a Ca-free medium, whether or not ouabain was present, the slice Ca content dropped to less than 30% of control values for slices in the usual media containing 1.25 μmoles/ml Ca. However, when cortical slices were incubated in the latter medium in the presence of 10^{-5} M ouabain, the Ca content of such slices was significantly elevated to 136% of levels in control slices incubated without ouabain. A striking enrichment in Ca content was found for mitochondria isolated from slices incubated in a Ca-containing medium in the presence of ouabain (Table X).[10] Thus, these findings provide strong circumstantial evidence in favor of an effect of ouabain on neural cells consisting of mobilization of cell Ca and, in the presence of adequate external Ca, its rise in the tissue cells—primarily in the mitochondrial compartment—a translocation which would be expected to stimulate mitochondrial respiration (Table IX).[81] The lack of effect on tissue Mg under these conditions contrasts strikingly with the lability of Ca levels and with effects on slice K and Na.

TABLE X

Effects of 10^{-5} M Ouabain on Ca and Mg in Slices of Cat Cerebral Cortex Incubated in Vitro[a]

Samples	Medium Ca^{2+} (μmoles/ml)	(μmoles/g fresh tissue wt)		Mg content ($\pm$ ouabain)
		Calcium content		
		Controls	+ Ouabain	
Incubated slices	1.25	3.6	4.9[b]	4.9 ($\pm$ 0.3)
	0.2	1.1	0.95	
		(mμmoles/mg protein)		
Mitochondria	1.25	13.4	43.6[b]	–

[a] Data are taken from Tower.[10] All incubations were carried out aerobically for 1 hr at 37°C in bicarbonate–saline–glucose media containing 27 mM K$^+$ plus other conditions as specified in the table. Mitochondria were isolated by the Stahl–Basford density gradient centrifugation techniques as detailed by Tower.[10]

[b] Significantly different from controls ($P < 0.001$).

C. Ouabain *in Vivo*

In adult animals, systemically administered ouabain does not penetrate into the CNS compartment. However, recent studies have been reported on the effects of ouabain injected directly into the brain[118] or by ventriculo-cisternal perfusion.[42,43,119] These studies *in vivo* confirm the basic observations of studies *in vitro* by demonstrating leakage of brain tissue K into the CSF and the development of severe, often fatal, convulsions. In unpublished studies on adult rats, D. M. Woodbury has observed under such conditions a depletion of brain tissue K leading to a very marked rise of K in the CSF without a concomitant rise of serum K. In very young rats, no brain barrier mechanisms operate for either ouabain or K, so that Woodbury observed that systemically administered ouabain rapidly depleted brain K and intracranially administered ouabain not only depleted brain K but led to rises of both CSF and serum K to levels of some 30 μmoles/ml.

V. PATHOLOGICAL CORRELATES

The foregoing observations on the effects of ouabain on electrolytes of neural tissues are pertinent to a number of aspects which relate to the functional significance of cerebral electrolytes and to some pathological correlates thereof. In the final analysis, the reception, conduction, and transmission of nerve impulses depend directly upon normal functioning of the membrane processes which maintain the differential distribution of Na and K across the excitable neural membranes. These processes depend in part upon a degree of membrane stability provided by membrane-bound $Ca^{[120]}$ and upon the operation of the Na–K transport or "pump" mechanisms, together with the energy-yielding mechanisms underlying them. In this sense, the ATPase supplying energy to these "pumps" is an enzyme system dependent upon Mg, Na, and K as cofactors for its optimal activity.[94] Such cofactor roles for various electrolytes are numerous, especially in relation to various cellular transport systems (as discussed in subsequent chapters of this *Handbook*).

A. Epileptogenicity

However, the majority of pathological problems relate more directly to failure of the "pump" systems and the consequent outward leakage and depletion of cellular K and its replacement by Na and Cl. The example provided by ouabain infused intracerebrally and the consequent development of convulsions may be closely analogous to the situation in clinical epilepsy, where samples of human epileptogenic cerebral cortex exhibit abnormal Na and K levels (Table XI).[53,121] Presumably these abnormalities reflect impairment of operation of the Na and K transport systems in epileptogenic neurons, and it is suggested[121] that the final common expression of epileptogenicity is likely to be such an impairment which creates the characteristic instability of the excitable membrane.

TABLE XI

Electrolytes in Incubated Slices of Normal and Epileptogenic Human Cerebral Cortex[a]

Observations	Normal	Epileptogenic
Slice swelling (%)	37.6	38.1
Slice K (μmoles/g)	100.2	80.6
Slice Na (μmoles/g)	113.4	137.5

[a] Data are taken from Tower[53,121] and are all referred to initial fresh weights of slices. All incubations were carried out aerobically for 1 hr at 37°C in bicarbonate–saline–glucose media containing 27 mM K^+. Except for the values for slice swelling, the differences between values for normal and epileptogenic slices are significant ($P < 0.01$). The degree of slice swelling and the high Na content suggest that some "preparatory" artifact[23,88] is present and is probably attributable to the difficulty of sampling and slicing human cerebral cortex excised from patients during neurosurgical procedures.

B. Cerebral Edema

Another important pathological problem is cerebral edema, recently the subject of an extensive symposium.[122] Contrary to the situation *in vitro* (Table VII) where tissue swelling (or edema) is characteristically associated

TABLE XII

Examples of Cerebral Edema *in Vivo*[a]

Species	Pathology	Dry wt. (%) Cerebral cortex Control	Exptl.	Subcortical white Control	Exptl.	Degree of swelling (%)
Man	Tumor	17.4	16.9	30.7	17.4*	81
Rabbit	Alkyl tin	20.2	20.0	30.8	22.9*	36
Cat	Cold lesion	17.9	18.5	31.2	22.1*	43
Cat	Circulatory arrest	16.0	13.5*	29.6	29.0	20

[a] Data are derived from reports by Stewart-Wallace,[123] Aleu *et al.*,[126] Pappius and Gulati,[125] and Tower.[50,130] The degree of swelling is calculated for the experimental samples which deviate significantly from controls and are denoted by an asterisk.

with gray matter (cerebral cortex slices), most forms of edema studied *in vivo* are characteristically localized in subcortical white matter (Table XII).[122] Examples are provided by cerebral tumors,[123] experimentally induced cold (freezing) lesions,[124,125] or alkyl tin intoxication.[126,127] In these forms of edema, the accumulation of extra fluid is primarily interstitial and between myelin lamellae, and the predominant ionic constituents in the edema fluid are characteristically Na and Cl. In the case of the cold lesion, the source of the edema fluid appears to be leakage from damaged blood vessels.[124] One of the few examples of edema in cerebral cortex is provided by circulatory arrest *in vivo* (Table XII).[50] The extra fluid here is primarily intracellular and is associated with influx of Na and loss of K, reflected by a significant rise of K levels in the CSF. In unpublished studies, R. S. Bourke has observed no comparable changes in corpus callosum after circulatory arrest; tissue water content and electrolyte composition were identical with controls. Hence, the process seems to be uniquely restricted to gray matter areas.

It is of interest that attempts to relieve edema of the white matter (of the types outlined above) by means of dehydrating procedures have failed to affect significantly the edematous brain (white matter). The relief afforded is a consequence of dehydration and shrinkage of the uninvolved brain (gray matter).[128–130] Such findings signal caution in the interpretation of results of clinical therapies without adequate and relevant analytical or experimental confirmation.

VI. CONCLUDING REMARKS

This chapter has not aimed for completeness; to do so would have required a book in itself, and several such have been referred to.[8,13,17,121,122] Instead a more selective yet broad sampling of major facets has been attempted. There is still much that is controversial or incomplete. A number of the problems and some of the areas of ignorance or promise have been indicated. It is perhaps fitting that a chapter on inorganic constituents should begin this *Handbook*. Much of what is discussed in subsequent chapters concerns structural components and organization, metabolic systems, and specialized agents, many, if not most, of which are devoted primarily to excluding, retaining, transporting, or sequestering the various inorganic constituents in order that neuronal and cerebral function can be maintained. The early investigators of brain chemistry knew little and understood less about these more complex aspects of brain composition and organization, but the presence of alkali salts, of iron, and especially of phosphorus has been known for almost 250 years.[131] And edema of the brain has probably been known since the days of prehistoric trepanations and certainly was familiar to Hippocrates. The length of the history and the persistence of so many unanswered questions attest to the complexities and to the fundamental importance of the subject. The constancy of the *milieu intérne* is nowhere more exquisitely demanded than in the nervous system and by the neuron.

And nowhere is that constancy so dependent upon the proper distribution and movement of inorganic constituents as in the central nervous system. Bioelectricity is one of Nature's most sophisticated and efficient developments, yet it operates from "batteries" and generators for which Na, K, Ca, Mg, Cl, and the other "simple" inorganic constituents are the essential components.

VII. REFERENCES

1. D. B. Tower, Chemical architecture of the central nervous system, in *Handbook of Physiology: Neurophysiology* (J. Field, H. W. Magoun, and V. E. Hall, eds.), Vol. 3, pp. 1793–1813, American Physiological Society, Washington (1960).
2. J. Folch, M. Lees, and G. H. Sloane-Stanley, The role of acidic lipides in the electrolyte balance of the nervous system of mammals, in *Metabolism of the Nervous System* (D. Richter, ed.), pp. 174–180, Pergamon Press, New York (1957).
3. R. Katzman and C. E. Wilson, Extraction of lipid and lipid cation from frozen brain tissue, *J. Neurochem.* **7**: 113–127 (1961).
4. J. Eichberg and R. M. C. Dawson, Polyphosphoinositides in myelin, *Biochem. J.* **96**: 644–650 (1965).
5. R. S. Bourke and D. B. Tower, Fluid compartmentation and electrolytes of cat cerebral cortex *in vitro*—II. Sodium, potassium and chloride of mature cerebral cortex, *J. Neurochem.* **13**: 1099–1117 (1966).
6. D. B. Tower and R. S. Bourke, Fluid compartmentation and electrolytes of cat cerebral cortex *in vitro*—III. Ontogenetic and comparative aspects, *J. Neurochem.* **13**: 1119–1137 (1966).
7. J. R. Wherrett and D. B. Tower, Glutamyl, aspartyl and amide moieties of cerebral proteins—I. Metabolic aspects, *Exptl. Brain Res.* (in press).
8. H. McIlwain, *Chemical Exploration of the Brain*, Elsevier Publishing Co., Amsterdam (1963).
9. H. McIlwain, *Biochemistry and the Central Nervous System* (3rd ed.), Little, Brown and Co., Boston (1966).
10. D. B. Tower, Ouabain and the distribution of calcium and magnesium in cerebral tissues *in vitro*, *Exptl. Brain Res.* **6**: 273–283 (1968).
11. R. C. Hanig and M. H. Aprison, Determination of calcium, copper, iron, magnesium, manganese, potassium, zinc, and chloride concentrations in several brain areas, *Anal. Biochem.* **21**: 169–177 (1967).
12. R. H. S. Thompson, The regional distribution of copper in human brain, in *Regional Neurochemistry* (S. S. Kety and J. Elkes, eds.), pp. 102–106, Pergamon Press, New York (1961).
13. J. N. Cumings, *Heavy Metals and the Brain*, C. C. Thomas Publisher, Springfield, Ill. (1959).
14. J. N. Cumings, The copper and iron content of brain and liver in the normal and in hepatolenticular degeneration, *Brain* **71**: 410–415 (1948).
15. B. Hallgren and P. Sourander, The effect of age on the non-haemin iron in the human brain, *J. Neurochem.* **3**: 41–51 (1958).
16. J. F. Manery, Water and electrolyte metabolism, *Physiol. Rev.* **34**: 334–417 (1954).
17. A. Van Harreveld, *Brain Tissue Electrolytes*, Butterworth, Inc., Washington (1966).
18. A. Walsh, The application of atomic absorption spectra to chemical analyses, *Spectrochim. Acta* **7**: 108–117 (1955).

19. J. B. Willis, The determination of metals in blood serum by atomic absorption spectroscopy, *Spectrochim. Acta* **16**: 259–278 and 551–558 (1960).
20. H. L. Kahn, Instrumentation for atomic absorption, *J. Chem. Educ.* **43**: A7–A40 and A103–A132 (1966).
21. R. N. Lolley, The calcium content of isolated cerebral tissues and their steady-state exchange of calcium, *J. Neurochem.* **10**: 665–676 (1963).
22. H. H. Hillman, The ion distribution in neural systems of different species in relation to their electrical properties, *in Comparative Neurochemistry* (D. Richter, ed.), pp. 249–260, Macmillan Co., New York (1964).
23. R. S. Bourke and D. B. Tower, Fluid compartmentation and electrolytes of cat cerebral cortex *in vitro*—I. Swelling and solute distribution in mature cerebral cortex, *J. Neurochem.* **13**: 1071–1097 (1966).
24. R. S. Bourke, E. S. Greenberg, and D. B. Tower, Variation of cerebral cortex fluid spaces *in vivo* as a function of species brain size, *Am. J. Physiol.* **208**: 682–692 (1965).
25. P. R. Lewis, The free amino-acids of invertebrate nerve, *Biochem. J.* **52**: 330–338 (1952).
26. B. A. Koechlin, On the chemical composition of the axoplasm of squid giant nerve fibers with particular reference to its ion pattern, *J. Biophys. Biochem. Cytol.* **1**: 511–529 (1955).
27. E. Cotlove, Determination of the true chloride content of biological fluids and tissues. II. Analysis by simple, nonisotopic methods. *Anal. Chem.* **35**: 101–105 (1963).
28. L. B. Flexner and J. B. Flexner, Biochemical and physiological differentiation during morphogenesis. IX. The extracellular and intracellular phases of the liver and cerebral cortex of the fetal guinea pig as estimated from distribution of chloride and radiosodium, *J. Cellular Comp. Physiol.* **34**: 115–127 (1949).
29. A. Lajtha, The development of the blood–brain barrier, *J. Neurochem.* **1**: 216–227 (1957).
30. A. Vernadakis and D. M. Woodbury, Electrolyte and amino acid changes in rat brain during maturation. *Am. J. Physiol.* **203**: 748–752 (1962).
31. A. Vernadakis and D. M. Woodbury, Cellular and extracellular spaces in developing rat brain, *Arch. Neurol.* **12**: 284–293 (1965).
32. H. Porter and J. Folch, Cerebrocuprein—I. A copper-containing protein isolated from brain, *J. Neurochem.* **1**: 260–271 (1957).
33. H. Porter and J. Folch, Brain copper-protein fractions in the normal and in Wilson's disease, *Arch. Neurol. Psychiat.* **77**: 8–16 (1957).
34. F. Hawking, T. J. Hennelly, and J. H. Quastel, Trypanocidal activity and arsenic content of the cerebrospinal fluid after administration of arsenic compounds, *J. Pharmacol. Exptl. Therap.* **59**: 157–175 (1937).
35. P. C. Baastrup and M. Schou, Lithium as a prophylactic agent: Its effect against recurrent depressions and manic-depressive psychosis, *Arch. Gen. Psychiat.* **16**: 162–172 (1967).
36. M. Schou and P. C. Baastrup, Lithium treatment of manic-depressive disorder, *J. Am. Med. Assoc.* **201**: 696–698 (1967).
37. A. C. Corcoran, R. D. Taylor, and I. H. Page, Lithium poisoning from the use of salt substitutes, *J. Am. Med. Assoc.* **139**: 685–688 (1949).
38. L. W. Hanlon, M. Romaine, F. J. Gilroy, and J. E. Deitrick, Lithium chloride as a substitute for sodium chloride in the diet, *J. Am. Med. Assoc.*, **139**: 688–692 (1949).
39. M. Schou, Lithium studies. 3. Distribution between serum and tissues, *Acta Pharmacol.* **15**: 115–124 (1958).
40. R. D. Keynes and R. C. Swan, Permeability of frog muscle fibres to lithium ions, *J. Physiol. (London)* **147**: 626–638 (1959).
41. H. Davson, A comparative study of the aqueous humor and cerebrospinal fluid in the rabbit, *J. Physiol. (London)* **129**: 111–133 (1955).
42. H. Cserr, Potassium exchange between cerebrospinal fluid, plasma and brain, *Am. J. Physiol.* **209**: 1219–1226 (1965).

43. R. Katzman, L. Graziani, R. Kaplan, and A. Escriva, Exchange of cerebrospinal fluid potassium with blood and brain, *Arch. Neurol.* **13**:513–524 (1965).
44. V. Fencl, T. B. Miller, and J. R. Pappenheimer, Studies on the respiratory response to disturbances in acid–base balance with deductions concerning ionic composition of cerebral interstitial fluid, *Am. J. Physiol.* **210**:459–472 (1966).
45. M. W. B. Bradbury and H. Davson, Transport of potassium between blood, cerebrospinal fluid, and brain, *J. Physiol. (London)* **181**:151–174 (1965).
46. J. S. Barlow and J. F. Manery, The changes in electrolytes, particularly chloride, which accompany growth in chick muscle, *J. Cellular Comp. Physiol.* **43**:165–191 (1954).
47. E. Cotlove, Mechanism and extent of distribution of inulin and sucrose in chloride space of tissues, *Am. J. Physiol.* **176**:396–410 (1954).
48. A. B. Hastings and L. Eichelberger, The exchange of salt and water between muscle and blood. I. The effect of an increase in total body water produced by the intravenous injection of isotonic salt solution, *J. Biol. Chem.* **117**:73–93 (1937).
49. F. L. Truax (1954), cited in Ref. No. 16.
50. D. B. Tower, Distribution of cerebral fluids and electrolytes *in vivo* and *in vitro*, in *Brain Edema* (I. Klatzo and F. Seitelberger, eds.), pp. 303–332, Springer-Verlag, New York (1967).
51. G. F. Bahr, G. Bloom, and U. Friberg, Volume changes of tissues in physiological fluids during fixation in osmium tetroxide or formaldehyde and during subsequent treatment, *Exptl. Cell Res.* **12**:342–355 (1957).
52. G. F. Bahr, G. Bloom, and E. Johannison, Further studies on fixation with osmium tetroxide, *Histochemie* **1**:113–118 (1958).
53. D. B. Tower, Problems associated with studies of electrolyte metabolism in normal and epileptogenic cerebral cortex, *Epilepsia* **6**:183–197 (1965).
54. A. Van Harreveld, J. Crowell, and S. K. Malhotra, A study of extracellular space in central nervous tissue by freeze-substitution, *J. Cell Biol.* **25**:117–137 (1965).
55. A. Van Harreveld and S. K. Malhotra, Demonstration of extracellular space by freeze-drying in the cerebellar molecular layer, *J. Cell Sci.* **1**:223–228 (1966).
56. M. W. Brightman, The distribution within brain of ferritin injected into cerebrospinal fluid compartments. II. Parenchymal distribution, *Am. J. Anat.* **117**:193–220 (1965).
57. C. F. Barlow, N. S. Domek, M. A. Goldberg, and L. J. Roth, Extracellular brain space measured by S^{35} sulfate, *Arch. Neurol.* **5**:102–110 (1961).
58. E. Streicher, Thiocyanate space of rat brain, *Am. J. Physiol.* **201**:334–336 (1961).
59. J. E. Richmond and A. B. Hastings, Distribution of sulfate in blood and between cerebrospinal fluid and plasma *in vivo*, *Am. J. Physiol.* **199**:814–820 (1960).
60. A. Van Harreveld, N. Ahmed, and D. J. Tanner, Sulfate concentration in cerebrospinal fluid and serum of rabbits and cats, *Am. J. Physiol.* **210**:777–780 (1966).
61. E. Streicher, D. P. Rall, and J. R. Gaskins, Distribution of thiocyanate between plasma and cerebrospinal fluid, *Am. J. Physiol.* **206**:251–254 (1964).
62. M. Pollay, Cerebrospinal fluid transport and the thiocyanate space of the brain, *Am. J. Physiol.* **210**:275–279 (1966).
63. M. Pollay and H. Davson, The passage of certain substances out of the cerebrospinal fluid, *Brain* **86**:137–150 (1963).
64. K. Welch, Concentration of thiocyanate by the choroid plexus of the rabbit *in vitro*, *Proc. Soc. Exptl. Biol. Med.* **109**:953–954 (1962).
65. D. J. Reed, D. M. Woodbury, and L. Holtzer, Brain edema, electrolytes and extracellular space, *Arch. Neurol.* **10**:604–614 (1964).
66. D. L. Woodward, D. J. Reed, and D. M. Woodbury, Extracellular space of rat cerebral cortex, *Am. J. Physiol.* **212**:367–370 (1967).
67. D. P. Rall, W. W. Oppelt, and C. S. Patlak, Extracellular space of brain as determined by diffusion of inulin from the ventricular system, *Life Sci.*, No. 2, 43–48 (1962).

68. I. M. Gibson and H. McIlwain, Continuous recording of changes in membrane potential in mammalian cerebral tissues *in vitro*, *J. Physiol. (London)* **176**:261–263 (1965).

69. B. Becker, Cerebrospinal fluid iodide, *Am. J. Physiol.* **201**:1149–1151 (1961).

70a. L. Z. Bito, M. W. B. Bradbury, and H. Davson, Factors affecting the distribution of iodide and bromide in the central nervous system, *J. Physiol. (London)* **185**:323–354 (1966).

70b. W. H. Oldendorf and H. Davson, Brain extracellular space and the sink action of cerebrospinal fluid, *Arch. Neurol.* **17**:196–205 (1967).

71. J. R. Pappenheimer, S. R. Heisey, and E. F. Jordan, Active transport of Diodrast and phenolsulfonphthalein from cerebrospinal fluid to blood, *Am. J. Physiol.* **200**:1–10 (1961).

72. S. B. Friedman, W. G. Austen, R. E. Rieselbach, J. B. Block, and D. P. Rall, Effect of hypochloremia on cerebrospinal fluid chloride concentration in a patient with anorexia nervosa and in dogs, *Proc. Soc. Exptl. Biol. Med.* **114**:801–805 (1963).

73. H. Waelsch, An attempt at integration of structure and metabolism in the nervous system, *in Structure and Function of the Cerebral Cortex* (D. B. Tower and J. P. Schadé, eds.), pp. 313–327, Elsevier Publishing Co., Amsterdam (1960).

74. O. H. Lowry, N. R. Roberts, K. Y. Leiner, M. L. Wu, A. L. Farr, and R. W. Albers, The quantitative histochemistry of brain. III. Ammon's horn, *J. Biol. Chem.* **207**:39–49 (1954).

75. J. G. Nicholls and S. W. Kuffler, Extracellular space as a pathway for exchange between blood and neurons in the central nervous system of the leech: Ionic composition of glial cells and neurons, *J. Neurophysiol.* **27**:645–671 (1964).

76. S. W. Kuffler, J. G. Nicholls, and R. K. Orkand, Physiological properties of glial cells in the central nervous system of amphibia, *J. Neurophysiol.* **29**:768–806 (1966).

77. H. M. Pappius and K. A. C. Elliott, Factors affecting the potassium content of incubated brain slices, *Can. J. Biochem. Physiol.* **34**:1053–1067 (1956).

78. R. Katzman and P. H. Leiderman, Brain potassium exchange in normal adult and immature rats, *Am. J. Physiol.* **175**:263–270 (1953).

79. W. C. Holland and G. V. Auditore, Distribution of potassium in liver, kidney and brain of rat and guinea pig, *Am. J. Physiol.* **183**:309–313 (1955).

80. J. L. Gamble, Retention of potassium by mitochondria, *Am. J. Physiol.* **203**:886–890 (1962).

81. C. S. Rossi and A. L. Lehninger, Stoichiometric relationships between accumulation of ions by mitochondria and the energy-coupling sites in the respiratory chain, *Biochem. Z.* **338**:698–713 (1963).

82. G. P. Brierly, Ion accumulation in heart mitochondria, *in Energy-Linked Functions of Mitochondria* (B. Chance, ed.), pp. 237–245, Academic Press, New York (1963).

83. B. Chance, Calcium-stimulated respiration in mitochondria, *in Energy-Linked Functions of Mitochondria* (B. Chance, ed.), pp. 253–269, Academic Press, New York (1963).

84. E. C. Weinbach and T. von Brand, The isolation and composition of dense granules from Ca^{++}-loaded mitochondria, *Biochem. Biophys. Res. Commun.* **19**:133–137 (1965).

85. A. L. Hodgkin and B. Katz, The effect of sodium ions on the electrical activity of giant axon of the squid, *J. Physiol. (London)* **108**:37–77 (1949).

86. C-L. Li, Cortical intracellular potentials and their responses to strychnine, *J. Neurophysiol.* **22**:436–450 (1959).

87. S. P. R. Rose, Preparation of enriched fractions from cerebral cortex containing isolated, metabolically active neuronal and glial cells, *Biochem. J.* **102**:33–43 (1967).

88. S. Varon and H. McIlwain, Fluid content and compartments in isolated cerebral tissues, *J. Neurochem.* **8**:262–275 (1961).

89. H. S. Bachelard, W. J. Campbell, and H. McIlwain, The sodium and other ions of mammalian cerebral tissues, maintained and electrically stimulated *in vitro*, *Biochem. J.* **84**:225–232 (1962).

90. J. C. Keesey, H. Wallgren, and H. McIlwain, The sodium, potassium and chloride of cerebral tissues: Maintenance, change on stimulation and subsequent recovery, *Biochem. J.* **95**: 289–300 and 301–310 (1965).

91. A. Ames and F. B. Nesbett, Intracellular and extracellular compartments of mammalian central nervous tissue, *J. Physiol. (London)* **184**: 215–238 (1966).

92. C-L. Li and H. McIlwain, Maintenance of resting membrane potentials in slices of mammalian cerebral cortex and other tissues *in vitro*, *J. Physiol. (London)* **139**: 178–190 (1957).

93. C. Yamamoto and H. McIlwain, Electrical activity in thin sections from the mammalian brain maintained in chemically defined media *in vitro*, *J. Neurochem.* **13**: 1333–1343 (1966).

94. G. Siegel and R. W. Albers, Nucleoside triphosphate phosphohydrolases, *in Handbook of Neurochemistry* (A. Lajtha, ed.), Vol. IV, Plenum Press, New York (1969).

95. J. Thomas and H. McIlwain, Chloride content and metabolism of cerebral tissues in fluids low in chlorides, *J. Neurochem.* **1**: 1–7 (1956).

96. T. Narahashi, J. W. Moore, and W. R. Scott, Tetrodotoxin blockage of sodium conductance increase in lobster giant axons, *J. Gen. Physiol.* **47**: 965–974 (1964).

97. Y. Nakamura, S. Nakajima, and H. Grundfest, Action of tetrodotoxin on electrogenic components of squid giant axons, *J. Gen. Physiol.* **48**: 985–996 (1965).

98. H. McIlwain, Tetrodotoxin and the cation content, excitability and metabolism of isolated mammalian cerebral tissues, *Biochem. Pharmacol.* **16**: 1389–1396 (1967).

99. C. Moore and B. C. Pressman, Mechanism of action of valinomycin on mitochondria, *Biochem. Biophys. Res. Commun.* **15**: 562–567 (1964).

100. B. C. Pressman, Induced active transport of ions in mitochondria, *Proc. Natl. Acad. Sci. (U.S.)* **53**: 1076–1083 (1965).

101. H. Yoshida, T. Nukada, and H. Fujisawa, The effect of ouabain on ion transport and metabolic turnover of phospholipids of brain slices, *Biochim Biophys. Acta* **48**: 614–615 (1961).

102. R. Whittam, Active cation transport as a pace-maker of respiration, *Nature (London)* **191**: 603–604 (1961).

103. R. Whittam, The dependence of the respiration of brain cortex on active cation transport, *Biochem. J.* **82**: 205–212 (1962).

104. A. Schwartz, The effect of ouabain on potassium content, phosphoprotein metabolism and oxygen consumption of guinea pig cerebral tissue, *Biochem. Pharmacol.* **11**: 389–391 (1962).

105. P. Joanny and J. Corriol, Influence de l'ouabaïne sur les mouvements ioniques, la respiration et la glycolyse aérobie du cortex cérébrale isolé de mammifère, *Arch. Sci. Physiol.* **18**: 325–337 (1964).

106. P. D. Swanson and H. McIlwain, Inhibition of the sodium-ion-stimulated adenosine triphosphatase after treatment of isolated guinea pig cerebral cortex with ouabain and other agents, *J. Neurochem.* **12**: 877–891 (1965).

107. J. C. Skou, Preparation from mammalian brain and kidney of the enzyme system involved in active transport of Na^+ and K^+, *Biochim. Biophys. Acta* **58**: 314–325 (1962).

108. S. L. Bonting, L. L. Caravaggio, and N. M. Hawkins, Studies on sodium-potassium-activated adenosinetriphosphatase. IV. Correlation with cation transport sensitive to cardiac glycosides, *Arch. Biochem. Biophys.* **98**: 413–419 (1962).

109. W. N. Aldridge, Adenosine triphosphatase in the microsomal fraction from rat brain, *Biochem. J.* **83**: 527–533 (1962).

110. R. Whittam, D. M. Blond, and M. Ruscak, The influence of ions on the metabolism of brain-cortex slices, *Biochem. J.* **96**: 47P–48P (1965).

111. E. T. Dunham and I. M. Glynn, Adenosinetriphosphatase activity and the active movement of alkali metal ions, *J. Physiol. (London)* **156**: 274–293 (1961).

112. P. C. Caldwell and R. D. Keynes, The effect of ouabain on the efflux of sodium from a giant squid axon, *J. Physiol. (London)* **148**: 8P–9P (1959).

113. W. Klaus, G. Kuschinsky, and H. Lüllman, Über den Zusammenhang zwischen positiv inotroper Wirkung von Digitoxigenin, Kaliumflux und intracellulären Ionenkonzentrationen im Herzmuskel, *Naunyn-Schmiedebergs Arch. Exptl. Pathol. Pharmakol.* **242**:480–496 (1962).

114. H. Lüllman and W. C. Holland, Influence of ouabain on an exchangeable calcium fraction, contractile force, and resting tension of guinea-pig atria, *J. Pharmacol. Exptl. Therap.* **137**: 186–192 (1962).

115. G. Gersmeyer and W. C. Holland, Influence of ouabain on contractile force, resting tension, Ca^{++} entry and tissue Ca content in rat atria, *Circulation Res.* **12**:620–622 (1963).

116. W. C. Govier and W. C. Holland, The relationship between atrial contractions and the effect of ouabain on contractile strength and calcium exchange in rabbit atria, *J. Pharmacol. Exptl. Therap.* **148**:284–289 (1965).

117. W. F. Keeton and A. H. Briggs, *In vivo* effects of ouabain on calcium metabolism in rabbit hearts and plasma, *Proc. Soc. Exptl. Biol. Med.* **118**:1127–1129 (1965).

118. A. Bignami and G. Palladini, Experimentally produced cerebral status spongiosus and continuous pseudorhythmic electroencephalographic discharges with a membrane-ATPase inhibitor in the rat, *Nature (London)* **209**:413–414 (1966).

119. A. Ames, K. Higashi, and F. B. Nesbett, Effects of P_{CO_2}, acetazolamide and ouabain on volume and composition of choroid-plexus fluid, *J. Physiol. (London)* **181**:516–524 (1965).

120. J. M. Tobias, A chemically specified molecular mechanism underlying excitation in nerve: A hypothesis, *Nature (London)* **203**:13–17 (1964).

121. D. B. Tower, *Neurochemistry of Epilepsy*, C. C. Thomas Publisher, Springfield, Ill. (1960).

122. I. Klatzo and F. Seitelberger (eds.), *Brain Edema*, Springer-Verlag, New York (1967).

123. A. M. Stewart-Wallace, Biochemical study of cerebral tissue, and changes in cerebral edema, *Brain* **62**:426–438 (1939).

124. I. Klatzo, A. Piraux, and E. J. Laskowski, The relationship between edema, blood–brain barrier and tissue elements in a local brain injury, *J. Neuropath. Exptl. Neurol.* **17**:548–564 (1958).

125. H. M. Pappius and D. R. Gulati, Water and electrolyte content of cerebral tissues in experimentally induced edema, *Acta Neuropathol.* **2**:451–460 (1963).

126. F. P. Aleu, R. Katzman, and R. D. Terry, Structure and electrolyte analyses of cerebral edema induced by alkyl tin intoxication, *J. Neuropathol. Exptl. Neurol.* **22**:403–413 (1963).

127. R. M. Torack, R. D. Terry, and H. M. Zimmerman, Fine structure of cerebral fluid accumulation. II. Swelling produced by triethyl tin poisoning compared with that in human brain, *Am. J. Pathol.* **36**:273–287 (1960).

128. H. M. Pappius, Biochemical studies on experimental brain edema, in *Brain Edema* (I. Klatzo and F. Seitelberger, eds.), pp. 445–460, Springer-Verlag, New York (1967).

129. R. A. Clasen, P. M. Cooke, S. Pandolfi, G. Carnecki, and G. Bryar, Hypertonic urea in experimental cerebral edema, *Arch. Neurol.* **12**:424–434 (1965).

130. D. B. Tower, Delineation of fluid compartmentation in cerebral tissues, in *Brain Barrier Systems* (A. Lajtha and D. H. Ford, eds.), pp. 465–480, Elsevier Publishing Co., Amsterdam (1968).

131. D. B. Tower, Origins and development of neurochemistry, *Neurology* **8**(Suppl. 1):3–31 (1958).

Chapter 2

CARBOHYDRATES

H. S. Bachelard

Institute of Psychiatry
Department of Biochemistry
De Crespigny Park, London, England

I. INTRODUCTION

Of possible sources of metabolic support for brain function *in vivo*, only glucose has been shown to be capable of maintaining cerebral tissues satisfactorily. Other sugars, e.g., mannose and maltose, have been found capable of supporting brain function in hepatectomized animals[1,2] but were considered to be converted to glucose elsewhere in the body before reaching the brain. Fructose will not support normal cerebral function in hepatectomized animals[1] and is utilized only very slowly by the brain in perfused preparations.[3] The brain relies on a rapid utilization of the glucose brought to it from the bloodstream and is particularly dependent on the circulating blood glucose, as it has a relatively small reserve of glycogen (less than that of muscle tissues).

The simple carbohydrate substances important in the brain involve, therefore, mainly glucose, sugar derivatives such as the hexose phosphates, and glycogen. Complex carbohydrate-containing substances will not be discussed—glycoproteins, mucopolysaccharides, and gangliosides are treated in subsequent chapters.

The levels reported in the literature for the two main carbohydrates, glucose and glycogen, have varied widely due partly to inherent difficulties involved in their estimation—in particular, due to the rapid changes which occur immediately after the blood supply to the brain is cut off. During the first few minutes after death, the levels of glucose, creatine phosphate, ATP, and glycogen fall rapidly, with a concomitant rise in the concentrations of lactate, inorganic phosphate, AMP, and ADP. The changes in these constituents are due to the rapid anaerobic glycolysis which occurs post-mortem, during which time replenishment from the bloodstream of glucose, the major substrate, cannot take place. Therefore, it is essential in estimating carbohydrate levels in the brain to use techniques designed to prevent or limit the rapid post-mortem changes.

II. POST-MORTEM AUTOLYSIS

The most satisfactory method of preventing post-mortem glycolysis is considered[4] to be the rapid-freezing technique described by Kerr.[5] Values for one of the most labile constituents, creatine phosphate, were determined in the brains of larger animals such as the cat or dog after pouring liquid air onto the exposed skulls and were found to be similar to those obtained using smaller animals (rats or mice) frozen whole by total immersion in liquid air. Richter and Dawson[6] inserted thermocouples at different depths into the cerebral cortex of living rats and followed the time course of the fall in temperature at the different depths after immersion of the animals in liquid air. The temperature of the surface of the cerebral cortex fell to below 0°C within 4 to 5 sec. In deeper parts of the brain, longer times were required—up to 20 sec—but the temperature was observed to remain close to 37°C until it dropped suddenly and rapidly to below zero. It appears feasible that circulation in deeper layers may be maintained until almost immediately before the sudden drop in temperature.

The required glycostasis has also been approached by injecting sodium iodoacetate into living cats and dogs.[7] Glycogen levels were found to be about 25 % lower than those obtained using the *in situ* freezing technique,[8] and the iodoacetate method was concluded to be a less satisfactory method, although an injection method could prove advantageous in experiments with larger animals.

III. GLUCOSE

Variations in reported levels of glucose in mammalian nervous tissues have been due partly to the rapid fall occurring during post-mortem autolysis and partly to the techniques which have been used for estimating glucose. Many reports have been based on the total reducing power of extracts, which is likely to yield elevated results. Gey[9] separated glucose by paper chromatography of deproteinized extracts prepared from rapidly frozen rat tissues and compared the glucose contents with the total reducing power of the extracts. The purified glucose accounted for only 40 % of the total reducing power of brain extracts in comparison with 45 % in muscle and 95 % in kidney and liver. The high proportion of non-glucose reducing activity in brain extracts is partly due to ascorbic acid, which occurs in relatively large amounts (1–2 μmole/g) in mammalian brain.[10,11]

During post-mortem autolysis in rat brain, the glucose content fell from an initial value in the frozen tissue of 0.45 μmole/g to less than 0.1 μmole/g within 2 min.[9] During this period, the total reducing activity fell less rapidly to about 50 % of the initial activity. Lowry *et al.*[12] recorded higher values for the glucose content, estimated enzymatically in rapidly frozen mouse brain extracts. These were 1.25–1.55 μmole/g, similar to those (1.5 μmole/g) found by Tews, Carter, and Stone[13] in rapidly frozen dog brain. In the mouse, over 85 % of the brain glucose was lost within 30 sec of the death of the animals.[12]

The estimations based on specific methods for the determination of glucose under conditions where minimal autolytic changes are likely to have occurred indicate that the glucose content of mammalian brain is close to 1.5 μmole/g (27 mg/100 g).

A. Intracellular Glucose Content

The total glucose content of the brain is, therefore, relatively low and must in part be contained in the fluids (blood and cerebrospinal fluid) associated with the tissue, so the cellular glucose content will be even lower. An approximate value for the intracellular content can be obtained by assessing the contributions of the blood, the cerebrospinal fluid, and the extracellular tissue space. The blood volume of the brain has been estimated at 3 %[14,15] and the cerebrospinal fluid at 9 %.[15] The extracellular space of the brain, based on studies using chemical tracers in vivo, may vary in different mammalian species, according to the size of the brain, from 15–20 % in the rat or guinea pig to over 30 % in the sheep or monkey.[16] Estimates using other techniques, such as chemical tracers in vitro,[17] impedance studies,[18] and freeze-substitution electron microscopy,[19,20] indicate that the extracellular space lies in the range 20–25 %. Use of the latter two techniques has shown that anoxia reduces the extracellular space considerably to correspond more closely to the small extracellular space observed in classical electron microscopy, where anoxia is likely to occur during specimen fixation.

Given a cerebrospinal fluid glucose content of 70 mg/100 ml[21,22] and assuming an extracellular space of 25 % which contains glucose at a concentration similar to that of the cerebrospinal fluid, the contribution of these to the total glucose content would be about 26.5 mg/100 g (Table I), i.e., it could account for the total glucose measured. The assumption that the extracellular glucose concentration is similar to that in the cerebrospinal fluid may not be valid; it may well be lower, but in the absence of established data it seems a reasonable approximation. The approximation also does not take into consideration the pronounced cellular heterogeneity of the tissue. Still, the

TABLE I

Calculated Intracellular Glucose Concentration

Space	Volume of brain (%)	Glucose content (mg/100 ml)	Possible contribution to total glucose (mg/100 g)
Blood	3	90	2.7
CSF	9	70	6.3
Extracellular space	25	70	17.5
		Total	26.5

intracellular glucose concentration is likely to be very low, and, if the concentration in the extracellular space is similar to that of the cerebrospinal fluid, it may approach zero, as has been found for muscle tissue under normal conditions.[23]

IV. SUGAR DERIVATIVES

Free simple carbohydrates other than glucose or glycogen occur in nervous tissue in small amounts. The levels of fructose and hexose phosphates range from 0.016 to 0.12 μmole/g (Table II), although higher fructose concentrations (0.97 μmole/g) have been found in peripheral nerve. The changes in concentration of the hexose phosphates during post-mortem autolysis may be as rapid as the change in glucose.[12] In mouse brain during the few seconds after death, hexose monophosphates fell 20–55% and fructose diphosphate concentrations increased by 50–60%.

V. GLYCOGEN

Glycogen, like glucose, is lost rapidly from nervous tissues within minutes of death unless precautions are taken to limit the extent of post-mortem anaerobic glycolysis. As discussed above, the most effective method of minimizing such changes is Kerr's rapid-freezing technique. After the blood supply has been cut off, glycogen (Table III) is initially not depleted as rapidly as glucose.[24] When glucose has been almost depleted, the rate of glycogen loss increases to rates of up to 3 μmole/min/g tissue.[12]

A. Techniques

The original method described for the isolation of glycogen[25] involved extraction of the glycogen with hot KOH followed by precipitation with

TABLE II

Content of Fructose and Sugar Phosphates[a]

Species	Tissue	F	F6P	FDP	G6P	GDP	Reference
Rabbit	Tibial nerve	0.97					36
Mouse	Brain (adult)	0.10	0.016	0.120	0.080		12, 36
	Brain (young)		0.023	0.109	0.091		12
Rat	Brain					0.032	37

[a] Values are expressed as μmole/g of rapidly frozen tissue.
  Notation: F, fructose; F6P, fructose 6-phosphate; FDP, fructose 1,6-diphosphate; G6P, glucose 6-phosphate; GDP, glucose 1,6-diphosphate.

TABLE III

Brain Glycogen Levels and Post-Mortem Changes

Species	Glycogen (as glucose, mg/100 g of rapidly frozen tissue)	Change post-mortem (% of initial value)			
		Time (min)	Glycogen	Glucose	Reference
Mouse (adult)	53				33
Mouse (adult)	40	0.5	−48	−86	12
Mouse (young)	45	0.5	−38	−75	12
Rat	92				35
Dog	120–160				33
Dog	86–144	3	−50	−90	24

ethanol; it is still the basis of the current method most frequently used. It was modified slightly by Pflüger[26] and later by Kerr,[27] who introduced an improved method for the removal of contaminating cerebrosides using chloroform–methanol extractions. The alcoholic–KOH method, like methods involving tissue extractions with perchloric acid or trichloroacetic acid, has recently been suspected of leading to partial degradation of the isolated glycogen. Extraction with dimethyl sulfoxide[28] yields glycogen of higher average molecular weight. However, the efficiency of the extraction was found to be lower, and while the possibility exists that selective extraction of higher molecular-weight glycogen molecules had occurred, it was argued that the neutral organic solvent produced glycogen in a more "native" state. The use of cold water in extraction was also concluded to yield more "native" glycogen. In a comparative study of extraction methods, the glycogen isolated by cold-water extraction was considerably higher in molecular weight than that extracted by other methods and showed extreme polydispersity.[29,30] Whether brain glycogen is, in fact, more highly branched than liver glycogen[31,32] may depend on factors such as the use of techniques which will prevent gross autolysis and the methods used in extraction. After isolation and purification, the glycogen is estimated as glucose after acid hydrolysis.

B. Glycogen Levels

Values for the glycogen content of brain tissue rapidly fixed by freezing *in situ* (Table III) range from 40–160 mg (glucose)/100g tissue, which is considerably lower than the content in normal nonfasted mammalian liver, but not much below (10–40%) the levels found in muscle tissue. The mild hypoglycemia which results from fasting has no effect on brain glycogen levels, but insulin-induced hypoglycemia can result in decreases of over 50%.[33,34] Anesthetics in some studies[35] resulted in decreases of 20–40% in brain glycogen, but other workers[12,33] found no change in glycogen content. Decreases in glycogen content during post-mortem autolysis of 30–50%

within the first few minutes have been observed (Table III); during the same time period, glucose levels fell by 75–90 %.

The chemistry of glycogen and its localization in nervous tissues will be treated by R. V. Coxon in "Glycogen Metabolism," Chapter 2 of Volume III of this series.

VI. REFERENCES

1. S. J. Maddock, J. E. Hawkins, and E. Holmes, The inadequacy of substances of the "glucose cycle" for maintenance of normal cortical potentials during hypoglucemia produced by hepatectomy with abdominal evisceration, *Am. J. Physiol.* **125**:551–565 (1939).
2. F. C. Mann and T. B. Magath, The effect of administration of glucose in the condition following total extirpation of the liver, *Arch. Internal Med.* **30**:171–181 (1922).
3. A. Geiger, J. Magnes, R. M. Taylor, and M. Veralli, Effect of blood constituents on uptake of glucose and on metabolic rates of the brain in perfusion experiments, *Am. J. Physiol.* **177**:138–149 (1954).
4. W. Thorn, H. Scholl, G. Pfleiderer, and B. Mueldener, Metabolic processes in the brain at normal and reduced temperatures and under anoxic and ischaemic conditions, *J. Neurochem.* **2**:150–165 (1958).
5. S. E. Kerr, Studies on the phosphorus compounds of the brain. I. Phosphocreatine, *J. Biol. Chem.* **110**:625–635 (1935).
6. D. Richter and R. M. C. Dawson, Brain metabolism in emotional excitement and in sleep, *Am. J. Physiol.* **154**:73–79 (1948).
7. A. Chesler and H. E. Himwich, The glycogen content of various parts of the central nervous system of dogs and cats at different ages, *Arch. Biochem.* **2**:175–181 (1943).
8. M. R. A. Chance and D. C. Yaxley, Central nervous function and changes in brain metabolite concentration. I. Glycogen and lactate in convulsing mice, *J. Exptl. Biol.* **27**:311–323 (1950).
9. K. F. Gey, The concentration of glucose in rat tissues, *Biochem. J.* **64**:145–150 (1956).
10. C. M. Damron, M. M. Monier, and J. H. Roe, Metabolism of L-ascorbic acid, dehydro-L-ascorbic acid and diketo-L-gulonic acid in the guinea pig, *J. Biol. Chem.* **195**:599–606 (1952).
11. S. Lin and H. P. Cohen, The effect of scorbutus and pentobarbital on the *in vivo* levels of "energy-rich" phosphates and their turnover in guinea pig cerebral tissue, *Arch. Biochem. Biophys.* **88**:256–261 (1960).
12. O. H. Lowry, J. V. Passonneau, F. X. Hasselberger, and D. W. Schulz, Effect of ischemia on known substrates and cofactors of the glycolytic pathway in brain, *J. Biol. Chem.* **239**:18–30 (1964).
13. J. K. Tews, S. H. Carter, and W. E. Stone, Chemical changes in the brain during insulin hypoglycaemia and recovery, *J. Neurochem.* **12**:679–693 (1965).
14. N. B. Everett, B. Simmons, and E. P. Lasher, Distribution of blood (Fe^{59}) and plasma (I^{131}) volumes of rats determined by liquid nitrogen freezing, *Circulation Res.* **4**:419–424 (1956).
15. H. L. Rosomoff, Method for simultaneous quantitative estimation of intracranial contents, *J. Appl. Physiol.* **16**:395–396 (1961).
16. R. S. Bourke, E. S. Greenberg, and D. B. Tower, Variation of cerebral cortex fluid spaces *in vivo* as a function of species brain size, *Am. J. Physiol.* **208**:682–692 (1965).
17. I. Gibson and H. McIlwain, Continuous recording of changes in membrane potential in mammalian cerebral tissues *in vitro*; recovery after depolarization by added substances, *J. Physiol.* **176**:261–283 (1965).

18. A. Van Harreveld, Water and electrolyte distribution in central nervous tissue, *Federation Proc.* **21**:659–664 (1962).
19. A. Van Harreveld, J. Crowell, and S. K. Malhotra, A study of extracellular space in central nervous tissue by freeze substitution, *J. Cell. Biol.* **25**:117–137 (1965).
20. A. Van Harreveld and S. K. Malhotra, Extracellular space in the cerebral cortex of the mouse, *J. Anat.* **101**:197–207 (1967).
21. R. G. Cooper and J. W. Archdeacon, Blood and cerebrospinal fluid glucose in the fasting state. *Am. J. Physiol.* **198**:260–262 (1960).
22. H. Davson, A comparative study of the aqueous humour and cerebrospinal fluid in the rabbit, *J. Physiol.* **129**:111–133 (1955).
23. P. J. Randle and G. H. Smith, Regulation of glucose uptake by muscle. 2. The effects of insulin, anaerobiosis and cell poisons on the penetration of isolated rat diaphragm by sugars, *Biochem. J.* **70**:501–508 (1958).
24. S. E. Kerr and M. Ghantus, The carbohydrate metabolism of brain. III. On the origin of lactic acid, *J. Biol. Chem.* **117**:217–225 (1937).
25. C. Bernard, Nouvelles recherches expérimentales sur les phénomes glycogéniques du foie, *Compt. Rend.* **44**:1325–1331 (1857).
26. E. Pflüger, Estimation of glycogen, *Arch. Ges. Physiol.* **93**:163–185 (1902).
27. S. E. Kerr, The carbohydrate metabolism of brain. I. The determination of glycogen in nerve tissue, *J. Biol. Chem.* **116**:1–8 (1936).
28. R. L. Whistler and J. N. BeMiller, Extraction of glycogen with dimethyl sulfoxide, *Arch. Biochem. Biophys.* **98**:120–121 (1962).
29. E. Bueding and S. A. Orrell, A mild procedure for the isolation of polydispersed glycogen from animal tissues, *J. Biol. Chem.* **239**:4018–4020 (1964).
30. S. A. Orrell and E. Bueding, A comparison of products obtained by various procedures used for the extraction of glycogen, *J. Biol. Chem.* **239**:4021–4026 (1964).
31. E. E. Goncharova, Some data on the structure of glycogen and polysaccharides of the brain synthesized *in vitro* by brain enzymes, in *Proc. 3rd All-Union Neurochem. Conf. Erevan* (A. V. Palladin and Ch. Buniatian, eds.), p. 455, Armenian C.C.P., Erevan (1962).
32. B. I. Khaikina and E. E. Goncharova, Metabolism and chemical structure of glycogen fractions in the brain, in *Problems of the Biochemistry of the Nervous System* (A. V. Palladin, ed.), pp. 87–95, Pergamon Press, Oxford (1964).
33. S. H. Carter and W. E. Stone, Effects of convulsants on brain glycogen in the mouse, *J. Neurochem.* **7**:16–19 (1961).
34. S. E. Kerr and M. Ghantus, The carbohydrate metabolism of brain. II. The effect of varying the carbohydrate and insulin supply on the glycogen, free sugar and lactic acid in mammalian brain, *J. Biol. Chem.* **116**:9–20 (1936).
35. A. W. Merrick, Encephalic glycogen differences in young and adult rats, *J. Physiol.* **158**:476–485 (1961).
36. M. A. Stewart and J. V. Passonneau, Identification of fructose in mammalian nerve, *Biochem. Biophys. Res. Commun.* **17**:536–541 (1964).
37. L. F. Leloir, Enzymic isomerization and related processes, *Advan. Enzymol.* **14**:193–218 (1953).

Chapter 3

AMINO ACIDS

Williamina A. Himwich and
Harish C. Agrawal*

Thudichum Psychiatric Research Laboratory
Galesburg State Research Hospital
Galesburg, Illinois

I. FREE AMINO ACID POOL OF CENTRAL NERVOUS SYSTEM

A. Origin of the Amino Acids

The amino acids of the brain arise from two sources. In both immature and adult animals some amino acids pass from the blood into the brain and join the amino acid pool. Other amino acids appear to be made solely within the brain substance itself, the carbon skeleton being derived essentially from glucose. If amino acids such as glutamate[1] or aspartate are administered during periods of insulin hypoglycemia, the amino acids appear to enter the brain more readily than under normal conditions. It has not yet been demonstrated whether this is due to the fact that the hypoglycemia *per se* injures the blood–brain barrier, whether the lack of glucose and of amino acids in the blood frees the transport systems (which, therefore, makes it more possible for the amino acids in the blood to move into the brain), or whether the metabolism of free amino acids and protein in the brain releases protein-binding sites which may be used by the incoming amino acids.

Amino acids are known to be synthesized in the brain from glucose fragments.[2] Whether this occurs only in the young animal or also in the older animal has not been completely demonstrated. However, it seems likely that carbon fragments are reused in the brain for the formation of amino acids as well as other compounds, since the entry of these materials and their exit seem to be limited.

B. Permeability of the Brain to Amino Acids in the Blood

Of the 26 or more amino compounds which are either amino acids or so closely related to them that we usually consider them with the amino acids,

* Present address: Charing Cross Hospital Medical School, London, England.

TABLE I

Amino Acid Levels in Various Species

Compound	Mouse (μmoles/g fresh tissue)		Rabbit (μmoles/g fresh tissue)		Guinea pig (μmoles/g fresh tissue)		Cat (μmoles/g fresh tissue)		Dog (μmoles/g fresh tissue)	
	Newborn[a]	Adult[a]	Newborn[b]	Adult[b]	Newborn[c]	Adult[c]	Newborn[d]	Adult[d]	Newborn[d]	Adult[d]
GSH	1.43[e]	1.00	0.75	0.58	1.17	0.93	0.36	0.49	0.57	1.00
Glycerophosphoryl-ethanolamine	0.25	0.47	0.35	0.57	0.97	0.66	0.19	0.77	0.28	0.49
Phosphoethanolamine	4.76	2.01	7.90	1.54	5.35	2.88	4.58	1.19	6.47	1.87
Taurine	16.50	9.13	5.94	1.04	3.42	1.61	9.20	2.30	6.82	1.28
Urea	6.72	4.13	8.32	2.63	3.50	7.73	9.73	3.81	5.36	2.95
Hydroxyproline	0.15	—	0.07	—	0.05	—	0.04	0.08	0.14	—
Aspartic acid	1.86	3.38	1.23	2.05	2.79	2.36	1.08	1.70	1.01	3.28
Threonine	0.68	0.39	0.43	0.14	0.31	0.26	0.58	0.17	0.86	0.29
Serine	1.21	0.78	1.30	0.76	1.79	0.68	1.08	0.48	0.79	0.58
Glutamine	3.78	4.83	3.50	3.08	4.87	3.88	3.85	2.83	2.53	4.66
Sarcosine	0.57	1.24	0.24	0.32	1.97	0.32	0.23	0.20	0.12	—
Proline	0.49	0.06	0.23	0.04	0.18	0.08	0.22	0.03	0.17	0.06
Glutamic acid	4.26	11.50	5.21	8.53	11.14	9.51	6.01	7.87	4.29	9.05
Citrulline	0.21[e]	—	0.06	—	—	—	0.04	0.04	0.03	—
Glycine	2.72	1.61	3.19	0.97	2.55	0.98	2.26	0.78	1.25	0.86
Alanine	3.26	0.66	0.94	0.42	0.94	0.65	2.24	0.48	0.76	0.99
a-Amino-n-butyric acid	0.02	0.03	0.03	0.01	0.03	0.02	0.02	0.02	0.04	0.08
Valine	0.32	0.08	0.21	0.08	0.10	0.06	0.21	0.06	0.09	0.17
Half cystine	0.007[e]	0.01[e]	—	—	—	—	—	—	—	—
Cystathionine	0.05	0.06	0.16	0.23	0.03	0.07	0.04	0.14	0.01	0.13
Methionine	0.02	0.03	0.02	0.01	0.02	0.10	0.01	0.02	—	0.04

TABLE I—*continued*

Compound	Mouse (μmoles/g fresh tissue)		Rabbit (μmoles/g fresh tissue)		Guinea pig (μmoles/g fresh tissue)		Cat (μmoles/g fresh tissue)		Dog (μmoles/g fresh tissue)	
	Newborn[a]	Adult[c]	Newborn[b]	Adult[b]	Newborn[c]	Adult[c]	Newborn[d]	Adult[d]	Newborn[d]	Adult[d]
Isoleucine	0.14	0.04	0.04	0.03	0.04	0.04	0.07	0.03	0.01	0.11
Leucine	0.24	0.08	0.15	0.06	0.08	0.07	0.17	0.07	0.05	0.17
Glucosamine	0.03[e]	—	—	—	—	—	—	—	—	—
Tyrosine	0.14	0.05	0.16	0.05	0.05	0.07	0.11	0.03	0.06	0.04
Phenylalanine	0.13	0.04	0.03	0.03	0.03	0.04	0.05	0.02	0.06	0.07
B-Alanine	0.05	0.05	0.04	0.03	0.04	0.05	0.07	0.02	0.02	0.02
GABA	1.35	2.52	1.18	1.46	1.91	1.88	1.25	1.37	0.78	3.29
Ornithine	0.05[f]	0.01	0.03	0.01	0.03	0.02	0.05	0.01	0.02	0.02
Ethanolamine	1.58	0.12	0.39	0.12	0.29	0.32	1.37	0.48	0.18	0.72
Lysine	0.15	0.18	0.06	0.04	0.02	0.07	0.07	0.08	0.19	0.23
Histidine	0.14[e]	0.05[g]	—	—	—	—	0.02	0.02	—	0.05
Ammonia	—	0.43[g]	—	—	—	—	—	—	—	—
Carnosine	0.04	0.24	—	—	0.07	0.07	—	—	—	0.12
Homocarnosine	—	—	—	—	—	0.059	—	—	—	—
Arginine	0.10[e]	0.13[g]	—	—	—	—	—	—	—	—
N-acetylaspartic acid	—	8.25[g]	—	5.4[f]	—	—	—	—	—	—

[a] Agrawal *et al.*[54]
[b] Agrawal, Davis, and Himwich.[43]
[c] Agrawal, Davis, and Himwich.[55]
[d] Unpublished data.
[e] Levi, Kandera, and Lajtha.[56]
[f] Jacobson.[57]
[g] Tews and Stone.[26]

only a few have been investigated extensively. Glutamate,[3] lysine,[4] leucine,[5] and proline[6] are not increased in brain if blood values are raised, while tyrosine,[7] glutamine,[3] methionine,[8] and phenylalanine[9] are. The work of Lajtha on the penetration of leucine[5] and lysine[4] stands alone as evidence of elegantly planned experiments to demonstrate the penetration of amino acids from blood. Flexner's[9] study on the mouse suggests that even in the adult there is a one to one exchange of tagged glutamic acid from the blood with the non-tagged in the brain; that is, although the concentration of glutamate does not increase, some radioactive glutamate does appear in the brain. In younger animals, it has been shown that brain concentration of glutamate can be increased by increasing the blood levels of glutamic acid.[10] The permeability of the brain to GABA has not been fully established. The statement is frequently made that glutamine enters the brain freely, based on a paper by Schwerin, Bessman, and Waelsch.[3] A confirmation of these data using current methods would be welcome. Waelsch[11] suggested that experiments in which the blood level of an amino acid is elevated are highly unphysiological and may demonstrate a blood–brain barrier function which is normally not operative.

C. Comparative Studies of Amino Acid Composition of the Brain

The free amino acid content of the whole brain does not vary markedly from species to species (Table I). It seems, therefore, that among the vertebrate animals studied the amino acid composition of whole brain presents a fairly constant picture, with some slight change in the pattern in the various species. The free amino acids, glutamic acid, glutamine, aspartic, and GABA in decreasing order of concentration, dominate the free amino acid pool.

D. Distribution in Cortical and Subcortical Areas

Differences of amino acid composition (Table II) appear when we break the brain down into various parts. For example, if we consider the brain on the basis of forebrain, midbrain, and hindbrain, we find that GABA is the dominant amino acid in the midbrain and hindbrain, whereas in the forebrain glutamic is predominant. If the brain is divided into structures which can be accurately sampled, such as neocortex, caudate nucleus, thalamus, and whose functions are more or less known, it is possible to see that, although glutamic acid is the predominant amino acid in the brain as a whole, its place is taken by GABA as we move from rostal structures to the caudal ones. The reason for this shift is not known, and as yet no reasonable explanation has been advanced.

E. Developmental Changes in Whole Brain and in Brain Parts

The generalization may be made that for most amino acids there is no change in concentration in whole brain during growth.[12] Two groups of

TABLE II

Amino Acid Levels in Rat Brain Areas

	Amino acid (μmoles/g fresh tissue)							
	Hemisphere		Midbrain		Pons medulla		Cerebellum	
Phosphorylethanolamine	2.23[a]	1.52[b]	1.08	1.47	0.514	0.84	0.469	0.81
Glycerophosphorylethanolamine	0.404		0.447		0.599		0.500	
Taurine	7.55	5.34	2.81	2.88	1.96	1.89	5.57	4.89
GSH	1.11	0.56	0.584	0.30	0.618	0.29	1.14	0.38
Aspartic acid	2.44	2.40	2.45	2.52	2.68	2.59	1.97	2.19
Glutamine and threonine	2.35	3.17	1.53	2.72	1.64	2.07	2.30	3.16
Serine and asparagine	1.16	1.32[c]	0.537	1.06[c]	0.630	1.33[c]	0.936	1.18[c]
Glutamic acid	11.57	9.26	7.50	6.39	6.85	4.66	10.33	8.42
Citrulline		0.03		0.03	0.139	0.06	0.096	0.07
Glycine	0.629	0.83	1.56	1.45	3.62	3.63	0.627	1.04
Alanine	0.666	0.77	0.395	0.52	0.560	0.64	0.591	0.88
Glucosamine		0.06	0.046	0.08	0.068	0.05	0.027	0.06
Valine	0.083	0.12	0.080	0.15	0.083	0.16	0.065	0.15
Cystine	0.006	0.01	0.012	0.02	0.030	0.02	0.024	0.02
Methionine	0.037	0.04	0.045	0.05	0.043	0.05	0.050	0.05
Cystathionine	0.015	0.02	0.025	0.03	0.053	0.07	0.268	0.23
Isoleucine	0.025	0.05	0.028	0.06	0.034	0.07	0.028	0.06
Leucine	0.064	0.10	0.071	0.11	0.064	0.12	0.065	0.12
Tyrosine	0.073	0.07	0.062	0.08	0.059	0.09	0.070	0.10
Phenylalanine	0.050	0.06	0.055	0.07	0.050	0.08	0.052	0.09
Ethanolamine	0.112	0.38	0.140	1.45	0.177	2.73	0.080	1.47
γ-Aminobutyric acid	2.33	1.52	3.79	3.06	1.77	1.32	1.54	1.56
NH$_3$	0.986	1.94	0.806	2.51	0.892	2.50	0.661	2.71
Ornithine	0.020	0.07	0.040	0.11	0.051	0.16	0.034	0.15
Lysine	0.207	0.22	0.363	0.36	0.435	0.52	0.366	0.49
Histidine	0.066	0.11	0.059	0.12	0.044	0.14	0.062	0.17
Carnosine	0.124	0.06	0.222	0.09	0.216	0.12	0.135	0.09
Arginine	0.111	0.12	0.186	0.15	0.200	0.23	0.145	0.19

[a] Kandera, Levi, and Lajtha.[58]
[b] Shaw and Heine.[59]
[c] Serine value only.

amino acids diverge from this general statement. Glutamic acid and its metabolic derivatives, aspartic acid, and GABA tend to show an increase in concentration as the brain matures. Glutamine is an exception to this change. Another group of amino acids, i.e., phenylalanine, glycine, leucine, taurine, alanine, phosphoethanolamine, proline, and phenylalanine, are high in the brain at birth or shortly thereafter, and fall rapidly during the period of maturation. This fall may be due to a number of things: if we assume the high concentration is a reflection of the concentration in the mother's blood, then the fall may be simply due to slow leakage of the amino acids out of the brain. On the other hand, it may simply mean that these amino acids do not enter the brain after birth and the end result is seen merely as the dilution effect of myelin and the growing tissue as a whole. This problem is of considerable interest but is not yet near solution. In the brain parts, essentially the same sort of patterns can be seen. It is of interest that the hippocampus tends to be closer to its mature composition of amino acids at birth even in animals born relatively immature, such as the dog, than other brain parts.[13] These data suggest a confirmation of Purpura's[14] electrophysiological finding for this structure. Berl,[15] however, found glutamine synthetase as low in the hippocampus as in the neocortex. It is also of interest that animals born as relatively mature as the guinea pig show only slight changes in amino acid composition in the brain parts as they mature.[16]

In general, attempts to correlate levels of free amino acids with cellular maturity measured either histologically, electrophysiologically, or behaviorally have been disappointing. It is possible that studies on the cellular level may be more rewarding. On the other hand, the chance that in such studies we are trying to compare two concomitant but unrelated phenomena should also be considered.

II. GLUTAMIC ACID AND ITS DERIVATIVES

A. Cellular Localization

Whittaker[17] using the methods of differential centrifugation developed in his laboratory concluded that there is no specific localization of the members of the glutamic acid family in the fraction rich in isolated presynaptic nerve terminals.

De Robertis and his colleagues,[18] however, have devised an ingenious method of cellular localization studies using the distribution of the enzymes involved in the metabolism of GABA, glutamate, and glutamine. From these data the localization of the amino acids can be inferred (Fig. 1). On this basis, GABA appears to have a possible role at the synaptic complex with two hypothetical mechanisms of action—one of which requires the transsynaptic diffusion of GABA to regulate excitability and the other depending upon accumulation of the amino acid within synaptic vesicles ready to be delivered when a nerve impulse arrives. The essential difference between an unspecific role for GABA and a true transmitter action may lie in these two mechanisms. De Robertis speculates on two metabolic paths for glutamate—the amino acid present in the mitochondria being metabolized through the citric acid cycle and that in the cytoplasm into GABA at the nerve endings and into glutamine in the neuronal perikarya. The aspects of glutamate compartmentation will be discussed in Chapter 19 of Volume II of this series.

B. Factors Affecting

The presence of GABA exclusively in the neural tissues and the preponderance of glutamic acid, glutamine, GABA, and aspartate (which form

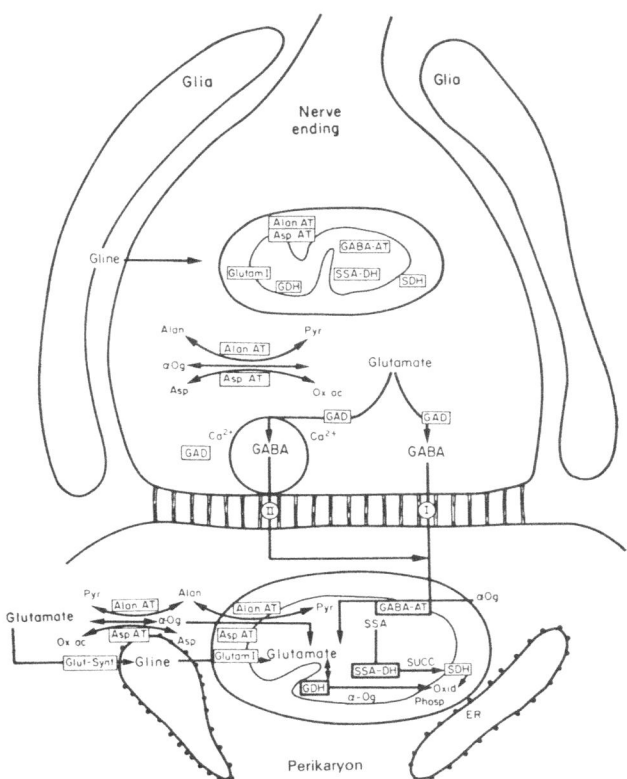

Fig. 1. General diagram of the structural and biochemical organization of the synaptic complex. (See description in the text.) Within the nerve ending there are the following subcompartments: (1) axoplasm, (2) mitochondrion, and (3) synaptic vesicle; within the perikaryon: (1) neuroplasm, (2) mitochondrion, and (3) endoplasmic reticulum (ER). The localization of the enzymes of the glutamate, glutamine, and GABA cycles in the different compartments and subcompartments is indicated. Within the mitochondrion the localization may be in the membranes or in the matrix. The different thicknesses of the lines around the enzymes give an idea of the Enz/SDH ratios. (For example, GDH, SSA–DH and GABA–AT in the perikaryon have the highest Enz/SDH ratio.) The various reactions, as well as the probable localization of the amino acids glutamate, GABA, and glutamine, are indicated. The two possible presynaptic mechanisms for GABA described in the text are indicated by I and II.

TABLE III

Factor	Animal	Brain parts	Amino acid				
			GABA	GLU-NH₂[a]	GA[a]	Asp[a]	GSH[a]
Psychotropic drugs							
CPZ[a](60)	Rat	Whole brain	↑	↑	↑		↑
CPZ(52)	Rat	Whole brain	→	(Acetylaspartic acid ↓)			
CPZ(61)	Rat	Cerebral cortex	→	—	→	→	↑
CPZ(62)	Monkey	Cortical and subcortical	→	(PE[a], serine, and glycine ↓)			
		Cortical and subcortical		(Cysteine, cystathionine, and alanine ↓)			—
		Midbrain	—	—	—	—	—
		Cerebellum	—	—	—	—	—
		Hypothalamus	—	—	—	—	—
Reserpine(63)	Monkey	Cortical and subcortical	→	→	→	→	—
			(Alanine ↑ except frontal lobe)				
(52)	Rat	Whole brain	→				
(64,65)	Mouse	Whole brain	→				
Ethanol(66–68)							
(69)	Rat	Whole brain	↑	→	↑	↑	—
(70)	Rat	Whole brain	←	—	←	←	—
(71)	Rat	Whole brain	→	—	—	—	—
	Rat	Cerebellum and hemispheres	→	→	→	—	—
Insulin							
hypoglycaemia(72–74)(23)	Rat	Whole brain	→	→	→	↑←	↑←
	Rat	Cerebral cortex, pons medulla	(Alanine and glycine ↓; NH₃ and GSH ↑)				
(20)	Rat	Cerebellum, midbrain					
	Dog	Cerebral cortex	→	(Alanine and glycine ↓; NH₃ and GSH ↑)			
Thyroidectomy(75)	Rat	Cerebral cortex					

TABLE III—*continued*

Factor	Animal	Brain parts	Amino acid				
			GABA	GLU-NH$_2$^{a}	GA^a	Asp^a	GSH^a
Insulin—*continued*							
Hibernation[76]	Garden dormice	Cerebral hemispheres	↑	↑ (Temporary)		↗ (Permanent)	—
[77]	Ground squirrel	Frontal cortex	↑	↑	↑	↗	—
				(Tyrosine and threonine ↓)			
		Hypothalamus	↑↓	—	—	↗	—
		Thalamus	↑	—	—	↗	—
Arousal[78]	Ground squirrel	Hypothalamus	→↓	—	—	—	—
		Caudate nucleus	→	—	—	—	—
		Pons	↑	—	—	—	—
Convulsions							
^{b}AOAA[26]	Dog	Cerebral cortex	↑	↑	—	↓(Tyrosine and NH$_3$↑)......	—
[79]	Mouse	Whole brain	←←	←	—	—	—
L-2,4-diaminobutyric acid[80]	Mouse	Whole brain	→→	→→	→	—	—
^{c}MSO[32]	Dog	Cerebral cortex	→	→→→	→→→	—	→
				(Methionine and cystathionine reduced)			
Local freezing[81] Strychnine[82] Pentylenetetrazol	Cat	Cortex	—	→	→	—	—
4-methoxy methyl-pyridoxine Hydrazides[20,30–32,83,84]	Rat	Whole brain	→	—	↑	—	→

TABLE III—*continued*

Factor	Animal	Brain parts	Amino acid				
			GABA	GLU-NH₂[a]	GA[a]	Asp[a]	GSH[a]
Convulsions—*continued*							
Hydroxylamine[33,35]	Rat	Whole brain and parts	↑	—	—	—	—
Environmental deprivation							
Isolation[85]	Dog	Cortical structures	—	—	—	—	—
		Thalamus–hypothalamus	←	→	←		
		Caudate nucleus	→	→	→		
		Superior colliculi		←	←		
Visual deprivation[86,87]	Rabbit	Occipital cortex	—	←	←	←	—
		Superior colliculi		←	←	→	
		Retina	→	—	—	—	—
Early weaning[88]	Rat	Cerebral cortex	←	←	←	←	
(15 day)		Midbrain	→	→	→	→	
(19 day)[88]	Rat	Cerebral cortex	—	←	—	←	
		Midbrain	→	—	→	→	—
Hypoxia							
Stagnant[89]	Rat						
5 days		Prosencephalon	←	—	→	→	
25 days		Prosencephalon	←	→	→	→	
		Rhombencephalon	—	—	→	→	
	Adult	Prosencephalon	—	—	→	—	
5% O₂[90]	Rat	Whole brain	←	—	—	—	
8% O₂[37]	Rat	Whole brain	↑	—	—	—	

[a] *Notation:* GLU-NH₂, glutamine; GA, glutamic acid; Asp, aspartic acid; GSH, glutathione; CPZ, chlorpromazine; PE, phosphoethanolamine; ↑ increasing concentration; ↓ decreasing concentration; — no change.

[b] AOAA denotes amino-oxyacetic acid.

[c] MSO denotes methionine sulfoxamine.

more than two-thirds of α-amino nitrogen of the cerebral pool) have induced many investigators to evaluate the effect of different psychotropic drugs, hormones, hypoxia, and environmental deprivation on members of the glutamic acid family in the brain in order to ascertain a direct role played by this group of amino acids in the physiological functioning of the fully integrated nervous system. The data are presented in a summary (Table III). Only those amino acids or related compounds which showed significant changes are given.

1. Psychotropic Drugs and Ethanol

Alterations in free amino acids of the brain after administration of psychotropic drugs is neither a consistent nor a reproducible phenomenon. Moreover, the changes observed in some amino acids of the brain after treatment with chlorpromazine, reserpine, imipramine, desmethylimipramine, and haloperidol cannot be correlated with the pharmacological actions of these drugs. The data suggest that some unappreciated factor in the experimental design is influencing the results.

The effect of ethanol intoxication on brain amino acids has been extensively studied by many investigators (Table III) and presents as confused a picture as the psychotropic drugs.

2. Insulin

The observations of amino acid changes during insulin hypoglycemia indicate that some of the dicarboxylic amino acids, especially glutamic acid, may also serve as a substrate for energy in lieu of glucose. The simultaneous decrease in glutamic acid accompanied by an increase in aspartic acid in neural tissues suggest that their carbon skeleton is largely derived from glucose and also indicate their intimate relationship to the TCA cycle. The increase in aspartate probably results from the reduced level of glucose, which leads to a limited production of citrate by the combination of acetyl coenzyme A and oxaloacetate. The oxaloacetate under these circumstances forms aspartate by transamination.[19] The prompt return of glutamate in brain to its normal level following the administration of glucose in the hypoglycemic animal suggests that normal brain function requires a certain level of free glutamic acid. The attainment of normal glutamate level, however, does not restore the EEG.[1]

It is tempting to speculate[20] that the convulsions which occur may be due to the fall in GABA as well as the release of lower centers by the hypoglycemia.[21] It would seem that the changes in amino acids are the result of the convulsions rather than their cause.[22] Moreover, it must also be noted that there is a significant increase in ammonia levels in all areas of the brain during hypoglycemic convulsions[23] and that the level of ammonia in the CNS has been frequently correlated with the functional activity of the CNS.[24-25] Therefore, the possibility cannot be ruled out that the increase in the ammonia content of the brain during hypoglycemia may be one of the contributing causes of convulsions.

3. Hibernation

Although the physiological significance of the findings in hibernation and arousal as well as the mechanism underlying arousal and the direct involvement of GABA and aspartate in the phenomenon of hibernation and arousal remain to be elucidated, it appears that during hibernation the changes in the concentrations of some amino acids may be a direct biochemical consequence of diminished activity of the Krebs cycle.

4. Convulsions

The effect of convulsive agents on the free amino acids of the brain have been extensively studied.[26] It appears from the data reported by various investigators that the changes in amino acids of the CNS are the result rather than the causes of convulsion.[22]

The effect of convulsant hydrazides and hydroxylamine on the free amino acid pool, in particular on brain GABA, have been reviewed by Roberts and Eidelberg[19] and Williams and Bain.[27] Hydrazides and hydroxylamine are carbonyl-trapping agents which react with the aldehyde groups of pyridoxal phosphate, and it must be pointed out that the enzymes responsible for the synthesis (glutamate decarboxylase, GAD) and degradation of GABA (GABA-α-keto-glutarate transaminase, GABA-T) are pyridoxal-phosphate-dependent. In addition, it is believed that GABA-T binds the coenzyme more firmly than does GAD and that hydroxylamine in all probability reacts with the bound form.[19] GABA is essentially formed in the central nervous system by decarboxylation of glutamic acid catalyzed by GAD and further metabolized by GABA-T. The hydrazides and hydroxylamine preferentially act on the synthesis and catabolism of GABA in the central nervous system. The former reduces its concentration by inhibiting GAD,[28-32] while the latter raises GABA levels by inhibiting the catabolizing enzyme, GABA-T.[33,34] No satisfactory explanation has been advanced to explain the convulsant action of hydrazides and hydroxylamine, however, it should be noted that, among the nitrogenous compounds of the brain, GABA is the only amino acid that shows a consistent pattern of change.

5. Environmental Deprivation

Experiments with animals reared in various environmental conditions have shown that social contact with other animals or the mother during postnatal ontogenesis is necessary for the integration of normal patterns of behavior and biochemical maturation of the brain. Studies have been made on the effect of isolation, visual deprivation, and early weaning on the free amino acids and protein content of discrete anatomical areas of the brain. Most of the investigators favor the view that a certain amount of physiological stimulation is needed for the normal development of the CNS and that alterations in environmental conditions interfere with the metabolism of protein and amino acids in the brain during development.

A plausible biochemical explanation for the increases in amino acids of the occipital cortex and superior colliculi, especially at 11 days of age, in animals visually deprived may be that the deprivation induced significant alterations in the normal rate of accumulation or in the metabolism of these amino compounds. The data for 11-day-old animals may be viewed as an animal's initial response to stress and the variation at 15 days (which returned to almost normal value) may suggest the animal's homeostatic adaptation to the environmental stress.[35]

6. Early Weaning

The increases in the free amino acids of the cortex of animals subjected to early weaning may reflect a change in the rate of protein synthesis, such that the amino acids, although accumulated at approximately the same rate as in the control, were not as rapidly incorporated into the protein.[36] On the other hand, the changes may reflect a non-specific stress response. If we accept either argument, we must assume that different anatomical areas of the brain respond differently to the same kind of stress. Different responses may also be seen at different developmental periods.

7. Hypoxia

A plausible explanation for the increase in brain GABA under hypoxic condition may be decreased utilization of the compound via GABA shunt pathway, i.e., reduced oxidative metabolism.[37]

C. Function

The function of this group of amino acids in the living brain is still not well established, in spite of much work on the subject. Krnjevic and his colleagues[38] using iontophoretic methods demonstrated that under the conditions of their experiments glutamic acid and aspartic acid are stimulators for neurons, while GABA as well as many other substances are inhibitory. Their work, however, does not appear to be generally accepted as relevant to brain function *in situ*. One problem is that in the intact animal brain GABA can be raised or lowered or glutamate decreased without producing clear-cut and reproducible behavioral effects. The most elegant and convincing data came from Jasper, Kahn, and Elliott,[39] who measured the amino acids released from the perforated pial cortex of the cat in different EEG states of "arousal." These data tend to confirm those of Krnjevic. However, confirmation from another laboratory has not yet appeared. Attempts to promote convulsions by lowering GABA[40] and to prevent them by increasing GABA[41,42] have been attempted by many workers, with widely divergent results. These results can be explained on many bases—part or all of which may be operant in any given experimental situation: (1) It may be that only the amino acid present in certain cellular structures is capable of influencing excitability. (2) If GABA, for example, is increased due to drug effect, the increase may be in neuroglia, or in the wrong place in the neurons for the GABA to be inhibitory (see Cellular Localization above). (3) The methods used to determine the amino acid concentration may not be sufficiently accurate to detect small differences at the lower end of the concentration gradient where they may be expected to have an influence. (4) The behavioral tests may be too gross or not of the proper kind to detect changes. The wealth of data accumulating on the concentrations and distribution of these substances in various areas of the brain is an "embarrassment of riches" when no functional role can be assigned to them.

III. AMINO ACIDS OTHER THAN GLUTAMIC ACID AND DERIVATIVES

Little is known of the function of these amino acids or of the factors which may affect them, other than those discussed above. Proline, valine, isoleucine, leucine, tyrosine, and phenylalanine are at relatively high levels in the brain during the early postnatal life of the rat and rabbit and fall to approximately adult levels in 2 to 3 weeks after birth.[12,43] With the exception of tyrosine and phenylalanine, these amino acids show a sharp rise after birth. The reasons for these changes are unknown. Several possible factors may be acting simultaneously; for example: (1) changes in protein synthesis after birth may result in differing amounts of these amino acids being used, (2) the initial high levels may reflect levels in the mother's blood just prior to delivery, and (3) the slow fall in concentration may be the result of the slow passage of these substances out of the brain. It is of interest that phenylalanine feeding to young animals does not appear to increase brain phenylalanine,[44] although it would seem that the blood–brain barrier is relatively permeable to phenylalanine. The presence of arginine in the CNS is not a consistent finding.

Taurine is another amino acid which is found in extremely high concentrations in the young animal and which falls during the period of maturation.[45] Marked species differences are also seen (Table I). The function of this amino acid is unknown in brain, although the levels may be affected by thyroidal function.[46] Low levels of taurine in the urine have been reported in children with mongolism, but it is difficult to equate the urinary level with the amount of taurine in brain.[47]

The current interest in cystathionine and the mental deficiency accompanying homocystinuria has resulted in analyses of human brain for this amino acid. Garritsen and Waisman[48] were the first to demonstrate the absence of cystathionine in the brains of patients with homocystinuria. In the dog, the values are normally low[8] but can be increased by methionine loading (personal observation). Werman, Davidoff, and Aprison[49] have suggested that this amino acid is an inhibitory transmitter similar to glycine.

Aprison and his colleagues[50] have presented an excellent case for glycine as an inhibitory transmitter in the cat spinal cord. These data have been confirmed by Curtis and his group.[51]

TABLE IV

Peptide	Animal or species	Comments
N-acetyl aspartyl glutamic acid[91]	Rabbit and horse brain	In horse brain rostral to caudal levels decreased.
N-acetyl-α-aspartyl glutamic acid[92,93]	Bovine brain Rabbit Dog Rat Guinea pig Man	N-acetyl aspartic acid serves as precursor for the synthesis of this peptide since the concentration of NAA is much higher than the N-acetyl-α-aspartyl glutamic acid. Rostral areas lowest.
N-acetyl aspartyl glutamic acid and N-acetyl-L-glutamic acid[94–96]	Human brain	
Homocarnosine[97]	All the mammalian species, including man	The distribution pattern of the peptide is similar to GABA in various parts of the brain.
Glutathione[52]	Found in almost all the species	See review by Tallan.[52]
γ-L-glutamyl L-glutamine[98]	All mammalian species, including man	Highest concentration in the chicken brain and lowest in the dog brain.
γ-L-glutamyl-L β-Aminoisobutyric acid[99]		
γ-L-glutamyl-L glutamic acid and γ-L-glutamyl-L-glutamine[100]	Bovine brain	
γ-Glutamylserine, γ-glutamylalanine γ-glutamylvaline and S-methyl glutathione[101]		
γ-L-Glutamyl glycine[102]		
N-acetyl aspartic acid (NAA)[52]	All the species	Concentration increased rostral to caudal in horse brain.[91]
Homocarnosine, homoanserine 3-methyl histidine 1-methyl histidine[103]	Bovine brain	
4 Hypothalamic peptides[104]	Dog brain	None of the peptide had pressor or ACTH releasing activity.
N-acetyl histidine[105]	Reptiles, amphibians, and fish	

IV. PEPTIDES

Many peptides have been identified in brains from various species of animals (Table IV). These peptides, in most cases, are peptides of glutamic acid. In spite of vigorous analytical work, the physiological significance of these peptides in the diverse functional activity of the CNS remains obscure. The fact that the levels of N-acetyl aspartic acid (NAA) increase with age[52] and its level in young can be influenced by thyrotoxic drugs[53] may offer a lead as to the functional role of this substance. Acetyl groups of the N-acetylated peptides in general may also serve as donors of acetyl group in the synthesis of brain lipids during myelination.

V. REFERENCES

1. W. A. Himwich and J. C. Petersen, Recovery from insulin coma and hematoencephalic exchange, *Diseases Nervous System* **19**:104–107 (1958).
2. A. Beloff-Chain, R. Catanzaro, E. B. Chain, I. Masi, and F. Pochiari, Fate of uniformly labelled 14C-glucose in brain slices, *Proc. Roy. Soc.* **B144**:22–28 (1955).
3. P. Schwerin, S. P. Bessman, and H. Waelsch, The uptake of glutamic acid and glutamine by brain and other tissues of the rat and mouse, *J. Biol. Chem.* **184**:37–44 (1950).
4. A. Lajtha, Amino acid and protein metabolism of the brain. II. The uptake of L-lysine by brain and other organs of the mouse at different ages, *J. Neurochem.* **2**:209–215 (1958).
5. A. Lajtha and J. Toth, The brain barrier system. II. Uptake and transport of amino acids by the brain, *J. Neurochem.* **8**:216–225 (1961).
6. W. Dingman and M. B. Sporn, The penetration of proline and proline derivatives into brain, *J. Neurochem.* **4**:148–153 (1959).
7. M. A. Chirigos, P. Greengard, and S. Udenfriend, Uptake of tyrosine by rat brain *in vivo*, *J. Biol. Chem.* **235**:2075–2079 (1960).
8. W. A. Himwich, unpublished data.
9. R. B. Roberts, J. B. Flexner, and L. B. Flexner, Biochemical and physiological differentiation during morphogenesis. XXIII. Further observations relating to the synthesis of amino acids and proteins by the cerebral cortex and liver of the mouse, *J. Neurochem.* **4**:78–90 (1959).
10. W. A. Himwich, J. C. Petersen, and M. L. Allen, Hematoencephalic exchange as a function of age, *Neurology* **7**:705–710 (1957).
11. H. Waelsch, *in Neurochemistry* (K. A. C. Elliott, I. H. Page, and J. H. Quastel, eds.), second edition, pp. 288–320, Charles C. Thomas, Springfield, Ill. (1955 and 1962).
12. H. C. Agrawal, J. M. Davis, and W. A. Himwich, Postnatal changes in free amino acid pool of rat brain, *J. Neurochem.* **13**:607–615 (1966).
13. A. R. Dravid, W. A. Himwich, and J. M. Davis, Some free amino acids in dog brain during development, *J. Neurochem.* **12**:901–906 (1965).
14. D. P. Purpura, *in Neurological and Electroencephalographic Correlative Studies in Infancy* (P. Kellaway, ed.), p. 117, Grune and Stratton, New York (1964).
15. S. Berl, Glutamine synthetase. Determination of its distribution in brain during development, *Biochemistry* **5**:916–922 (1966).
16. J. M. Davis, W. A. Himwich, and H. C. Agrawal, Free amino acids of newborn and adult guinea pig brain, *Develop. Psychobiol.* (in press).
17. V. P. Whittaker, *in Progress in Biophysics and Molecular Biology* (J. A. V. Butler and H. E. Husley, eds.), Vol. 15, pp. 39–96, Pergamon, Oxford (1965).
18. L. Salganicoff and E. DeRobertis, Subcellular distribution of the enzymes of the glutamic acid, glutamine and γ-aminobutyric acid cycles in rat brain, *J. Neurochem.* **12**:287–309 (1965).
19. E. Roberts and E. Eidelberg, *in International Review of Neurobiology* (C. C. Pfeiffer and J. R. Smythies, eds.), Vol. 2, pp. 279–332, Academic Press, New York–London (1960).

20. J. K. Tews, S. H. Carter, and W. E. Stone, Chemical changes in the brain during insulin hypoglycaemia and recovery, *J. Neurochem.* **12**:679–693 (1965).
21. H. E. Himwich, *in Brain Metabolism and Cerebral Disorders*, Williams and Wilkins Co., Baltimore (1951).
22. A. Lajtha, *in International Review of Neurobiology* (C. C. Pfeiffer and J. R. Smythies, eds.), Vol. 6, pp. 1–98, Academic Press, New York–London (1964).
23. R. K. Shaw and J. D. Heine, Effect of insulin on nitrogeneous constituents of rat brain, *J. Neurochem.* **12**:527–532 (1965).
24. H. Weil-Malherbe, *in Neurochemistry* (K. A. C. Elliott, I. H. Page, and J. H. Quastel, eds.), second edition, pp. 321–330, Charles C. Thomas, Springfield, Ill. (1962).
25. G. Pintillie, N. Mison-Crighel, and G. Badiu, Convulsive effect elicited by topical application of penicillin on glutamate–glutamine system of brain, *Nature* **214**: 1131–1132 (1967).
26. J. K. Tews and W. E. Stone, *in Progress in Brain Research, Horizons in Neuropsychopharmacology* (W. A. Himwich and J. P. Schade, eds.), Vol. 16, pp. 135–163, Elsevier, Amsterdam (1965).
27. H. L. Williams and J. A. Bain, Convulsive effect of hydrazides: relationship to pyridoxine, *Intern. Rev. Neurobiol.* **3**:319–348 (1961).
28. K. F. Killam and J. A. Bain, Convulsant hydrazides. I. *In vitro* and *in vivo* inhibition of vitamin B_6 enzymes by convulsant hydrazides, *J. Pharmacol. Exptl. Therap.* **119**:255–262 (1957).
29. C. F. Baxter and E. Roberts, Demonstration of thiosemicarbazide-induced convulsions in rats with elevated brain levels of γ-aminobutyric acid, *Proc. Soc. Exptl. Biol.* **104**:426–427 (1960).
30. K. A. C. Elliott and N. M. Van Gelder, The state of Factor I in rat brain: the effects of metabolic conditions and drugs, *J. Physiol.* **153**:423–432 (1960).
31. E. W. Maynert and H. K. Kaji, On the relationship of brain γ-aminobutyric acid to convulsions, *J. Pharmacol. Exptl. Therap.* **137**:114–121 (1962).
32. J. K. Tews and W. E. Stone, Effects of methionine sulfoxamine on levels of free amino acids and related substances in brain, *Biochem. Pharmacol.* **13**:543–545 (1964).
33. E. Roberts, C. F. Baxter, and E. Eidelberg, *in Structure and Function of the Cerebral Cortex* (D. B. Tower and J. P. Schade, eds.), p. 392, Elsevier, Amsterdam (1960).
34. C. F. Baxter and E. Roberts, Elevation of γ-aminobutyric acid in rat brain with hydroxylamine, *Proc. Soc. Exptl. Biol.* **101**:811–815 (1959).
35. W. A. Himwich, Multi-disciplined studies of the visual system in developing rabbits, *International Symposium Neuro-ontogeneticum*, Prague (1967), in press.
36. W. A. Himwich, J. M. Davis, and H. C. Agrawal, Effects of early weaning on some free amino acids and acetylcholinesterase activity of rat brain, *in Recent Advances in Biological Psychiatry*, Vol. 10, pp. 266–270, Plenum Press, New York (1968).
37. J. D. Wood, A possible role for gamma-amino butyric acid in the homeostatic control of brain metabolism under conditions of hypoxia, *Exptl. Brain Res.* **4**:81–84 (1967).
38. K. Krnjevic and J. W. Phillis, Iontophoretic studies of neurones in the mammalian cerebral cortex, *J. Physiol.* **165**:274–304 (1963).
39. H. H. Jasper, R. T. Khan, and K. A. C. Elliott, Amino acids released from the cerebral cortex in relation to its state of activation, *Science* **147**:1448–1449 (1965).
40. E. W. Maynert and H. K. Kaji, On the relationship of brain γ-aminobutyric acid to convulsions, *J. Pharmacol. Exptl. Therap.* **137**:114–121 (1962).
41. R. Tapia, H. Pasantes, M. Perez de la Mora, B. G. Ortega, and G. H. Massieu, Free amino acids and glutamate decarboxylase activity in brain of mice during drug-induced convulsions, *Biochem. Pharmacol.* **16**:483–496 (1967).
42. C. F. Baxter and E. Roberts, Elevation of γ-aminobutyric acid in rat brain with hydroxylamine, *Proc. Soc. Exptl. Biol. Med.* **101**:811–815 (1959).

43. H. C. Agrawal, J. M. Davis, and W. A. Himwich, Postnatal changes in free amino acid pool of rabbit brain, *Brain Res.* 3:374–380 (1966–1967).
44. W. A. Himwich, A. R. Dravid, and T. F. C. Berk, in *Regional Development of the Brain in Early Life* (A. Minkowski, ed.), pp. 221–238, William Clowes and Sons Ltd., London (1967).
45. H. C. Agrawal, J. M. Davis, and W. A. Himwich, Maturational changes in amino acids in CNS of different mammalian species, in *Recent Advances in Biological Psychiatry*, Vol. 10, pp. 258–265, Plenum Press, New York (1968).
46. A. E. Ramirez De Guglielmone and C. J. Gómez, Influence of neonatal hypothyroidism on amino acids in developing rat brain, *J. Neurochem.* 13:1017–1025 (1966).
47. J. J. Thomas, H. O. Goodman, J. S. King, Jr., and A. Wainer, Taurine excretion and intelligence in mongolism, *Proc. Soc. Exptl. Biol. Med.* 119:832–833 (1965).
48. T. Gerritsen and H. A. Waisman, Homocystinuria: absence of cystathionine in the brain. *Science* 145:588 (1964).
49. R. Werman, R. A. Davidoff, and M. H. Aprison, The inhibitory action of cystathionine, *Life Sci.* 5:1431–1440 (1966).
50. R. Werman, R. A. Davidoff, and M. H. Aprison, Is glycine a neurotransmitter? *Nature* 214:680–681 (1967).
51. D. R. Curtis, L. Hosli, G. A. R. Johnston, and I. H. Johnston, Glycine and spinal inhibition, *Brain Res.* 5:112–114 (1967).
52. H. H. Tallan, in *Amino Acid Pools* (J. T. Holden, ed.), pp. 471–485, Elsevier, Amsterdam–London–New York (1962).
53. E. Mussini, F. Marcucci, and S. Garattini, Postnatal changes in brain N-acetyl-L-aspartic acid content of normal and hypothyroid suckling rats, *J. Neurochem.* 14:551–554 (1967).
54. H. C. Agrawal, J. M. Davis, and W. A. Himwich, Developmental changes in mouse brain: weight, water content and free amino acids, *J. Neurochem.* 15:917–923 (1968).
55. H. C. Agrawal, J. M. Davis, and W. A. Himwich, Changes in some free amino acids of guinea pig brain during postnatal ontogeny, *J. Neurochem.* (in press).
56. G. Levi, J. Kandera, and A. J. Lajtha, Control of cerebral metabolite levels. I. Amino acid uptake and levels in various species, *Arch. Biochem. Biophys.* 119:303–311 (1967).
57. K. B. Jacobson, Studies on the role of N-acetylaspartic acid in mammalian brain, *J. Gen. Physiol.* 43:323 (1959).
58. J. Kandera, G. Levi, and A. Lajtha, Control of cerebral metabolite levels. II. Amino acid uptake and levels in various areas of the rat brain, *Arch. Biochem. Biophys.* (1968), in press.
59. R. K. Shaw and J. D. Heine, Ninhydrin positive substances present in different areas of normal rat brain, *J. Neurochem.* 12:151–155 (1965).
60. N. Okumura, S. Otsuki, and H. Nasu, The influence of insulin hypoglycemic coma, repeated electroshock and chlorpromazine or B-phenylisopropyl methylamine administration on the free amino acids in the brain, *J. Biochem. (Tokyo)* 46:247–249 (1959).
61. R. S. Piha, S. S. Oja, and A. J. Uusitalo, The effect of chloropromazine on free amino acids in the rat brain, *Ann. Med. Exptl. Biol. Fenniae (Helsinki)* 40:5–26 (1962).
62. S. I. Singh and C. L. Malhotra, Amino acid content of monkey brain. IV. Effects of chlorpromazine on some amino acids of certain regions of monkey brain, *J. Neurochem.* 14:135–140 (1967).
63. S. I. Singh and C. L. Malhotra, Amino acid content of monkey brain—effects of reserpine on some amino acids of certain regions of monkey brain, *J. Neurochem.* 11:865–872 (1964).
64. H. Balzer, P. Holtz, and D. Palm, Untersuchungen über die biochemischen Grundlagen der konvulsiven Wirkung von Hydraziden, *Naunyn-Schmiedebergs Arch. Exptl. Pathol. Pharmakol.* 239:520–552 (1960).
65. H. Balzer, P. Holtz, and D. Palm, Reserpin und γ-Aminobuttersäuregehalt des Gehirns, *Experientia* 17:38–40 (1961).

66. H.-M. Hakkinen and E. Kulonen, Increase in the γ-aminobutyric acid content of rat brain after ingestion of ethanol, *Nature* **184**:726 (1959).

67. H.-M. Hakkinen and E. Kulonen, The effect of ethanol on the amino acids of rat brain with a reference to the administration of glutamine, *Biochem. J.* **78**:588–593 (1961).

68. H.-M. Hakkinen and E. Kulonen, Comparison of various methods for the determination of γ-aminobutyric acid and other amino acids in rat brain with reference to ethanol intoxication, *J. Neurochem.* **10**:489–494 (1963).

69. R. A. Ferrari and A. Arnold, The effect of central nervous system agents of rat-brain γ-aminobutyric acid level, *Biochim. Biophys. Acta* **52**:361–367 (1961).

70. E. S. Higgins, The effect of ethanol on GABA content of rat brain, *Biochem. Pharmacol.* **11**:394–395 (1962).

71. E. R. Gordon, The effect of ethanol on the concentration of γ-aminobutyric acid in the rat brain, *Can. J. Physiol. Pharmacol.* **45**:915–918 (1967).

72. R. M. C. Dawson, Studies on the glutamine and glutamic acid content of the rat brain during insulin hypoglycaemia, *Biochem. J.* **47**:386–391 (1950).

73. R. M. C. Dawson, Cerebral amino acids in fluoroacetate-poisoned, anaesthetised, and hypoglycaemic rats, *Biochim. Biophys. Acta* **11**:548–552 (1953).

74. R. O. Cravioto, G. Massieu, and J. J. Izquierdo, Free amino acids in rat brain during insulin shock, *Proc. Soc. Exptl. Biol. (N.Y.)* **78**:856–858 (1951).

75. A. E. R. DeGuglielmone and C. J. Gomez, Influence of neonatal hypothyroidism on amino acids in developing rat brain, *J. Neurochem.* **13**:1017–1025 (1966).

76. P. Mandel, Y. Godin, J. Mark, and Ch. Kayer, The distribution of free amino acids in the central nervous system of garden dormice during hibernation, *J. Neurochem.* **13**:533–537 (1966).

77. Lj. T. Mihailovic, Lj. Krzalic, and D. Cupic, Changes of glutamine, glutamic acid and GABA in cortical and subcortical brain structures of hibernating and fully aroused ground squirrels *(Citellus citellus)*, *Experientia* **21**:1–5 (1965).

78. Lj. Krzalic, D. Cupic, and Lj. Mihailovic, Changes of aspartic acid in various cerebral structures of hibernating and aroused ground squirrels, *Arch. Intern. Physiol. Biochim.* **73**:817–825 (1965).

79. D. P. Wallach, Studies on the GABA pathway. I. The inhibition of γ-aminobutyric acid–γ-ketoglutaric acid transaminase *in vitro* and *in vivo* by U-7524 (amino-oxyacetic acid), *Biochem. Pharmacol.* **5**:323–331 (1961).

80. K. Kuriyama, E. Roberts, and M. K. Rubinstein, Elevation of γ-aminobutyric acid in brain with amino-oxyacetic acid and susceptibility to convulsive seizures in mice: A quantitative re-evaluation. *Biochem. Pharmacol.* **15**:221–236 (1966).

81. S. Berl, D. P. Purpura, M. Girado, and H. Waelsch, Amino acid metabolism in epileptogenic and non-epileptogenic lesions of the neocortex (cat), *J. Neurochem.* **4**:311–317 (1959).

82. Y. Aelony, J. Logothetis, B. Bart, F. Morell, and M. Bovis, Free amino acid concentrations in cerebral cortex of guinea-pigs with epileptogenic lesions, *Exptl. Neurol.* **5**:525–532 (1962).

83. C. F. Baxter and E. Roberts, Demonstration of thiosemicarbazide-induced convulsions in rats with elevated brain levels of γ-aminobutyric acid, *Proc. Soc. Exptl. Biol. (N.Y.)* **104**:426–427 (1960).

84. R. S. DeRopp and E. H. Snedeker, Effect of drugs on amino acid levels in the rat brain: Hypoglycemic agents, *J. Neurochem.* **7**:128–134 (1961).

85. H. C. Agrawal, M. W. Fox, and W. A. Himwich, Neurochemical and behavioral effects of isolation-rearing in the dog, *Life Sci.* **6**:71–78 (1967).

86. W. A. Himwich, J. M. Davis, and H. C. Agrawal, Biochemical substrates for the development of the matured evoked potential, *in Recent Advances in Biological Psychiatry* (J. Wortis, ed.), Vol. 9, pp. 271–279, Plenum Press, New York (1967).

87. W. A. Himwich, Multi-disciplined studies of the visual system in developing rabbits, *International Symposium Neuro-ontogeneticum*, Prague (1967), in press.

88. W. A. Himwich, J. M. Davis, and H. C. Agrawal, Effects of early weaning on some free amino acids and acetylcholinesterase activity of rat brain, in *Recent Advances in Biological Psychiatry* (J. Wortis, ed.), Vol. 10, pp. 266–270, Plenum Press, New York (1968).

89. A. R. Dravid and L. Jilek, Influence of stagnant hypoxia (oligaemia) on some free amino acids in rat brain during ontogeny, *J. Neurochem.* **12**:837–843 (1965).

90. R. A. Lovell and K. A. C. Elliott, The γ-aminobutyric acid and factor I content of brain, *J. Neurochem.* **10**:479–488 (1963).

91. A. Curatolo, P. D'Arcongelo, A. Lino, and A. Brancati, Distribution of N-acetyl-aspartic and N-acetyl-aspartyl-glutamic acids in nervous tissue, *J. Neurochem.* **12**:339–342 (1965).

92. E. Miyamoto, Y. Kakimoto, and I. Sano, Identification of N-acetyl-α-aspartylglutamic acid in the bovine brain, *J. Neurochem.* **13**:999–1003 (1966).

93. E. Miyamoto and T. Tsujio, Determination of N-acetyl-α-aspartylglutamic acid in the nervous tissue of mammals, *J. Neurochem.* **14**:899–903 (1967).

94. J. V. Auditore and H. Hendrickson, Investigation of unidentified pharmacologically active substances in the human central nervous system, *Intern. J. Neuropharmacol.* **3**:1–7 (1964).

95. J. V. Auditore, E. J. Olson, and L. Wade, Isolation, purification and probable structural configuration of N-acetyl aspartyl glutamate in human brain, *Arch. Biochem. Biophys.* **114**:452–458 (1966).

96. J. V. Auditore, L. Wade, and E. J. Olson, Occurrence of N-acetyl-L-glutamic acid in the human brain, *J. Neurochem.* **13**:1149–1155 (1966).

97. A. Kanazawa and I. Sano, A method of determination of homocarnosine and its distribution in mammalian tissues, *J. Neurochem.* **14**:211–214 (1967).

98. A. Kanazawa and I. Sano, The distribution of γ-L-glutamyl-L-glutamine in mammalian tissues, *J. Neurochem.* **14**:596–598 (1967).

99. Y. Kakimoto, A. Kanazawa, T. Nakajima, and I. Sano, Isolation of γ-L-glutamyl-L-β-aminoisobutyric acid from bovine brain, *Biochim. Biophys. Acta* **100**:426–431 (1965).

100. Y. Kakimoto, T. Nakajima, A. Kanazawa, M. Takesada, and I. Sano, Isolation of γ-L-glutamyl-L-glutamine from bovine brain, *Biochim. Biophys. Acta* **93**:333–338 (1964).

101. A. Kanazawa, Y. Kakimoto, T. Nakajima, and I. Sano, Identification of γ-glutamylserine, γ-glutamylalanine, γ-glutamylvaline and S. methylglutathione of bovine brain, *Biochim. Biophys. Acta* **111**:90–95 (1965).

102. A. Kanazawa, Y. Kakimoto, T. Nakajima, H. Shimizu, M. Takesada, and I. Sano, Isolation and identification of γ-L-glutamylglycine from bovine brain, *Biochim. Biophys. Acta* **97**:460–464 (1965).

103. T. Nakajima, F. Wolfgram, and W. G. Clark, The isolation of homoanserine from bovine brain, *J. Neurochem.* **14**:1107–1112 (1967).

104. B. Shome and M. Saffran, Peptides of the hypothalamus, *J. Neurochem.* **13**:433–448 (1966).

105. V. Erspamer, M. Roseghini, and A. Anastasi, Occurrence and distribution of N-acetylhistidine in brain and extracerebral tissues of poikilothermal vertebrates, *J. Neurochem.* **12**:123–130 (1965).

Chapter 4

PEPTIDES

John J. Pisano

Laboratory of Clinical Biochemistry
National Heart Institute
National Institutes of Health
Bethesda, Maryland

I. INTRODUCTION

This review is intended primarily to include those peptides of known structure having a wide distribution in the central nervous system. This obviously excludes localized peptides such as the pituitary hormones and hypothalamic releasing factors which are reviewed in later chapters. Substance P would fall within the scope of the review, but the special interest in this peptide also merits a separate chapter. Most of this review deals with the tripeptide glutathione and several dipeptides. Each will be discussed under three general headings: Isolation and Characterization, Occurrence in the Central Nervous System, and Metabolism and Role in the Central Nervous System. This order is intended to facilitate the reading, expose gaps in our knowledge, and indicate those areas which merit further research. A recent treatise is highly recommended as a comprehensive review of all facets of the peptide field.[1]

II. GLUTATHIONE

A. Isolation and Characterization

Glutathione, or γ-glutamylcysteinylglycine, holds a special place in any discussion of biologically active peptides. It is probably the most ubiquitous peptide known, having been found in virtually every living cell. It was one of the earliest natural peptides isolated whose structure was proven by synthesis, and it was the first peptide whose biosynthesis was accomplished *de novo* in an *in vitro* cell-free system.[2] Wieland[3] accurately recounts the early history of glutathione research, and, of the many excellent reviews, those by Knox[4] and Waley[5] are the most recent and authoritative. They should be consulted for a balanced survey of the tremendous body of literature on glutathione, as the present review stresses the neurochemical aspects.

B. Occurrence in the Central Nervous System

Although glutathione has never been isolated from brain and formally characterized, there is no reason to doubt its presence in this tissue. Highly specific enzyme assays employing glyoxylase and glutathione reductase have been used to measure it in brain. Furthermore, the amount found by an ion-exchange procedure which has very impressive resolving power[6] is similar to that obtained by enzyme assay.

An important study in brain[7] deals with the relative amounts of the oxidized and reduced forms of the peptide *in situ*. In analyses preserving *in situ* conditions obtained by dropping live animals into liquid nitrogen, only 2–5 % of the peptide was found in the oxidized form. By contrast, when animals were killed by a blow or by exsanguination, and the brains removed, weighed, and ground in a fixing agent (solid CO_2 or acid) approximately 3 min after death, as much as 50 % of the peptide was found in the oxidized form. The method of tissue preparation is obviously of paramount importance when the relative amounts of oxidized and reduced forms of the peptide are determined.

Three basic approaches have been employed to assay glutathione in tissues.[8] Enzymatic assays employing glyoxylase or glutathione reductase are usually considered most reliable because they are completely specific. Less common amperometric procedures have also been employed with confidence. Chemical procedures based on measurement of the sulfhydryl group of the peptide have been troublesome because of the common occurrence of other reducing substances in tissues.

The glyoxylase method[7] is generally regarded as most specific. Lactic acid production from methyl glyoxal is catalyzed by the ubiquitous enzymes, glyoxylase I and II, readily obtainable from yeast. The former enzyme requires reduced glutathione (GSH).

$$CH_3COCHO + GSH \xrightarrow{\text{glyoxylase I}} CH_3CHOHCOSG \xrightarrow{\text{glyoxylase II}}$$

$$CH_3CHOHCOOH + GSH$$

Lactic acid production may be measured by a variety of procedures. Manometric assays were very popular but there is now interest in spectrophotometric techniques. They are based on the reaction of GSH with chromogenic disulfides or on the disappearance of NADPH. In the latter case, all GSH is oxidized by the addition of a disulfide. The oxidized glutathione (GSSG) is then reduced by the addition of glutathione reductase and NADPH.[9,10] A very sensitive and easy spectrophotofluorometric assay of GSH seems promising.[11,12] It is based on a unique reaction of GSH with ortho-phthalaldehyde in an alkaline medium to give a highly fluorescent product.

Representative levels of total glutathione found in brain vary from 50 to 110 mg/100 g fresh tissue (Table I). In one comparative study in rats, brain

TABLE I

Total Glutathione of Cerebral and Other Tissues

Source	Glutathione (mg/100 g of fresh tissue)
Sheep cortex	88
Ox cortex	110
Ox cerebellum	77
Rat cortex	110
Rat whole brain	100
Rat midbrain	97
Guinea pig cortex	90
Guinea pig midbrain	67
Guinea pig cerebral hemisphere	90
Guinea pig whole brain (age, 3 days)	67
Rat liver[a]	190
Rat brain	54
Rat muscle	32
Rat kidney	129

[a] The last four entries are taken from Davison and Hird[9]; all others, from Martin and McIlwain.[7] The values are averages of 2–7 animals. Oxidized glutathione was reduced with glutathione reductase prior to assay with glyoxylase or spectrophotometrically. The data are considered representative of the amounts of glutathione in brain. Although numerous assays of brain have been reported, it is meaningless to average the data because of the many variables encountered.

was found to contain one-fourth as much peptide as liver, but almost twice that found in muscle.[9] Insufficient data are available to attach significance to the observed regional and species differences of brain glutathione.

C. Metabolism and Role in the Central Nervous System

The half-life of GSH was reported soon after the application of isotopic methods in biological research.[13] It is about 4 hr in liver, 70 hr in brain, and 65 days in rat erythrocytes.[14,15] GSH is labeled within minutes following intravenous injection of glutamic acid-^{14}C into rats and mice.[16] Similar results were obtained in an earlier study in which rats were injected intracisternally with methionine-^{35}S.[17] Methionine was shown to be rapidly converted to cystine, glutathione, and taurine *in vitro* as well as *in vivo*. Incorporation of glycine-^{14}C into glutathione of rat cerebral slices was inhibited about 50% when they were incubated with 5×10^{-4} M isomytal, azacylonol, chlorpromazine, or acetylpromazine.[18]

The biosynthesis of GSH, elucidated by Bloch, Snoke, and co-workers, is accomplished in two steps, both of which·require ATP.[19,20]

$$\text{Glu} + \text{CysH} + \text{ATP} \xrightarrow{\text{Mg}^{2+}} \gamma\text{-Glu-CysH} + \text{ADP} + \text{Pi}$$

$$\gamma\text{-Glu-CysH} + \text{Gly} + \text{ATP} \xrightarrow{\text{Mg}^{2+}, \text{K}^+} \gamma\text{-Glu-CysH-Gly}$$
$$+ \text{ADP} + \text{Pi}$$

The first step is catalyzed by γ-glutamylcysteine synthetase and the second by glutathione synthetase. This latter enzyme has been purified 5000-fold from baker's yeast.[21] The overall mechanism of biosynthesis appears to be similar in animals, plants, yeasts, and bacteria. Studies with brain *in vitro* have not been reported but there is no reason to expect a different mechanism in this tissue.

Glutathione may also be synthesized by a γ-glutamyl transpeptidation reaction starting with cysteinylglycine and a γ-glutamyl donor, typically a peptide having an N-terminal γ-glutamyl residue. However, the reaction is readily reversible and GSH is, in fact, an excellent γ-glutamyl donor to amino acids and peptides. It is more generally believed that γ-glutamyl transpeptidation is the first step in the catabolism of glutathione. Peptidases do not hydrolyze the γ-glutamyl link of glutathione, but a potent cysteinylglycinase activity has been demonstrated in tissues. γ-Glutamyl transpeptidation and the probable reaction products, γ-glutamyl dipeptides, have been demonstrated in brain.[22-24]

At one time it was thought that GSH directly participated in protein synthesis because of its abundance in tissues, rapid turnover, and active participation in γ-glutamyl transpeptidation.[25] However, in the early thinking about the physiological roles of GSH, its involvement in protein synthesis, in the control of carbohydrate metabolism,[26] and in cell division all were overshadowed by another unconfirmed theory that it played an important role in cellular respiration.[4,5] Even though all cells actively reduce GSSG and oxidize GSH wlth molecular oxygen, the necessary coupling factors which must transport electrons to oxygen have never been demonstrated in animals. However, glutathione is known to have an important function in the respiration of certain plants where it is readily oxidized by dehydroascorbic acid.[27,28] For example, in pea cotyledons one-fourth to one-half of the oxygen uptake has been observed to proceed by the following multienzyme system which transfers electrons from NADPH to oxygen:

1. $\text{ascorbic acid} + [\text{O}] \xrightarrow[\text{oxidase}]{\text{ascorbic acid}} \text{dehydroascorbic acid} + \text{H}_2\text{O}$

2. $\text{dehydroascorbic acid} + 2\text{GSH} \xrightarrow[\text{reductase}]{\text{dehydroascorbic acid}} \text{GSSG} + \text{ascorbic acid}$

3. $\text{GSSG} + \text{NADPH} + \text{H}^+ \xrightarrow[\text{reductase}]{\text{glutathione}} 2\text{GSH} + \text{NADP}^+$

A similar enzyme system may be of importance (but has not been demonstrated) in the respiration of the nuclei of mammalian as well as plant cells, as nuclei have substantial concentrations of GSH, ascorbic acid, and glutathione reductase.

It has also often been noted that GSH can protect reactive SH groups and remove inhibitory heavy metals from purified enzymes. Since many other sulfhydryl compounds have the same action, it is possible that all these agents are merely reversing the insults imposed on the enzymes by the isolation procedure. Not to be overlooked, however, is the abundance of GSH in cells where other sulfhydryl compounds are not found in significant concentrations. Studies with erythrocyte triosephosphate dehydrogenase indicate a direct *in vivo* protective action of GSH. In erythrocytes, GSH peroxidase, and not catalase, appears to be reponsible for the decomposition of any H_2O_2 produced by the coupled oxidation of various hydrogen donors with oxyhemoglobin.[29] Addition of H_2O_2 to duck erythrocytes (which are deficient in catalase) or azide-treated (to inhibit catalase) human cells causes a corresponding decrease in GSH.[30]

$$H_2O_2 + 2GSH \longrightarrow GSSG + 2H_2O$$

Red blood cells are particularly rich in GSH peroxidase, but tissues also contain appreciable levels.[29] Brain has not been examined. Elucidation of the physiological role of this enzyme still awaits the identification of the source of the peroxide it reduces.

An equally interesting proposal for an *in vivo* action of GSH involves the relationship of glutathione reductase and the hexose monophosphate shunt, which is a major source of NADPH in the cell. The NADPH produced in the shunt pathway is required for the reductase reaction

$$GSSG + NADPH + H^+ \longrightarrow 2GSH + NADP^+$$

In red blood cells that have a reduced glucose 6-phosphate dehydrogenase activity, inherited or drug-induced, there is insufficient production of NADPH, a concomitant reduction in GSH, and elevation in peroxides. This condition sensitizes cells to hemolysis by a number of agents.[31]

Brain also contains GSH reductase,[32] and it has been suggested that the hexose monophosphate shunt also plays a major role in maintaining glutathione in the reduced state in this tissue.[33] Incubation of guinea pig brain minces with 2 mM chloropromazine resulted in a two-fold increase in the amount of added GSSG reduced. However, similar levels of amytal caused a 20% diminution in GSSG reduction, and chlordiazepoxide hydrochloride (Librium), phenobarbital, serotonin, epinephrine, and norepinephrine had no appreciable effect. These experiments with brain are not as convincing as those reported with erythrocytes, but observations made with other tissues will obviously continue to suggest experiments with brain.

Recent and more convincing experiments that support the traditional theory that GSH regulates the level of essential sulfhydryl groups has been obtained with the enzyme glyceraldehyde-3-phosphate dehydrogenase[34] and in studies on the phenomenon of mitochondrial swelling which appears to be associated with marked changes in membrane sulfhydryl content.[35] Specific glutathione transhydrogenases such as glutathione-insulin transhydrogenase[36] may have important roles in biosynthesis or in controlling the relative levels of oxidized and reduced material in cells.

Despite the lack of definitive results in many experiments implicating GSH, there is no question about the importance of the tripeptide in certain enzymatic reactions. It is an essential cofactor for the following enzymes: glyoxylase, formaldehyde dehydrogenase, maleylacetoacetate isomerase, maleic acid isomerase, indolpyruvic acid tautomerase, and DDT-dehydrochlorinase. Of this group of enzymes, only glyoxylase has been studied in brain.[37] The significance of this enzyme remains obscure as methylglyoxal has never been found in tissues. An examination of the above enzymatic reactions reveals the interesting fact that in no case is there oxidation of GSH. In the glyoxylase and formaldehyde dehydrogenase reactions, GSH adds to a carbonyl group to form a thiohemiacetal which then tautomerizes to a thioester. Reaction of GSH with an olefin group is its mechanism of action with the remaining enzymes mentioned.[4,5]

Another unique action of GSH which does not involve its oxidation is its ability to initiate a feeding reaction in *Hydra litoralis* and *Physalia* gastrozooids. It was thought that this reaction was absolutely specific for GSH until it was shown that ophthalmic acid (γ-glutamyl-α-amino-*n*-butyrylglycine), the glutathione analogue isolated from calf lens, was somewhat more active.[5] The mechanism of action is unknown, but the observation indicates again the diversity of action of glutathione.

III. γ-GLUTAMYL DIPEPTIDES

A. Isolation and Characterization

In addition to the previously mentioned γ-glutamyl polypeptides, Kakimoto and collaborators have identified several γ-glutamyl dipeptides in bovine brain.[38–42] Trichloroacetic acid extracts were chromatographed on cationic and anionic ion-exchange columns developed with volatile buffers. In every case the peptide structure was deduced by degradation and by comparison with the synthetic peptide in a variety of enzymatic, chemical, and physical tests.

B. Occurrence in the Central Nervous System

Table II lists the γ-glutamyl dipeptides found in whole bovine brain. Even the most abundant, γ-glutamylglutamic acid and γ-glutamylglutamine, are only present to the extent of about 1.0 mg/100 g fresh tissue. γ-Glutamylglutamine has also been found in many other tissues in varying amounts, but the levels in brain are comparatively high. The peptide is rather evenly distributed in the different regions of human brain.[43]

TABLE II

γ-Glutamylpeptides of Bovine Brain[a]

Peptide	Concentration (mg/100 g fresh wt.)	Reference
γ-Glutamylglutamic acid	0.7–1.0	38
γ-Glutamylglutamine	0.7–1.0	38
γ-Glutamylglycine	0.07–0.1	39
γ-Glutamylalanine	0.05–0.1	41
γ-Glutamyl-L-β-aminoisobutyric acid	0.03–0.04	40
γ-Glutamylserine	0.02–0.04	41
γ-Glutamylvaline	0.01–0.02	41
S-Methylglutathione	0.01–0.02	41

[a] Peptides were estimated by visual comparison of spots with synthetic standards of electropherograms.

C. Metabolism and Role in the Central Nervous System

The physiological significance of γ-glutamyl dipeptides is unknown, but they appear to be widely distributed in nature. Of those listed in Table II, γ-glutamylalanine and γ-glutamylvaline have been isolated from plants.[5] The latter peptide, as well as several other γ-glutamyl dipeptides and β-aspartyl di- and tripeptides, have been isolated from human urine.[44] Iris bulbs (*Iris tingitana*) contain relatively huge amounts of γ-glutamyl-β-aminoisobutyric acid, which accounts for about 15% of the nonprotein amino nitrogen in extracts. The difference between the iris peptide and that isolated from brain is that the plant peptide contains the D isomer of β-aminoisobutyric acid. It is also the D isomer of the amino acid which is found in human urine.[45] The origin and significance of the L isomer is unknown, and its presence in the brain peptide is the first report of its occurrence in tissues. S-Methylglutathione is another new peptide, but γ-glutamyl-S-methylcysteine has been isolated from kidney bean seed and lima bean seed.[5] The Japanese workers can also lay claim to the first reports of γ-glutamylglutamine, γ-glutamylglycine, and γ-glutamylserine in nature. Although free γ-glutamylglutamic acid was not observed previously, it is a known constituent of microbial growth factors of the folic acid group and a constituent of the capsular substance of certain microorganisms which contain poly-γ-glutamylglutamic acid.[5,25]

The most likely pathway for the formation of γ-glutamyl dipeptides in brain is γ-glutamyl transpeptidation. The transpeptidase in brain,[22,23] as in other tissues, catalyzes the reversible transfer of the γ-glutamyl moiety of GSH to acceptor amino acids as well as peptides and certain other compounds. Since amino acids differ in acceptor activity and the resulting γ-glutamyl dipeptides formed likewise differ in donor activity, this may account for the occurrence of only a few of the possible 20 γ-glutamyl dipeptides.

Current concepts of protein biosynthesis preclude the possibility that γ-glutamyl dipeptides are intermediates in protein synthesis or products of their degradation. Their significance may be related to GSH metabolism. The only known mechanism for removal of the γ-glutamyl moiety of GSH in tissues is γ-glutamyl transpeptidation. Virtually all tissues contain the transpeptidase but none has been found to contain a γ-glutamyl peptidase. Further speculation on the significance of γ-glutamyl transpeptidase has centered around the other product of transpeptidation, cysteinylglycine, which is a stronger reducing agent than GSH.[4] γ-Glutamyltranspeptidation is freely reversible and cysteinylglycine is an excellent γ-glutamyl acceptor. This reaction forms GSH without the expenditure of energy, but its occurrence appears to be of minor importance because of the small amounts of γ-glutamyl dipeptides present in brain and other tissues. In plant seeds and bulbs, γ-glutamyl dipeptides occur in abundance, and they could be considered as reservoirs of amino acids which serve the plant during the early stages of growth. It is not known if their disappearance during growth involves γ-glutamyl transpeptidation. γ-Glutamyl dipeptides were once implicated in amino acid transport in kidney, which is the richest source of γ-glutamyl transpeptidase.[4]

It is interesting that more β-aspartyl di- and tripeptides than γ-glutamyl peptides have been identified in human urine;[44] nevertheless, the Japanese workers were not able to find any in brain or other tissues, nor have other investigators found β-aspartyltranspeptidase activity. On the other hand, it has been elegantly demonstrated[46] that the α-aspartyl peptide link can isomerize to the β-aspartyl form by a nonenzymatic mechanism at physiological pH and temperature. The urinary excretion of β-aspartyl peptides could be the consequence of this α–β conversion as the β-aspartyl peptides are resistant to enzymatic hydrolysis and once formed would be rapidly excreted.

IV. HOMOCARNOSINE AND RELATED PEPTIDES

A. Isolation and Characterization

At the beginning of this decade, when the role of γ-aminobutyric acid in the central nervous system was under intensive study, it was thought that the amino acid might be a precursor of other more active substances. The simple experiment of looking for γ-aminobutyric acid-containing substances in acid hydrolysates led to the discovery of γ-amino-n-butyryl histidine, homocarnosine.[47] The dipeptide was isolated from acetic acid extracts of bovine brain and purified by chromatography on Dowex-50 columns. The structure of the peptide was deduced by chemical degradation and confirmed in comparative studies with synthetic peptide. In chromatographic, electrophoretic, and enzymatic tests, the natural and synthetic compounds were indistinguishable. Homocarnosine has also been crystallized from other bovine brain extracts.[48,49]

Brain also contains homoanserine, γ-amino-n-butyryl-1-methylhistidine, a finding not unexpected in view of the similar structure and distribution of carnosine and anserine. Homoanserine was isolated in crystalline form from perchloric acid extracts of bovine brain by several column chromatographic steps in an ambitious and fruitful study.[49] Its structure was established by elemental and degradation analyses, and by comparison with synthetic standard in chromatographic and electrophoretic procedures. The same authors also isolated homocarnosine, 3-methylhistidine and 1-methylhistidine, the first two substances in crystalline form.

B. Occurrence in the Central Nervous System

It was estimated that bovine brain contains about four times more homocarnosine than homoanserine.[49] However, it is certain that the isolated 1 mg homoanserine/kg wet weight of brain is a minimal value as losses were undoubtedly considerable.

Of the known imidazolyl peptides, only homocarnosine has been quantitatively measured in brain. Using an isotope dilution assay, Abraham, Pisano, and Udenfriend[50] studied the distribution of homocarnosine in several different animals. Human brain appears to contain the highest levels of this peptide (about 8 mg/100 g), but bovine, monkey, guinea pig, and rabbit brains have comparable levels (Tables III and IV). Because of the limited number of analyses, generalizations are not possible, but it seems that the rat and chicken contain intermediate levels (about 1 mg/100 g) and, surprisingly, duck, swine, cat, and dog contain very little (<0.1 mg/100 g). Essentially the same results were obtained by Kanazawa and Sano,[51] who measured homocarnosine by a paper electrophoretic method. They also detected low levels of homocarnosine, <0.1 mg/100 g, in fish brain. Perhaps

TABLE III

Homocarnosine in Whole Brains of Animals ^{a}

	Abraham, Pisano, and Udenfriend[50]	Kanazawa and Sano[51]
	(mg/100 g wet wt.)	
Guinea pig	4.4 (2)	1.4 ± 0.4 (4)
Rabbit	4.3 (3)	2.7 ± 0.5 (4)
Rat	1.4 (5)	1.3 ± 0.4 (3)
Chicken	0.72 (1)	0.77 ± 0.2 (3)
Dog	<0.1 (1)	<0.1 (1)
Cat	<0.1 (1)	<0.1 (1)
Swine	<0.1 (1)	—
Duck	<0.1 (2)	—
Crucian carp	—	<0.1 (1)
Mackerel	—	<0.1 (2)

^{a} Numbers of animals tested are indicated within parentheses.

TABLE IV

Gross Anatomical Distribution
of Homocarnosine[50]a

	Cerebrum	Cerebellum	Brainstem
		(mg/100 g wet wt.)	
Human (1)	7.9	3.8	5.0
Bovine (1)	3.5	2.4	5.6
Monkey (2)	2.0	2.4	3.7

a Numbers of animals tested are indicated within parentheses. In more detailed studies by Kanazawa and Sano[51] the thalamus, hypothalamus, and cerebellum seemed to contain higher levels than other areas but no consistent pattern was evident. Five brains were examined; in some as much as a fivefold difference between regions was observed.

those animal brains which have little homocarnosine contain relatively high levels of homoanserine, a relationship not uncommon between anserine and carnosine levels in muscle.[52]

The gross regional distribution of homocarnosine has been studied (Table IV). Some differences were observed, but no consistent pattern was evident. Possibly the thalamus, hypothalamus, and cerebellum contain more than the cerebral cortex and brainstem. The higher levels observed in the white matter of the frontal cortex of the cerebrum by Abraham, Pisano, and Udenfriend[50] were not observed by Kanazawa and Sano[51] (Table V). Postmortem changes were also investigated by the latter authors. They found no significant difference between rabbit brain left at room temperature for 2 hr and for 15 hr.

In one study, homocarnosine was not detected in the spleen, kidney, liver, or skeletal muscle of the duck, guinea pig, rat, and rabbit, nor was it found in the sciatic nerve of the rabbit.[50] Other investigators did not find it in the intestine, kidney, liver, lung, spleen, or muscle of the rat and guinea pig, but they did find 0.1–0.2 mg/100 g in rabbit liver.[51] This later finding is in contrast to the previous report[50] and, if confirmed, is the first demonstration of homocarnosine in tissue other than the central nervous system.

Although the early reports[52,53] on the occurrence of carnosine (β-alanylhistidine) in brain left much to be desired from a chemical viewpoint, it now appears more certain that both carnosine and anserine are present in brain. Employing methods that distinguish carnosine from homocarnosine, the former peptide was detected in all species examined in one study.[50] The levels found were in all cases less than homocarnosine and ranged from a trace to 0.7 mg/100 g. Carnosine was also detected in bovine brain during the course of homocarnosine isolation[48] but not during homoanserine

TABLE V

Homocarnosine in Human Frontal Cortex[a]

Age, sex	Hours after death	Cause of death	Gray matter	White matter
			(mg/100 g wet wt.)	
78, male	7	Subdural hemorrhage	5.1	6.9
45, male	7	Carbon monoxide poisoning	4.8	11.3
66, female	19	Myocardial infarction	5.1	4.6
44, male	19	Myocardial infarction	9.7	8.2
52, male	<12	—	—	12.1
52, male	<12	—		12.6
29, female	<12	—	1.3	5.4
9 weeks, male	<12	—	0.9	3.0
9 weeks, female	<12	—	0.8	
46, female	<12	—	0.7	2.4
45, male	<12	—	1.1	2.8
20, female	<12	—	0.5	2.7

[a] The first five cases were taken from Kanazawa and Sano,[51] who employed a paper electrophoretic method and the remainder from Abraham, Pisano, and Udenfriend[50] who used an isotope dilution method. Therein might be the reason for the different results from the two laboratories in the analysis of white and gray matter.

isolation.[49] Carnosine was also reported in human cerebrospinal fluid in an excellent study,[54] but apparently not seen in another study where elevated CSF homocarnosine was noted.[55]

The first good evidence that anserine (β-alanyl-1-methylhistidine) occurs in brain was reported by Tsunoo et al.[56] who isolated 0.4 g of the dipeptide from 10 kg of chicken brains. The crystalline material (isolated from water extracts by column chromatography) and its hydrolysis products (1-methylhistidine and β-alanine) were compared to the authentic substances by paper chromatography using a variety of solvents. It is somewhat surprising that the authors did not observe carnosine or homocarnosine. However, apparent discrepancies in reports from different laboratories should not be viewed with undue concern as it is quite easy to overlook these peptides when they are not the main compounds of interest.

C. Metabolism and Role in the Central Nervous System

The mechanism of carnosine biosynthesis has been elucidated independently in two laboratories.[57,58] Active carnosine synthetase preparations have never been obtained, but weak preparations from chick pectoral muscle, the best source, are satisfactory. The partially purified enzyme catalyzes the following reactions:

1. β-Alanine + ATP + enzyme $\rightleftharpoons$ enzyme-β-alanyl adenylate + PPi

2. Enzyme-β-alanyl adenylate + histidine $\rightleftharpoons$ carnosine + AMP
   + enzyme

The enzymatic reaction is surprisingly nonspecific. Both histidine and β-alanine may be replaced by a number of amino acids. The enzyme not only catalyzes the synthesis of the four natural peptides, carnosine, anserine, ophidine, and homocarnosine, but will also act with several other β- or ω-amino acids in place of β-alanine and with lysine, ornithine, arginine, and other amino acids in place of histidine.

The low rate of synthesis observed in the carnosine studies was also encountered in attempts to demonstrate the biosynthesis of homocarnosine. However, *in vivo* biosynthesis was readily demonstrated when γ-aminobutyric acid was administered or fed to rats at levels that allowed its accumulation in muscle and other tissue.[59] It appears that the absence of homocarnosine in muscle and other peripheral tissues is due to the absence of the precursor, γ-aminobutyric acid. Thus, brain has abundant γ-aminobutyric acid, relatively little β-alanine, and contains more homocarnosine than carnosine. However, it is also possible that even less carnosine biosynthesis occurs in brain than supposed from the low levels of β-alanine in this tissue since carnosine and not homocarnosine is taken up by brain slices incubated in a medium containing the peptide.[60] It would be interesting to see if brain carnosine levels could be elevated by injection of carnosine into rats. If so, then uptake from blood could account for some (if not all) brain carnosine.

Discussion of the role of imidazolyl dipeptides in brain brings to mind the subject of their metabolism in skeletal muscle. For half a century muscle has been known to be a uniquely rich source of carnosine and anserine; yet, despite numerous studies with this tissue, the somewhat unsatisfying proposal for their metabolic role is that they act as buffers which help control the pH of muscle, particularly during anaerobic glycolysis when lactic acid is produced in quantity.[61,62] Extension of the buffer concept to brain must accommodate the fact that this tissue contains much less (about 1/20 as much) of the imidazolyl dipeptides. However, the rapid generation of lactic acid observed in muscle is not characteristic of brain metabolism and a smaller buffer capacity would suffice. Inadequate as it may seem, there is currently no more plausible explanation of the biological role of these dipeptides. Hopefully, new insights will be possible with the accumulation of more information. One obvious approach is to measure the peptide in blood and urine from persons with a variety of pathological conditions. This approach has already led to some interesting findings.

It appears that carnosine, anserine, homocarnosine, and presumably homoanserine are in some way related to two types of progressive disorders of the central nervous system. In 1962, Bessman and Baldwin[63] noted an increased urinary excretion of carnosine and anserine, as well as histidine and 1-methylhistidine, in five patients with juvenile amaurotic idiocy (also called cerebromacular degeneration, juvenile Tay-Sachs disease, and Vogt's or Sjogren's disease) who came from three separate families. Levenson, Lindahl-Kiessling, and Rayner[64] 2 years later reported similar results in four of fifteen patients with the same disease. It is quite possible that more members of this group had the generalized imidazolyl dipeptide and amino aciduria, for the methods employed detected only those cases which were grossly (ca. > tenfold) elevated. Recently, Perry et al.[55] reported increased urinary excretion of carnosine and anserine in two unrelated patients with another kind of progressive neurological disease. Myoclonic and grand mal seizures were observed during the first 3 months of life and the patients subsequently showed severe psychomotor retardation. This syndrome is clinically different from that reported by the other laboratories and it is probable that the biochemical disorders are also distinct. Perry et al. did not observe the elevated 1-methylhistidine excretion seen by Bessman and Baldwin. The former patients also did not even excrete the amino acid after anserine loading, unlike normal persons who excreted large quantities of 1-methylhistidine. This and elevated serum levels of carnosine and anserine led to the proposal that their patients have a carnosinase deficiency. The patients of Bessman and Baldwin, on the other hand, seem to have a renal condition, but additional studies of serum, particularly after carnosine loading, seem warranted to exclude a generalized disturbance of imidazole metabolism.

It is worthy of note that Perry et al. observed about 0.25 mg homocarnosine per 100 ml CSF in their patients. This is more than 10 times the highest normal level previously seen.[50] The Canadian workers also observed 0.25 to 0.17 mg/100 ml in 13 other children with mental deficiency but none in the CSF of most neurologically normal adults. It is unfortunate that neither these workers nor Bessman and Baldwin measured homocarnosine in urine. As previously noted, this substance is found exclusively in the central nervous system [one unconfirmed exception is its presence in rabbit liver[51]], and its occurrence in urine offers a unique opportunity to study specifically central nervous system metabolism. Few, if any, other known substances in urine are found exclusively in the central nervous system. While normal CSF contains a trace to 0.025 mg/100 g, 24-hr urine samples have 0.5 to 2.5 mg.[50]

Even though there is no evidence to indicate that the biochemical defects are the cause of the two types of progressive nervous system disorders mentioned, widespread analysis of urine for imidazole compounds (dipeptides and amino acids) seems worthy of consideration as they could be useful genetic markers. In the cases reported by Bessman and Baldwin, but not those of Perry et al., the urinary defect appears to be transmitted as a dominant trait and the cerebromacular degeneration as a recessive trait. The fact that the two traits have been found in three unrelated families makes it likely that they are manifestations of the same gene.

V. *N*-ACETYL-α-ASPARTYLGLUTAMIC ACID

A. Isolation and Characterization

The isolation of *N*-acetylaspartylglutamic acid from horse and rabbit brain was first reported in an abstract which contained little documentation of its structure.[65] A year later it was reported that the peptide was indistinguishable from synthetic *N*-acetyl-β-aspartylglutamic acid and clearly separable from *N*-acetyl-α-aspartylglutamic acid when chromatographed on paper or on columns of Dowex-2-formate.[66] Also, in 1964, Auditore and Hendrickson[67] isolated a substance from extracts of human brain which very preliminary studies indicated was *N*-acetylaspartylglutamic acid. Subsequently they presented a detailed report of the isolation and characterization of the peptide from human brain.[68] A combination of paper and column chromatographic techniques was used without the aid of synthetic standards. Hydrazinolysis of the pure peptide gave acetyl hydrazide and glutamic acid, indicating an *N*-acetyl peptide in which the C-terminal amino acid was glutamic acid. Hydrolysis of the peptide in acid gave equimolar amounts of aspartate and glutamate. These data led to the proposal that the human brain peptide was *N*-acetylaspartylglutamic acid. Still undetermined was the nature of the aspartylglutamate bond. Miyamoto, Kakimoto, and Sano settled the question when they isolated the peptide from bovine brain and proved the structure to be *N*-acetyl-α-aspartylglutamic acid.[69] The key reaction was hydrazinolysis, which gave them α- and no β-aspartylhydrazide. They also made comparisons with synthetic standards and showed that their brain peptide was readily separable from synthetic *N*-acetyl-β-aspartylglutamate but indistinguishable from *N*-acetyl-α-aspartylglutamate in chromatographic tests. The peptide was found only in brain of the rabbit, guinea pig, and rat, but less than 1 μg/g in other tissues would not have been detected.

In view of these studies, it appears that the previous report of *N*-acetyl-β-aspartylglutamic acid in rabbit and horse brain is incorrect. Perhaps an explanation of the discrepancy lies in the fact that aspartyl peptides undergo a well-known α-β interconversion even under physiological conditions of *p*H and temperature.[70] Conceivably, the β-peptide formed during isolation or storage and was preferentially purified.

B. Occurrence in the Central Nervous System

N-Acetyl-α-aspartylglutamate has been found exclusively in the central nervous system.[71,72] Perchloric acid extracts[72] corresponding to approximately 1 g of brain were passed through a 20 × 0.9 Dowex-1 formate column which was then developed by gradient elution with formic acid. Fractions containing the peptide were free of contaminants. They were hydrolyzed in NaOH and the quantity of peptide determined by reaction with ninhydrin.

In the most complete study to date,[71] the regional distribution of N-acetyl-α-aspartylglutamate was determined in several different animals. Also reported are analyses of whole brain and postmortem changes. The peptide was extracted from 2–3 g nervous tissues by homogenization in ethyl alcohol and isolated by chromatography on Dowex-1 formate columns. Eluates were dried and the residue hydrolyzed with HCl. Since N-acetyl-α-aspartylglutamate was the only glutamic acid-containing peptide in the Dowex-1 eluate, glutamic acid determination by paper electrophoresis and reaction with ninhydrin was a valid measure of the peptide.

TABLE VI

Distribution of N-Acetyl-α-aspartylglutamic Acid in the Central Nervous System

Neuraxis levels[a]	Rabbit	Dog	Rat	Guinea pig (mg/100 g)	Horse[b]	Mouse	Monkey
Whole brain	19.2		10.7	23.0		8.0 (8)	8.0 (8)
	16.0		14.4	23.0			
	16.7		15.2				
Brain cortex					19.1		
Cerebrum	12.0	12.0					
	11.9	15.2					
Midbrain		21.1					
Thalamus					21.3		
Cerebellum	24.2	23.2			30.4		
	29.3						
Mesencephalon					47.1		
Brainstem	35.1	29.0					
	33.3						
	40.3						
	35.3						
Pons					54.7		
Medulla					45.6		
Spinal cord	38.3						
	31.9						
	27.9						
Spinal cord, thoracic					39.5		
Spinal cord, lumbar					27.4		
N. ischiadicus	37.6						
	27.4						

[a] Author's designation.
[b] Values for the horse were taken from reference 72, mouse and monkey (eight pooled brains) from 74, and the other animals from reference 71.

It can be seen in Table VI that slightly higher levels of N-acetyl-α-aspartylglutamate were found in horse brain. It is not yet known if this is related to the different analytical procedure employed. In general, there is a gradual increase in N-acetyl-α-aspartylglutamate toward the more caudal areas. The lowest value was 11.9 mg/100 g in the cerebellum of the rat; the highest, 54.7 mg/100 g in the pons of the horse. The opposite distribution was observed for N-acetylaspartate in the horse. Cortex contained 112.0 and pons 28.0 mg/100 g.

In postmortem human brain, no remarkable differences in the concentration between white and gray matter were observed, and no areas showed unusually high levels. The lowest mean value of four brains was 12.4 mg/100 g in frontal gray matter and the highest, 24.2 mg/100 g in thalamus.[71]

A gradual increase in N-acetyl-α-aspartylglutamate has been observed in rabbit brain held at 20°C. The first increase, about 3%, was observed after 3 hr of incubation but a maximal 50% increase was noted at 7, 14, or 20 hr. Three areas showed different rates of postmortem changes. Measured 15 hr after death, the cerebrum, cerebellum, and brainstem showed 120, 65, and 44% increases, respectively.

C. Metabolism and Role in the Central Nervous System

Mention of N-acetyl-α-aspartylglutamate immediately calls to mind the well-known and more abundant N-acetylaspartic acid. The current postulated function of N-acetylaspartic acid in the central nervous system is that it partly satisfies the anion deficit of brain; the same could be said of the peptides. In a recent study,[73] neither the α- nor β-isomer of aspartylglutamate nor their N-acetyl derivatives had activity on the frog rectus or guinea pig ileum and they are probably not responsible for the above activities previously noted in crude extracts of human brain.[67] It was also shown that neither the free nor acetylated peptide behaved as central transmitter substances when tested on the cat's piriform cortex.

Metabolism of N-acetyl-α-aspartylglutamate, N-acetylaspartate, and N-acetylglutamate were simultaneously studied in mouse and monkey brain.[74] Pooled brain from eight mice contained 88 mg N-acetylaspartate, 8 mg N-acetyl-α-aspartylglutamate, and 1.6 mg N-acetylglutamate per 100 g brain. Incubation of monkey brain slices and ^{14}C precursors indicated that these substances are not metabolically inert as supposed from work in other laboratories. Incorporation of glutamate-^{14}C into N-acetyl-α-aspartylglutamate, while small, was still of the same order of magnitude as incorporation into glutathione. With aspartate-^{14}C there was greater incorporation of aspartate into N-acetylaspartate than into N-acetyl-α-aspartylglutamate, suggesting that the former compound is a precursor of N-acetyl-α-aspartylglutamate. Other data with acetate-^{14}C raise questions about this biosynthetic pathway which can only be answered by more detailed study.

It is interesting to note that N-acetyl-α-aspartylglutamate is at the N-terminus of the muscle protein actin.[75] A recently reported[76] "contractile protein" in brain may also have the same N-terminus. Possibly related are the observed higher levels of the peptide seen in postmortem rabbit brain which may be due to proteolysis of brain protein and release of the peptide rather than decreased catabolism of peptide synthesized during postmortem storage.

VI. PEPTIDES OF UNDETERMINED STRUCTURE

An early examination of bovine and swine hypothalamic and posterior pituitary tissues indicated that hypothalamus does not contain a significant peptide pool, but the posterior pituitary contains very high concentrations of polypeptides, several times greater than can be attributed to the known peptides.[77] It is now apparent, however, that the hypothalamus contains several polypeptides which are releasing factors of pituitary hormones (see M. Reiss, Chapter 26 of Volume IV of this series). Other peptides have also been characterized in extracts of swine hypothalamus tissue.[78] Of four which have been purified from fresh tissue, one had properties of a lipopeptide; the remaining three were apparently simple peptides. The smallest had a molecular weight of approximately 1400; the other two, 9000. Extraction of an acetone powder of swine hypothalamic tissue yielded several other peptides or small proteins, one of which was purified. It was a basic protein which had a molecular weight of 14,000. None of the peptides or small proteins isolated had pressor or ACTH-releasing activity and their biological activity, if any, is completely unknown. The amounts isolated ranged from 25 to 200 mg per kilogram fresh tissue.

In the latest report of a continuing study of phosphopeptides in nervous tissue, 28 phosphopeptides were isolated from bovine brain.[79] The peptides may contain 50 to 100 amino acid residues and some could be regarded as small proteins. They all yielded much ammonia on mild acid but not alkaline hydrolysis, suggesting the presence of phosphoramido compounds.[80] Analysis of subcellular fractions of rat brain indicated that the amidated phosphopeptides are present in high concentration in myelin sheaths, in membranous structures, and in the nuclei. The phosphorus of these phosphopeptides has an extremely fast turnover and the authors suggest that amidated phosphopeptides may participate in transport phenomena of cellular membranes. Phosphoserine was the only phosphorylated amino acid obtained by mild acid hydrolysis of the peptides.

In another study of brain phosphopeptides which were similarly extracted (but not further characterized), pentylenetetrazol (Metrazol) did not affect the *in vivo* incorporation of ^{32}P despite the high turnover of the phosphorus.[81] Cell-free incorporation of phosphate into a bovine phosphopeptide fraction prepared by a different procedure has also been observed.[82] O-Phosphoserine isolated from the phosphopeptides appeared to be the sole radioactive phosphate in the peptide fraction. It has been known for some time that brain phosphoproteins contain sequences of phosphorylserine residues similar to those found in casein and phosvitin.[83] Brain phospho-proteins are phosphorylated and dephosphorylated by cerebral enzymes and these reactions have been thought to be involved in ion transport.[83–85]

Highly encephalitogenic basic peptides (mol. wt. 3500) isolated from bovine spinal cord[86] have recently been shown not to be normal consti-tuents of nervous tissue.[87] They were formed during the extraction procedure through the action of proteases on a basic protein (mol. wt. 16,500) whose encephalitogenic activity has been known for some time.[88] Nevertheless, it is interesting that such peptide degradation products of the protein appear to have all the activity of the native material. It is also noteworthy, particu-larly to those interested in peptides, that a great deal of time and effort was spent before it was realized that acid extraction of acetone powder of spinal cord and brain simultaneously removed the basic protein and the active protease. Chloroform–methanol extraction of fresh tissue prior to acid extraction apparently inactivates the protease, as only the basic protein and no peptides are found.[87,88]

Ungar, Galvan, and Clark claim to have isolated a peptide from brain concentrates of trained rats which affects the behavior of untrained rats. These experiments have not yet been verified in other laboratories nor has the peptide nature of the "active" factor been established.[89]

VII. CONCLUSION

Recent developments and refinements of chromatographic and chemical techniques now make it possible to isolate and characterize readily peptides from any source. If this were the sole aim of an investigation, it would seem that success is assured. But such an approach is a relative waste of time. Why spend time accumulating peptides which have no function if the same effort could be directed to the characterization of peptides which have a definable biological role?

In his discussion of the quest for new biologically active substances, E. B. Chain reviewed past discoveries and emphasized the point that the successful search follows a characteristic pattern: biological observation; isolation and determination of the structure; synthesis of the active com-pound.[90] To continue this successful pattern, the biologist must accelerate his efforts and expose new opportunities to the chemist who, because of his new powerful techniques, is rapidly exhausting the old list of unknown biologically active compounds. Cooperation of biologists and chemists is likely to lead to mutually satisfying research. When each chooses a totally independent path, he may have to content himself with a catalogue of obser-vations or a list of compounds which will lay idle for many years.

When compared to the pituitary hormones and the hypothalamic pep-tides (releasing factors), studies on the metabolic roles of the peptides discussed in this review seem very shallow. With the exception of glutathione (which has many functions), none has a clearly definable role. Most promising is the work with the imidazolyl dipeptides where elevated levels in serum, cerebrospinal fluid, and urine have been related to two types of progressive disorders of the central nervous system. However, it remains to be proved that the biochemical disturbance is the cause of the neurological disorder.

VIII. REFERENCES

1. E. Schröder and K. Lübke, *The Peptides*, *Vol. 1*, *Methods of Peptide Synthesis*, *Vol. 2*, *Synthesis, Occurrence, and Action of Biologically Active Peptides*, Academic Press, New York (1965).
2. K. Bloch, The synthesis of glutathione in isolated liver, *J. Biol. Chem.* **179**:1245–1254 (1949).
3. T. Wieland, in *Glutathione* (S. Colowick, A. Lazarow, E. Racker, D. R. Schwarz, E. Stadtman, and H. Waelsch, eds.), pp. 45–57, Academic Press, New York (1954).
4. W. E. Knox, in *The Enzymes* (P. D. Boyer, H. Lardy and K. Myrbäck, eds.), Vol. 2, pp. 253–294, Academic Press, New York (1960).
5. S. G. Waley, Naturally occurring peptides, in *Advances in Protein Chemistry* (C. B. Anfinsen, Jr., M. L. Anson, John T. Edsall, and Frederic M. Richards, eds.), Vol. 21, pp. 2–112, Academic Press, New York (1966).
6. H. H. Tallan, S. Moore, and W. H. Stein, Studies on the free amino acids and related compounds in the tissues of the cat, *J. Biol. Chem.* **211**:927–939 (1954).
7. H. Martin and H. McIlwain, Glutathione, oxidized and reduced, in the brain and in isolated cerebral tissue, *Biochem. J.* **71**:275–280 (1959).
8. C. G. Thomson and H. Martin, Techniques for determining glutathione in animal tissues, in *Biochemical Society Symposia*, Vol. 17, pp. 17–25, Cambridge University Press, Cambridge (1959).
9. B. E. Davidson and F. J. R. Hird, The estimation of glutathione in rat tissues. A comparison of a new spectrophotometric method with the glyoxylase method, *Biochem. J.* **93**:232–236 (1964).
10. C. W. I. Owens and R. V. Belcher, A colorimetric micro-method for the determination of glutathione, *Biochem. J.* **94**:705–711 (1965).
11. V. H. Cohn and J. Lyle, A fluorometric assay for glutathione, *Anal. Biochem.* **14**:434–440 (1966).
12. T. L. McNeil and L. V. Beck, Fluorometric estimation of GSH-OPT, *Anal. Biochem.* **22**:431–441 (1968).
13. H. Waelsch and D. Rittenberg, Glutathione. II. The metabolism of glutathione studied with isotopic ammonia and glutamic acid, *J. Biol. Chem.* **144**:53–58 (1942).
14. G. W. Douglas and R. A. Mortensen, The rate of metabolism of brain and liver glutathione in the rat studied with C^{14}-glycine, *J. Biol. Chem.* **222**:581–585 (1956).
15. R. A. Mortensen, M. I. Haley, and H. A. Elder, The turnover of erythrocyte glutathione in the rat, *J. Biol. Chem.* **218**:269–273 (1956).
16. A. Lajtha, S. Berl, and H. Waelsch, Amino acid and protein metabolism of the brain. IV. The metabolism of glutamic acid, *J. Neurochem.* **3**:322–332 (1959).
17. M. K. Gaitonde and D. Richter, in *Metabolism of the Nervous System* (D. Richter, ed.), pp. 449–455, Pergamon Press, New York (1957).
18. Y. Takahashi and Y. Akabane, Incorporation of [^{14}C] glycine into glutathione in rat cerebral cortex slices, *J. Neurochem.* **7**:89–96 (1961).
19. J. E. Snoke and K. Bloch, in *Glutathione* (S. Colowick, A. L. Lazarow, E. Racker, D. R. Schwarz, E. Stadtman, and H. Waelsch, eds.), pp. 129–144, Academic Press, New York (1954).

20. J. E. Snoke and K. Bloch, Studies on the mechanism of action of glutathione synthetase, *J. Biol. Chem.* **213**:825–835 (1955).
21. E. D. Mooz and A. Meister, Tripeptide (glutathione) synthetase. Purification, properties, and mechanism of action, *Biochemistry* **6**:1722–1734 (1967).
22. F. Binkley and C. K. Olson, Metabolism of glutathione. IV. Activators and inhibitors of the hydrolysis of glutathione, *J. Biol. Chem.* **188**:451–457 (1951).
23. P. J. Fodor, A. Miller, and H. Waelsch, Quantitative aspects of enzymatic cleavage of glutathione, *J. Biol. Chem.* **202**:551–565 (1953).
24. G. G. Glenner, J. E. Folk, and P. J. McMillan, Histochemical demonstration of a gamma-glutamyl transpeptidase-like activity, *J. Histochem. Cytochem.* **10**:481–489 (1962).
25. J. S. Fruton, in *The Proteins* (H. Neurath, ed.), Vol. I, pp. 189–310, Academic Press, New York (1963).
26. H. McIlwain, Thiols and the control of carbohydrate metabolism in cerebral tissues, *Biochem. J.* **71**:281–285 (1959).
27. L. W. Mapson, Enzyme systems associated with the oxidation and reduction of glutathione in plant tissues, in *Biochemical Society Symposia*, Vol. 17, pp. 28–41, Cambridge University Press, Cambridge (1959).
28. L. W. Mapson, Photo-oxidation of ascorbic acid in leaves, *Biochem. J.* **85**:360–369 (1962).
29. G. C. Mills, Glutathione peroxidase and the destruction of hydrogen peroxide in animal tissues, *Arch. Biochem. Biophys.* **86**:1–5 (1960).
30. G. Cohen and P. Hochstein, Glutathione peroxidase: The primary agent for elimination of hydrogen peroxide in erythrocytes, *Biochemistry* **2**:1420–1428 (1963).
31. E. Beutler, in *The Metabolic Basis of Inherited Disease* (J. B. Stanbury, J. B. Wyngaarden, and D. S. Fredrickson, eds.), pp. 1060–1089, McGraw-Hill, New York (1966).
32. T. W. Rall and A. L. Lehninger, Glutathione reductase of animal tissues, *J. Biol. Chem.* **194**:119–130 (1952).
33. S. S. Hotta, Effect of chlorpromazine and amytal on the reduction of glutathione, *Arch. Biochem. Biophys.* **113**:395–398 (1966).
34. R. N. Perham and J. I. Harris, Amino acid sequences around the reactive cysteine residues in glyceraldehyde-3-phosphate dehydrogenases, *J. Mol. Biol.* **7**:316–320 (1963).
35. M. V. Riley and A. L. Lehninger, Changes in sulfhydryl groups of rat liver mitochondria during swelling and contractions, *J. Biol. Chem.* **239**:2083–2089 (1964).
36. H. M. Katzen, F. Tietze, and DeWitt Stetten, Jr., Further studies on the properties of hepatic glutathione-insulin transhydrogenase, *J. Biol. Chem.* **238**:1006–1011 (1963).
37. A. Geiger, XCV. Role of glutathione in anaerobic tissue glycolysis, *Biochem. J.* **29**:811–823 (1935).
38. Y. Kakimoto, T. Nakajima, A. Kanazawa, M. Takesada, and I. Sano, Isolation of gamma-L-glutamyl-L-glutamic acid and gamma-L-glutamyl-L-glutamine from bovine brain, *Biochim. Biophys. Acta* **93**:333–338 (1964).
39. A. Kanazawa, Y. Kakimoto, T. Nakajima, H. Shimizu, M. Takesada, and I. Sano, Isolation and identification of gamma-L-glutamylglycine from bovine brain, *Biochim. Biophys. Acta* **97**:460–464 (1965).
40. Y. Kakimoto, A. Kanazawa, T. Nakajima, and I. Sano, Isolation of gamma-L-glutamyl-L-beta-aminoisobutyric acid from bovine brain, *Biochim. Biophys. Acta* **100**:426–431 (1965).
41. A. Kanazawa, Y. Kakimoto, T. Nakajima, and I. Sano, Identification of gamma-glutamyl-serine, gamma-glutamylalanine, gamma-glutamylvaline and *S*-methylglutathione of bovine brain, *Biochim. Biophys. Acta* **111**:90–95 (1965).
42. I. Sano, Y. Kakimoto, A. Kanazawa, and T. Nakajima, Identification of glutamylpeptides in brain (Ger), *J. Neurochem.* **13**:711–719 (1966).
43. A. Kanazawa, I. Sano, and H. Shimizu, The distribution of γ-L-glutamyl-L-glutamine in mammalian tissues, *J. Neurochem.* **14**:596–598 (1967).

44. D. L. Buchanan, E. E. Haley, and R. T. Markiw, Occurrence of β-aspartyl and γ-glutamyl oligopeptides in human urine, *Biochemistry* 1:612–620 (1962).
45. Y. Kakimoto and M. D. Armstrong, The preparation and isolation of D-(−)-β-aminoisobutyric acid, *J. Biol. Chem.* 236:3283–3286 (1961).
46. E. E. Haley and B. J. Corcoran, β-Aspartyl peptide formation from an amino acid sequence in ribonuclease, *Biochemistry* 6:2668–2672 (1967).
47. J. J. Pisano, J. D. Wilson, L. Cohen, D. Abraham, and S. Udenfriend, Isolation of γ-aminobutyrylhistidine (homocarnosine) from brain, *J. Biol. Chem.* 236:499–502 (1961).
48. A. Kanazawa, Y. Kakimoto, E. Miyamoto, and I. Sano, Isolation and identification of homocarnosine from bovine brain, *J. Neurochem.* 12:957–958 (1965).
49. T. Nakajima, F. Wolfgram, and W. G. Clark, The isolation of homoanserine from bovine brain, *J. Neurochem.* 14:1107–1112 (1967).
50. D. Abraham, J. J. Pisano, and S. Udenfriend, The distribution of homocarnosine in mammals, *Arch. Biochem. Biophys.* 99:210–213 (1962).
51. A. Kanazawa and I. Sano, A method of determination of homocarnosine and its distribution in mammalian tissues, *J. Neurochem.* 14:211–214 (1967).
52. C. L. Davey, The significance of carnosine and anserine in striated skeletal muscle, *Arch. Biophys.* 89:303–308 (1960).
53. E. A. Hosein and M. Smart, The presence of anserine and carnosine in brain tissue, *Can. J. Biochem. Physiol.* 38:569–573 (1960).
54. J. C. Dickinson and P. B. Hamilton, The free amino acids of human spinal fluid determined by ion exchange chromatography, *J. Neurochem.* 13:1179–1187 (1966).
55. T. L. Perry, S. Hansen, B. Tischler, R. Bunting, and K. Berry, Carnosinemia, a new metabolic disorder associated with neurologic disease and mental defect, *New Engl. J. Med.* 277:1219–1227 (1967).
56. S. Tsunoo, K. Horisaka, M. Kawasumi, K. Aso, and S. Tokue, Concerning the free amino acids in chicken brain. The isolation and identification of histidine and anserine (Ger), *J. Biochem. (Tokyo)* 54:355–362 (1963).
57. R. E. Winnick and T. Winnick, Carnosine-anserine synthetase of muscle. I. Preparation and properties of a soluble enzyme from chick muscle, *Biochim. Biophys. Acta* 31:47–55 (1959).
58. G. D. Kalyankar and A. Meister, Enzymatic synthesis of carnosine and related β-alanyl and γ-aminobutyryl peptides, *J. Biol. Chem.* 234:3210–3218 (1959).
59. D. Abraham, J. J. Pisano, and S. Udenfriend, Synthesis of homocarnosine in muscle *in vivo*, *Biochim. Biophys. Acta* 50:570–572 (1961).
60. D. Abraham, J. J. Pisano, and S. Udenfriend, Uptake of carnosine and homocarnosine by rat brain slices, *Arch. Biochem. Biophys.* 104:160–165 (1964).
61. C. L. Davey, The effects of carnosine and anserine on glycolytic reactions in skeletal muscle, *Arch. Biochem. Biophys.* 89:296–302 (1960).
62. Y. Qureshi and T. Wood, The effect of carnosine on glycolysis, *Biochim. Biophys. Acta* 60:190–192 (1962).
63. S. P. Bessman and R. Baldwin, Imidazole aminoaciduria in cerebromacular degeneration, *Science* 135:789–791 (1962).
64. J. Levenson, K. Lindahl-Kiessling, and S. Rayner, Carnosine excretion in juvenile amaurotic idiocy, *Lancet* 2:756 (1964).
65. A. Curatolo, N-Acetyl-aspartyl-glutamic acid in nervous tissue, *Abstracts VI Intern. Cong. Biochem.*, New York, V-E-98 (1964).
66. E. Marchetti and G. Mattalia, On the structure of the natural dipeptide N-acetyl-aspartyl-glutamic acid (NAAGA), *Experientia* 21:687–688 (1965).
67. J. V. Auditore and H. Hendrickson, Investigation of unidentified pharmacologically active substances in the human central nervous system, *Intern. J. Neuropharmacol.* 3:1–7 (1964).

68. J. V. Auditore, E. J. Olson and L. Wade, Isolation, purification, and probable structural configuration of N-acetyl-aspartylglutamate in human brain, *Arch. Biochem. Biophys.* **114**:452–458 (1966).

69. E. Miyamoto, Y. Kakimoto, and I. Sano, Identification of N-acetyl-α-aspartylglutamic acid in the bovine brain, *J. Neurochem.* **13**:999–1003 (1966).

70. D. L. Buchanan, E. E. Haley, R. T. Markiw, and A. A. Peterson, Studies on the *in vivo* metabolism of α and β-aspartylglycine-1-C^{14}, *Biochemistry* **1**:620–623 (1962).

71. E. Miyamoto and T. Tsujio, Determination of N-acetyl-α-aspartyl-glutamic acid in the nervous tissue of mammals, *J. Neurochem.* **14**:899–903 (1967).

72. A. Curatolo, P. D'Arcangelo, A. Lino, and A. Brancati, Distribution of N-acetyl-aspartic and N-acetyl-aspartyl-glutamic acids in nervous tissue, *J. Neurochem.* **12**:339–342 (1965).

73. D. Morris and D. W. Straughan, The interaction of alpha- and beta-L-aspartyl-L-glutamic acid and of their N-acetyl derivatives on central neurones and certain isolated peripheral tissues, *Biochem. Pharmacol.* **14**:1679–1681 (1965).

74. K. L. Reichelt and E. Kvamme, Acetylated and peptide bound glutamate and aspartate in brain, *J. Neurochem.* **14**:987–996 (1967).

75. R. E. Alving and K. Laki, N-Terminal sequence of actin, *Biochemistry* **5**:2597–2601 (1966).

76. S. Puszkin, D. D. Clarke, and S. Berl, Contractile protein in brain, *Federation Proc.* **27**(2):464, Abs. 1392 (1968).

77. T. Winnick, R. E. Winnick, R. Archer, and C. Fromageot, Amino acids and peptides of posterior pituitary and hypothalamus tissues, *Biochim. Biophys. Acta* **18**:488–494 (1955).

78. B. Shome and M. Saffran, Peptides of the hypothalamus, *J. Neurochem.* **13**:433–448 (1966).

79. P. Mandel and M. Ledig, Amidated phosphopeptides in nervous tissue, *Biochem. Biophys. Res. Commun.* **24**:275–279 (1966).

80. T. Rathlev and T. Rosenberg, Non-enzymatic formation and rupture of phosphorus to nitrogen linkages in phosphoramido derivatives, *Arch. Biochem. Biophys.* **65**:319–339 (1956).

81. B. S. Ramarao, G. Hauser and F. N. LeBaron, Further studies on the incorporation of ^{32}P into the phosphatidopeptides of brains from normal and metrazol-treated rats, *Biochim. Biophys. Acta* **84**:348–350 (1964).

82. E. Orikabe, H. Matsui, S. Ishikawa, and N. Shimazono, Studies on phosphoprotein in brain. I. Phosphorylation of bound phosphopeptide and its isolation, *J. Biochem. (Tokyo)* **57**:346–354 (1965).

83. P. J. Heald, Studies on the phosphoproteins of brain. III. Phosphorylserine sequences in brain phosphoprotein, *Biochem. J.* **80**:510–514 (1961).

84. P. J. Heald, *Phosphorus metabolism of brain*, Pergamon Press, New York (1960).

85. R. W. Albers, Biochemical aspects of active transport, *Ann. Rev. Biochem.* **36**:727–756 (1967).

86. P. R. Carnegie and C. E. Lumsden, Fractionation of encephalitogenic polypeptides from bovine spinal cord by gel filtration in phenol–acetic acid–water, *Immunology* **12**:133–145 (1967).

87. P. R. Carnegie, B. Bencina, and G. Lamoureux, Experimental allergic encephalomyelitis: Isolation of basic proteins and polypeptides from central nervous tissue, *Biochem. J.* **105**:559–568 (1967).

88. M. W. Kies, Chemical studies on an encephalitogenic protein from guinea pig brain, *Ann. N.Y. Acad. Sci.* **122**:161–170 (1965).

89. G. Ungar, L. Galvan, and R. H. Clark, Chemical transfer of learned fear, *Nature* **217**:1259–1261 (1968).

90. E. B. Chain, in *Reflection on Research and the Future of Medicine* (C. E. Lyght, ed.), pp. 129–169, McGraw-Hill, New York (1967).

Chapter 5

PROTEINS

Samuel Bogoch

*Foundation for Research on the Nervous System
and Boston University School of Medicine
Boston, Massachusetts*

I. INTRODUCTION

This chapter will be concerned with general principles of and illustrative observations on central nervous system proteins. It is hoped that it will serve as a useful orienting introduction to other chapters in the *Handbook* that will deal in greater detail with particular classes of proteins, such as phosphoproteins, or particular proteins extracted from certain nervous system structures, such as the myelin proteins, or with individual proteins with specific enzymatic properties. The principles discussed will be illustrated by examples from various studies done on individual proteins of the nervous system.

Knowledge of the proteins of the nervous system has been meager until very recent years. The rapid development of general protein biochemistry during the past 50 years has not been accompanied by parallel development for nervous system proteins. There are many reasons for this lag, and some of these reasons are interrelated. Thus, the general tendency to exclude the brain from the list of sampled organs for isolation of proteins in radioactive exchange studies is not simply due to the fact that the brain is harder to get to. The generally used aqueous methods for extraction of proteins in other body organs do not yield adequate amounts of brain proteins. Only 10–20 % of the total brain proteins are extracted by one or two homogenizations in aqueous buffer, and that which is extracted is frequently inextricably associated with lipids, readily loses its solubility, and precipitates on standing or when placed onto ion exchange columns.

A. Extractability and Total Content

The use of organic solvents to first remove lipids from brain tissue to permit protein determination has the doubly negative effect of solubilizing a small fraction of the total protein, while at the same time denaturing the remainder. Thus what is left after organic solvent extraction is insoluble protein residues enmeshed in stringy form. Because of this difficulty it is not possible by this method to state exactly what is the total protein content of nervous tissue. Such values obtained from the organic solvent insoluble residue must be determined either by dry weight, which is inaccurate because of the presence of other large molecules bound to proteins in the insoluble residue; by nitrogen content, which is inaccurate because of the considerable concentrations of amino sugars bound to the brain proteins; or by Folin–Lowry protein determination, which is relatively inaccurate in grossly insoluble material. All these methods, however, give approximate ideas of the total content of protein. Biuret and α-amino nitrogen determinations are additional useful measures.

From the approximate total content, it is clear that single or double extraction with aqueous buffer does not yield more than 10–20% of the total protein content of nervous tissue. The use of strong salt concentrations, or extremes of pH, may yield somewhat more of the total or may provide relatively specific fractions that are characterized by the particular extreme method. However, there are remaining problems of continuing solubility after extraction and of the definition of the native state of the proteins thus extracted.

B. Prevalence of Conjugated Proteins

A further complication in the extractability and definition of central nervous system protein is the increasingly apparent fact that most of these are conjugated proteins in the classical sense, i.e., are covalently bound or otherwise strongly bound to carbohydrates, to residues rich in phosphorus, or to large lipid or nucleic acid residues. In addition, this conjugated material is not readily separable by ordinary means from the proteins to which they are joined. Thus, glycoliponucleoprotein fractions are not infrequently observed. Indeed, in recent studies that started by finding that many of the proteins isolated by aqueous buffers contain appreciable amounts of tightly bound carbohydrate residues,[1] the search has now turned to the question of whether there are indeed any proteins in the group thus extracted that do not contain bound carbohydrate to some significant degree.[2]

C. Membranous Origin

The above problems of extractability, solubility, estimate of total concentration, ionic binding, and covalent conjugation of proteins to other types of substances all come together in one significant way: It is becoming increasingly apparent that all these problems relate to the fact that the tissue of origin of these proteins is distinctive in one special way; i.e., the brain is perhaps the most complex membranous tissue in biology. Recent electron micrographs of a few millimicrons of a synaptic region indicate the high degree of membranous structural differentiation. Because of this tremendous ordering of macromolecular constituents at the level of fine structure, it is apparent that many of the proteins of the nervous system are involved in a three-dimensional geometric membranous array. Attempts to isolate "aqueous" vs. "membranous" proteins from separated subcellular organelles[2-4] support this notion. The usual differentiation between "structural" and "enzymatic" protein may be of less value in the case of nervous system protein until a more exact separation of individual protein residues from individual specific membranes in various subcellular organelles can be made.

D. Variation in Structure with Functional State of Nervous System

A final general complexity may be referred to in terms of the increasing evidence that the degree of conjugation with other substances and the actual concentration of individual protein fractions both may vary markedly with the functional state of the nervous system at a given moment in time. This phenomenon is illustrated in later parts of this chapter with reference to the concentration of hexose bound to brain proteins, as well as to the absolute concentration of individual brain protein fractions, in resting as compared with training states in pigeon brain.[4]

E. The Native State of CNS Proteins

From all of the foregoing, some of the usual prohibitions about the physicochemical means of obtaining information on proteins in general may not be so profitably invoked for brain proteins. Thus, whereas it is well known that the use of organic solvents causes denaturation of many proteins, certain nervous system proteins are extracted in reasonably reproducible fashion by the use of such solvents. That the majority of brain proteins may not be soluble in aqueous form is regrettable, especially to those who are accustomed to studying proteins that are essentially aqueous in form, i.e., the serum proteins. However, it should be noted that because of the highly membranous structure of certain brain proteins, there is no *a priori* reason for assuming that they should be able to exist in a "native" state in an aqueous medium. The only state to which they are accustomed throughout their existence may well be a state which has only a small degree of hydration and few degrees of freedom of rotation because of the close packing and conjugation with other organic-soluble substances in the membrane. Thus, *in situ*, these proteins may exist in a form more akin to a semisolid state. The concepts of denaturation are largely derived from studies of serum proteins. These concepts therefore may require considerable modification when membranous proteins of the nervous system are studied. It seems reasonable, at this state of development of our knowledge of nervous system proteins, to state that no method of extraction or solubilization should be excluded if it yields *reproducible* protein fractions that can be studied at least in terms of primary protein structure, if not readily in terms of secondary or tertiary structure, and enzymatic or other function.

II. CLASSIFICATION OF NERVOUS SYSTEM PROTEINS ACCORDING TO DIFFERENT PROPERTIES

Table I summarizes some of the properties of nervous system proteins that are useful now, and may be increasingly useful as data accumulates, for the classification of nervous system proteins as they are purified and described in the future. In a sense, it is a summary of the specific properties to be considered in defining a particular protein as discrete from other proteins of the nervous system. While the total information on the proteins of the nervous system is still meager, it can be seen from Table I and from the discussion that follows that there are sufficient examples at present of particular protein fractions or individual purified proteins that need to be distinguished in terms of all of the listed parameters. At the same time it must be stated that there is no single protein of the nervous system for which data is available on all the parameters listed in Table I. (See Table II for examples.)

Another feature of these parameters is the fact that each by itself often provides the first approach to the definition of a given central nervous system protein. From a practical point of view it then will be increasingly important to try to gather data on each of the other properties in as systematic a manner as possible so that early differentiation of discrete proteins can be made.

A. Conditions for Optimal Extractability

1. Solvent, pH, and Ionic Strength

This property has been perhaps most neglected in studies to date on nervous system proteins. The classical separation of proteins on the basis of the

TABLE I

Classification of Nervous System Proteins According to Properties

A. Conditions for optimal extractability
   1. Solvent; pH; ionic strength
   2. Method and number of homogenizations
B. Properties on column chromatography and subsequent solubility
C. Properties on electrophoresis
D. Specific substances constantly conjugated (e.g., carbohydrates, lipids, nucleic acids, copper, sulfur, and phosphorus)
E. Amino acid composition
F. Physicochemical properties
G. Response to enzymatic and other hydrolysis
H. Organ, regional, species, cell, and subcellular organelle specificity
I. Enzymatic properties
J. Immunochemical specificity
K. Developmental phase specificity
L. Pathological state specificity
M. Functional specificity

TABLE II

Some Specified and Partially Characterized Nervous System Proteins

A. Central nervous system proteins
    1. Posterior pituitary proteins[42]
      a. Neurophysin
      b. "Granule protein"
    2. Proteins of brain nuclei: histones[43]
    3. 10B and 11A: human and pigeon brain proteins
    4. Membrane proteins
      a. 12 and 13
      b. Sheep brain[44] and rat brain[36] microsomal proteins
    5. Dendritic proliferation proteins (1 and 2)
    6. S-100
    7. Copper proteins
    8. Myelin proteins
      a. Proteolipids
      b. Neurokeratin
      c. Neurosclerin
      d. Basic myelin proteins
      e. Encephalitogenic proteins
    9. Retinal opsins
    10. Cerebrospinal fluid proteins[45]
B. Peripheral nervous system proteins
    1. Nerve growth factor protein
    2. Neurofilament proteins of squid axoplasm

solvent which will extract them has limited use, however, because in addition to the type of solvents, aqueous, detergent, organic, etc., having importance, there is evidence that other features such as the pH and the type and molarity of salt concentration are also of crucial importance. Thus, for example, while studies employing detergents for the extraction of brain protein rich in carbohydrate constituents yield particular fractions,[5] there is little doubt that many of these fractions can be extracted by nondetergent phosphate buffer solution[1,2] and that many of the glycoproteins extracted by the aqueous methods are not apparently extracted by the detergent methods. It has not yet been determined whether proteins extracted by lysolecithin[6] can also be extracted by aqueous buffers. That a simple sequential application of different solvents to the tissue and its residues does not answer this question is made clear in the next section, which illustrates that the state of physical disruption of the membranes is crucial to what proportion of the total protein is made available to the solvent.[2,3] The type and concentration of detergent used also may be critical. An example is seen in some studies[7] that show that sodium deoxycholate at 0.1 % solubilizes only 45 % of brain protein, whereas 0.3 % solubilizes 81 % of protein as measured by Biuret.

Similarly, other studies[8] on the solubility of brain proteins in chloroform–methanol show that the amount of chloroform–methanol soluble protein increased two- to sixfold upon the removal of the inorganic salt from the tissue during the process of centrifugation or dialysis. An example of the effect of pH on the ability of chloroform–methanol to extract nervous system white matter proteolipids is provided by studies[9] that showed that at acid pH more proteolipid is solubilized than at neutral or alkaline pH. Phosphoproteins have been extracted from brain microsomes at pH 10.5, clarified by centrifugation at pH 7.0, and precipitated at pH 4.5. After precipitation, only part could be redissolved at pH 9.[10,11] The effect of extremes of pH and salt concentration on protein extractability has been examined,[12] but as noted in the use of detergents, subsequent separation of protein species may be difficult for fractions originally solubilized under extreme conditions. In addition, the formation of complexes by extracted proteins may occur at subsequent separation states,[13] so that not only initial pH but subsequent pH at various stages of separation may be critical.

2. Method and Number of Homogenizations

The method of disruption of tissue is now increasingly recognized as important, not only for total concentration of neural proteins extracted, but also with regard to the type of protein that can be so extracted. Thus, using aqueous phosphate buffers at near neutrality, it was demonstrated that two homogenizations in the Waring blender yielded 7.7 mg protein/g. By simply increasing the number of homogenizations up to 9 to 14, 19.3 mg/g was obtained. When separated subcellular organelles were used as the starting material for the extraction, the yield was further increased to 34 mg/g. Not only was the total amount of protein extracted increased by these means, but also a group of proteins (11B, 12, and 13) extracted with difficulty were obtained whose subsequent history showed them to be tightly binding to DEAE (Cellex-D) columns and to require detergent for elution from those columns.[2,3]

B. Properties on Column Chromatography and Subsequent Solubility

These last-mentioned proteins, 11B, 12, and 13, although perfectly soluble before being placed on the column, clearly interacted with the column in some way to make their elution without detergent impossible by any means tested to date. These very proteins have been increasingly shown to be in high concentration in membranous portions of the nervous system such as isolated microsomes and glial cell membrane ghosts.[2,4] These observations taken together with the comments made underline the importance of considering many of the proteins of the nervous system as tightly bound in membrane structure and thus requiring special methods for extraction and solubilization. The separation on carboxymethyl cellulose of seven subfractions of a white matter protein fraction obtained by preliminary separation on Biogel[13] further illustrates the importance of utilizing more than one carrier medium to differentiate individual protein fraction on the basis of size or charge, or both.

C. Properties on Electrophoresis

Electrophoresis on various carriers including paper, starch gel, agar gel, and polyacrylamide gel has been used in the past fifteen years to define anywhere from 7 to 30 fractions from various specimens of brain tissue or peripheral nerve, and most of these studies have been recently reviewed.[14] Electrophoretic techniques also have been applied with some success to clinical pathological diagnostic use.[15] Due to the great heterogeneity of the brain proteins, the resolution obtained in the above studies has been less satisfactory than that obtained in more recent studies where protein fractions are first delineated by differential precipitation or by column chromatography. Thus direct application of aqueous brain proteins to disc gel electrophoresis gives evidence for only between 15 and 30 proteins.[14] However, when column chromatography is first applied to the aqueous extracts of gray matter of human brain, 50 major fractions can be designated by means of tube-to-tube analysis of the carbohydrate content of the eluates. Each of these 50 major fractions when further tested on disc gel electrophoresis was shown to be heterogeneous, with 3–11 distinct proteins in each major fraction. While the overlap between the adjacent fractions is apparent, it is nonetheless possible to demonstrate well over 100 and perhaps closer to 200 quite discrete brain protein fractions on this basis.[1]

It should be noted from Fig. 1 that the proteins shown here are those eluted by aqueous solvents from Cellex-D chromatography. At the time the study was done, Groups 11B, 12, and 13 proteins were neither extracted in the original homogenizations, since only two homogenizations were used, nor were they eluted from the columns, since Triton was not used. It must be concluded that the number of discrete brain proteins that can be demonstrated on disc gel electrophoresis after preliminary separation by chromatographic properties, and by carbohydrate composition, is far too great to expect an adequate separation when any single known carrier medium is used alone. It becomes increasingly apparent that two or three sequential separations by different means, such as those employed in the above study, are necessary before one can begin to reach a level of discreteness in fractions that permits their comparison between different parts of brain or in different subcellular organelles. That discrete fractions very sharply defined by the above methods, such as 10B, may still contain one major and two or three minor subfractions is suggested by the evidence that three distinct antibodies are produced to whole 10B.[4] The improvement in column chromatography has delineated sharply two distinct subpeaks, 10BI and 10BII (see Fig. 2),[4] and these are now being compared with the evidence of constituent antibodies.

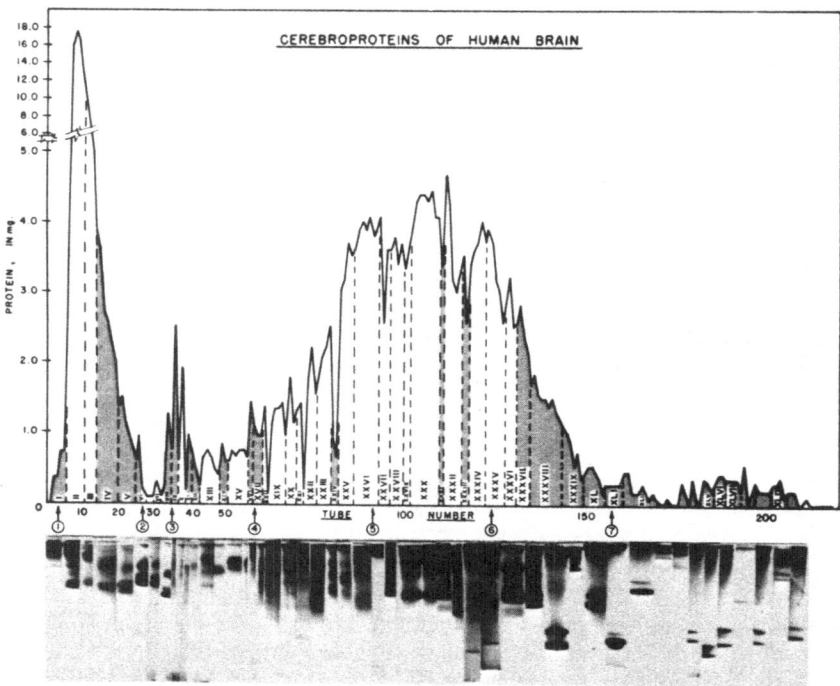

Fig. 1. Three-dimensional separation of brain proteins. Column chromatography followed by protein and carbohydrate analysis tube to tube, and then by disc gel electrophoresis of 50 subfractions.[1] Tube number and volume of effluent are shown on abscissa. Cerebroprotein fractions are designated by Roman numerals I through L. Shaded areas represent cerebroprotein fractions with hexose content of 2.5% or greater. Numbered arrows on abscissa indicate points at which new solvents were introduced into solvent mixing chamber: (1) 75 cm^3 of 0.005 M sodium phosphate buffer, pH 7; (2) mixture of 12.5 cm^3 of 0.005 M sodium phosphate buffer, pH 7, and 12.5 cm^3 of 0.05 M sodium phosphate buffer, pH 4.7; (3) 75 cm^3 of 0.05 M sodium phosphate buffer, pH 4.7; (4) mixture of 50 cm^3 of 0.05 M sodium phosphate buffer, pH 4.7, and 50 cm^3 of 0.3 M sodium phosphate buffer, pH 4.3; (5) 100 cm^3 of 0.3 M sodium phosphate buffer, pH 4.3; (6) 100 cm of 1.0 M sodium phosphate buffer, pH 4.1; and (7) 100 cm^3 of 1.0 M phosphoric acid, pH 1.0. Electrophoretic pattern for each major cerebroprotein fraction is shown under the respective column fraction.[4]

D. Specific Substances Constantly Conjugated

As noted in Table I, such substances can be carbohydrates, lipids, or nucleic acids, or the complex may contain readily identifiable substances such as copper, sulfur, and phosphorus. Examples where the presence of carbohydrates is useful as a means of separating nervous system proteins according to their absolute concentration of carbohydrate attached has already been given.[1,16] In the case of the proteolipids, not only was the

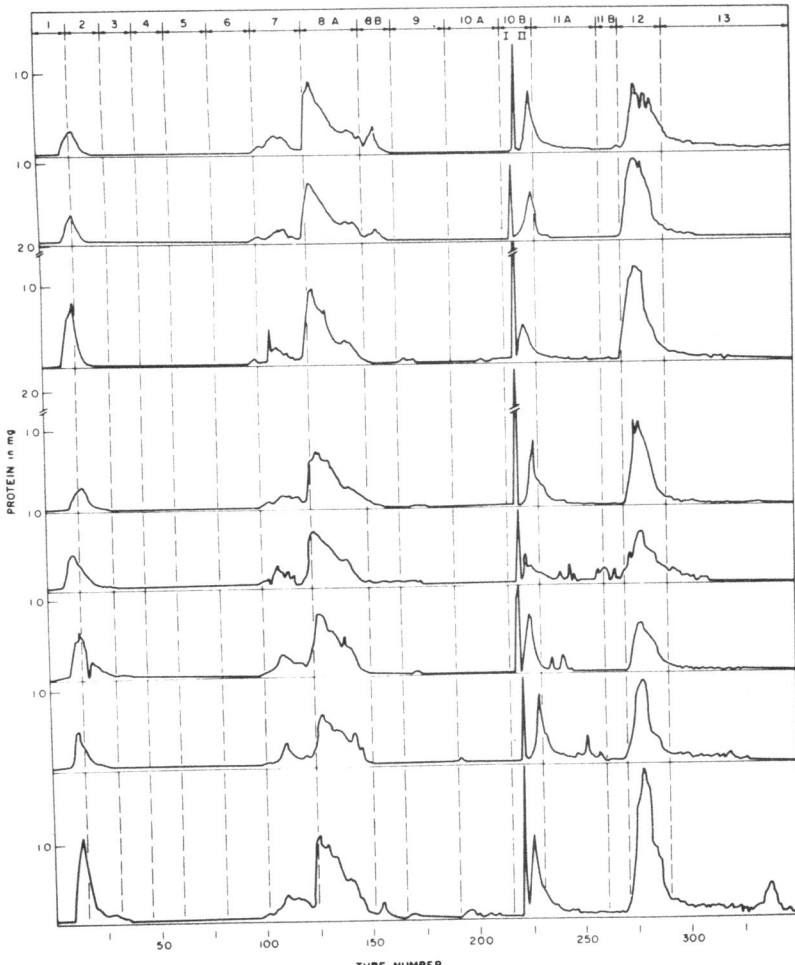

Fig. 2. Column chromatography of pigeon brain proteins. The method used to obtain the above data differs from that used to obtain the data given in Fig. 1 in the following ways: (1) Two homogenizations only were employed in the extraction of Fig. 1; exhaustive homogenization (4–13 times) was used in Fig. 2. (2) Extract was concentrated against Carbowax in Fig. 1, rather than by perevaporation as in Fig. 2. (3) Purified preparation of DEAE (Cellex D) was used in Fig. 2. (4) The same eluting solvents (1 through 7 in Fig. 1) also were employed in Fig. 2, but the spacing between these was increased in order to permit more complete elution with each solvent. The total number of tubes was increased to 259 for the first 7 eluting solvents in Fig. 2. (5) Triton X-100, not used in Fig. 1, was routinely employed later, as in Fig. 2, following solvent (7), i.e., beginning with tube 260 and ending with tube 350.

presence of bound lipid apparently relevant to the extractability of this group of proteins, but the high content of sulfur containing amino acid aided in their distinction from other components.[17] Similarly, the presence of a high concentration of bound copper in both normal tissue and in Wilson's disease permitted the separation and identification of several copper proteins of the nervous system.[18,19]

The presence of lipids bound to protein requires definition in terms of the ease of dissociation of the lipid from the protein fragments. Thus, for example, some of these bonds are dissociated by alcohol and ether extraction, some cannot be so dissociated but can be separated by chromatography on silicic acid, and some cannot be separated by either of these two methods but can be separated by extraction with chloroform–methanol when acidified with hydrochloric acid.[17]

The high degree of association of phosphorus with a number of protein fractions in brain has been recently demonstrated for ox brain microsomes.[11] While it is not yet clear how many of these fractions that are rich in phosphate are so because of the accidental adsorptive association of phosphorus with protein throughout the fractionation procedure, it is unlikely that this accounts for all of the phosphoproteins. Certainly the demonstration that phosphopeptides occur in some brain proteins, containing what would appear to be polyphosphoryl serine, suggests some specificity of conjugation of phosphorus within the protein. Phosphoinositide bound to brain protein also has been a useful marker in the definition of fractions.[20,21]

E. Amino Acid Composition

While it is clear that an accurate determination of the amino acid composition of the protein is one of the critical differentiating parameters for its specification, this method can provide misleading similarities between two protein fractions that have not been completely purified. Under these conditions, minor variations in overall amino acid composition may be quite critical in illustrating the presence to quite an appreciable extent of some contaminating protein. On the other hand, the presence of high concentrations of sulfur-containing amino acids already has been referred to as a useful marker in one series of separation steps. A further example is found in the demonstration of the high content of lysine and basic amino acids in acid-extractable proteins of bovine brain white matter that are inducers of experimental allergic encephalomyelitis.[22] It should not be forgotten that with minor differences in amino acid composition between two proteins, structural sequence studies may show them to be quite different. There are a number of examples of brain proteins for which amino acid analyses have been reported.[9,17]

F. Physicochemical Properties

The physicochemical properties of certain nervous system proteins can be used for their separation. This is illustrated by the fact that high-speed centrifugation prior to column chromatography produces a greater yield of proteins and a greater ability to separate these, presumably because of the diminution in the amount of associated lipids.[1–4] Similarly, centrifugation of chloroform–methanol soluble nervous system protein has aided in the separation of lipoproteins or proteolipids from free lipids on the one hand and from proteins on the other.[17] The use of dialysis and separation on exclusion gels are, of course, fundamental methods in the separation of brain proteins, as for proteins elsewhere. The demonstration of water-soluble lipoprotein complexes as presumably distinct from those that are soluble in organic solvents (proteolipids) has been made recently.[23] These contain cholesterol, glycolipids, and phospholipids and appear in the $105,000 \times g$ supernate of rat brain homogenates. While their physicochemical properties in the centrifuge have been of aid in their preparation, necessary attempts to separate the lipid and protein by different solvents, by silicic acid chromatography, and by acidified organic solvents have not been reported as yet. It is not clear, therefore, to what degree and in what manner the lipids are bound to the particular brain proteins *in situ*.

Definitive physicochemical tests of purity applied to highly purified proteins elsewhere, such as their sedimentation behavior under varying ionic conditions, have seldom been used to date on nervous system proteins with any clear benefit, largely because the nervous system proteins have not yet been purified to a sufficient degree. An example of the problems met in the use of this one parameter as a distinguishing characteristic is seen in the work on the S-100 protein fraction.[24] Whereas the S-100 protein was eluted in a single peak from DEAE cellulose, G-100 sephadex, and hydroxyapatite columns and migrated in a single band well ahead of other brain soluble proteins when submitted to starch gel electrophoresis in the discontinuous buffer system at several pH values between 6.5 and 9, electrophoresis on mixed agarose–acrylamide gel in discontinuous glycine buffer system demonstrated that S-100 separated into several bands.[25] It has been possible to distinguish a minimum of two bands by both electrophoretic and immunological studies.[26,27] As repeated throughout this chapter, the use of a single parameter, no matter how sophisticated, is not in itself sufficient for the demonstration of homogeneity of a given nervous system protein; indeed, the use of three or more such parameters sequentially may not suffice. This applies only to what might be called macroheterogeneity and still leaves the problem of microheterogeneity to be wrestled with, as in the most purified fractions of serum α-1 glycoprotein.[28]

G. Response to Enzymatic and Other Hydrolysis

The fact that neurokeratin, a fraction of brain protein extracted by Ewald and Kuhn 90 years ago, was resistant to the action of gastric and pancreatic juice formed the basis for the definition of the first crude myelin protein fraction. Further purification of similar material resistant to pepsin and trypsin, obtained by different extraction procedures, the proteolipids,[17] led to the definition of a major component of the myelin proteins. This method of differentiating proteins doubtless will find further useful application as organelle-specific neuroproteins are isolated.

H. Organ, Regional, Species, Cellular, and Subcellular Organelle Specificity

The early studies demonstrating the presence of organ-specific substances in brain were published in 1928.[29] Since that time, an organ-specific antigen in brain that is thermostable, and is referred to as "BE," has been demonstrated.[30] This has been shown to be heterogeneous in recent studies.[31–33] One of these thermostable brain-specific antigens, AP, has been purified and shown to be a glycoprotein from which neuraminic acid can be removed with neuraminidase. Other proteins reported to be brain specific, of unknown relation to the above and to each other, have been described.[24,34–36] Some indication is available of regional specificity for particular proteins in terms of particular brain protein fractions. Thus 11A occurs in higher concentration in the hippocampus than in the caudate of the cat (S. Bogoch and R. Galambos, unpublished results).

Some degree of species specificity is suggested by the work on 10B and 11A.[4] There is now considerable evidence that 10B is predominately or exclusively a glial protein.[4] Subcellular organelle specificity of proteins only recently has been approached.[2–4] There also are indications of fairly specific groups of proteins that appear to be related to particular membranous fractions such as synaptosomes and microsomes.[2–4,35,36] Most of the proteins so far extracted still exhibit heterogeneity. One major problem for all of these studies is the purity of the isolated subcellular organelle fraction as determined by electron microscopy. The use of biochemical enzyme markers for determination of such purity will be increasingly critical in the definition of organelle-specific proteins.[37]

I. Enzymatic Properties

Investigators whose primary interest is in a particular enzyme frequently proceed to attempt to localize it in particular subcellular fractions, and most of the chapters in the *Handbook* dealing with specific enzymes will contain data of relevance to subsidiary localization. However, when particular subcellular fractions are first extracted for total protein, it then may be useful to examine individual enzymatic activity in particular proteins fractionated from these extracts. Thus, for example, in preliminary enzyme studies on the fractionation of constituent proteins of glioblastoma tumor cells, neutral proteinase, acid proteinase, and several other selected dipeptidases were found to be present and to be unevenly distributed between the brain protein groups 1 through 13 (A. Lajtha, N. Marks, and S. Bogoch, unpublished results).

J. Immunochemical Specificity

The extraction of a given protein from a particular species, brain region, or subcellular fraction has already been referred to above. In addition, once a particular protein is extracted and purified, the ability to prepare antisera to it can afford a further useful method of chemical classification.

Thus, for example, it has been possible to prepare specific antisera in rabbits to protein 10B of both normal human and Tay–Sachs disease human brain, as well as from normal pigeon brain.[2,4,16] Disc gel electrophoresis of the antigen always has shown it to contain between two and three distinct components, depending on the functional state of the brain.[2,4] The immunochemical data correspond, in that one major and two minor bands are demonstrated by Ochterlony plate reactions. In pigeon brain, antisera to 10BI, 10BII, and 11A have been prepared (S. Bogoch and B. R. Das, unpublished results). Preliminary evidence shows all three to share at least one antigenic determinant, although each also has further unique determinants. The nature and concentration of carbohydrate moieties glycosidically bound to these three pigeon brain proteins also differs.[2,4] The correlation of these quantitative chemical observations on bound sugars, with the physical properties on column chromatography and gel electrophoresis and with the immunochemical data mentioned above, thus provides a four-dimensional analysis of discreteness or heterogeneity of each purified protein fraction.

K. Developmental Phase Specificity

In addition to the description of certain proteins that appear to influence the development of nerve cells, as nerve growth factor protein, there are undoubtedly certain nervous system proteins whose appearance at a particular developmental point reflects the appearance of a particular histological structure or heralds the onset of a particular function dependent upon it. One example of this is the dendritic proliferation protein of groups 1 and 2.

There is a suggestion of a "critical period" between 15 and 20 gestational weeks, in which the protein of groups 1 and 2 increases markedly. In contrast, the stability of other groups during this period is noteworthy.[4] The change in groups 1 and 2 correlates in time with the massive appearance of new dendritic cell processes, and with the rapid growth of the cortical gray mantle, beginning in the insular area and spreading over both occipital and frontal poles of the cerebral cortex.[4] Discrete specimens from the insular region as distinct from the poles, in a 20-wk-old human brain, showed the insular area to control 1.3 mg/g of groups 1 and 2, while the poles contained only 0.1 mg/g. This temporal relationship to nerve cell process growth has led to the use of the term dendritic proliferation protein for the relevant proteins of groups 1 and 2.

Figure 3 shows the protein patterns on column chromatography of human fetal brain protein groups 1–3 (group 3 is the terminal portion of of group 2).[4] From top to bottom, the specimens are 14-, 15-, 16-, and 22-

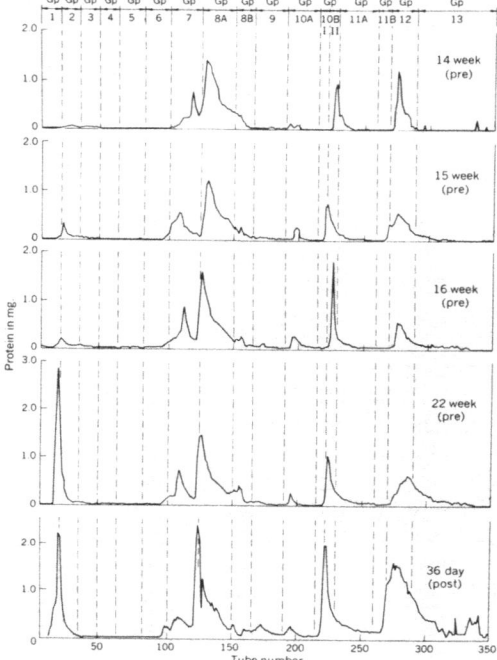

Fig. 3. Column chromatography of brain proteins in human cerebral hemispheres of normal prenatal human fetuses obtained at therapeutic interruptions of pregnancy and one normal 36-day postnatal brain. Methods as in Fig. 2.

week prenatal gestational age, followed by a 36-day-old postnatal specimen. The accumulation of groups 1 and 2 in appreciable amounts can be seen in Fig. 4 to occur between the sixteenth and twenty-second week.

   While this clearly represents a group of proteins,[16] and there is as yet no direct evidence that they are actually located in the dendritic processes, the abrupt increase in the concentration of these proteins has a developmental phase specificity and is distinct from the minimal changes in total protein and other protein groups during this period.[4]

L. Pathological State Specificity

   The change in concentration in certain normal brain proteins in a given disease state may be specific to it or a reflection of a pathological process shared by that disease and others. For example, the increase in concentration in human 10B appears to relate to glial proliferation since it is seen in both Tay–Sachs disease, marked by gliosis,[16] and in glial tumors.[2,4] However, there may be subtle differences in chemical structure as evidenced by the

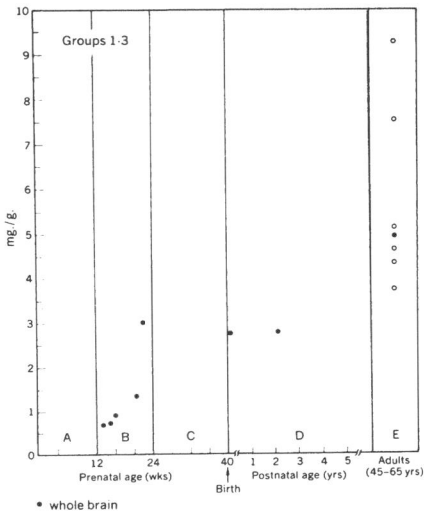

Fig. 4. Concentration in milligrams per gram wet tissue of groups 1–3 brain proteins in human cerebral hemispheres of normal prenatal human fetuses obtained at therapeutic interruptions of pregnancy. Normal postnatal children and adults.

essential absence of bound carbohydrate units in tumor 10B[4] and their marked excess in Tay–Sachs 10B.[16] The fact that the "regressed" glial cell of tumor type has little carbohydrate attached is of interest to the mucoid–memory hypothesis of coding of experimental information.[4] Change in overall brain protein metabolism in several pathological states recently has been well reviewed.[38,39]

M. Functional Specificity

Brain protein changes in excitation and inhibition have been recently reviewed.[38] Some relationship to function in the sodium–potassium–ATPase system has been suggested for the brain phosphoproteins.[11,40] By means of indirect evidence obtained in pharmacological studies with inhibitors of protein synthesis, recently reviewed,[4] total brain proteins have been implicated in learning and memory functions. Three specific glycoprotein fractions of brain have been shown to increase in training and learning situations.[4]

Thus, data obtained by comparing resting with training pigeon brain have suggested that not only is there a quantitative difference in particular brain proteins (2, 10B, and 11A) in terms of the (Folin–Lowry) polypeptide chains, but also that the concentration of the bound carbohydrates in these conjugated proteins differs in the two different functional states.[4]

The assignment of functional significance is after all the end goal of understanding in the chemical study of the nervous system proteins. While it may not be necessary, as Thudichum suggested,[41] to know all the chemistry of the brain in its utmost detail before its functional disorders can be treated, it is clear that our present knowledge is so far from the ideal that the ability to make rational modification of existing brain processes based upon chemical knowledge is not frequently at hand.

Nonetheless, methodology for the study of the nervous system proteins has made appreciable advances in the past few years, and with a much increased effort, and time, the most complex organ system in nature may yet permit understanding.

III. REFERENCES

1. S. Bogoch, P. C. Rajam, and P. C. Belval, Separation of cerebroproteins of human brain, *Nature* **204**:73–78 (1964).
2. S. Bogoch, P. C. Belval, W. H. Sweet, W. Sacks, and G. Korsh, in *Protides of the Biological Fluids*, Proc. XVth Colloq., Brugge (H. Peeters, ed.), Vol. 15, pp. 129–131, Elsevier, Amsterdam (1968).
3. A. Quamina and S. Bogoch, in *Protides of the Biological Fluids*, Proc. XIIIth Colloq., Brugge (H. Peeters, ed.), Vol. 14, pp. 211–216, Elsevier, Amsterdam (1966).
4. S. Bogoch, *The Biochemistry of Memory: With an Inquiry into the Function of the Brain Mucoids*, Oxford University Press, New York (1968).
5. E. G. Brunngraber and E. A. Bejnarowicz, in *Protides of the Biological Fluids*, Proc. XIIIth Colloq., Brugge (H. Peeters, ed.), Vol. 14, pp. 201–205, Elsevier Press, Amsterdam (1966).
6. H. Bauer, D. Matzelt, and I. Schwarz, Untersuchungen über Hirnproteine bei Einwirkung von Lysolecithin auf Hirnhomogenate, *Klin. Wochschr.* **40**:251–255 (1962).
7. M. K. Johnson, personal communication (1967).
8. M. B. Lees, Effect of ion removal on the solubility of rat brain proteins in chloroform–methanol mixtures, *J. Neurochem.* **15**:153–159 (1968).
9. F. Wolfgram, A new proteolytic fraction of the nervous system. I. Isolation and aminoacid analyses, *J. Neurochem.* **13**:461–470 (1966).
10. K. Kothary, as quoted in Rodnight,[11] PhD Thesis, London University (1964).
11. R. Rodnight, in *Protides of the Biological Fluids*, Proc. XIIIth Colloq., Brugge (H. Peeters, ed.), Vol. 13, pp. 39–52, Elsevier, Amsterdam (1966).
12. F. N. LeBaron and J. Folch, The effect of pH and salt concentration on aqueous extraction of brain proteins and lipoproteins, *J. Neurochem.* **4**:1–8 (1959).
13. R. E. Martenson and F. N. LeBaron, Studies on the acid-extractable proteins of bovine brain white matter, *J. Neurochem.* **13**:1469–1479 (1966).
14. K. Warecka and H. Bauer, Studien über Hirnproteine. Immunochemische Untersuchungen der Wasserlöslichen Fraktionen, *Deut. Z. Nervenheilkunde* **189**:53–66 (1966).
15. A. Lowenthal, *Agar Gel Electrophoresis in Neurology*, Elsevier, New York (1964).
16. S. Bogoch and P. C. Belval, in *Inborn Disorders of Sphingolipid Metabolism* (S. M. Aronson and B. W. Volk, eds.), pp. 273–287, Pergamon Press, New York (1966).
17. J. Folch-Pi, in *Protides of the Biological Fluids*, Proc. XIIIth Colloq., Brugge (H. Peeters, ed.), Vol. 13, pp. 21–34, Elsevier, New York (1966).
18. H. Porter and S. Ainsworth, The isolation of the copper-containing protein cerebrocuprein I from normal human brain, *J. Neurochem.* **5**:91–98 (1959).

19. H. Porter and J. Folch-Pi, Cerebrocuprein I. A copper containing protein isolated from brain, *J. Neurochem.* **1**:260–271 (1957).
20. F. M. LeBaron, G. Hauser, and E. Ruiz, The occurrence and metabolism of protein-bound phosphoinositides in several lipid-protein complexes from brain, *Biochim. Biophys. Acta* **60**:338–349 (1962).
21. F. N. LeBaron, The nature of the linkage between phosphoinositides and proteins in brain, *Biochim. Biophys. Acta* **70**:658–669 (1963).
22. M. W. Kies, E. C. Alvord, R. E. Martenson, and F. N. LeBaron, Encephalitogenic activity of bovine basic proteins, *Science* **151**:821–822 (1966).
23. N. Herschkowitz, E. F. Shooter, and G. M. McKhann, Isolation and characterization of soluble brain lipoproteins, *Abstracts, 1st Intern. Meeting Intern. Soc. Neurochem.*, *Strasbourg*, p. 94 (1967).
24. B. W. Moore and D. McGregor, Chromatographic and electrophoretic fractionation of soluble proteins of brain and liver, *J. Biol. Chem.* **240**:1647–1653 (1965).
25. G. Gombos, G. Vicendon, G. Tardy, and B. Mandel, Heterogeneite electrophoretique et preparation vapide de la fraction proteique S-100, *Compt. Rend. Sci. Paris* (D): 1533–1535 (1966).
26. G. Gombos, K. Uyemuze, J. Tardy, and G. Vincedon, Heterogeneity of beef brain S-100 protein, *Abstracts, 1st Intern. Soc. Neurochem., Strasbourg*, p. 86 (1967).
27. G. Vincendon, A. Waksman, K. Uyemuza, G. Tardy, and G. Gombos, Ultracentrifugal behavior of beef brain S-100 protein fraction, *Arch. Biochem. Biophys.* **120**:233–235 (1967).
28. K. Schmid, A. Polis, K. Hunziker, R. Fricke, and M. Yayoshi, Partial characterization of the sialic acid-free forms of α-acid glycoprotein from human plasma, *Biochem. J.* **104**:361–363 (1967).
29. E. Witebsky and J. Steinfeld, Untersuchungen über spezitische Antigenfunktionen von organen, *Z. Immunitaltsforsch.* **58**:271–296 (1928).
30. F. Milgram, M. Tuggac, and E. Witebsky, Immunological studies on adrenal glands, *Immunology* **6**:105–118 (1963).
31. P. C. Rajam and S. Bogoch, Antigenic constituents of human cortical gray matter, *Nature* **211**:1200–1201 (1966).
32. P. C. Rajam and S. Bogoch, Brain antigens: Components of subfractions from human gray matter, *Immunology* **11**:211–215 (1966).
33. P. C. Rajam, S. Bogoch, M. A. Rushworth, and P. C. Forrester, Antigenic constituents of basic proteins from human brain, *Immunology* **11**:217–221 (1966).
34. K. Warecka and H. Bauer, Studies on "brain-specific" proteins in aqueous extracts of brain tissue, *J. Neurochem.* **14**:783–787 (1967).
35. E. Mehl, International symposium on the metabolism of the nucleic acids and proteins and the function of neurone (May 1967), *Excerpta Med.*, in press.
36. C. W. Cotman and H. R. Mahler, Resolution of insoluble proteins in rat brain subcellular fractions, *Arch. Biochem. Biophys.* **120**:384–396 (1967).
37. V. P. Whittaker, Some properties of synaptic membranes isolated from the central nervous system, *Ann. N.Y. Acad. Sci.* **137**:982–998 (1966).
38. A. Lajtha, Alteration and pathology of cerebral protein metabolism, *Intern. Rev. Neurobiol.* **7**:1–40 (1964).
39. D. Richter, Protein metabolism in pathological states, *in Protides of Biological Fluids* (H. Peeters, ed.), Vol. 13, pp. 137–143, Elsevier, Amsterdam (1966).
40. P. J. Heald, Phosphoprotein metabolism and ion transport in nervous tissue; a suggested connexion, *Nature* **193**:451–454 (1962).
41. J. L. Thudichum, Treatise on the Chemical Constitution of the Brain (D. L. Drabkin, trans.), first publication 1884, Archon Books, Hamden, Connecticut (1962).

42. C. R. Dean and D. B. Hope, The isolation of purified neurosecretory granules from bovine pituitary posterior lobes, *Biochem. J.* **104**:1082–1088 (1967).
43. A. Neidle, Basic proteins, *in Protides of the Biological Fluids* (H. Peeters, ed.), Vol. 13, pp. 35–38, Elsevier, Amsterdam (1966).
44. K. Got, G. M. Polya, J. B. Polya, and L. M. Cockerill, Water-insoluble proteins from subfractions of sheep brain microsomes, *Biochem. Biophys. Acta* **135**:225–235 (1967).
45. S. Bogoch, Studies on cerebrospinal fluid. Quantitative fractionation of carbohydrate constituents, *J. Biol. Chem.* **235**:16–22 (1960).

Chapter 6

ACIDIC PROTEINS

Blake W. Moore

Department of Psychiatry
Washington University School of Medicine
St. Louis, Missouri

I. INTRODUCTION

The importance of soluble, acidic proteins in the nervous system recently has been recognized, in terms of quantity and possibly of function. This has come about partly as a result of application of new techniques of protein fractionation such as chromatography and zone electrophoresis. "Acidic proteins" can be defined as those proteins which move faster than serum albumin on zone electrophoresis (starch or acrylamide gel) and which bind most strongly to the basic ion exchangers used in protein chromatography, such as DEAE-cellulose.

Although earlier papers described the fractionation of nervous system proteins by zone electrophoresis,[1,2] the first paper[3] explicitly to show the importance, in quantitative terms, of soluble acidic proteins in brain was based on the two-dimensional "protein map" technique (DEAE-cellulose × starch gel). For example, in rat brain 26% of the soluble protein was tightly bound to DEAE-cellulose (i.e., it required higher than 0.2 M chloride for elution), while in liver only 16% was tightly bound. The maps showed a greater number and amount of acidic proteins (based on both chromatographic and electrophoretic properties) in brain than in liver.

II. THE S-100 PROTEIN

A. Preparation and Properties

The "protein maps" demonstrated in brain an electrophoretically fast-moving protein which was absent in liver.[3] This protein was named "S-100" since it was soluble at pH 7 in saturated ammonium sulfate and it was prepared in pure form[4] by the following sequence of steps:

1. Extraction of soluble proteins by homogenizing in 5 mM tris phosphate, pH 7.2.
2. Ammonium sulfate fractionation, S-100 being precipitated at pH 4 by saturated ammonium sulfate.
3. DEAE-cellulose chromatography.
4. Sephadex G-100 chromatography.
5. DEAE-Sephadex chromatography.

It was homogeneous by the following criteria: starch and acrylamide gel electrophoresis; chromatography on DEAE-cellulose, G-100 Sephadex, DEAE-Sephadex, and hydroxyapatite; sucrose density gradient centrifugation; double agar diffusion of the protein against its antiserum; and analytical ultracentrifugation.[5]

The protein had a molecular weight of about 24,000 by sucrose density gradient centrifugation and by chromatography on Sephadex G-100. The amino acid analysis is given in Table I and is characterized by a high proportion of glutamic and aspartic acid residues. It contained no carbohydrate or lipid.

Kessler, Levine, and Fasman[6] have recently done an ORD study of S-100 and find that it contains about 40% α-helix in the native form. Heat treatment reduces the amount of helix and this conformational change is slower in the presence of Ca^{2+} or EDTA.

Recently we have found in our laboratory that S-100 shows a specific interaction with Ca^{2+} ion. The native tryptophan fluorescence (it contains one tryptophan residue) is increased by about twofold by low levels of Ca^{2+} and this effect is inhibited by Na^+ and even more so by K^+ (Table II). The effect is specific for Ca^{2+}, the Mg^{2+} ion not giving any enhancement of fluorescence. We have also found a change in the difference spectrum of S-100 (with Ca^{2+} vs. without Ca^{2+}) involving tryptophan, tyrosine, and phenylalanine, which occurs in the same range of Ca^{2+} concentrations as the fluorescence effect. Here also Mg^{2+} has no effect. The data for effect of Ca^{2+} on

TABLE I

Amino Acid Analysis of Beef S-100[a]

Glu	36	Tyr	3
Asp	21	Pro	1
Lys	17	Phe	16
His	8	Leu	17
Arg	3	Ileu	6
Ser	10	Val	13
Thr	8	Ala	12
Cys	3	Gly	9
Try	1	Met	4

[a] Values are nearest integer numbers of residues per 24,000 mol. wt.

TABLE II

Effect of Various Ions on Fluorescence[a] of S-100

Divalent cation	Monovalent cation	Relative fluorescence
None	60 mM K$^+$	1.00
5 mM Ca^{2+}	60 mM K$^+$	1.85
5 mM Mg^{2+}	60 mM K$^+$	1.00
5 mM Sr^{2+}	60 mM K$^+$	1.04
None	None	1.00
0.5 mM Ca^{2+}	None	1.68
None	60 mM Na$^+$	1.00
0.5 mM Ca^{2+}	60 mM Na$^+$	1.29
None	60 mM K$^+$	1.00
0.5 mM Ca^{2+}	60 mM K$^+$	1.25

[a] Fluorescence was measured in an Aminco Bowman Spectrofluorimeter at a primary wavelength of 290 mμ and secondary of 355 mμ in 20 mM tris chloride, pH 8.2, in the presence of added metal ions.

tryptophan fluorescence (in 60 mM KCl at pH 8.2) can be fitted by a curve derived from the assumption of two consecutive binding sites for Ca^{2+} with dissociation constants of 1 and 0.2 mM. Equilibrium dialysis with ^{45}Ca shows eight sites with $K = 1$ mM and two sites with $K = 0.2$ mM. It is possible that these changes may be conformational changes. Since they are specific for Ca^{2+} and K$^+$, it is tempting to speculate that they are involved in the function of S-100 in the nervous system.

Gombos et al.[7] obtained multiple bands when S-100 was electrophoresed on acrylamide gel. We have found that only one band is obtained when EDTA is added to the gel buffer or when Ca^{2+} is carefully excluded, but when small amounts of Ca^{2+} are added, six or eight bands can be seen. This splitting is very sensitive to small amounts of Ca^{2+} since the effect of about 2 to 5 μM Ca^{2+} can be detected. These changes are probably connected with the possibly conformational changes described above.

B. Specificity to the Nervous System

Levine was able to prepare antiserum to S-100[8] and, using the sensitive and specific method of microcomplement fixation, we have shown[4] that, in the rat, the concentration in brain is at least 1000 times more than that in any other organ.

C. Species Similarity

The S-100 protein has been shown to be present in an immunologically cross-reacting form in all vertebrate species examined[4,8,9] when they were

TABLE III

S-100 Contents of Areas of the Nervous System

	Micrograms S-100 per gram wet weight			
	Rat[a] (60 days)	Rabbit	Beef	Monkey[a]
Whole brain	94	—	253	—
Cerebrum	90	161	—	123
Cerebellum	115	708	—	321
Spinal cord	—	243	—	—
Sciatic nerve	—	352	—	—

[a] In terms of beef S-100 equivalents.

tested by C'-fixation using the anti-beef serum. In addition, we have recently been able to detect a cross-reaction in invertebrates: lobster, crayfish, octopus, cockroach, and drosophila.

D. Distribution Within the Nervous System

The S-100 protein, in vertebrates, is present in both the peripheral and central nervous systems (Table III), in both myelinated and unmyelinated nerve. There is a characteristic distribution in areas of human brain, generally being higher in white matter than in gray (Table IV). It was highest in the cerebellum, in the molecular layer.[9,10]

Hyden[11] has published evidence, based on wet dissection of neurons followed by immunodiffusion, and on immunofluorescence measurements, that S-100 is localized primarily in glial cytoplasm and neuronal nuclei. However, S-100 was not detectable in purified nuclei from brain.[12]

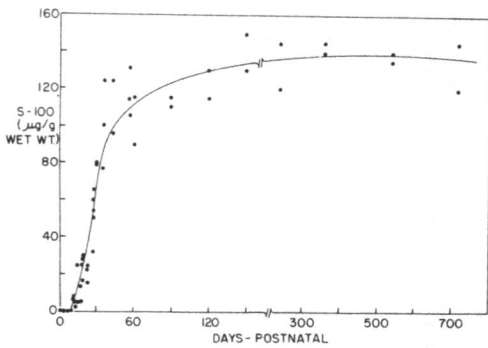

Fig. 1. Content of S-100 (in terms of beef brain equivalents) in whole rat brain at various ages after birth.

TABLE IV

S-100 and 14-3-2 Contents of Various Parts of Human Brain

	μg/g wet weight[a] S-100			μg/g wet weight[a] 14-3-2		
	Mean	SEM	N[b]	Mean	SEM	N[b]
1. Frontal gray	45	± 3	16	191	± 21	14
2. Frontal white	177	± 16	16	118	± 10	14
3. Parietal gray	53	± 4	16	208	± 20	14
4. Parietal white	190	± 22	16	116	± 11	14
5. Occipital gray	61	± 4	16	239	± 16	14
6. Occipital white	208	± 18	16	131	± 14	14
7. Temporal gray	49	± 4	16	170	± 13	14
8. Temporal white	134	± 14	16	101	± 10	14
9. Corpus callosum	154	± 12	16	101	± 8	14
10. Globus pallidus	131	± 13	15	196	± 12	14
11. Putamen	59	± 6	16	271	± 27	14
12. Hippocampus	105	± 8	16	169	± 11	14
13. Head of caudate nuc.	61	± 7	16	237	± 19	14
14. Hypothalamus	138	± 8	16	158	± 16	14
15. Thalamus, cent. nuc.	126	± 10	16	251	± 22	14
16. Thalamus, lat. nuc.	133	± 10	16	213	± 16	14
17. Cing. gyrus, rost.	54	± 7	16	186	± 20	14
18. Amygdaloid body	68	± 9	16	201	± 10	14
19. Corp. quad. inf.	143	± 8	12	205	± 21	11
20. Corp. quad. sup.	131	± 8	13	180	± 16	11
21. Cerebellar gray	190	± 10	16	282	± 19	14
22. Cerebellar white	197	± 16	16	199	± 10	14
23. Mesen. teg.	182	± 23	13	188	± 24	11
24. Teg. of pons	221	± 17	15	150	± 10	13
25. Ventral pons	208	± 11	16	142	± 12	14
26. Pineal	111	± 14	11	307	± 27	11

[a] In terms of beef S-100 or 14-3-2 equivalents.
[b] Number of brains assayed.

A study of Wallerian degeneration in rabbit sciatic nerve[12] suggests, on the other hand, that it may be localized in the axon in peripheral nerve since it disappears in parallel with breakdown of the axon.

E. Changes During Development

S-100 is nearly absent at birth in rats (less than 1 μg per gram wet brain). It begins to appear at 12 to 15 days postnatally (Fig. 1), then rises rapidly, and begins to level off at 60 to 80 days at 100 to 120 μg/g. There is also a statistically significant increase with age from 16 to 82 years in 21 of 26 areas of human brain.

F. Turnover

McEwen and Hyden[14] concluded that S-100 is turned over very rapidly in brain. In their experiments S-100 was isolated as a band after acrylamide gel electrophoresis. Rubin and Stenzel,[15] using a cell-free protein-synthesizing system from brain, concluded that S-100 represents about 15 % of the protein synthesis in that system. S-100 was isolated by precipitation with the antiserum. In both these experiments there is some possibility that the S-100 may have been contaminated by other components which could have contributed to the labeling, i.e., aminoacyl-RNA.

III. OTHER ACIDIC PROTEINS

Davison and Schmitt[16,17] have studied an acidic protein from axoplasm of the giant squid axon which they conclude is the monomer of the neurotubule system. This protein does not cross-react immunologically with S-100.

Four other acidic proteins from brain have also been prepared in our laboratory[9] and we have obtained antiserum to two of them. On the basis of double agar diffusion we have shown that at least one of these (called 14-3-2) is at least 32 times more concentrated in brain than in other organs and that there is immunological cross-reactivity among species. It is present in even higher concentration in brain than is S-100 but shows a somewhat different distribution in human brain (Table IV).

IV. REFERENCES

1. B. Bailey and P. J. Heald, The separation of the cytoplasmic proteins of brain by electrophoresis in a starch gel medium, *J. Neurochem.* **6**:342 (1961).
2. J. Vas and H. J. Van Der Helm, Electrophoresis of brain proteins in polyacrylamide gel, *J. Neurochem.* **11**:209 (1964).
3. B. W. Moore and D. McGregor, Chromatographic and electrophoretic fractionation of soluble proteins of brain and liver, *J. Biol. Chem.* **240**:1647 (1965).
4. B. W. Moore, A soluble protein characteristic of the nervous system, *Biochem. Biophys. Res. Commun.* **19**:739 (1965).
5. G. Vincendou, A. Waksman, K. Uyemura, J. Tardy, and G. Gombos, Ultracentrifugal behavior of beef brain S-100 protein fraction, *Arch. Biochem. Biophys.* **120**:233 (1967).
6. D. Kessler, L. Levine, and G. Fasman, Some conformational and immunological properties of a bovine brain acidic protein (S-100), *Biochemistry* **7**:758 (1968).
7. G. Gombos, G. Vincendou, J. Tardy, and P. Mandel, Hétérogénéité électrophorétique et préparation rapide de la fraction protéique S-100, *Compt. Rend. Acad. Sci. (Paris)* **263**:150 (1966).
8. L. Levine and B. W. Moore, Structural relatedness of a vertebrate brain acidic protein as measured immunochemically, *Neurosci. Res. Progr. Bull.* **3**:18 (1965).
9. B. W. Moore and V. J. Perez, Specific acidic proteins of the nervous system, *Soc. Gen. Phys. Symp.*, Wood's Hole, Mass., 1967 (in press).

10. B. W. Moore, V. J. Perez, and M. Gehring, Assay and regional distribution of a soluble protein characteristic of the nervous system, *J. Neurochem.* (in press).

11. H. Hyden and B. McEwen, A glial protein specific for the nervous system, *Proc. Natl. Acad. Sci. (U.S.)* **55**:354 (1966).

12. A. R. Dravid and J. A. Burdman, Acidic proteins in rat brain nuclei: Disc electrophoresis, *J. Neurochem.* **15**:25 (1968).

13. V. J. Perez and B. W. Moore, Wallerian degeneration in rabbit tibial nerve, changes in amounts of the S-100 protein, *J. Neurochem.* (in press).

14. B. S. McEwen and H. Hyden, A study of specific brain proteins on the semi-micro scale, *J. Neurochem.* **13**:823 (1966).

15. A. L. Rubin and K. H. Stenzel, *In vitro* synthesis of brain protein, *Proc. Natl. Acad. Sci. (U.S.)* **53**: 963 (1963).

16. P. Davison, F. Huneeus-Cox, D. Lusted, and F. O. Schmitt, The subunit structure of neurofilament protein (Abstract), *in First Intern. Meeting, Intern. Soc. Neurochem.*, p. 50 (1967).

17. F. Huneeus-Cox, Electrophoretic and immunological studies of squid axoplasm proteins, *Science* **143**:1036 (1964).

Chapter 7

NUCLEIC ACIDS

D. A. Rappoport, R. R. Fritz, and J. L. Myers

Department of Pediatrics
The University of Texas Medical Branch
Galveston, Texas

I. INTRODUCTION

Nucleic acids in the central nervous system, as in other organs, are characterized by their size, composition, and role in protein synthesis. However, the regional tissue heterogeneity in the brain and the diversity of the cell types and their processes[1] complicate both the acquisition and interpretation of experimental data. Despite these difficulties, recent reports suggest that, in addition to the established role of nucleic acids in the biosynthesis of proteins,[2] RNA and protein synthesis may be involved in the accrual of sensory information in the brain, thus indicating a possible approach to elucidation of brain function on a molecular basis.[3]

A. Location and Types

Intracellularly, DNA is almost entirely localized in the cell nuclei, however, small amounts have been isolated from mitochondria.[4,5] In cell nuclei, the DNA is complexed with histones and acidic proteins in the form of a nucleoprotein, and this constitutes the chromatin (interphase chromosomes).[6] Microscopically, the chromatin is dispersed throughout the nucleoplasm, and it surrounds and extends into the interior of the nucleolus.[7] This arrangement can be visualized after staining with Feulgen reagent, a DNA-specific stain,[8] or acid Fast Green, which specifically combines with histones.[9]

The ribonucleic acids exist in three well differentiated forms and are distributed in all parts of the cell. The three major groups of RNA within the cell are designated as soluble or transfer RNA (s or tRNA), ribosomal RNA (rRNA), and messenger RNA (mRNA). Of the total RNA in somatic cells, approximately 15–20% is tRNA, 75–80% is rRNA, and less than 5% is mRNA.[10] The tRNA's are low-molecular-weight polynucleotides (3 to

4×10^4 Dalton) mainly localized in the cytoplasm, but small amounts are always found in the nuclei and nucleoli. Isolated individual tRNA's from various organisms are single-stranded polynucleotides, terminating at one end with guanine (G) and at the other end with cytidylcytidyladenylate (CCA).[11] The terminal adenylate group can combine with amino acids to form amino acyl tRNA's through the mediation of amino acyl synthetases. For each of the amino acids, there is a specific tRNA with which it can combine.[12]

The tRNA polynucleotides contain some unusual bases, such as pseudouridine, hypoxanthine, thymine (2.5–4.6% of total tRNA) and also methylated bases as four methyl adenines, four methyl guanines, one methyl cytosine and one methyl uracil (2–5% of total tRNA), as well as some 2'-O-methylated sugars in all bases (0.1–0.2% of total tRNA).[12–14]

The ribosomes are a particulate RNA nucleoprotein (RNP) consisting of almost equal quantities of RNA (rRNA) and protein, and, under electron microscopy, appear as a sphere 240 Å in diameter.[15–18] The ribosomes, which have a sedimentation constant of 70–80 S, consist of two RNA nucleoprotein subunits, a 45–50 S particle containing a 27–28 S RNA (mol. wt. 1.6×10^6 Dalton) complexed with protein and a 28–30 S particle containing a 16–18 S RNA (mol. wt. 6×10^5 Dalton) also combined with protein.[19] Some of the bases in both the 18 S and 28 S rRNA's are methylated, similar to those in tRNA, but only one-quarter as many as in tRNA, some of the sugar in all the nucleotides are methylated in the 2'-position.[13] In general, the bulk of the cellular RNA is ribosomal RNA and is localized in the cytoplasm.

The association of ribosomes with mRNA forms polysomes, which, under the electron microscope, can be seen as rosettes or clusters of ribosomes in the cytoplasm of both glia and neurons.[16,20,21] The polysomes are the structures on which the various amino acyl tRNA's are assembled for the synthesis of a specific protein. Each amino acyl tRNA is attached to a specific nucleotide triplet on the mRNA, thus the sequence of amino acids in the protein is determined by the succession of nucleotide triplets in the primary structure of the mRNA. A characteristic feature of neurons is the presence of a Nissl substance in the cytoplasm, consisting of parallel rows of endoplasmic reticulum (cytomembranes) lined at the outer surface with ribosomes.[22,23]

Messenger RNA is found in small quantities in the nuclei and it can be detected, when it is added to a ribosomal preparation, by its ability to enhance the incorporation of labeled amino acids into ribosomal proteins.[24–26] The molecular weight of mRNA is approximately 7×10^5 Dalton, corresponding to a sedimentation constant of 18 S.[24] The base composition of mRNA is complementary to DNA, determined by hybridization experiments.[26,27] Both tRNA and rRNA have a high G + C content in contrast to mRNA, which is rich in A + U (Table I).[10,28]

It is now established that all RNA's are synthesized in the nuclei.[19,29,30] Both rRNA's (18 S and 28 S) originate in the nucleoli, and it is uncertain whether tRNA is also synthesized in this organelle or in the nucleoplasm.

TABLE I

Base Composition of "Rapidly Labeled" RNA Compared to Ribosomal RNA, Transfer RNA, and DNA

Type of nucleic acid	C	A	G	U(T)	A + U(T)/ G + C	Reference
tRNA (rat brain cortex)	28.3	19.8	32.2	19.7	0.653	10
rRNA (rat brain cortex)	31.2	18.4	32.3	18.1	0.576	10
DNA	29.1	20.7	19.7	30.5	1.48	10
DNA[a]	20.0	29.1	21.5	29.4	1.41	45
rRNA[a]	28.6	18.0	31.6	21.8	0.661	45
"Rapidly labeled" RNA	24.7	24.7	22.2	28.5	1.13	45

[a] Analysis made on the combination of the rat cerebral cortex, cerebellum, and the brain stem.

Messenger RNA is formed on the chromatin in the nucleoplasm[24,26] and is then associated with a 35 S preribosomal particle before it is extruded in the cytoplasm.[26]

B. Deoxyribonucleic Acid

Since the majority of the cells in brain are diploid, there is generally a fixed quantity of DNA per cell. Some polyploidy has been reported in brain cells; however, the number of these cells is relatively small.[31] In general, the DNA content in mammalian brains is between 6.1–7.1 pg/cell.[32–34] In other vertebrate brains, where data is available, the DNA content per cell varies greatly.[35] Since the DNA content per diploid cell and the diploid chromosome number are known for many animals, the average DNA content per chromosome can be calculated. The calculated values for DNA per chromosome varied for a group of mammals, but a range of 0.10–0.15 pg of DNA per chromosome emerges as an average estimate. The reliability of the reported DNA analytical values are questionable, hence the above calculation of DNA per chromosome is only an approximation.[36]

Base composition of pure DNA isolated from brain showed that it is richer in adenine and thymine than in guanine and cytosine (Table II). The DNA extracted from human cerebral gray matter had an A + T/G + C ratio of 1.48, while the white matter had a ratio of 1.41. Similar values for the base ratios were found in DNA from human liver, thymus, sperm, and endometrium.[37–40] The actual content of DNA was reported to be 0.50 mg/g fresh weight in human gray matter and 0.72 mg/g fresh weight in white matter.[37] The base ratio for DNA does not change in various regions of the rat brain nor even among the various mammalian brains (Table II).

The finding that mitochondria actually contain DNA has raised the question of how it compares with nuclear DNA.[4,5,28] Recently, DuBuy, Mattern, and Riley[5] have isolated mouse brain mitochondrial DNA and

TABLE II

Brain DNA Base Composition from Various Vertebrates

	A	G	C	T	$\dfrac{A + T}{G + C}$	Reference
Human						
Gray matter	29.1	20.7	19.7	30.5	1.48	37
White matter	28.8	20.4	20.9	29.7	1.41	37
Rat						
Whole brain	29.1	21.5	20.0	29.4	1.41	45
Cerebrum	29.9	20.0	19.3	30.8	1.54	90
Cerebellum	29.1	21.1	20.0	29.6	1.43	90
Diencephalon	29.3	21.3	19.1	30.3	1.47	90
Brain stem	29.3	21.5	19.3	29.9	1.45	90
Cattle						
Cerebrum	28.1	21.5	20.2	30.3	1.40	40
Pig						
Whole brain	30.0	20.0	20.7	29.4	1.46	89

studied some of its physical characteristics. They found that the buoyant density of mouse brain mitochondrial DNA is 1.701 g/cc, which compares well with the buoyant density of mouse brain nuclear DNA of 1.702 g/cc. The melting temperatures (T_m) of the two are 86.2 and 82.8°C, respectively. Using the buoyant density, they calculated the G + C content in mitochondrial DNA as 42% and that of nuclear DNA as 43%. Calculation of G + C from T_m values gave 41% for nuclear DNA and 33% for mitochondrial DNA. These data establish that mitochondrial DNA is very similar to nuclear DNA. Presence of DNA in mitochondria implies that these organelles are capable of self-replication. However, the existing evidence indicates that mitochondrial DNA does not contain a sufficient number of codons for the synthesis of the entire mitochondrion, thus additional codons from nuclear DNA are transmitted via nuclear mRNA.[28]

C. Ribonucleic Acids

Edstrom[41] and Edstrom and Pigon[42] have reported that there is a proportionality between RNA content and the surface area of the cell body. In addition, they found a correlation between the volume of the cell and the surface of the cell body. The perikarya contain more RNA than the axons; however, when the total axoplasm is contrasted to the perikaryon, the former occupies four times the volume of the latter. Edstrom showed that there is a decrease in RNA concentration along two-thirds of the axon distal to the hillock, then the RNA increases toward the distal end of the axon. In contrast to the glia, the neurons contain more RNA by a factor of approximately thirteen.[43] Hyden has demonstrated that the content of RNA in neurons varies over a wide range, from 200 pg/cell in hypoglossal cells and 530 pg/cell in the anterior horn cells to 1550 pg/cell in Deiters' giant cells of the rabbit. In the latter, the nucleus contains 580 pg of RNA. In man 40–50 years of age, Hyden found that the anterior horn cells contain 607 pg of RNA/cell, which decreases to 540 pg/cell in man 60–70 years of age.[43]

In an attempt to characterize brain RNA, many investigators have reported the base composition of total RNA extracted from a variety of animal brains and from various regions of the brain.[10,15,18,44,45] They found that the base ratio of this RNA was essentially that of rRNA (A + U/G + C = 0.65). A more detailed analysis of brain RNA was reported by Mahler, Moore, and Thompson,[10] who isolated RNA from rat cerebral cortex and subjected it to fractionation by sucrose gradient centrifugation and with methylated albumin kieselguhr (MAK) columns. Three RNA fractions were isolated with sedimentation coefficients of 4, 17, and 28 S. The 4 S RNA fraction had a base ratio (A + U/G + C) of 0.653, representing tRNA. Both 17 and 28 S fractions were ribosomal RNA's with a base ratio (A + U/G + C) of 0.633. Isolated ribosomal RNA from cerebral cortex had a base ratio identical to that of the 17 and 28 S RNA fractions.

Studies of RNA synthesis in liver and ascites cells have established that "rapidly labeled" RNA formed in cell nuclei from labeled precursors is mRNA,[46] since its base composition resembled DNA and it enhanced amino acid incorporation into protein in an isolated ribosomal system. Egyhazi and Hyden[47,48] separated rapidly labeled RNA from microsurgically isolated Deiters' cells and the surrounding glial cells of the rat brain. Analysis of the base ratio of this RNA showed that it was rich in A + U and closely resembled the base composition of DNA. Jacob et al.[45] isolated rapidly labeled RNA from rat brain (combined sample of cerebral cortex, cerebellum, and brain stem) by means of sucrose gradient centrifugation and also by means of methylated albumin kieselguhr columns. They characterized this RNA as messenger-like RNA from its base composition and by competitive hybridization with homologous DNA. Its sedimentation coefficient differed from those reported by other investigators, since it was heterogeneous in composition and a considerable portion sedimented above the 28 S region.

II. DISTRIBUTION

The nucleic acid content in the entire brain and various brain parts from different animals have been determined in an attempt to gain some insight into its regional composition, developmental changes, and species differences.

A. Total and Regional Content

Analyses of nucleic acids in the whole brain have been used to estimate the average content of RNA and DNA per cell. Such estimates have been

TABLE III

Nucleic Acid[a] Composition in the Brain from Man, Monkey, Dog, and Cat

Species	RNA[b]	DNA[b]	RNA[c]	DNA[c]	RNA/DNA	Reference
Man	—	—	26.3	6.80	3.87	91
Man	—	—	—	7.10	—	37
Monkey white matter	64	124	—	—	0.52	92
Monkey gray matter	115	96	—	—	1.20	92
Dog	—	—	—	6.5	—	32
Dog	—	—	7.9	6.7	1.18	93
Dog white matter	53	63	—	—	0.84	75
Dog gray matter	111	53	—	—	2.10	75
Cat	—	—	—	7.10	—	32
Cat	—	—	7.90	6.90	1.14	93

[a] Reported RNA-P and DNA-P values multiplied by ten to obtain RNA and DNA values.
[b] Values in μg/100 mg fresh tissue.
[c] Values in pg/avg. cell.

used by some investigators to assess comparative or phylogenetic relationships between mammals and other vertebrates.[34,35,49,50] However, such data do not show any correlation among the various animals tested, as shown in Tables III, IV, and V. Comparisons of nucleic acid content based

TABLE IV

Brain Nucleic Acid[a] Composition from Various Mammals

Species	RNA[b]	DNA[b]	RNA[c]	DNA[c]	RNA/DNA	Reference
Rat	368	138	—	—	2.67	91
	188	200	—	—	0.94	94
	135	94	—	—	1.43	91
	175	123	—	—	1.42	95
	—	—	7.2	6.1	1.18	93
	111	85	—	—	1.30	35
	(901)[d]	(642)	—	—	1.40	96
Rabbit	104	75	—	—	1.39	35
	(644)	(412)	—	—	1.56	96
Guinea pig	—	—	16.1	6.9	2.33	93
	115	93	—	—	1.35	35
Mouse	(1038)	(928)	—	—	1.10	96

[a] Reported RNA-P and DNA-P values multiplied by ten to obtain RNA and DNA values.
[b] Values in μg/100 mg fresh tissue.
[c] Values in pg/avg. cell.
[d] The numbers in parentheses indicate values based on μg/100 mg dry wt.

TABLE V

Brain Nucleic Acid[a] Composition from Lower Vertebrates

Species	RNA[b]	DNA[b]	RNA[c]	DNA[c]	RNA/DNA	Reference
Trout	134	152	—	—	0.88	35
Carp	146	142	—	—	1.02	35
Carp	—	—	4.50	3.50	1.28	93
Newt	100	343	—	—	0.29	35
Frog	73	94	—	—	0.78	35
Turtle	131	62	—	—	2.11	35
Grass snake	130	63	—	—	2.06	35
Chicken	—	—	4.50	2.20	2.05	91
Chicken	88	24	—	—	2.00	35

[a] Reported RNA-P and DNA-P values multiplied by ten to obtain RNA and DNA values.
[b] Values in μg/100 mg fresh tissue.
[c] Values in pg/avg. cell.

on the analysis of large portions of the brain are of limited value, since any differences that may exist on a cellular level are masked by such gross analyses.

The regional distribution of nucleic acids in the brains from the human and other animals is listed in Tables VI and VII. The neurons contain Nissl bodies rich in RNA, hence their content of RNA per cell is higher than glia. Since all the cells contain the same amount of DNA, the areas in the brain with an abundance of neurons will have a higher RNA/DNA ratio than those parts constituted mainly of glial cells.

In the human brain, the highest content of nucleic acids is in the subfornical organ and pineal gland. Both tissues contain many small neurons, with an average content of RNA/DNA in the subfornical organ of 2.05 and in the pineal gland of 1.58. The lowest content of the nucleic acids is in the corpus callosum, fornix, peduncle, and the red nucleus. These tissues contain many axons and attached oligodendroglia, hence the RNA/DNA content of one or less. The other brain areas listed under telencephalon, mesencephalon, and diencephalon contain many neurons, hence the RNA/DNA ratio is well above 1.4, and these areas, constituting gray matter, contain more than 3 μg of RNA and 1.8 μg of DNA/mg dry wt. (Table VI). The heterogeneity of cell types and sizes within each of the above regions of the human brain does not permit correlation of the gross analysis of nucleic acid content with the morphological characteristics of these regions.

In the rat, guinea pig, and rabbit brains, the predominance of neuronal to glial cells in the hypothalamus, thalamus, cerebral cortex, corpus striatum, and hippocampus is reflected in the RNA/DNA ratios numerically above 1.2 (Table VII). The exception is the cerebellum gray matter, which has a high RNA and a very high DNA, indicating the presence of many small

TABLE VI

Regional Distribution of Nucleic Acids in Human Brain[51]

Region	μg/mg dry wt.		RNA/DNA
	RNA	DNA	
Telencephalon			
Frontal cortex	5.47	2.56	2.14
Motor cortex	4.38	2.56	1.72
Calcarine cortex	6.41	4.34	1.47
Caudate nucleus	4.40	3.09	1.44
Mesencephalon			
Periaqueductal	4.69	2.60	1.89
Red nucleus	1.80	1.72	1.07
Substantia nigra	3.29	1.81	1.83
Diencephalon			
Mammillary bodies	4.17	2.73	1.54
Subfornical organ	17.9	8.75	2.05
Pineal gland	15.9	10.1	1.58
Corpus callosum	1.60	1.83	0.87
Fornix	2.45	2.68	0.92
Peduncle	1.05	1.36	0.77

neurons. Thus the resulting RNA/DNA is very low. In the cerebellar and cerebral white matter, the predominance of glia cells is reflected in the RNA/DNA ratio of less than 1.10. In those areas listed as "mixed," there is a predominance of glial cells and the RNA/DNA ratio has an intermediate value. Here, the olfactory bulbs show a very high content of RNA and DNA, hence the RNA/DNA ratio is numerically low, indicating the presence of numerous small neurons.

In the non-mammalian vertebrates, the newt had the highest content of DNA in the hemispheres and the olfactory lobes, with a moderate amount of RNA (Table VIII). Although these areas do contain numerous small neurons, this resulted in a very low RNA/DNA ratio. The frog brain also has a higher concentration of DNA in the hemispheres and olfactory lobes than in the turtle, grass snake, or chicken, indicating the presence of many small neurons and glia, hence the RNA/DNA ratio was below that of the aforementioned animals. The hemispheres of the trout, turtle, grass snake, and chicken have a RNA/DNA ratio above two, indicating the presence of numerous large neurons.

From the above data, it can be concluded, as did Mandel and co-workers,[35,49] that there are no phylogenetic correlates from these analytical studies of nucleic acids.

TABLE VII

Regional Distribution of RNA and DNA in Various Vertebrates[50]

Region	μg/100 mg wet wt.								
	RNA			DNA			RNA/DNA		
	Rabbit	Guinea pig	Rat	Rabbit	Guinea pig	Rat	Rabbit	Guinea pig	Rat
Metencephalon									
Cerebellar gray	183	165	159	260	370	484	0.70	0.45	0.33
Cerebellar white	65	82	103	85	97	264	0.77	0.85	0.39
Diencephalon									
Hypothalamus	123	127	139	60	95	103	2.04	1.34	1.34
Thalamus	108	107	108	52	67	73	2.05	1.59	1.47
Telencephalon									
Corpus striatum	123	117	112	52	67	84	2.33	1.74	1.33
Cerebral gray	160	187	148	67	67	84	2.37	2.78	1.77
Cerebral white	62	87	99	55	90	96	1.14	0.97	1.03
Hippocampus	145	150	142	62	90	85	2.32	1.67	1.66
Olfactory bulb	195	208	163	185	218	223	1.05	0.95	0.72
Myelencephalon									
Medulla oblongata	85	105	85	50	75	68	1.70	1.40	1.24
Mesencephalon	95	115	90	57	75	75	1.65	1.53	1.19

TABLE VIII

Nucleic Acid Distribution in Various Areas of the Brain of Some Vertebrates[35]

	Trout	Newt	Frog	Turtle	Grass snake	Chicken
RNA[a] (μg/100 mg wet wt.)						
Hemispheres	138	106	115	137	111	117
Olfactory lobes	—	106	111	120	121	141
Cerebellum	124	—	—	124	146	141
DNA[a] (μg/100 mg wet wt.)						
Hemispheres	68	365	137	29	41	42
Olfactory lobes	—	365	166	65	69	53
Cerebellum	348	—	—	128	70	138
RNA/DNA						
Hemispheres	2.03	0.29	0.84	4.73	2.71	2.78
Olfactory lobes	—	0.29	0.67	1.85	175	2.66
Cerebellum	0.36	—	—	0.96	2.08	1.02

[a] Data for RNA-P and DNA-P multiplied by ten to obtain the above values for RNA and DNA.

B. Cell Density

Determination of the DNA content in human, dog, and cat cerebral cortex by Heller and Elliott[32] showed that there was 7.1, 6.5, and 7.1 pg/nucleus, respectively. From the above data, the average cell densities in this tissue were calculated. In human dog, and cat, the corpus callosum contained approximately the same total number of cells as the cerebral cortex, while the cerebellar cortex contained 4–5 times this number of cells. In examining three distinct layers of the cerebellum of rabbits and monkeys, Kissane and Robins[33] determined the DNA content as well as the cell density of these layers. The three layers are designated as the molecular layer, the granular layer, and the subcortical white matter. In both species, the nuclear density in the cerebellar granular layer was found to be approximately 20 times that in the molecular layer or in the subcortical white matter, and the cellularity of the molecular and subcortical layers was approximately equal. Other investigators have examined the DNA content and cellularity in different areas of human brain cortex,[51] in various cortical layers in the motor and visual cortex of the monkey,[52] and in assorted regions of rat brain.[53]

III. DEVELOPMENT

A. Total and Regional Changes

The single biochemical process which can be considered to represent growth is the biosynthesis of protein, and both DNA and RNA have an essential role in this process. In the intact brain, both DNA and RNA increase during cell multiplication (mitosis); however, when mitosis ceases, there is no further increase in DNA, but RNA continues to increase with cell enlargement. When all growth ceases, DNA and RNA in the tissue remain constant throughout the remaining life span of the animal, although RNA undergoes a continuous turnover in this period.[54,55] The studies by Mandel and coworkers[34,35,49] illustrate that the temporal changes of the nucleic acids in the developing brain varies with different animals; thus in the guinea pig brain, the RNA content increases to the tenth day postnatally, while the DNA remains constant from birth and remains unaltered through adulthood.[49] In the rat, both nucleic acids increase up to the 10–15th day postnatally,[35] and subsequently remain constant. The changes in nucleic acid content during development in the fowl, dog, cat, rabbit, rat, guinea pig, and human are shown in Tables IX and X.

In the rabbit brain, the nucleic acids increase rapidly for 30 days after birth, then they increase slowly until the 90th day, at which time RNA and DNA reach adult levels. Similar findings were reported by other investigators.[56–60] The dog and cat brains show a rapid increase in RNA and DNA in the first 30 days of the postnatal period, then RNA increases slowly for four more months, while DNA remains constant. In the chick embryo,

TABLE IX

RNA and DNA Content per Brain of Various Species During Development[a]

Species	Age	RNA (mg/brain)	DNA (mg/brain)	RNA/DNA
Fowl	10 Days	0.30	0.30	1.00
	1 Month	3.50	2.30	1.52
	Adult	5.10	2.30	2.55
Dog	10 Days	25	20	1.25
	1 Month	40	35	1.14
	Adult	50	40	1.25
Cat	10 Days	12	9	1.33
	1 Month	27	20	1.35
	Adult	33	20	1.65

[a] Data were extrapolated from graphs by Mandel et al.[35] When comparing with the original data, there were some discrepancies with respect to some of the DNA and RNA content. The RNA/DNA ratio was also different, but the general trends of these ratios were comparable.

TABLE X

RNA and DNA Content per Brain of Various Species During Development[a]

Species	Age	RNA (mg/brain)	DNA (mg/brain)	RNA/DNA
Rabbit	10 Days	6	4	1.50
	1 Month	8	6	1.33
	Adult	13	10	1.30
Rat	10 Days	1.7	1.2	1.41
	1 Month	2.3	1.8	1.27
	Adult	2.3	1.8	1.27
Guinea pig	10 Days	5.2	3.8	1.36
	1 Month	5.2	3.8	1.36
	1 Year	5.2	3.8	1.36
Human	4 Days	400	300	1.33
	1 Year	1400	850	1.69
	Adult	2550	1000	2.55

[a] Data were extrapolated from graphs by Mandel et al.[35] When comparing with the original data, there were some discrepancies with respect to some of the DNA and RNA content. The RNA/DNA ratio was also different, but the general trends of these ratios were comparable.

both RNA and DNA increase rapidly and both continue to increase after hatching until the chick is 30 days old. Then RNA continues to increase slowly for another 30 days, while DNA remains constant. At 60 days after birth, RNA and DNA attain adult levels. Both nucleic acids in the human brain increase rapidly, and at the same rate, until ten months postnatally; then DNA continues to increase slowly, while RNA increases more rapidly for approximately two years, and subsequently adult levels are attained.

In the rat cerebral cortex, Keup[61] has found that the content of DNA is higher than that of RNA during the prenatal period. This relation is reversed shortly after birth and the RNA/DNA ratio is 1.5–2.2 for the total brain of adult rats. Recently, Adams[62] has also shown an increase in total DNA, nuclear, and transfer RNA's in rat brains from birth up to 18 days postnatally, and then they remain relatively constant through the adult. The total microsomal and ribosomal RNA content of the brain also increased up to the 18-day-old rat, but then declined. With this decrease, there was a concomitant change in the composition of ribosomal RNA.

In a study carried out by Bernsohn and Norgello,[63] the base composition and concentration of ribosomal RNA was compared in whole brains of newborn and adult rats. Their results showed a decrease in ribosomal RNA based on unit wet weight of brain; but on the basis of total brain, there was a three-fold increase in the adult. There was also a change in base composition, an increase in A (17.5–20.0), and a commensurate decrease in G (34.5–32.9) as the brain matured. What occurs in ribosomal RNA to account for the observed changes in A and G content is not clear at present.

During the postnatal development in mouse brain, DNA and RNA diminish slowly in relation to total nitrogen or fresh weight of brain tissue.[64] The values for RNA were 570 mg/100 g at the age of 1–2 days and 300 mg/100 g at the age of 3–6 months, and the corresponding values for DNA were 380 and 240 mg/100 mg. Other investigators[35,49,65,68] also report an increase in total RNA and DNA in the brain during postnatal growth of the rat. Uzman and Rumley[67] showed that in one-day mouse brain, total nucleic acid content was 1.2%, which decreased to 0.6% in the adult.

Flexner and Flexner[68] found a decrease from about 70 mg of DNA-P/ 100 g wet brain cortex at the 25th day *in vivo* in the guinea pig embryo to about 10 mg/100 g at the 48th day, with little change in the adult level (6 mg/100 g). The decrease in RNA-P was from about 35 to 15 mg/100 g at the same ages, respectively.

Lestie and Davidson[55] studied the changes in brain nucleic acid in the developing chick embryo from the eighth day to two days after hatching. They found that, on the basis of fresh weight of tissue, RNA concentration fluctuated while DNA decreased; but on the basis of unit cell, RNA increased steadily ($41–118 \times 10^{-7}$ μg/cell), while DNA content was constant (23.5×10^{-7} μg/cell). Similar findings were reported by Szepsenwol, Mason, and Shontz.[69]

B. Chromatin

Dingman and Sporn[70] isolated the chromatin material from embryonic and adult brains of the chicken and studied the changes in brain nucleic acid, histones, and proteins during development of the embryo and in adult chicken brain. They found that the ratio of total protein/DNA remained relatively constant between the range of 2.7–3.1 in both embryonic and adult brain; however, the histone/DNA ratio showed a slight decrease from embryo (0.88–0.98) to adult (0.76–0.80). In addition to this, when they examined the RNA/DNA ratio in the chromatin material, they found there was a progressive decrease in this ratio from a range of 0.36–0.23 in the embryo from 2–8 days. Subsequently, the ratio remained relatively constant from the eighth day embryonic brain through the adult in a range of 0.19–0.17. They found that the content of DNA increased from 11 μg of DNA per brain in the four-day embryo to 140–145 μg per brain in the eight-day embryo, and to 390–555 μg per brain in the adult brain of the chicken.

C. DNA Nucleoprotein

A recent study by Kurtz and Sinex[71] has focused attention on the thermal stability of DNA nucleoprotein, which apparently shows changes with increasing age of the animal and, hence, reflects a quantitative change between DNA and protein. They found that the protein/DNA ratio in nucleoprotein of the mouse brain was high at birth and dropped until the age of six months. Subsequently, the ratio remained relatively constant until the 14th month and then continued to rise up to the 30th month. Concomitant with this observation, the melting temperature of the nucleoprotein complex was measured. The changes in T_m followed approximately the changes in protein per DNA ratios observed during the growth of the mouse. Thus, in the mouse, the T_m of the nucleoprotein at birth was 80 and it gradually dropped to 72 in the 13th month of age. Subsequently, it reached a value higher than that observed at birth. These studies indicate that there is a considerable change in the DNA nucleoprotein (DNAP) during various phases of the developing mouse brain, although the exact significance of this phenomenon remains to be explored.

IV. METHODOLOGY

A. Analyses

The use of classical procedures for the separation and analysis of nucleic acids from brain, namely, the Schmidt–Thannhauser,[72] Schneider,[73] and the Ogur and Rosen[74] procedures, resulted in analytical errors of which many investigators are unaware. Evaluation of these procedures by Logan[75] and Munro[76] revealed that in the presence of high concentrations of acids for the precipitation of tissue components and with the subsequent extraction of lipids with alcohol a portion of the nucleoprotein is solubilized. This loss of both protein and RNA can be avoided by the addition of sodium acetate to the alcohol in order to neutralize the acid in the precipitate or by bypassing this step in the procedure.[77] Many of the difficulties in the above classical procedures can be avoided by the modifications proposed in the procedure by Shibko et al.[78]

Analysis of phosphorus as a means of nucleic acid determination has the advantage that its great sensitivity permits the use of very small amounts of samples.[79] However, the possibility of phospholipid and phosphoprotein contaminants in the nucleic acid extracts render this method unreliable.[75,76] A detailed evaluation of these procedures has been reported by Munro.[76]

B. Fractionation

A serious problem has been encountered in the rapid degradation of nucleic acids during homogenization of the brain. Addition of bentonite, which selectively adsorbs RNase and DNase[80,81] does not completely prevent this degradation, nor does the addition of sodium lauryl sulfate (SDS) or polyvinylsulfonate completely prevent nucleic acid degradation. It has been shown that brain and liver contain an RNase inhibitor.[82,83] The liver inhibitor has been isolated and addition of this inhibitor to a brain homogenate has effectively minimized RNA degradation. The most effective procedures for extraction of intact nucleic acids are those using aqueous phenol as devised by Kirby[84] and modified by Georgiev[85] and Sabatini.[86,87] Phenol effectively prevents degradation of the nucleic acids and has permitted isolation of RNA with sedimentation coefficients of 35 S or higher.

In separation of the RNA components by sucrose gradient centrifugation or by means of MAK columns, three distinct fractions of RNA are obtained: 4, 18, and 28 S, each of which contains a group of RNA's. Recently, Peacock and Dingman[88] subjected the above subfractions of RNA to polyacrylamide gel electrophoresis and they found that each of these fractions was a heterogeneous mixture of RNA's. These observations emphasize that the present methods of nucleic acid fractionation yield heterogeneous mixtures of RNA, in which unique RNA components remain undetected. Development of the polyacrylamide separation of homologous RNA's may permit better interpretation of nucleic acid metabolism in the brain.

V. CONCLUSION

The past studies on the isolation, fractionation, and analysis of the brain nucleic acids have now evolved so that reliable detailed data can be obtained. This will enhance future interpretation of the role of nucleic acids in the cytoarchitecture and metabolism of the brain, and possibly facilitate clarification of brain function on a biochemical basis. It appears that the accrual of detailed information on the nucleic acids in the central nervous system is now well initiated.

VI. REFERENCES

1. R. L. Friede, *in Topographic Brain Chemistry*, pp. 2–7, Academic Press, New York (1966).
2. P. N. Campbell, *in Progress in Biophysics and Molecular Biology* (J. A. V. Butler and H. E. Huxley, eds.), Vol. 15, pp. 1–36, Pergamon Press, New York (1965).
3. H. Hyden, *in Brain Function* (M. A. B. Brazier, ed.), Vol. II, pp. 29–68, University of California Press, Berkeley and Los Angeles (1964).
4. D. R. Dahl, R. Jacobs, and F. E. Samson, Jr., Characterization of two mitochondrial particulates from rat brain, *Am. J. Physiol.* **198**:467–70 (1960).
5. H. G. DuBuy, C. F. T. Mattern, and L. Riley, Comparisons of the DNA's obtained from brain nuclei and mitochondria of mice and from the nuclei and kinetoplasts of *Leishmamia enrietti*, *Biochim. Biophys. Acta* **123**:298–305 (1966).
6. H. Busch, *in Histones and Other Nuclear Proteins*, pp. 91–119, Academic Press, New York (1965).
7. H. Busch, R. Desjardins, D. Grogan, K. Higashi, S. T. Jacob, M. Muramatsu, T. S. Ro, and W. J. Steele, *in International Symposium: The Nucleolus, Its Structure and Function*, Monograph 23, pp. 193–222, National Cancer Institute, Bethesda, Maryland (1966).
8. M. Alfert and I. I. Geschwind, A selective staining method for basic proteins of cell nuclei, *Proc. Natl. Acad. Sci. (U.S.)* **39**:991–999 (1953).
9. M. L. Bernstiel, M. I. H. Chipchase, and W. G. Flamm, On the chemistry and organization on neucleolar proteins, *Biochim. Biophys. Acta* **87**:112 (see R. C. Huang) (1964).
10. H. R. Mahler, W. J. Moore, and R. J. Thompson, Isolation and characterization of RNA from cerebral cortex of rat, *J. Biol. Chem.* **241**:1283–1289 (1966).
11. R. W. Holley, G. A. Everett, J. T. Madison, and A. Zamir, Nucleotide sequences in the yeast alanine transfer RNA, *J. Biol. Chem.* **240**:2122–2128 (1965).
12. G. L. Brown, *in Progress in Nucleic Acid Research* (J. N. Davidson and W. E. Cohn, eds.), Vol. 2, pp. 260–305, Academic Press, New York (1963).
13. E. Borek and P. R. Srinivasan, The methylation of nucleic acids, *Ann. Rev. Biochem.* **35**:275–298 (1966).
14. L. N. Simon, A. J. Glasky, and T. H. Rejal, Enzymes in the CNS. I. RNA methylase, *Biochim. Biophys. Acta* **142**:99–104 (1967).
15. R. K. Datta, Brain ribosomes, *Brain Res.* **2**:301–322 (1966).
16. C. E. Zomzely, S. Roberts, D. M. Brown, and C. Provost, Cerebral protein synthesis. I. Physical properties of cerebral ribosomes and polysomes, *J. Mol. Biol.* **20**:455–468 (1966).
17. M. R. V. Murthy and D. A. Rappoport, Biochemistry of the developing rat brain. VI. Preparation and properties of ribosomes, *Biochim. Biophys. Acta* **95**:132–145 (1965).
18. S. Yamagami, M. Masui, and Y. Kawakita, Preparation and properties of ribosomes from guinea pig brains, *J. Neurochem.* **10**:849–850 (1963).
19. R. P. Perry, *in Progress in Nucleic Acid Research and Molecular Biology* (J. N. Davidson and W. E. Cohn, eds.), Vol. 6, pp. 220–253, Academic Press, New York (1967).
20. H. R. Mahler and A. T. Campagnoni, Isolation and properties of polyribosomes from cerebral cortex, *Biochemistry* **6**:956–967 (1967).
21. A. Rich, J. R. Warner, and H. M. Goodman, *in Synthesis and Structure of Macromolecules*, Cold Spring Harbor Symp. Quant. Biol. **XXVIII**:269–286 (1963).
22. S. L. Palay and G. E. Palade, The fine structure of neurons, *J. Biophys. Biochem. Cytol.* **1**:69–88 (1955).
23. G. E. Palade and P. Siekevitz, Liver microsomes, an integrated morphological and biochemical study, *J. Biophys. Biochem. Cytol.* **2**:171–200 (1956).
24. F. Lipmann, *in Progress in Nucleic Acid Research* (J. N. Davidson and W. E. Cohn, eds.), Vol. 1, pp. 135–158, Academic Press, New York (1963).

25. S. Yamagami, R. R. Fritz, and D. A. Rappoport, Biochemistry of the developing rat brain. VII. Changes in the ribosomal system and nuclear RNAs, *Biochim. Biophys. Acta* **129**:532–547 (1966).

26. G. P. Georgiev, in *Progress in Nucleic Acid Research and Molecular Biology* (J. N. Davidson and W. E. Cohn, eds.), Vol. 6, pp. 259–351, Academic Press, New York (1967).

27. S. Spiegelman, in *Recent Progress in Microbiology*, *Symp.*, *8th Intern. Congr. Microbiol.*, 1962, *Montreal, Canada*, pp. 95–115 (1963).

28. S. Granick and A. Gibor, in *Progress in Nucleic Acid Research and Molecular Biology* (J. N. Davidson and W. E. Dohn, eds.), Vol. 6, pp. 143–183, Academic Press, New York (1967).

29. D. M. Prescott, in *Progress in Nucleic Acid Research and Molecular Biology* (J. N. Davidson and W. E. Cohn, eds.), Vol. 3, pp. 33–54, Academic Press, New York (1964).

30. R. M. S. Smellie, in *Progress in Nucleic Acid Research* (J. N. Davidson and W. E. Cohn, eds.), Vol. 1, pp. 27–55, Academic Press, New York (1963).

31. M. P. Viola, Histochemical differences between glia nuclei of the rat spinal cord, *Sperimentale* **113**:317–333 (1963).

32. I. H. Heller and K. A. C. Elliott, Deoxyribonucleic acid content and cell density in brain and human brain tumors, *Can. J. Biochem. Physiol.* **32**:584–592 (1954).

33. J. M. Kissane and E. Robins, The fluorometric measurement of DNA in animal tissues with special reference to the CNS, *J. Biol. Chem.* **233**:184–188 (1958).

34. P. Mandel, R. Bieth, and E. Stoll, Biochemical evolution of the brain of a chicken embryo. The nucleic acids, *Compt. Rend. Soc. Biol.* **142**:1020–1022 (1948).

35. P. Mandel, H. Rein, S. Harth-Edel, and R. Mardell, in *Comparative Neurochemistry* (D. Richter, ed.), pp. 149–163, Pergamon Press, New York (1964).

36. M. J. D. White, in *The Chromosomes* (K. Mellanby, ed.), pp. 14–15, 40–42, John Wiley & Sons, New York (1963).

37. N. Robinson, Composition of deoxyribonucleoprotein in human brain, *Clin. Chim. Acta* **14**:429–434 (1966).

38. E. Chargaff, S. Zamenhof, and L. B. Shettles, Human deoxypentose nucleic acid, *Nature* **165**:756–757 (1950).

39. N. I. Gold and S. H. Sturgis, Cytosine–thymidine ratios of endometrial DNA's, *J. Biol. Chem.* **196**:143–150 (1952).

40. C. F. Emanuel and I. L. Chaikoff, Deoxyribonucleic acid of the CNS, kidney, and spleen: A comparison of some chemical and physical properties, *J. Neurochem.* **5**:236–244 (1960).

41. J.-E. Edstrom, The content and the concentration of RNA in motor anterior horn cells from the rabbit, *J. Neurochem.* **1**:159–165 (1956).

42. J.-E. Edstrom and A. Pigon, Relation between surface, RNA content, and nuclear volume in encapsulated spinal ganglion cells, *J. Neurochem.* **3**:95–99 (1958).

43. H. Hyden, in *Neurochemistry* (K. A. C. Elliott, I. H. Page, and J. H. Quastel, eds.), pp. 331–375, Chas. C. Thomas, Springfield, Illinois, 2nd edition (1962).

44. A. Yajima, The nucleic acid content of the brain tissue of rats as influenced by age, *Tohoku J. Exptl. Med.* **85**:252–255 (1965).

45. M. Jacob, J. Stevenin, R. Jund, C. Judes, and P. Mandel, Rapidly-labelled RNA in brain, *J. Neurochem.* **13**:619–628 (1966).

46. G. P. Georgiev, O. P. Samarina, M. I. Lerman, M. N. Smirnov, and A. N. Severtzov, Biosynthesis of messenger and ribosomal RNA's in the nucleolochromosomal apparatus of animal cells, *Nature* **200**:1291–1294 (1963).

47. E. Egyhazi and H. Hyden, RNA with high specific activity in neurons and glia, *Brain Res.* **2**:197–200 (1966).

48. E. Egyhazi and H. Hyden, Biosynthesis of rapidly labelled RNA in brain cells, *Life Sci.* **5**:1215–1223 (1966).

49. P. Mandel and R. Bieth, Comparative study of the biochemical development of the brain in various mammals, *Compt. Rend. Acad. Sci.* **235**:485–487 (1952).

50. M. Jacob and P. Mandel, *in Protides of the Biological Fluids* (H. Peeters, ed.), Vol. 13, pp. 63–80, Elsevier Publishing Company, Amsterdam (1966).

51. R. Landolt, H. H. Hess, and C. Thalhemier, Regional distribution of some chemical structural components of the human nervous system, *J. Neurochem.* **13**:1441–1452 (1966).

52. E. Robins, D. E. Smith, and K. M. Eydt, The quantitative histochemistry of the cerebral cortex. I. Architectonic distribution of ten chemical constituents in the motor and the visual cortices, *J. Neurochem.* **1**:54–67 (1956).

53. L. May and R. G. Grenell, Nucleic acid content of various areas of the rat brain, *Proc. Soc. Exptl. Biol. Med.* **102**:235–239 (1959).

54. M. Winick and A. Noble, Quantitative changes in DNA, RNA, and protein during prenatal and postnatal growth in the rat, *Develop. Biol.* **12**:451–466 (1965).

55. I. Leslie and J. N. Davidson, The chemical composition of the chick embryonic cell, *Biochim. Biophys. Acta* **7**:413–428 (1951).

56. V. I. Krasil'nekova, Cytochemical study of nucleic acids in rabbit brain cells in ontogenesis, *Tsitologiya* **2**(1):29–36 (1960); *Ref. Zh. Biol.* **1961**, No. 2A253 (Transl.).

57. K. G. Manukyan, Nucleic acids and phospholipids of the rabbit brain in ontogenesis, *Dokl. Akad. Nauk SSSR* **101**:1085–1088 (1955).

58. K. G. Manukyan, Exchange of phosphorus of nucleic acids and phospholipids in the brain of rabbit during ontogenesis, *Dokl. Akad. Nauk SSSR* **102**:567–570 (1955).

59. E. B. Skvirskaya and T. B. Silich, Nucleic acids in the various parts of the brain, *Dokl. Akad. Nauk SSSR* **93**:1073–1075 (1953).

60. E. B. Skvirskaya and O. P. Chepinogu, Metabolism of nucleic acids in tissue of brain and liver in ontogenesis, *Dokl. Akad. Nauk SSSR* **92**:1007–1010 (1953).

61. W. Keup, *in Progress in Neurobiology, Vol. II, Ultrastructure and Cellular Chemistry of Neural Tissue* (H. Waelsch, ed.), pp. 215–223, Hoeber-Harper Book, New York (1957).

62. D. H. Adams, The relationship between cellular nucleic acids in the developing rat cerebral cortex, *Biochem. J.* **98**:636–640 (1966).

63. J. Bernsohn and H. Norgello, Base composition of ribosomal RNA in newborn and adult rat brain, *Proc. Soc. Exptl. Biol. Med.* **122**:22–24 (1966).

64. W. Albrecht, Changes in the content of the different phosphate fractions in the brain and muscles of the mouse during growth, *Z. Naturforsch.* **116**:248–252 (1956).

65. D. R. Dahl and F. E. Samson, Jr., Metabolism of rat brain mitochondria during postnatal development, *Am. J. Physiol.* **196**:470–472 (1959).

66. S. S. Oja, Postnatal changes in the concentration of nucleic acids, nucleotides, and amino acids in the rat brain, *Ann. Acad. Sci. Fennicae Ser. A. V. Med.* **125**:1–69 (1966).

67. L. L. Uzman and M. K. Rumley, Changes in the composition of the developing mouse brain during early myelination, *J. Neurochem.* **3**:179–184 (1958).

68. J. B. Flexner and L. B. Flexner, Biochemical and physiological differentiation during morphogenesis. XIV. The nucleic acids of the developing cerebral cortex and liver of the fetal guinea pig, *J. Cellular Comp. Physiol.* **37**:1–16 (1951).

69. J. Szepsenwol, J. Mason, and M. E. Shontz, Phospholipids and nucleic acids in embryonic tissues of the chick, *Am. J. Physiol.* **180**:525–529 (1955).

70. C. W. Dingman and M. B. Sporn, Studies on chromatin. I. Isolation and characterization of nuclear complexes of DNA, RNA, and protein from embryonic and adult tissues of the chicken, *J. Biol. Chem.* **239**:3483–3492 (1964).

71. D. I. Kurtz and F. M. Sinex, Age related differences in the association of brain DNA and nuclear protein, *Biochim. Biophys. Acta* **145**:840–842 (1967).

72. G. Schmidt and S. J. Thannhauser, A method for the determination of DNA, RNA, and phosphoproteins in animal tissues, *J. Biol. Chem.* **161**:83–89 (1945).
73. W. C. Schneider, Phosphorous compounds in animal tissues. I. Extraction and estimation of DNA and of RNA, *J. Biol. Chem.* **161**:293–303 (1945).
74. M. Ogur and G. Rosen, The nucleic acid of plant tissues. I. The extraction and estimation of DNA and RNA, *Arch. Biochem.* **25**:262–276 (1950).
75. J. E. Logan, W. A. Mannell, and R. J. Rossiter, Estimation of nucleic acids in tissue from the nervous system, *Biochem. J.* **51**:470–480 (1952).
76. H. N. Munro and A. Fleck, *in Methods of Biochemical Analysis* (David Glick, ed.), Vol. XIV, pp. 113–176, Interscience Publishers, New York (1966).
77. G. R. Barker and J. A. Hollinshead, Nucleotide metabolism in germinating seeds, *Biochem. J.* **93**:78–83 (1964).
78. S. Shibko, P. Korvistoinen, C. A. Tratnyek, A. R. Newhall, and L. Friedman, A method for sequential quantitative separation and determination of protein, RNA, DNA, lipid, and glycogen from a single rat liver homogenate or from a subcellular fraction, *Anal. Biochem.* **19**:514–528 (1967).
79. O. Lindberg and L. Ernster, *in Methods of Biochemical Analysis* (David Glick, ed.), Vol. III, pp. 1–17, Interscience Publishers, New York (1956).
80. B. Singer and H. Fraenkel-Conrat, Effects of bentonite on infectivity and stability of TMV-RNA, *Virology* **14**:59–65 (1961).
81. T. J. Brownhill, A. S. Jones, and M. Stacey, The inactivation of ribonuclease during the isolation of ribonucleic acids and ribonucleoproteins from yeast, *Biochem. J.* **73**:434–438 (1959).
82. K. Shortman, Studies on cellular inhibitors of ribonuclease. I. The assay of the ribonuclease–inhibitor system, and the purification of the inhibitor from rat liver, *Biochim. Biophys. Acta* **51**:37–49 (1961).
83. Y. Takahashi, K. Mase, and H. Sugano, Preparation of polysomes from rat brain tissue, *Biochim. Biophys. Acta* **119**:627–629 (1966).
84. K. S. Kirby, A new method for the isolation of RNA's from mammalian tissues, *Biochem. J.* **64**:405–408 (1965).
85. G. P. Georgiev, V. L. Mant'eva, and I. B. Zbarskii, RNA fractions in cell nuclei isolated by phenol and by sucrose–glycerophosphate, *Biochim. Biophys. Acta* **37**:373–374 (1960).
86. A. Sibatani, K. Yamana, K. Kimura, and H. Okagaki, Fractionation with phenol of RNA of animal cells, *Biochim. Biophys. Acta* **33**:590–591 (1959).
87. K. Yamana and A. Sibatani, Fractionation of RNA's with phenol, *Biochim. Biophys. Acta* **41**:295–303 (1960).
88. A. C. Peacock and C. W. Dingman, Resolution of multiple RNA species by polyacrylamide gel electrophoresis, *Biochemistry* **6**:1818–1827 (1967).
89. I. Pechan, Nucleic acids of the brain. I. Isolation of DNA from the pig brain and its arrangement with preparations from other tissues, *Biologia* **16**:292–295 (1961).
90. A. Yajima, The decrepitude of the rat brain with special references to changes in nucleic acids, *Tohoku J. Exptl. Med.* **89**:235–244 (1966).
91. I. Leslie, *in The Nucleic Acids*, Vol. II, pp. 1–44, Academic Press, New York (1955).
92. D. Bodian and D. Dziewiatkowski, The disposition of radio-active phosphorus in normal, as compared with regenerating and degenerating nervous tissue, *J. Cellular Comp. Physiol.* **35**:155–177 (1950).
93. R. Bieth and P. Mandel, A comparative study of adult brain constituents in various classes of vertebrates, *Experientia* **9**:185–186 (1953).
94. W. C. Schneider, Phosphorus compounds in animal tissues. III. A comparison of methods for the estimation of nucleic acids, *J. Biol. Chem.* **164**:747–751 (1946).

95. W. C. Schneider and H. L. Klug, Phosphorus compounds in animal tissues. IV. The distribution of nucleic acids and other phosphorus containing compounds in normal and malignant tissues, *Cancer Res.* **6**:691–694 (1946).

96. C. H. Ti, C. K. Wang, and Y. Chen, The nucleic acid contents of animal brains as determined by two wavelength spectrophotometry, *Sheng Wu Li Hsuek Pao* **4**:551–557 (1964).

Chapter 8

LIPIDS

George Rouser and Akira Yamamoto

Section of Lipid Research
Division of Neurosciences
City of Hope Medical Center
Duarte, California

I. INTRODUCTION

In this chapter the lipid composition of vertebrate and invertebrate nervous systems is considered. Changes during development and aging and in pathological states in humans are outlined, and the general role of lipids as cellular membrane components of the nervous system is emphasized. When possible, reference is made to recent comprehensive treatments of subjects which we consider more briefly here. In general, basic concepts and available information are summarized, and the controversial nature of some areas is pointed out. Major limitations of procedures and present information are noted, with recommendations for further study. Procedures for determination of lipid composition and their limitations are considered first since data and concepts must be judged in the light of the methods used.

II. DETERMINATION OF LIPID COMPOSITION

A. Lipid Classes Present in the Vertebrate Nervous System

Most of the lipid of the vertebrate nervous system is accounted for by cholesterol and well-characterized phospholipids and glycolipids (see Figs. 1–17). The amounts of each lipid class change with age (see Figs. 19–23, Section III).

Sterols other than cholesterol are trace components found in highest concentration shortly after birth.[1] Fatty acid esters of cholesterol (cholesterol esters) are found in normal brain as minor components (0.06 to 0.60 % of the total lipid). Although the presence in brain of another type of cholesterol ester, cholesterol sulfate, has been suggested, we have not found this substance in human brain at any age.

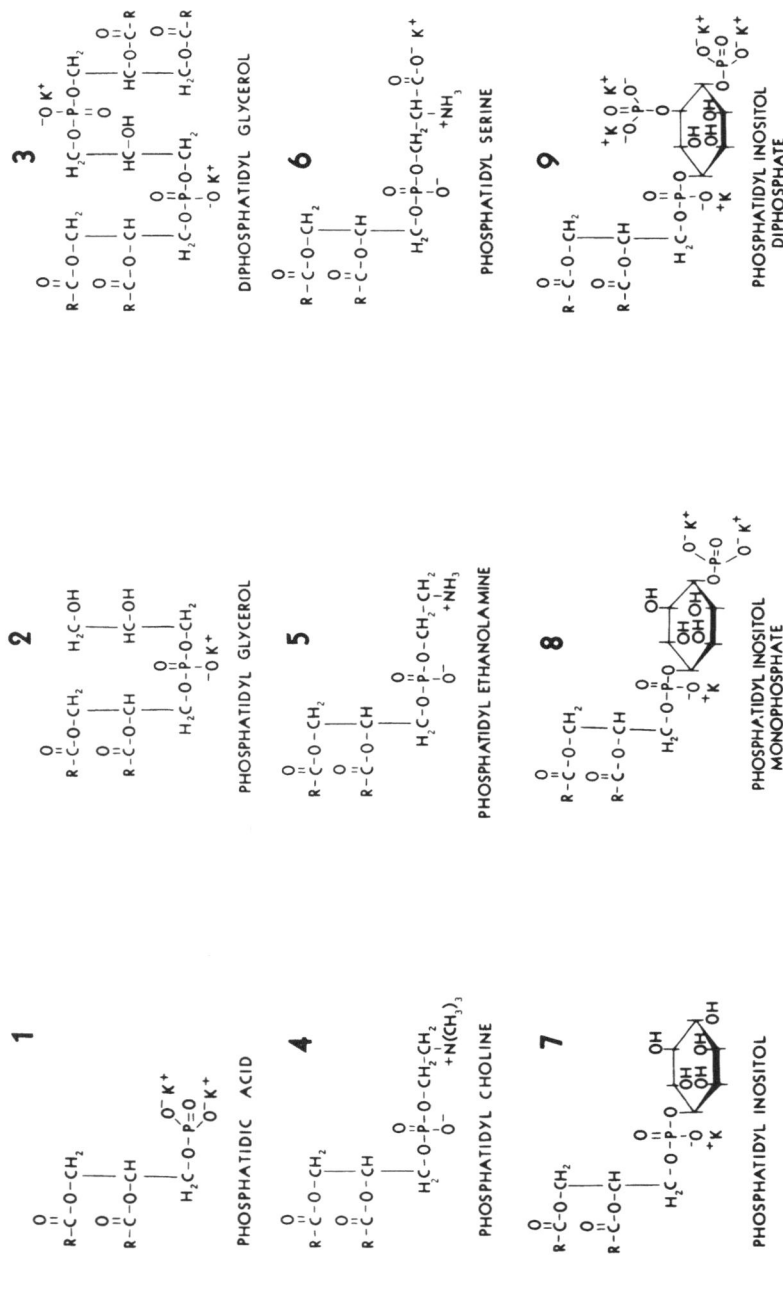

Figs. 1–9. Lipids of the vertebrate nervous system.

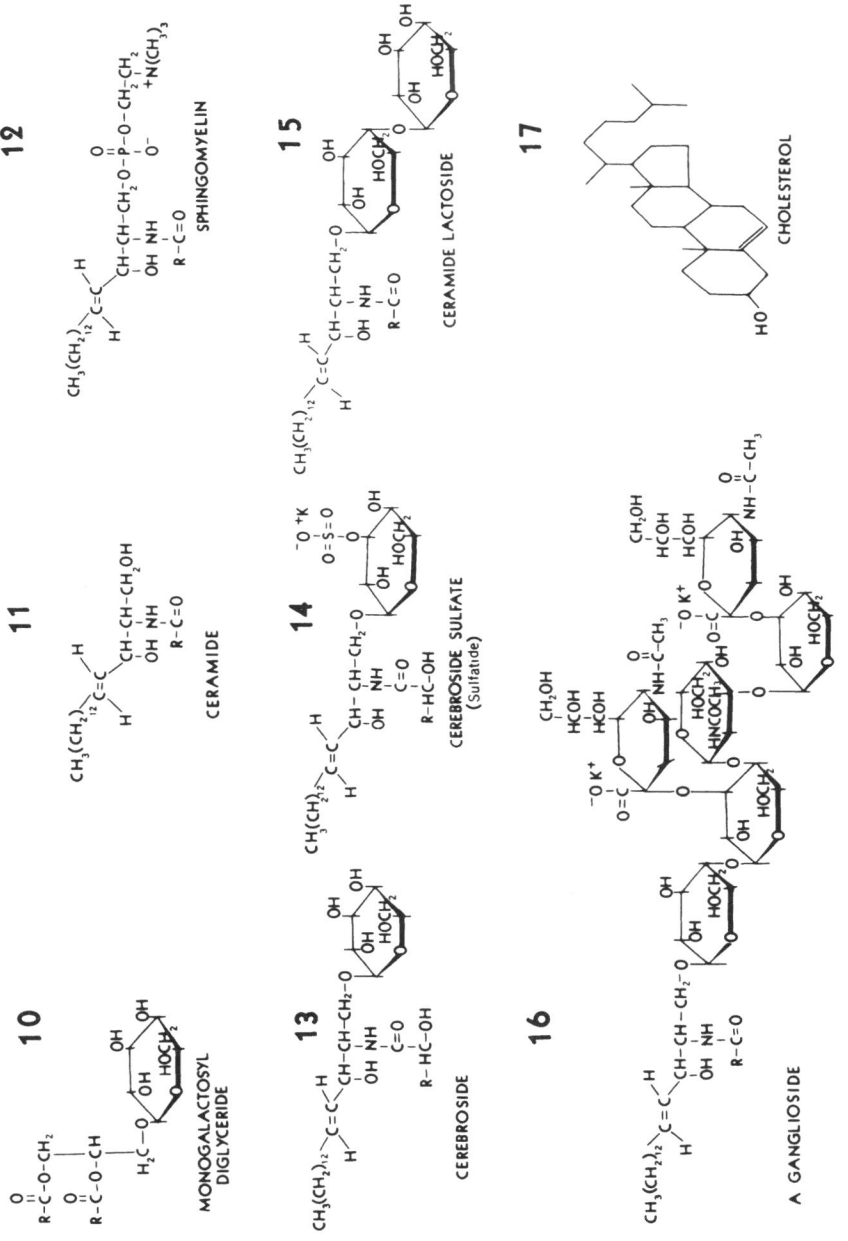

Figs. 10–17. Lipids of the vertebrate nervous system.

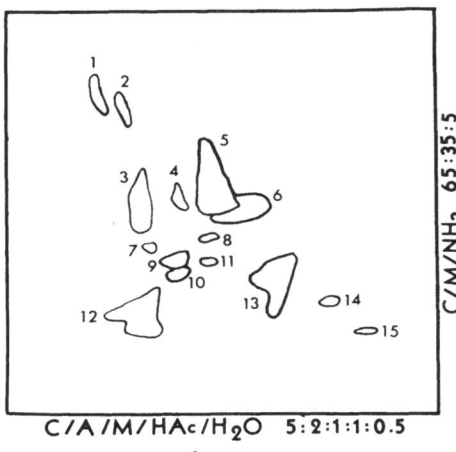

C/A/M/HAc/H$_2$O   5:2:1:1:0.5

Fig. 18. Diagram of the acidic lipid spots seen by two-dimensional TLC of the final fraction eluted from a DEAE cellulose column. Sample spotted at the lower right. Chloroform–methanol–28 % aqueous ammonia 65:35:5 used first for development in the vertical direction followed by chloroform–acetone–methanol–acetic acid–water 5:2:1:1:0.5 in the horizontal direction. Components are: (1) N-acetyl phosphatidyl ethanolamine; (2) lyso bisphosphatidic acid; (3) diphosphatidyl glycerol (cardiolipin); (4) phosphatidyl glycerol; (5) sulfatide (normal fatty acids); (6) sulfatide (hydroxy fatty acids); (7–11) uncharacterized acidic phospholipids; (12) phosphatidic acid; (13) phosphatidyl inositol; and (14–15) uncharacterized.

The major phospholipid classes of the vertebrate nervous system are well characterized, although some of the minor components are not. The uncharacterized minor components are acidic lipids isolated together in a concentrated fraction by ion exchange cellulose column chromatography[2,3] after which they can be separated by two-dimensional thin layer chromatography (Fig. 18). Minor components that have been fairly well characterized are phosphatidyl glycerol, N-acetylphosphatidyl ethanolamine,[4] and lyso bisphosphatidic acid.[5] Evidence for the presence of phosphatidyl glycerol phosphate has been presented.[6]

The two most abundant glycolipids of higher vertebrates are cerebroside and cerebroside sulfate (sulfatide). Examination of different vertebrate species has shown that, in the mature animal, the cerebroside and sulfatide content of brain declines as species progressively lower on the vertebrate scale are examined. This is correlated with a decrease in the amount of myelin. The most complex glycolipids, the gangliosides, are a complex mixture of substances differing in the number and position of sialic acid moieties and the type and number of carbohydrate moietes. The exact number of ganglioside types is still uncertain. One of the most abundant types is shown in Fig. 16. Ceramide dihexosides, monoglycosyldiglyceride, diglycosyldiglyceride and acyl ceramides[7,8] are trace glycolipids.

Some lipid appears to become oxidized in vivo to become a part of an insoluble deposit called ceroid or lipofuscin. Little is known of the chemistry of this oxidized material and the amount can be judged only roughly by visual examination of tissue sections.

Each lipid class (Figs. 1–16), while having the same polar groups, is actually a mixture of molecular species that differ in length, degree of unsaturation and, in some cases, mode of linkage of the hydrocarbon chains. Ultimately these must be separated and determined to provide a complete lipid analysis. Progress is being made in this area by chromatographic separation of derivatives (diglyceride acetates or dimethyl phosphatidates). At present, however, determination of the fatty acids and aldehydes released by hydrolysis of the mixture of molecular species from each lipid class is the only available, generally applicable, accurate, and precise quantitative procedure giving an indication of the nature of molecular species variations.

B. Lipid Classes Present in the Invertebrate Nervous System

Qualitatively there are two major differences between the lipids found in brains of vertebrates and invertebrates. These were disclosed in recent unpublished studies (G. Rouser, C. Baxter, G. Simon). Invertebrate brains do not contain cerebroside or sulfatide detectable by thin layer chromatography. The other major qualitative difference is absence of sphingomyelin in some invertebrates which is associated with the presence of a related substance, either ceramide aminoethylphosphonate or ceramide phosphorylethanolamine. In some invertebrates (e.g., lobster) only sphingomyelin is present as in vertebrates. In others, a small amount of sphingomyelin and a large amount of ceramide aminoethylphosphonate are found, whereas in some species only ceramide aminoethylphosphonate or ceramide phosphorylethanolamine are found.

C. Approaches to Quantitative Analysis

Accurate determination of lipid composition is possible only when reliable procedures for sample selection and extraction as well as specific, accurate analytical procedures are employed.

The *first step* in quantitative analysis is careful selection of the most meaningful sample. The problems of sampling of nervous tissue have been recently considered in detail.[3,9] Since almost all of the lipid of the nervous system is present in membranes, a fundamental approach to sampling is isolation of pure preparations of individual membranes. In the ideal case, each different membrane or organelle from each different cell type would be isolated in pure form with quantitative yield. It would then be possible to determine the changes in relative proportions of the membranes as well as any changes in composition of individual membranes. Unfortunately, procedures available for isolation of subcellular particles are not adequate for quantitative isolation and may fail to yield pure particles or membranes even with large sacrifices in yield. Furthermore, light and electron microscopy have demonstrated the presence of complex, abnormal structures in pathological states, and thus the study of subcellular structures in these conditions is even more complicated. Since experience, skill, and special equipment are required for subcellular particle work, these structures cannot be studied in a routine manner in many laboratories. Most studies must employ a different approach to sampling. Reproducible, larger anatomical structures can be taken for analysis. It must be kept in mind that lipids occur in membranes which differ in composition and that the relative proportions of membranes vary in different parts of the nervous system.

Whole brain is the most common sample for small animal studies. Analysis of a representative sample of whole brain is the only way that changes in the brain as a whole can be determined since different parts of the brain differ in composition. Whole brain samples provide valuable data for understanding normal human brain development and the aging process, as well as changes in various pathological states. In principle, any reproducible morphological unit of the nervous system obtainable in its entirety can be used to provide accurate values for the unit at different ages, etc. Such samples include the cerebrum, cerebellum, brain stem, and some individual nuclei.[9] Gray and white matter samples when separated quantitatively (from an entire morphological unit) by aspiration[10] can be used, but difficulties are encountered in some species during development and in some pathological states where the line between gray and white matter is not distinct. The practice of taking small samples of gray and white matter for analysis is not acceptable for quantitative studies since such samples removed from different parts of one brain vary in composition, not infrequently by a factor of two or more.

The *second step* in quantitative analysis is complete extraction. Most investigators use the Folch solvent, chloroform–methanol 2:1. Important modifications of the procedure of Folch, Lees, and Sloane-Stanley[11] are the use of the multiple extraction[2,12–14] rather than single extraction, addition of an antioxidant,[2] and extraction in a nitrogen atmosphere.[12,13] Lipid extracts should be evaporated in the cold in a nitrogen atmosphere[2,12–14] with care not to evaporate to complete dryness, which causes decomposition of some lipids.[2,14] The procedure for brain has been described recently in detail.[3]

The *third step* in lipid analysis is usually separation of lipids from nonlipid contaminants. This has been accomplished by solvent partition[11] or partition column chromatography.[10,15] The widely used solvent partition of procedure of Folch, Lees, and Sloane–Stanley[11] does not remove all nonlipid contaminants and some lipid is commonly lost into the nonlipid phase. The most satisfactory procedure for quantitative analysis is partition column chromatography with a cross-linked dextran gel (Sephadex) which separates water-soluble nonlipid contaminants quantitatively from lipids and provides total gangliosides as a separate fraction.[2,3,15] The procedure has been applied with equal success to nervous tissue of vertebrates and invertebrates at all stages of development. A small amount of protein may be present in the lipid fraction from Sephadex. The protein content can be determined by the method of Lowry *et al.*[16]

The *fourth step* in quantitative analysis is usually chromatographic separation, which provides a high degree of specificity in analysis. Most modern methods of analysis employ chromatographic separation of intact lipids, although degradation and chromatography of the water-soluble products released is used occasionally. The degradation procedures have several limitations, the most severe of which is that fatty acids are split from the mixture of lipids and thus the fatty acid composition of each lipid class cannot be determined. Procedures based on chromatographic separation of intact lipids do not possess this limitation and also combine speed, accuracy, precision, and specificity.

Chromatographic separation of lipid classes is usually accomplished by column or thin layer chromatography (TLC) or a combination of the two procedures.[2,3,14] TLC has almost completely replaced paper chromatography because TLC is more versatile and provides better separations. Since TLC is more rapid than column chromatography, it is commonly used in preference to column chromatography whenever possible. All of the lipid classes that are major components and some of the minor components can be determined after separation by two-dimensional TLC without column chromatography.[2,3,14] The speed of the TLC procedure makes it ideal for studies in which many samples must be analyzed. The procedures for lipids of the nervous system were recently described in detail.[3]

Column chromatography prior to TLC provides several advantages. Larger amounts of lipids can be isolated in less time by column chromatography than by TLC, and columns can be used to concentrate minor components not detectable by TLC alone.[2,3] Diethylaminoethyl (DEAE) cellulose column chromatography[2,3,12,13] has become one of the most commonly used chromatographic procedures for separation of lipids of the nervous system. Florisil, a synthetic magnesium silicate, in combination with silicic acid has been used for column chromatographic isolation of cerebroside,[2] and silicic acid has been used to advantage for quantitative recovery of all glycolipids except gangliosides.[17] More recently, triethylaminoethyl (TEAE) cellulose has been shown to possess advantages for lipid class separations.[2,3] The advantages and limitations of the various procedures were described in recent publications.[2,3]

The *fifth and final step* in lipid class analysis is quantitative determination of each lipid class by weighing or measuring the intensity of color produced by an appropriate reaction, most commonly the latter. Procedures recently described in detail[2,3] include phosphorus analysis for phospholipids, carbohydrate assay for glycolipids, hydrolysis and determination of long-chain base of glycosphingolipids by the color produced with trinitrobenzenesulfonic acid, determination of cholesterol by the zinc chloride–acetyl chloride or ferric chloride procedures, determination of cholesterol esters, triglycerides, and free fatty acids (after conversion to methyl esters) by the hydroxamic acid procedure for ester groups, and plasmalogen determination by hydrolysis and aldehyde determination with *p*-nitrophenylhydrazine or gas phase chromatography. Gas phase chromatography is also used for analysis of fatty acids after conversion to their methyl esters.[3,18] The procedure is applicable to nanogram amounts of methyl esters of both nonhydroxy and hydroxy fatty acids.

III. BRAIN LIPID COMPOSITION AT DIFFERENT AGES

The lipid changes in brain during development and aging have been studied fairly extensively. Early work in both man and animals indicated the developmental process to consist primarily of progressive deposition of myelin with increase in the lipids present largely or entirely in myelin. The early conclusions were based on comparisons of composition of gray matter (low in myelin) and white matter (high in myelin), and these have been supported for the most part by recent studies in which myelin was isolated and analyzed. There is general agreement that deposition of myelin in animals and man is associated with increase of cerebroside, sulfatide, sphingomyelin, and ethanolamine plasmalogen. Most studies of human brain have been conducted with samples of gray and white matter. Since such samples obtained from different parts of one brain vary in composition, precise determination of age-dependent changes was not possible.

The compilation by Bürger[19] of human brain weight, total solids, total lipid, and cerebrosides was enlightening. Using averages from many brains, it was shown that males and females differ. Brain weight continued to increase in males up to about age 35, but little change in brain weight of females was noted after about 8 years of age. Average brain weight in both sexes declined in later life. In both sexes, total solids, total lipids and cerebrosides increased into the twenties and decreased in later life.

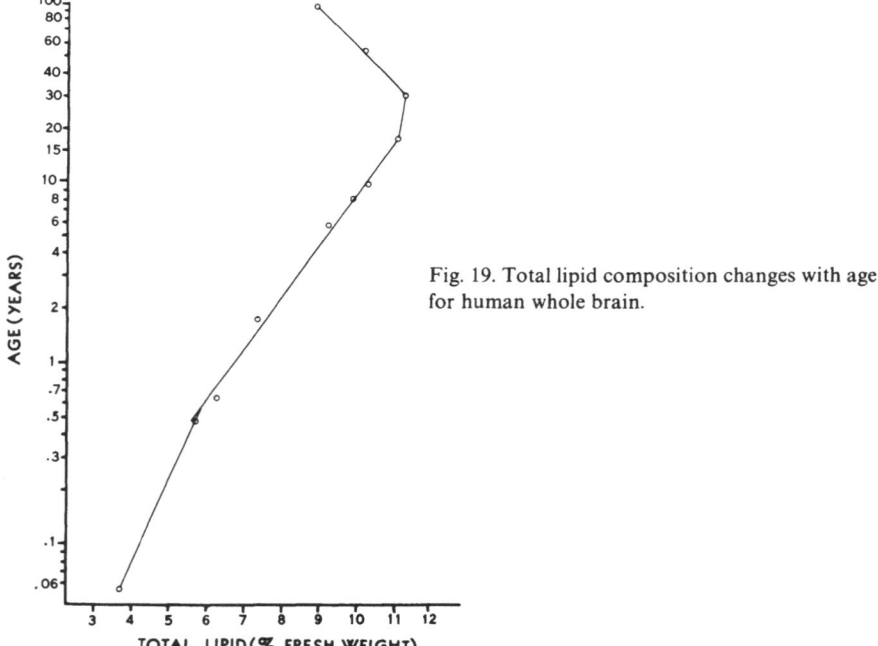

Fig. 19. Total lipid composition changes with age for human whole brain.

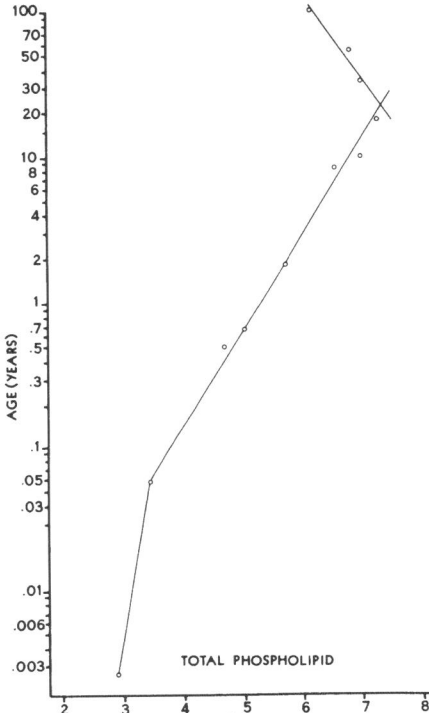

Fig. 20. Total phospholipid changes with
age in human whole brain.

The lipid composition data from recent studies of male, human, whole brain[20–22] give the most complete quantitative picture available for changes during development and aging (fetus to 98 years). The data for total solids, total lipids, and lipid class composition, summarized in graphic form in Figs. 19–23, disclose important features not recognized in earlier investigations employing samples of gray and white matter rather than whole brain. Human brain lipid composition changes continuously throughout life, the changes being greatest during the first few years. The changes in later life (aging) are generally the reverse of those during development. The changes are very predictable and straight line periods are obtained by plotting on semilogarithmic or logarithmic paper. The logarithmic plots show that sphingomyelin and sulfatide continue to increase in absolute amount (millimoles/-100 g fresh wt.) up to age 35. The equations of Table I were calculated for straight-line periods observed from semilogarithmic plots.[20] Calculated values have thus far been within $\pm 2\%$ of measured values. Equations for other expressions are presented elsewhere.[3] These equations can be used to calculate the lipid composition of normal, male, human whole brain at any age. Values calculated from the equations can be used where data by direct

TABLE I

Equations for Calculation of Normal, Male, Human, Whole Brain Water and Lipid Composition[a]

Component	Period	Equation
Water	(1) 1 D–3 W	(1) 85.92–1.374 (log age)
(% fresh wt.)	(2) 3 W–18 Y	(2) 82.69–3.931 (log age)
	(3) 18–33 Y	(3) 76.36 + 0.9514 (log age)
	(4) 33–98 Y	(4) 72.36 + 3.551 (log age)
Total lipid	(1) 1 D–6 M	(1) 5.94 + 1.30 (log age)
(% fresh wt.)	(2) 6 M–26 Y	(2) 6.69 + 3.50 (log age)
	(3) 26–98 Y	(3) 18.6–4.85 (log age)
Lipid phosphorus	(1) 1 D–3 W	(1) 3.93 + 0.393 (log age)
	(2) 3 W–23 Y	(2) 5.30 + 1.55 (log age)
	(3) 23–98 Y	(3) 10.1–1.94 (log age)
Cholesterol	(1) 1 D–3 W	(1) 2.379 + 0.2567 (log age)
	(2) 3 W–6 Y	(2) 3.988 + 1.633 (log age)
	(3) 6–8.5 Y	(3) 8.330 (log age) − 1.202
	(4) 8.5–33 Y	(4) 5.894 + 0.6288 (log age)
	(5) 33–98 Y	(5) 11.27 − 2.900 (log age)
Cholesterol	(1) 1 D–3 W	(1) 0.109 + 0.031 (log age)
ester	(2) 3 W–6 M	(2) 0.0051–0.0530 (log age)
	(3) 6 M–33 Y	(3) 0.0176 − 0.0077 (log age)
	(4) 33–98 Y	(4) 0.0520 (log age) − 0.0713
Cerebroside	(1) 1 D–3 W	(1) 0.1056 + 0.0365 (log age)
	(2) 3 W–6 M	(2) 0.5921 + 0.4273 (log age)
	(3) 6 M–2 Y	(3) 0.6848 + 0.7173 (log age)
	(4) 2–6 Y	(4) 0.5819 + 1.1034 (log age)
	(5) 6–21 Y	(5) 2.171 (log age) − 0.2312
	(6) 21–98 Y	(6) 4.511 − 1.384 (log age)
Sulfatide	(1) 1 D–3 W	(1) 0.0235 + 0.0076 (log age)
	(2) 3 W–6 M	(2) 0.1749 + 0.1293 (log age)
	(3) 6 M–6 Y	(3) 0.2008 + 0.2100 (log age)
	(4) 6–10 Y	(4) 0.8687 (log age) − 0.3230
	(5) 10–40 Y	(5) 0.2427 + 0.3212 (log age)
	(6) 40–98 Y	(6) 2.332–0.9606 (log age)
Ganglioside	(1) 1 D–8 M	(1) 0.154 + 0.0181 (log age)
	(2) 8 M–98 Y	(2) 0.146 − 0.0298 (log age)
Ceramide	(1) 1 D–3 W	(1) constant at 0.046 ± 0.05
	(2) 3 W–6 M	(2) 0.142 + 0.0731 (log age)
	(3) 6 M–1 Y	(3) (−) 0.008 − 0.401 (log age)
	(4) 1–33 Y	(4) 0.0354 + 0.0284 (log age)
	(5) 33–98 Y	(5) 0.281 − 0.123 (log age)
Triglyceride	Birth–98 Y	(1) 0.016 + 0.0038 (log age)
Free fatty acid	(1) 1 D–10 Y	(1) 0.429 + 0.120 (log age)
	(2) 10–18 Y	(2) 0.684 − 0.157 (log age)
	(3) 18–33 Y	(3) 1.26 (log age) − 1.10
	(4) 33–98 Y	(4) 2.59 − 1.17 (log age)

TABLE I (continued)

Component	Period	Equation
Sphingomyelin	(1) 1 D–3 W	(1) 0.1608 + 0.0264 (log age)
	(2) 3 W–38 Y	(2) 0.5040 + 0.3244 (log age)
	(3) 38–98 Y	(3) 1.782 − 0.4903 (log age)
Phosphatidyl	(1) 1 D–33 Y	(1) 1.892 + 0.2121 (log age)
choline	(2) 33–98 Y	(2) 2.760 − 0.4231 (log age)
Phosphatidyl	(1) 1 D–3 W	(1) 1.221 − 0.1291 (log age)
ethanolamine	(2) 3 W–30 Y	(2) 1.804 + 0.6071 (log age)
	(3) 30–98 Y	(3) 4.358 − 1.186 (log age)
Phosphatidyl	(1) 1 D–3 W	(1) 0.5969 + 0.1109 (log age)
ethanolamine	(2) 3 W–30 Y	(2) 1.031 + 0.4614 (log age)
plasmalogen	(3) 30–98 Y	(3) 3.303–1.101 (log age)
Phosphatidyl	(1) 1 D–3 W	(1) 0.5145 + 0.0687 (log age)
serine	(2) 3 W–6 Y	(2) 0.7549 + 0.2603 (log age)
	(3) 6–10 Y	(3) 0.0408 + 1.160 (log age)
	(4) 10–33 Y	(4) constant at 1.23 ± 0.02
	(5) 33–98 Y	(5) 1.960 − 0.4798 (log age)
Phosphatidyl	(1) 1 D–6 Y	(1) 0.1327 + 0.0194 (log age)
inositol	(2) 6–13 Y	(2) 0.0223 + 0.1603 (log age)
	(3) 13–37 Y	(3) 0.3459 − 0.1290 (log age)
	(4) 37–98 Y	(4) 0.1673 (log age) − 0.1161
Phosphatidic acid	(1) 1 D–98 Y	(1) 0.0268 + 0.0098 (log age)
Diphosphatidyl	(1) 1 D–2 Y	(1) 0.0490 + 0.0167 (log age)
glycerol	(2) 2 Y − 98 Y	(2) 0.0619 − 0.0159 (log age)

[a] Millimoles/100 g fresh weight. Abbreviations: D, days; W, weeks, M, months; Y, years. Log age refers in each case to $\log_{10}$ age in years.

analysis are not available for comparison with values for various pathological states.

Earlier studies on the age-dependent changes in fatty acid composition of brain were confined almost exclusively to the sphingolipids. The studies were in general agreement, i.e., the changes in cerebroside, sulfatide, and sphingomyelin consisted of an increase in the longer chain fatty acids, particularly those with 24 carbons.

Recent comprehensive data[22] defined more precisely the general nature of the changes of human brain glycerolphospholipid fatty acids (Fig. 24–28). Discussion of these changes is facilitated using a shorthand designation for chain length and number of double bonds. Thus, e.g., 20:0 designates a 20-carbon acid without a double bond, 18:1 designates an 18-carbon acid with one double bond, etc. In human brain, changes in fatty acid composition of glycerolphospholipids were found to follow a general pattern. The pattern consists of a decrease in 16:0 and 18:0 and polyunsaturated acids (three or more double bonds) with an increase in 18:1 and fatty aldehydes (present almost exclusively in ethanolamine phosphoglycerides, i.e., phosphatidyl

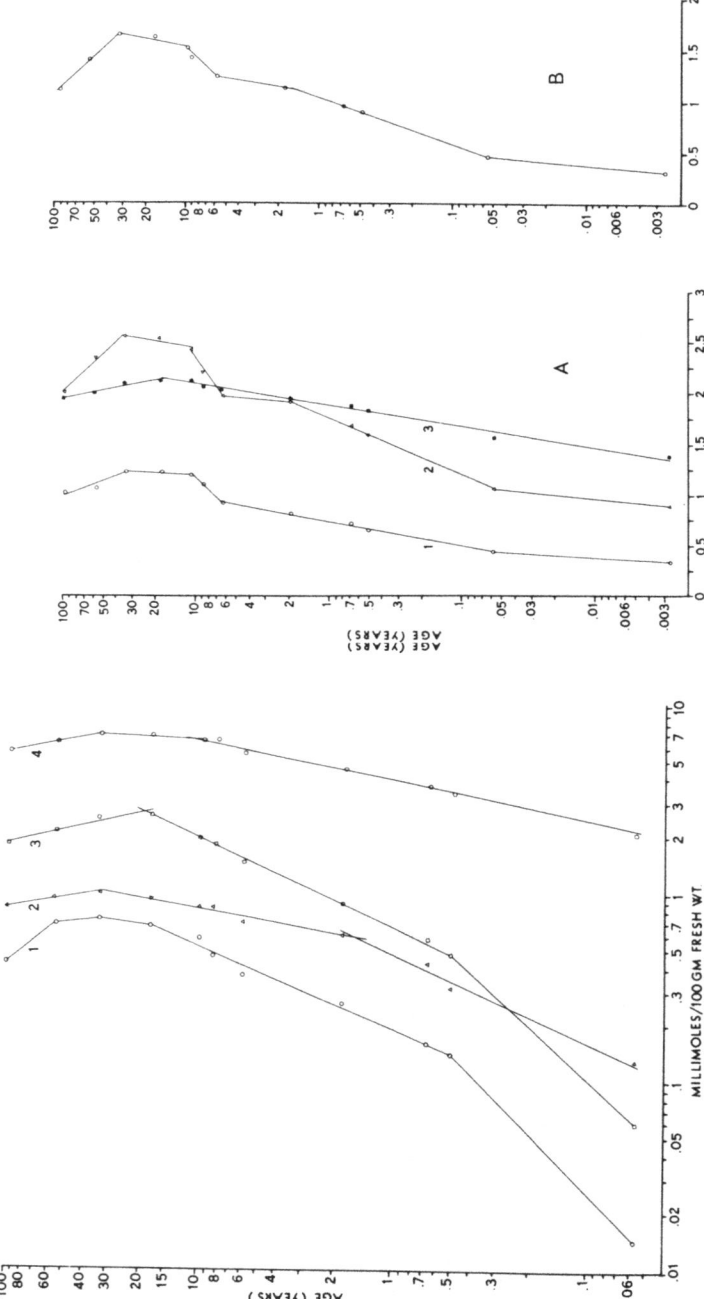

Fig. 22. Changes in brain composition with age. A: (1) phosphatidyl serine; (2) phosphatidyl ethanolamine; and (3) phosphatidyl choline. B: phosphatidyl ethanolamine plasmalogen.

Fig. 21. Changes in human brain composition with age. (1) Sulfatide, (2) sphingomyelin, (3) cerebroside, and (4) cholesterol.

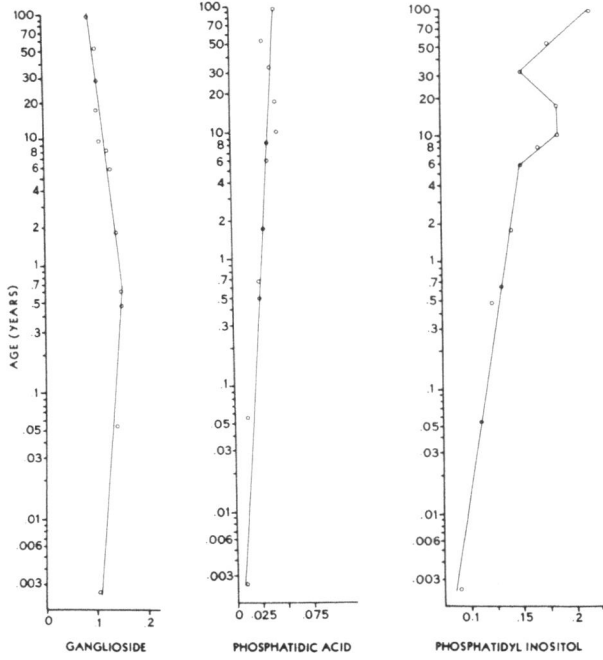

Fig. 23. Changes in brain composition with age for ganglioside, phosphatidic acid, and phosphatidyl inositol.

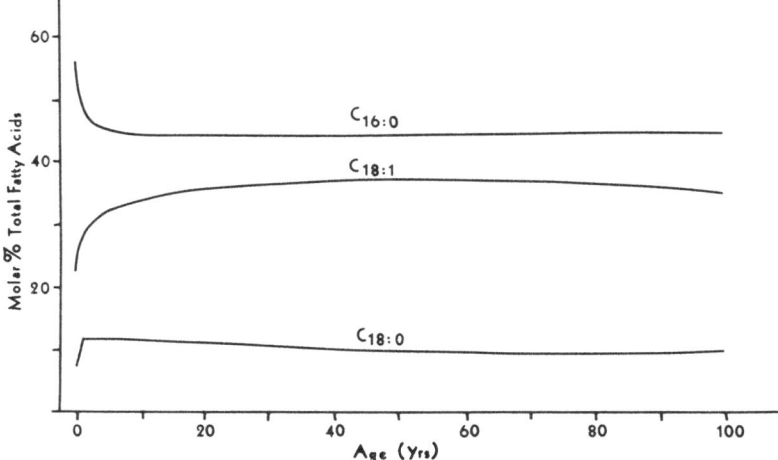

Fig. 24. Changes in human brain phosphatidyl choline—major fatty acids, birth to 100 years.

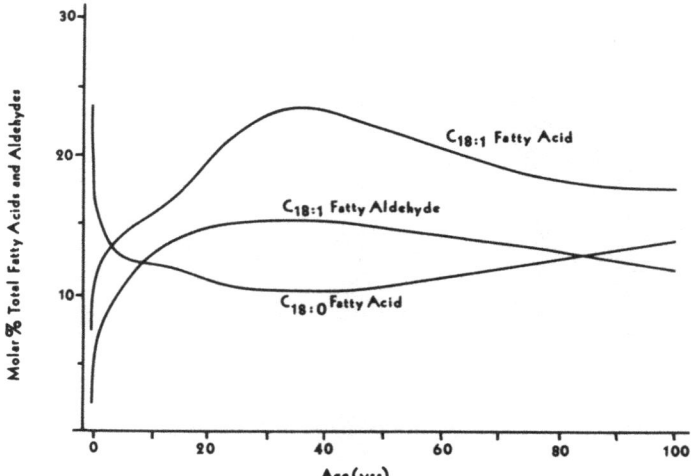

Fig. 25. Changes in human brain phosphatidyl ethanolamine—C_{18} fatty acids and aldehydes, birth to 100 years.

ethanolamine plasmalogen). Rat brain was found[23] to show a different pattern since 18:0 in phosphatidyl choline increased rather than declined.

The changes in human brain sphingolipid fatty acids (Figs. 29–33) were found to follow a general pattern with cerebroside, sulfatide, and sphingomyelin all showing a decrease of shorter chain (up to C_{22}) and an increase of

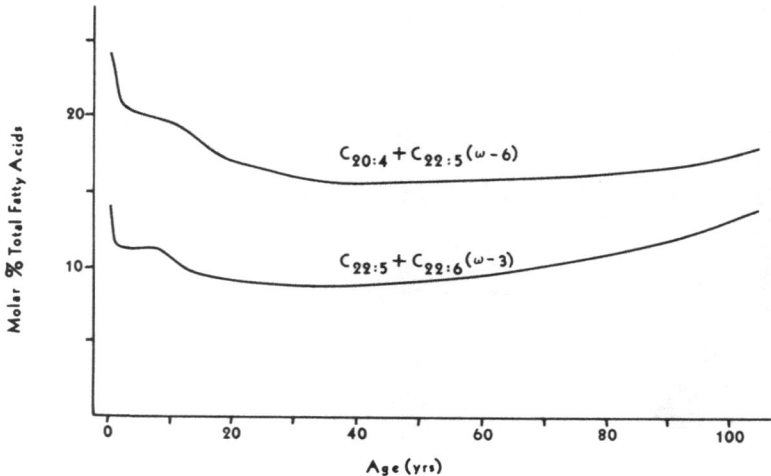

Fig. 26. Changes in human brain phosphatidyl ethanolamine—polyunsaturated fatty acids, birth to 100 years.

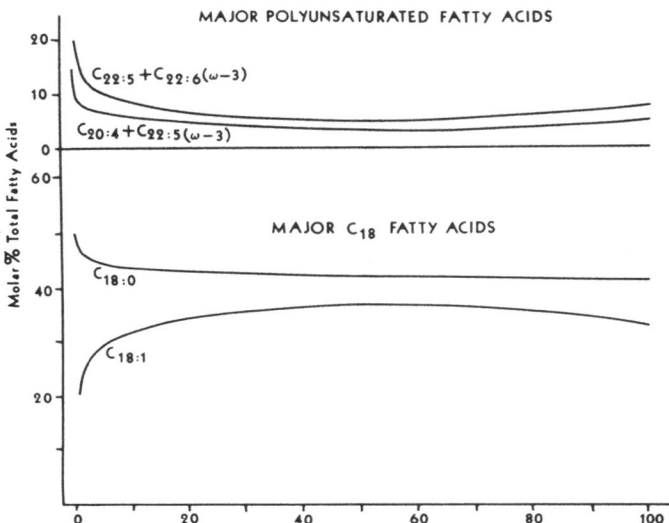

Fig. 27. Changes in human brain phosphatidyl serine—major polyunsaturated fatty acids and major C_{18} fatty acids, birth to 100 years.

longer chain (up to C_{26}) acids, particularly $24:1$. These changes were seen in both the hydroxy and nonhydroxy acids of cerebroside and sulfatide. In sphingomyelin, the increase of $24:0$ in the early years is maintained, but in cerebrosides and sulfatides $24:0$ first rises and then falls. The hydroxy-nonhydroxy fatty acid ratio increases with age.

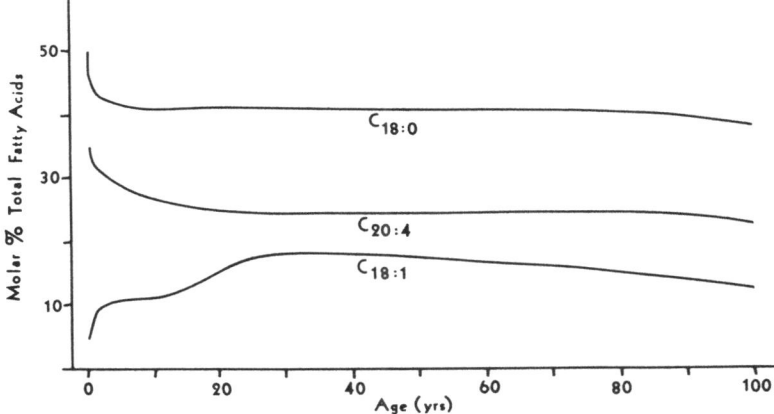

Fig. 28. Changes in human brain phosphatidyl inositol—major fatty acids, birth to 100 years.

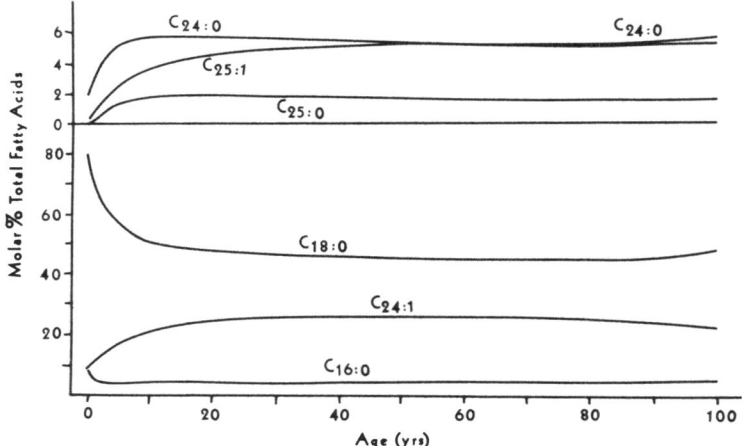

Fig. 29. Changes in human brain sphingomyelin—major fatty acids, birth to 100 years.

The predictable course noted from semilogarithmic plots of lipid class composition changes was also noted for fatty acids of each lipid class. Figure 34 illustrates such a semilogarithmic plot where it can be seen that the line for increase of 18:1 is almost exactly balanced by the line for the sum of

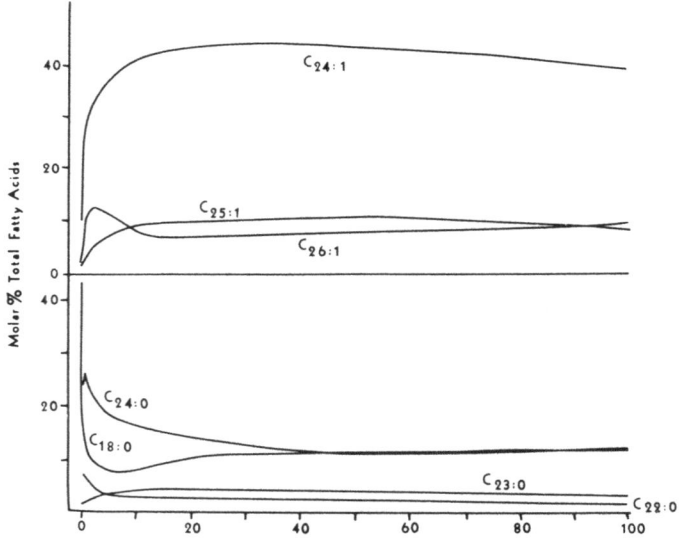

Fig. 30. Changes in human brain cerebroside—normal (nonhydroxy) fatty acids, birth to 100 years.

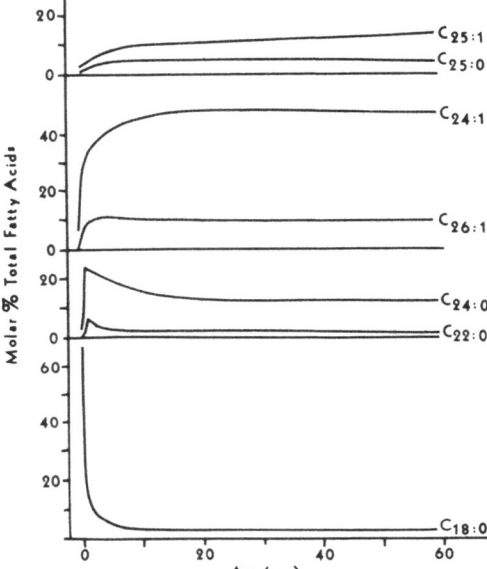

Fig. 31. Changes in human brain sulfatide—normal (nonhydroxy) fatty acids, birth to 60 years.

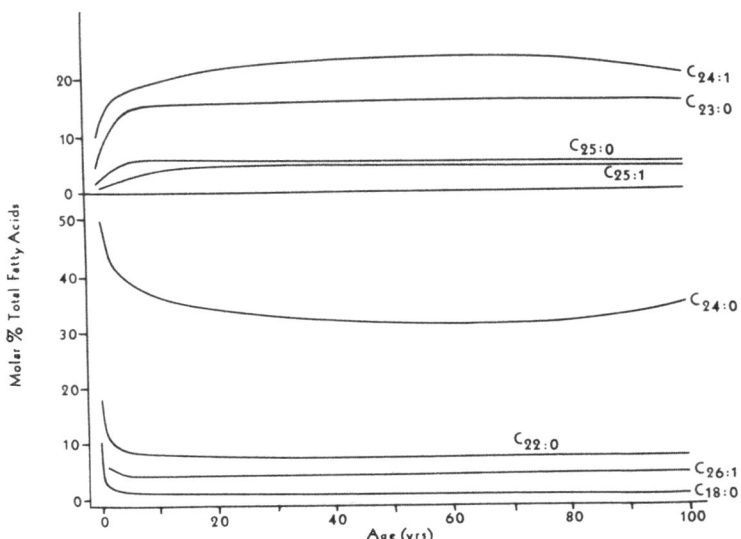

Fig. 32. Changes in human brain cerebroside—hydroxy fatty acids, birth to 100 years.

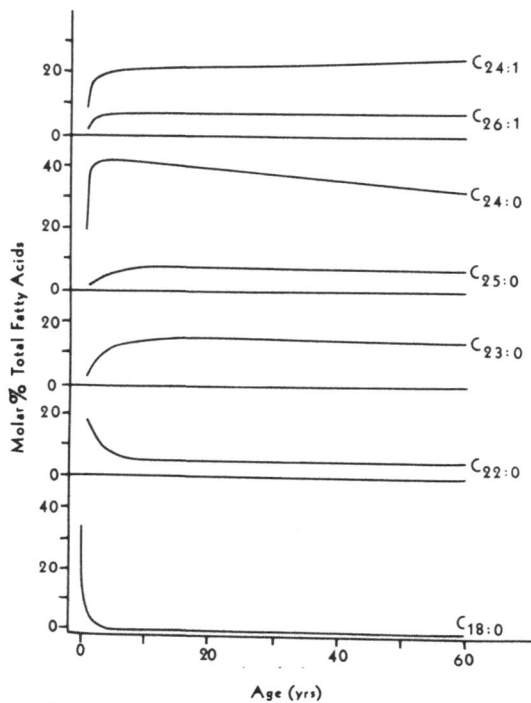

Fig. 33. Changes in human brain sulfatide—2-hydroxy fatty acids, birth to 60 years.

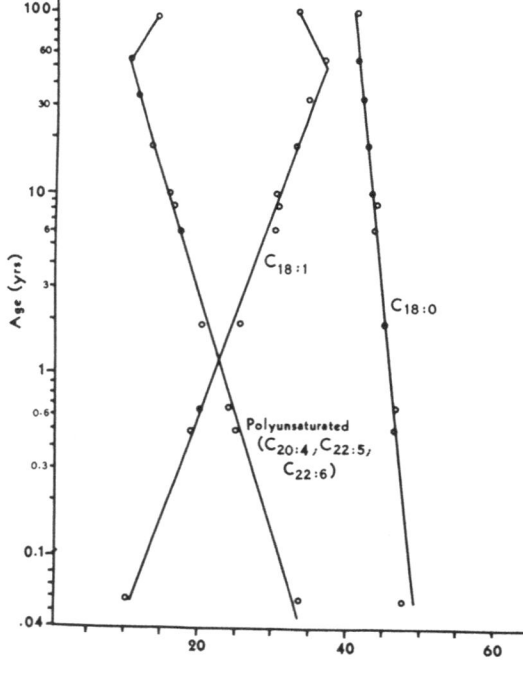

Fig. 34. Semilog plot of human brain phosphatidyl serine—major fatty acids.

the polyunsaturated fatty acids of phosphatidyl serine. Regression equations for the fatty acid changes have not been calculated.

The precision of the developmental and aging processes, when carefully determined with due regard to proper sampling and analytical procedures, provides a firm basis for the contention that accurate and precise procedures should be used. Precise measurement of brain lipid composition now provides a good quantitative measure of the extent to which brain development may have become distorted in pathological states.

IV. LIPIDS OF SUBCELLULAR PARTICLES

Most of the lipid of the nervous system is present in cellular membranes which serve as selective permeability barriers at the surface of cells and within cells at the organelle level. All membranes are composed of structural protein to which lipid is attached. Most membranes have a variety of associated enzyme proteins in addition to structural protein, although myelin is largely devoid of associated enzymes. The membrane-associated enzymes catalyze many reactions, some of which are related to membrane permeability. All cellular membranes contain phospholipid, and some contain, in addition, cholesterol and glycolipid. Triglyceride, a storage form of lipid, is not a true membrane lipid and is found in cells as fat droplets (vacuoles) as are cholesterol esters. An important lipid cofactor, coenzyme Q, clearly occurs in mitochondrial membranes and it seems probable that the lipid vitamins (tocopherols and vitamins A, D, and K) as well as lipid hormones (steroids and prostaglandins) are at least in part on or within membranes. Lipids such as squalene are important metabolites occurring in very small amount in nervous tissue.

Determination of the lipid composition of subcellular particles of brain and other parts of the nervous system is a difficult and challenging task. Procedures must be devised for isolation of pure preparations which must then be shown to possess morphological characteristics at least similar to and preferably identical with those of the same structural unit in the fresh, intact tissue. This must be done by electron microscopy which also gives an indication of the nature and extent of contaminating structures, although the technique is not at present a precise quantitative tool. Morphological observations must be supplemented with chemical and enzymatic demonstration of the presence of characteristic components and the absence of enzymes and other substances characteristic of possible contaminating structures. Following this, careful lipid analysis is required. With such extensive requirements, and despite rapid and important advances, it is thus not surprising that the status of the field is relatively primitive and that opinions regarding lipid composition differ rather widely.

No comprehensive review of the literature on subcellular particles of the nervous system has been presented, although much of the literature on the lipids of subcellular particles of other organs has been reviewed critically,[24] and in a recent treatment of the general subject with extensive new data, a number of important generalizations were presented.[25] At present, the most profitable course appears to be consideration of the general findings with all organs from different species, followed by consideration of special features of the nervous system.

Data obtainable from vertebrate organs indicate the following:

1. Each different subcellular particle or membrane has a characteristic, reproducible composition that differs from other organelles and membranes.

2. Among vertebrates, mitochondrial lipid class composition (i.e., the molar ratios of lipids) does not show organ or species variability except to a minor extent arising from differences in the relative amounts of inner (cristae) and outer membranes which differ in composition, although the total amount of lipid (mg/mg protein) is variable.[26]

3. The relative amounts of the lipid classes of nuclei from different vertebrate organs are similar, if not identical, and little or no species variation is apparent.

4. Rough and smooth endoplasmic reticulum (microsomes) from any one organ appear to have the same lipid class ratios among vertebrates.

5. Endoplasmic reticulum from different organs shows two types of variability. Diphosphatidyl glycerol (cardiolipin), generally found only in mitochondria, is present in endoplasmic reticulum of heart. Endoplasmic reticulum preparations of different organs have different amounts of sphingomyelin with three categories being apparent: high (kidney and brain), medium (heart), and low (liver). The endoplasmic reticulum from any one organ does not appear to show species variation among vertebrates.

6. Plasma membranes show extensive variation from organ to organ and species to species, and it seems probable that glycolipids (cerebrosides, sulfatides, ceramide polyhexosides, and gangliosides) are found only in plasma membranes or membranes derived from the plasma membrane (e.g., myelin).

7. The fatty acid composition of each lipid class shows wide organ and species variability and is influenced by changes in diet.

Caution in interpretation of data for subcellular particles of the nervous system is indicated in view of the multiplicity of cell types and the fact that studies of other organs have shown the lipid class composition of microsomes and plasma membranes from different cell types to be different. Purified fractions of myelin, microsomes (derived from endoplasmic reticulum), ribosomes, mitochondria, nuclei, nerve endings and synaptic vesicles (released from ruptured nerve endings) have been reported. A plasma membrane fraction has not been recognized. On the basis of studies of other organs, little or no variation is to be expected for the lipid class composition (molar ratios) of mitochondria or nuclei from different cell types in the nervous system despite large differences in morphology and enzymatic activity. The other fractions may, however, be expected to vary in composition within one species as well as from species to species. Recently, rat brain myelin was reported to vary in lipid class composition with age.[27] Differences in composition were also noted when different isolation procedures were used. If this expected general variability of lipid class composition does occur, it should be pointed out that even when very reproducible results are obtained by careful application of a given procedure, it may not be possible to determine whether the preparation consists of only one membrane or is a mixture, and whether the same type of membrane with a different composition exists but was not isolated.

Results from different laboratories differ both qualitatively and quantitatively. Confusion exists as to whether or not cerebroside and sulfatide are true components of membranes other than myelin. Most subcellular particle preparations contain both of these glycolipids and most investigators appear to accept these as true components, even of mitochondria.[28,31] A different interpretation has been expressed[3] on the basis of analysis of preparations isolated by procedures described in detail. Mitochondria free of cerebroside and sulfatide were isolated from bovine brain and the phospholipid distribution of the preparations was found to be the same as reported for bovine heart, liver, and kidney mitochondria.[26] The diphosphatidyl glycerol (cardiolipin) content of the mitochondria was much higher than reported from other laboratories, in keeping with the contention[24] that this lipid class is easily underestimated by several analytical procedures. Other structures had low and variable amounts of cerebroside and sulfatide and the amount appeared to correlate rather well with the extent of myelin contamination observed by electron microscopy. It was concluded that cerebroside and sulfatide are present in preparations other than myelin as a result of contamination with myelin. In agreement with this conclusion is the fact that unmyelinated brain is devoid of these glycolipids. Since the unmyelinated brain contains the other subcellular structures, it may be concluded that cerebroside and sulfatide are present only in myelin.

Another controversial matter is the localization of gangliosides. Some authors have interpreted their results as indicating that gangliosides are localized in nerve endings, but this has been denied by others. In our studies[3] gangliosides were found in all subcellular fractions analyzed but the amounts were variable and always much lower than in whole brain lipid extracts. It was concluded that gangliosides occur as contaminants in the subcellular fractions thus far isolated and that they are characteristic and major components of a membrane not yet isolated, very possibly the plasma membrane of the neuron. Neurons contain ganglioside since in Tay–Sachs disease the cells are swollen and contain membranous cytoplasmic bodies of which gangliosides are major components. In a recent microdissection study[32] it was found that gangliosides were more abundant in neurones than glial cells and that the neuropil obtained from immediately around the neuron cell body and dendrites contained a large amount of ganglioside.

It is clear that use of different procedures for isolation and analysis have produced quantitative data of a widely different nature for brain subcellular particles. Some recent values[3] are presented in Table II. From the data it

TABLE II

Phospholipid Composition of Brain Subcellular Particles[a]

	Myelin[b]	Microsomes[c]	Mitochondria[d]	Nuclei[c]	Nerve-ending particles[c]	Synaptic vesicles[c]
Phosphatidyl choline	21.6	47.1	36.1	52.9	38.2	45.6
Phosphatidyl ethanolamine	36.1	25.6	35.8	25.7	33.8	24.9
Phosphatidyl serine	21.6	12.4	Trace	7.2	13.8	13.2
Sphingomyelin	11.3	11.1	ND[e]	5.2	6.2	10.6
Phosphatidyl inositol	1.9	3.4	3.6	7.4	4.8	5.1
Phosphatidic acid	0.8	ND	4.1	0.5	1.6	0.5
Diphosphatidyl glycerol	ND	ND	18.2	ND	1.7	ND
Uncharacterized	6.9		2.0			

[a] Data from reference 3.
[b] Human brain.
[c] Bovine brain.
[d] Bovine and human brains, the values being about identical.
[e] ND denotes not detected.

appears that each subcellular particle has a different, characteristic, and reproducible composition with the reservations noted above for purified preparations from brain. In general, the latter values were different from those reported from earlier studies,[35–37] although better agreement was noted with data in one report.[40]

Bovine and human brain mitochondrial phospholipid distributions were found to be very similar,[3] although the composition found was in contrast to work previously reported in the literature.[28–31] Bovine and human brain nuclei were also found to have similar phospholipid distribution,[3] although the values were different from those reported by most investigators.[33–40]

Despite differences in methodology, it seems clear that myelin lipid composition shows species variation and that myelin from the central nervous system differs from myelin of the peripheral nervous system.[34–48] Of special interest with regard to the composition of myelin is the reported presence of desmosterol,[49,50] cholesterol esters,[51] and polyphosphoinositides.[52] That the polyphosphoinositides are components of plasma membranes or their elaborations (myelin) is also indicated by the finding of the site of diphosphoinositide biosynthesis in rat liver to be the plasma membrane.[53]

V. LIPID CHANGES IN PATHOLOGICAL STATES

A. General Comments

It seems reasonable to suppose that mutations can arise for all enzymes. In general, determination of the levels of blood lipids and demonstration in organs of lipid deposits by histological or histochemical means or both, and to a lesser extent, observation of cutaneous deposits, have directed attention to changes in lipid metabolism. Subsequently, detailed chemical analysis has followed and, when abnormal levels of tissue lipids were found, *in vitro* enzymatic assay has been used to determine the specific enzymatic defect. In the future it is expected that analysis of *in vitro* cell cultures and subcellular particles isolated from organs will play a more prominent role and provide valuable data.

Some of the disorders have been shown or strongly indicated to arise from a genetically determined change in the level of an enzyme of lipid metabolism. In other cases, lipid changes appear to be secondary to a known or unknown directly genetically determined enzyme defect in some other area of metabolism. Secondary (not directly determined by gene mutation) changes may arise from changes in metabolic balance, degeneration, and death of cells. The conditions produced by mutation of a structural gene for an enzyme of lipid metabolism or a control or regulatory gene affecting an enzyme of lipid metabolism have commonly been of most interest to investigators. Even though requiring much time and effort, study of lipid changes secondary to another genetically determined change in another area of metabolism should receive more attention in the future. Careful determination of the secondary changes can disclose the relationship of lipid metabolism to general metabolism and suggest approaches to the control of disorders. From the clinical standpoint, determination of secondary changes may be as important in some cases as knowledge of the primary enzymatic defects. Another important consideration is the fact that, although secondary, some lipid changes may prove to be so characteristic that they may serve to differentiate a disease from other related conditions grouped together as a syndrome and in general perhaps provide confirmation of a diagnosis even when the specific enzymatic defect is not known.

Differentiation of primary defects from secondary changes presents numerous problems. Mutations of genes that control the levels of several enzymes can give rise to multiple enzyme changes. Complex changes may arise from a mutation affecting one enzyme of a multienzyme complex that is active only when all enzyme proteins are together (such as α-keto acid dehydrogenase and some fatty acid synthetases). Such complexes have been considered in detail in a recent publication.[54] An illustration of a complex situation is provided by the "mucopolysaccharide" disorders in which lipid changes are commonly found. Until specific enzyme abnormalities have been demonstrated in the disorders usually classified as mucopolysaccharidoses, it seems unwise to consider the lipid changes as strictly secondary.

In contrast, specific defects for enzymes of glycogen metabolism have been demonstrated in the glycogen deposition disorders and thus the lipid changes are clearly of a secondary nature.

Progress in the understanding of hereditary neuropathological states is slowed by the absence of chemical data defining and separating the many and varied conditions. The great expansion of knowledge and laboratory procedures has made further progress in the area even more dependent upon the cooperation of chemists, clinicians, pathologists, and electron microscopists. Furthermore, some form of classification of the various disorders is desirable as an aid to chemical studies. Although numerous classifications of pathological states have been presented, a general classification based primarily upon consideration of lipid changes in hereditary neuropathological states does not appear to have been reported previously.

The classification presented below was prepared with the major objectives being to summarize and categorize data with regard to lipid changes with pertinent references to the literature. The arrangement of the disorders was chosen in order to emphasize changes in membranes and organelles whether or not lipids prove to be directly or indirectly involved. Membranes which are composed of structural protein, lipid, and membrane-associated enzymes are a complex, the operation of which may be affected by genetically determined changes in any of its components. Abnormal membrane structure may be detected microscopically, or abnormal permeability or excretion may be observed clinically.

Listing of disorders with little or no neurological involvement along with those in which the nervous system is affected seemed essential for several reasons. With this arrangement, emphasis is placed upon the possibilities for study of the mechanisms leading to neurological disturbances by comparing the lipid changes in these with changes in conditions in which the nervous system is not affected. Since, for example, neurological involvement is pronounced in the acute (infantile) form of cerebrosidase deficiency (Gaucher's disease), but is not seen in the less severe chronic form of the disorder, it is possible that other disorders exist in both forms but the acute form with neurological involvement has not been associated with the chronic form of the same disorder. Neurological involvement may not appear immediately and is greatly delayed in some disorders, e.g., abetalipoproteinemia. Thus, chemical studies must at times be conducted without proof of neurological involvement and in the light of lipid changes that may appear without neurological involvement ever becoming apparent. Changes observed in organs other than brain can be observed during life and are often more readily interpreted than those in the nervous system in which degenerative conditions may obscure the specific change in a lipid as is encountered in the more chronic forms of Niemann–Pick disease.

It seemed desirable from the practical standpoint to include follicular lipidoses of the spleen, despite the fact that it appears to arise from ingestion of mineral oil. The condition is frequently observed and could thus be an acquired change superimposed upon and complicating the study of changes in hereditary abnormalities.

The literature can be judged from any one of several viewpoints. If the most critical standards are adopted, many disorders cannot be designated as disease entities. Thus, decisions must often be based upon incomplete information on only one or a few cases which may differ in one or more respects. There is justification therefore for the viewpoint that such disorders should not be admitted to a definite category despite the fact that the information available, although incomplete, would appear to justify such placement on a tentative basis. Procedures differ in their reliability and the skill with which they are used varies. Thus, the validity of some data may be doubted, but complete rejection of it cannot be fully justified until more data become available. Even in the undesirable situation where formalin-fixed tissue was used for chemical analysis, it must be admitted, at least in some cases, that the data may have pointed to the correct answer. Thus, while critical evaluation demands that such data be considered as very probably incorrect, its presence in the literature cannot be denied. The reader is advised to become acquainted with the limitations of procedures and to make his own value judgments which must at present be a matter of opinion and interpretation in many cases. We have sought to present an outline indicating in a sense "the state of the art" without attempting to present detailed evaluations requiring more space and proving in the final analysis to be opinions based upon inconclusive data. Without doubt, better and more extensive data will be obtained and progress will be most rapid when methods with severe limitations are discarded in favor of better procedures.

Several comprehensive books[54–61] provide a great deal of general information on normal and abnormal structure and function of the nervous system and should be useful in conjunction with the literature references cited.

B. Classification of Hereditary Disorders with Reference to Neuropathology and Lipid Metabolism

1. Specific Lipid or Lipoprotein Abnormalities Indicated as the Genetically Determined Defect

a. Fatty Acid Metabolism

(i) *Short chain fatty acid metabolic defect of Sidbury, Smith, and Harlan.*[62] A hereditary condition with nervous system involvement characterized by lethargy, seizures, acidosis, dehydration, an unusual odor similar to that of sweaty feet, and death within the first month of life. Butyric and hexanoic acids occur in body fluids.

(ii) *Branched-chain fatty acid oxidation defect, Refsum's disease* (heredo-ataxica hemeralopia polyneuritiformis; heredopathia atactica polyneuritiformis). A hereditary condition with nervous system involvement in which the accumulation of phytanic acid (3, 7, 11, 15-tetramethylhexade-canoic acid) occurs in tissue lipids[63] and is considered to arise from a partial deficiency of branched-chain fatty acid oxidizing enzyme.[64] Note that phytanic acid was not found in a case that may be considered an example of Refsum's syndrome rather than Refsum's disease.[65]

b. Glycolipid Metabolism

(i) Cerebroside degradation defect, Gaucher's disease. A heredi-tary condition (involvement of the nervous system in the acute, infantile form) with deposition of cerebroside in tissues, arising from a deficiency of the degradative enzyme cerebrosidase[66,67] which hydrolyzes cerebroside to ceramide and free hexose.

(ii) Cerebroside sulfate degradation defect, metachromatic leuko-dystrophy. A hereditary condition with nervous system involvement from sulfatide accumulation in brain and kidney which arises from a deficiency of the sulfatase hydrolyzing sulfatide to cerebroside.[68]

(iii) Cerebroside sulfate biosynthetic defect, globoid cell leuko-dystrophy (Krabbe disease). A hereditary disorder with nervous system in-volvement in which a relative decrease of brain sulfatide has been attributed to deficiency of the enzyme cerebroside sulfotransferase catalyzing sulfate transfer to cerebroside to form sulfatide.[69]

(iv) Ceramide polyhexoside degradation defect, Fabry's disease (angiokeratoma corporis diffusum). A hereditary condition in which the nervous system is not involved. The findings in this disorder were recently described in detail.[70] Accumulation of ceramide trihexoside, dihexoside, and sulfated dihexoside was found in formalinized kidney and a deficiency of ceramide polyhexoside degradative enzyme is indicated.[71] A tenfold increase in ceramide di- and trihexosides was found in urine.[72] The condi-tion merits additional study with complete lipid analysis of all organs em-ploying fresh-frozen specimens, particularly in view of earlier reports of abnormal amounts of phospholipids in some organs.

(v) Monosialoganglioside degradation defect, Tay–Sachs disease (infantile amaurotic familial idiocy). A large accumulation of an abnor-mal monosialoganglioside in brain and to a lesser extent in other organs. The specific enzyme defect is not known, but is presumably for ganglioside degradation.[73]

(vi) Hexosaminidase deficiency leading to deposition of N-acetylhexosamine-containing ceramide tetrasaccharide and the Tay–Sachs type ganglioside.[74] A disorder similar to Tay–Sachs disease was shown to be different in that there was visceral involvement from accumula-tion of a ceramidepolyhexoside, apparently arising from a deficiency of hexosaminidase. This disorder appears to be different from two other forms of systemic late infantile lipidosis in which accumulation of a normal (rather than Tay–Sachs type) ganglioside is found (see Section V, B, 2, a).

c. Phospholipid Metabolism

(i) Sphingomyelin degradation defect, Niemann–Pick disease. A hereditary condition with nervous system involvement in which accumulation of sphingomyelin[75] arises from a deficiency of sphingomyelinase, the degradative enzyme converting sphingomyelin to ceramide and phosphoryl choline.[76,77] Recently, the accumulation of lyso bisphosphatidic acid has been reported.[5] The significance of this observation is not known, although it indicates the disease to be more complex than previously recognized.

Note: Although genetically determined disorders of glycerolphospholipid metabolism are expected, none have been proven to exist. The recent finding[5] of lyso bisphosphatidic acid elevation in Niemann–Pick disease and several other disorders may represent a secondary accumulation or perhaps in some cases a genetically determined defect. The failure to observe glycerolphospholipid abnormalities could arise from failure of glycerolphospholipids to be deposited as such. These lipids, in contrast to sphingolipids, contain polyunsaturated fatty acids and it is thus possible that they are converted to insoluble oxidation products (ceroid, lipofuscin, etc.; see pigment deposition disorders, Section V, B, 2, v). It is also possible that genetic changes in enzymes of glycerolphospholipid metabolism have gone unrecognized because such disorders are of a less severe nature, or conversely that these represent highly lethal mutations fatal at an early age, perhaps in fetal life.

d. Cholesterol Ester Storage.

A familial asymptomatic condition with hepatomegaly associated with deposition of cholesterol ester in liver.[78] Absence of large blood lipid and lipoprotein changes except for elevated bile acids suggests an abnormality of bile metabolism or secretion, or both.

e. Plasma Lipoprotein Abnormalities

(i) Abetalipoproteinemia. A hereditary condition with involvement of the nervous system (also referred to as acanthocytosis or Bassen–Kornzweig syndrome) that is characterized by a deficiency of betalipoprotein.[79] Recent evidence suggests that the protein moiety is present but the lipoprotein is not formed,[80] and the fatty acid composition of erythrocytes suggested the possibility of an essential fatty acid deficiency.[81]

(ii) Familial high density lipoprotein deficiency, Tangiers disease. In this condition without nervous system involvement, hypocholesterolemia and storage of cholesterol esters in foam cells is noted and is associated with a deficiency of high density lipoprotein.[82,83]

f. Plasma Lecithin-Cholesterol Acyltransferase Deficiency.

A hereditary condition without nervous system involvement in which a deficiency of the plasma enzyme catalyzing the transfer of an acyl group from lecithin (phosphatidyl choline) to cholesterol to form cholesterol ester has been noted in patients who also have low plasma cholesterol ester, anemia, proteinuria, foam cells in marrow and glomerular tufts of the kidney, and α-lipoprotein deficiency.[84] A comprehensive review of the literature pertaining to this enzyme was presented recently.[85]

g. Hydrocarbon Deposition

(i) Follicular lipidosis of the spleen.[86] Although this condition is apparently acquired by ingestion of mineral oil, its high incidence makes its appreciation and distinction from other conditions essential since it may be encountered in association with hereditary disorders.

2. Disorders with Lipid Changes of Less Certain Significance

a. Amaurotic Familial Idiocies. It is now apparent that the group of disorders designated as the amaurotic familial idiocies is very heterogeneous and that the original assumption that all forms are disturbances of ganglioside metabolism is not proven. The various clinical and anatomical entities are commonly subdivided according to age of onset and then further subdivided by special features.

(i) Infantile form (Tay–Sachs disease). A proven ganglioside defect (see section V, B, 1, b).

(ii) Late infantile form (Bielschowsky–Jansky disease). Ganglioside deposition[87] has been described as less extensive than in the infantile form but as involving the same type of ganglioside. The tissue was, however, formalin fixed and it is known that this treatment alters ganglioside distribution. A sequence of changes beginning with altered mitochondria and progressing through membrane body formation, and finally lipofuscin has been described.[88] In a myoclonic variant of the late infantile group[89,90] no lipid abnormality was apparent.[89]

(iii) Systemic infantile and late infantile lipidosis. Until recently, cases described as generalized gangliosidosis, Tay–Sachs disease with visceral involvement, neurovisceral lipidosis, and Hurler's syndrome variant were thought to be a single disease entity with abnormal ganglioside metabolism and to be similar to Tay–Sachs disease. It now appears that there are at least three conditions. In one, accumulation of abnormal amounts of Tay–Sachs disease type monosialoganglioside and a ceramide tetrahexoside has been reported to arise from a deficiency of hexosaminidase (see Section V, B, 1, b). In two similar conditions, accumulation of a normal type of monogialoganglioside (rather than the Tay–Sachs type) has been reported. The two conditions are differentiated as follows.[91] Type 1 (infantile) is characterized by neurological deterioration with onset between birth and 5 months of age, visceromegaly, and radiological bone changes similar to those found in Hurler's syndrome. In Type 2, neurological impairment begins somewhat later (7–14 months of age) and hepatosplenomegaly and radiological bone changes are absent. By these criteria, the cases of Landing *et al.*[92] and O'Brien *et al.*[93] are the Type I disorder. Similar cases were described by others.[94,95] Excessive accumulation of mucopolysaccharide has also been

reported[96] and provides additional evidence for the similarity of this condition to Hurler's syndrome. The Type 2 disorder was described by Gonatas and Gonatas[97] with accumulation of a normal type monosialoganglioside in brain and other organs and ceramide tetrahexoside in brain.[98] Another case of Type 2 appears to have been reported[99] and another report[100] in which infantile amaurotic idiocy was attributed to accumulation of an incompletely characterized disialoganglioside may represent the Type 2 disease.

(iv) Juvenile form   (Spielmeyer–Vogt disease). See pigment deposition disorders (Section, V, B, 2, v) for consideration of the disorder usually termed juvenile amaurotic idiocy. It is not certain whether or not one type of juvenile disorder with imidazole aminoaciduria[101] is a different disease entity. The aminoaciduria may simply not be observed in all patients with the same disease.

(v) Adult form   (Kufs disease). Chemical data do not appear to have been reported in the rare adult form.[102]

b. Chediak–Higashi Syndrome.   A familial condition with minimal involvement of the nervous system. The characteristic clinical features are[103] early recognized photophobia, retinal albinism, light skin, gray or light brown hair, recurrent infections, fevers unrelated to recognizable infections, and terminally at least moderate to marked splenomegaly and less striking hepatomegaly. Neurological manifestations are not always observed but can include convulsions, cranial or peripheral neuropathies, and rarely, mental retardation. Characteristic enlarged ("giant") granulations (possibly lyosomes) in leukocytes, widespread histiocytic infiltration of spleen, liver, lymph nodes, brain, and in some cases almost all tissues. Acidic lipid deposition in tissues was observed by staining[103] and reduction of sphingomyelin and glycolipid in leukocytes reported.[104]

c. Coat's Disease.[105,106]   Although the condition involves the eye, it may be generalized. Deposition of cholesterol and cholesterol esters and an abnormal amount of acid mucopolysaccharide were indicated by tissue-section staining procedures.

d. Dementias[107]

(i) Alzheimer's disease (presenile dementia) and senile dementia. These diseases are differentiated primarily on the basis of age of onset. The primary clinical manifestation of both is a slowly progressive mental deterioration. The presence of senile plaques and Alzheimer's neurofibrillary changes in cerebral cortical neurons in a generally atrophic brain is characteristic and commonly complicated by arteriosclerotic vascular disease. These changes are found in other conditions and perhaps particularly noteworthy is their presence in mongoloids without the marked dementia of Alzheimer's disease.[108] Chemical studies[109–113] have shown changes in total lipids, gangliosides, phospholipids, and fatty acids as well as mucopoly-

saccharides, but the changes appear to accompany rather than cause degeneration. Data obtained recently in the authors' laboratory for senile and presenile dementia whole brain lipids demonstrated a uniform pattern of change which consisted of an increase in the phospholipid–glycolipid and cerebroside–sulfatide ratios. The greater decrease of sulfatide could arise from more rapid conversion of sulfatide to cerebroside than breakdown of cerebroside (possibly defective cerebroside degradation) or from defective sulfatide biosynthesis.

(ii) Creutzfeldt–Jacob disease. A general decrease of lipid with a large decrease of gangliosides, a small elevation of cerebroside, and increase of cholesterol have been reported.[114] Further study is important. This disorder has been suggested to be a "slow-virus" disease.

(iii) Pick's disease. No specific information on lipid changes has been reported.

e. Encephalopathy with Fatty Changes in Liver and Spleen. This group of disorders is heterogeneous[115–118] and some forms appear in association with virus infections.

f. Epilepsy Responsive to High Fat Diet. The improvement noted in some epileptic patients on a high fat diet[119] suggests the possibility of a lipid disorder.

g. Familial Hypolipidemia with Retarded Development but Without Steatorrhea. An apparently new disorder[120] requiring further study.

h. Farber's Disease. This relatively rare familial neurological disorder was originally described as a lipogranulomatosis.[121] Recent biochemical and histochemical studies have disclosed accumulation and urinary excretion of mucopolysaccharide, apparently chondrotin sulfate B, and it was suggested that the lipid accumulation is secondary, i.e., arises from degenerative changes.[122] In tissues, an increase of sphingomyelin, ceramide, and total gangliosides with a normal distribution of ganglioside types was found.[123] The large accumulation of ceramide in particular appears to be unique but does not arise from a ceramidase deficiency.

i. Hashimoto's Disease. Although, as far as is known, this disorder is limited to the thyroid, it is included here since lipid changes have been reported as important.[124] A search for lipid changes in other organs could be informative.

j. Heller's Disease. With onset at 2–6 years, neurological involvement develops in a subacute and progressive manner. From chemical studies of body fluids in one case, the condition was postulated to be a variant of Tay–Sachs disease.[125]

k. Huntington's Chorea.[126] Normal blood lipids and brain lipid changes postulated to be secondary to an unknown gene-controlled abnormality were reported.[127,128]

l. Hypocholesterolemia of Lifelong Standing in Idiopathic Steatorrhea with Diffuse Central Nervous System Involvement.[129] An abnormality of lipid metabolism is indicated, but the nature of the abnormality has not been investigated.

m. Kinky Hair Disease. The condition described by Menkes *et al.*[130] as a "sex-linked recessive disorder with retardation of growth, peculiar hair, and focal cerebral and cerebellar degeneration" was later referred to as kinky hair disease[131] and a decrease of only one polyunsaturated fatty acid (22-carbon six double bond) in brain lipid reported. The significance of the observation is unknown.

n. Leukodystrophies. The classification of these disorders has been considered in detail in a general classification of demyelinative diseases.[132] The defects in two conditions, metachromatic leukodystrophy and Krabbe globoid cell leukodystrophy, appear to have been determined (see Section V, B, 1, b). It is apparent from recent publications[133–141] that the complete acceptance of any one classification has not been obtained. The term sudanophilic leukodystrophy is commonly used to designate a rather heterogeneous group of disorders that are referred to by others as Pelizaeus–Merzbacher disease, which is considered by others to be one special type of sudanophilic leukodystrophy. Lipid data presented in the same publications has failed to disclose specific abnormalities, although frequent reference is made to the accumulation of cholesterol esters. No chemical information has been reported for two rare leukodystrophies, spongy degeneration of the white matter (Canavan's disease)[132,142] and megalencephaly with hyaline panneuropathy.[132]

o. Lipidosis of Jervis, Harris, and Menkes.[143] An unusual fatty acid distribution (presence of a large amount of 22-carbon two double bond fatty acid and a low level of 24-carbon fatty acid) was found for spleen glycolipid.[144]

p. Lipidosis of Sawitsky, Human, and Hyman.[145] The condition does not affect the nervous system and there is no indication that it is familial.[146] A characteristic feature of the disorder is the presence of blue (Wright stain) granules in the cytoplasm of histiocytes. The condition may thus be called blue-granule histiocyte disease. Patients may first be examined because of X-ray disclosure of lung abnormalities. Hepatosplenomegaly is present and in some cases thrombocytopenia disappearing after splenectomy has been reported. Mucopolysaccharides were indicated as present in the blue-staining granules[145] which were also reported to contain bound lipid and protein[147] with an elevation of total lipid and sphingomyelin in spleen in one case,[148] and in another elevated cerebroside and sphingomyelin in spleen and increased excretion of acid mucopolysaccharides.[149]

q. *Lipodystrophies.*[150–153] Disturbances of fat metabolism (adipose tissue) that do not involve the nervous system and occur mostly in females. Progressive lipodystrophy is a rare, idiopathic condition characterized by a progressive loss of subcutaneous fat from tissues of the upper part of the body, whereas below the waist fat may increase. There are types which appear as gross accumulations of fat in certain symmetrical regions of subcutaneous tissues. Atrophy of adipose tissue may occur in conjunction with diabetes mellitus. Adipose tissue triglyceride appears to be normal in composition.

r. *Multicentric Reticulohistiocytosis* (lipoid dermatoarthritis). This disorder does not affect the nervous system and does not appear to be familial. It is characterized by mucocutaneous nodules, arthritis, a variety of local and systemic signs and symptoms, and typical histopathology highlighted by multinucleated histiocytic giant cells.[154] Presence in storage cells of triglyceride, cholesterol, cholesterol esters and phospholipid but no glycolipid or mucopolysaccharide has been reported.[155] The disorder has been classified[153] as a granulomatous disease with lipid storage along with histiocytosis X, lipoid proteinosis (Urbach-Wiethe disease), and Farber's disease.

s. *Multiple Sclerosis.* This disorder is thought by some to be of viral etiology and by others to be a disorder of lipid metabolism. An allergic component is known. Brain lipid changes reported include the presence of cholesterol esters and a deficiency of sphingomyelin, plasmalogen, and long chain saturated and unsaturated fatty acids.[156–158] Recently, it has been proposed that lesions are produced from fat embolisms[159–161] and that the cause of the disorder is a deficiency of 22-carbon six double bond fatty acid (22:6), a member of the ω-3 family of acids derived from α-linolenic acid (18:3ω:3).[162]

t. *Necrobiosis Lipoidica.* A skin condition found associated with diabetes (necrobiosis lipoidica diabeticorum) and in nondiabetics (necrobiosis lipoidica) who may, however, be in the prediabetic stage. The lesions and course are similar in both forms.[163]

u. *Neuroaxonal (Infantile) Dystrophy.*[164] The condition is thought by some to be the nonpigmented, infantile form of Hallervorden–Spatz disease, but the differences in the conditions are great enough to make this questionable. Staining indicated the occurrence of lipids in axons (in spheroids) and as deposits in cells of lymph nodes and spleen.

v. *Pigment Deposition Conditions.* Little is known of the chemical nature of the insoluble pigments variously referred to as ceroid or lipofuscin. Lipid pigment deposits occur in a variety of experimentally produced states, are generally greater in older animals, and are prominent in several hereditary metabolic disorders.[165] The deposits react with fat stains and have long been regarded as containing oxidized lipid. Pigment deposits have been produced by antioxidant (tocopherol) deficiency and other means[165] and unsaturated lipid seems always to be implicated in some manner. Thus, even the fatty liver conditions induced by ethanol and carbon tetrachloride[166,167] are prevented by antioxidants. A reasonable hypothesis is that many agents can damage cell membranes, the unsaturated fatty acids of which then undergo more rapid autoxidation and polymerization, giving rise to insoluble pigment deposits. Lipid oxidation products are very reactive toward many substances, including proteins, nucleic acids, sulfhydryl compounds, etc. It seems unlikely that all pigments are the same chemically. Although grouped together here, disorders in which an outstanding feature is ceroid or lipofuscin deposition should not necessarily be thought of as clinically or pathologically similar or related to a genetically determined defect of lipid metabolism. Rather, pigment deposition may be prominent as the result of damage to membranes that may arise in many ways.

(i) Ceroid storage disease in childhood. Slow general development, hepatosplenomegaly, and death by 5 years of age with deposition in various organs of ceroid, particularly massive in one of two cases, was reported.[168]

(ii) Diffuse lipofuscinosis of the adult central nervous system.[169] At onset (49 years of age), progressive cerebellar ataxia, action myoclonus, and epilepsy followed by dementia and death at 55 years. Light and electron microscopic studies of the nervous system disclosed lipofuscin pigment in all neurons and most glial cells. The phospholipids and glycolipids of brain appeared to be normal.[169] It was suggested[169] that other progressive dementias and cerebellar disorders, particularly dyssynergia cerebellaris myoclonica,[170] may be related and show lipofuscin deposition.

(iii) Pigmented lipid histiocytosis and susceptibility to infection.[171-174] A progressive septic granulomatosis in which a pigment different from lipofuscin appears. Evidence has been presented indicating the condition to be a failure to destroy or digest phagocytized bacteria.[174]

(iv) Hallervorden–Spatz disease.[175,176] A familial condition of variable clinical picture with abnormal posture and muscle tone, involuntary movements, and progressive dementia. Accumulation of large amounts of pigmented material in the globus pallidus and zona reticulata of the substantia nigra (grossly visible as brownish discolorations in these regions) is characteristic.

(v) Refsum's syndrome. A condition with the typical features of Refsum's disease (see Section V, B, 1, a) was described[65] in which phytanic acid was not found. In liver, abnormal membranes and elevated lipofuscin were present. The changes observed by electron microscopy suggested lipofuscin to be the end product of mitochondrial degeneration.

(vi) Certain types of infantile and juvenile disorders commonly classified as amaurotic familial idiocies (see Section V B2a). Opinions differ as to nomenclature and differentiation of various forms of some of the disorders occurring in the early years of life. Pigment deposition is a prominent feature of the condition commonly designated as juvenile amaurotic familial idiocy (Spielmeyer–Vogt disease), the features of which have recently been described in detail.[177] Another group of investigators[178] has preferred to use the term Batten's disease or Batten–Spielmeyer–Vogt disease to include conditions considered separately by others. Emphasis has been placed[179] upon pigment deposition and differentiation of a pigment variant of amaurotic familial idiocy from Hallervorden–Spatz disease.

w. Reticuloendothelial Granulomas (Reticuloendothelioses, Histiocytosis X). These conditions are tissue proliferative disorders sometimes accompanied by lipid deposition. The several conditions may be different stages or forms of the same disease. Letterer-Siwe disease and xanthoma disseminatum may be the two extremes with eosinophilic granuloma and Hand–Schüller–Christian disease representing intermediate and special forms.[153] The type of lipid deposited is still somewhat questionable.[180] Lipid values in some organs appear to be normal.

x. Splenic Lipidosis Associated with Thrombocytopenic Purpura. A secondary lipidosis postulated[181] to require for development a genetically determined predisposition involving lipid degradative enzymes, accelerated destruction of the formed elements of the blood (especially platelets), and steroid therapy, especially in massive or prolonged dosage (although possibly not essential in some cases particularly prone to the condition). Accumulation in spleen of phospholipid with an excess of sphingomyelin was reported as characteristic[182,183] but found in only one case in another series.[181] A different but apparently related condition with cardiac amyloidosis has been observed.[184] Splenic lipidosis also occurs in diabetes.[185]

y. Subacute Sclerosing Leukoencephalitis. Suspected to be of viral etiology, viruslike particles are seen by electron microscopy.[186] In a crude myelin fraction, cholesterol was found to be elevated and cerebroside, sulfatide, and phosphatidyl ethanolamine to be decreased.[187] The brain ganglioside pattern was found to be abnormal,[188] and this was suggested to be the result of degenerative changes.

z. Whipple's Disease.[189,190] In this condition, originally described as a lipodystrophy, lipid deposition is noted in the intestinal tract and lymph nodes and blood cholesterol is low. Recent work has placed emphasis upon the presence of mucopolysaccharides and glycoprotein and the possibility that the disorder may be a bacterial infection. Bacilli in the intestinal wall have been observed to grow well in an extracellular matrice that is thought to contain lipid.[191]

a'. Wolman's Disease.[192,193] A hereditary xanthomatosis with calcification of the adrenals. Deposition of cholesterol and triglyceride is found in adrenals, spleen, liver, lymph nodes, bone marrow, small intestine, lungs, and thymus. Death occurs during the first half year of life, presumably from intestinal malabsorption.

3. Disorders of Mucopolysaccharide Metabolism with Lipid Changes

Disorders in which excessive urinary excretion of mucopolysaccharide is found are considered to be primary (genetically determined) mucopolysaccharide defects, although specific enzyme deficiencies have not been demonstrated. Lipid changes are regularly found in these as in some of the glycogen storage diseases. Progress in the classification of polysaccharide disorders has been summarized in recent reviews.[194–196] Six distinct entities have been proposed. The syndromes are[196]: (1) Hurler or gargoylism (chondroitin sulfate B, heparin monosulfate); (2) Hunter (chondroitin sulfate B, heparin monosulfate); (3) Sanfilippo (heparin monosulfate); (4) Morquio–Brailsford–Ullrich (keratosulfate); (5) Schere (chondroitin sulfate B); (6) Maroteaux–Lamy (chondrotin sulfate B). Variants of the disorders have been observed, but their relationships to the other disorders are still uncertain. A useful general classification of connective tissue diseases was reported.[197] An increase of total ganglioside in brain in Hurler's syndrome has been established and the ganglioside distribution has been shown to be different from normal brain,[198] also the level of ceramide polyhexosides is abnormally high.[199]

Other conditions suspected to be disorders of mucopolysaccharide metabolism are psoriasis,[200] lipoid proteinoisis,[201] and hypertrophic interstitial neuritis.[102]

4. Conditions that Merit Examination

Many pathological states have not been examined for possible lipid changes. Some that may be of particular interest are pointed out in this section.

a. Miscellaneous Human Neuropathological States. The group of disorders known as the ataxias present interesting challenges and have not been investigated for possible lipid abnormalities. Ataxia telangiectasia[203–205] can be singled out as of special interest in view of the fact that, as in Fabry's disease (Section V, B, 1, b), proliferation of capillaries of the skin is found. It is thus possible that ataxia telangiectasia is also a disorder of glycolipid metabolism. Vascular abnormalities are commonly found in disorders in which lipid metabolism is altered. In ataxia telangiectasia, a γ-globulin deficiency associated with a defective thymus has been described.[206–208]

Disorders in which peripheral nerve changes are prominent merit careful study, particularly since one well-known condition, Refsum's disease, has been shown to be a fatty acid metabolism defect (see Section V, B, 1, a). Hypertrophic interstitial neuritis of Dejerine and Sottas and the familial tumors (von Recklinghausen's disease and tuberose sclerosis) are of interest. Histological techniques have suggested one form of hypertrophic interstitial neuritis to be a mucopolysaccharide disorder.[209] Classification of the neuropathies has presented various problems,[210–212] and different disease entities may perhaps be distinguished only by careful chemical studies. In this regard it is to be noted that a condition with the general features of Refsum's disease has been studied and phytanic acid was not found to accumulate.[65] The distinction between a disease and a syndrome may become apparent only after chemical analysis.

Kuru[213] is a condition presenting such interesting and complex features as restriction to a part of New Guinea, appearing in male and female children but only in adult females, and an abrupt onset of neurological disturbance with rapid progression to a fatal end. The disorder is a major cause of death in the area in which it is found. Although thought perhaps to be of viral etiology, the condition has been described as a hormonally programmed central nervous degeneration that might involve membranes and enzymes of lipid metabolism.[214]

Examination of brain lipids in developmental anomalies such as the microencephalies should provide interesting data and might disclose new lipid disorders. Examination of lipids in nervous system disorders induced by nutritional deficiencies and drugs may provide important data, allowing better interpretation of the role of lipids in normal structure and function and in hereditary metabolic diseases. Study of disorders, such as mongolism (Down's syndrome), now known to be chromosomal abnormalities might also disclose important deviations in membrane lipid composition. The finding[215] by electron microscopy of structural abnormalities of axons and synapses in neuropsychiatric disorders suggest membrane abnormalities.

b. Amino Acid Metabolism and Epithelial Transport Abnormalities.[216] Abnormalities of excretion and transport in general may well be associated with or arise from defects in membranes and hence possibly of lipid metabolism. Disorders of amino acid metabolism, including phenylketonuria, hyperglycinurias, maple syrup urine disease, i.e., branched-chain ketonuria, histidinemia, proline and hydroxyproline disorders, and sulfur amino acid defects, produce neurological disorders that are unexplained.

A second group of disorders recently grouped together[216] as abnormal epithelial transport states includes hereditary vitamin D-resistant rickets with hypophosphatemia, the Fanconi syndrome, renal glycosuria, renal tubular acidosis, vasopressin-resistant diabetes insipidus, cystinuria, Hartrup disease, and cystic fibrosis of the pancreas.

c. Muscle Disorders. Aside from the muscular dystrophies and myofibril disorders (central core diseases and nemaline myopathy), several disorders in which organelle and membrane changes are prominent are of special interest. Mitochondrial abnormalities[217] include Luft's disease in which defective coupling of respiration and oxidative phosphorylation gives rise to a high basal metabolic rate in the absence of thyroid abnormalities, megaconial myopathy with giant (10^3 times normal) mitochondria, and pleoconial myopathy with increased numbers of mitochondria. A new disorder was described[218] as "a generalized disorder of the nervous system, skeletal muscle and heart resembling Refsum's disease and Hurler's syndrome," but no accumulation of phytanic acid or mucopolysaccharide was found. Abnormal inclusions (zebra bodies, perhaps abnormal lysomes) and mitochondria[219] were observed, however. In still another variant,[220] mitochondrial enzyme hyperactivity was reported. It was noted that abnormally large mitochondria have been observed in heart muscle of rats on polyunsaturated fatty acid-deficient diets.[221] It should be noted also that mitochondrial inclusions are found in the Sanfilippo type of mucopolysaccharidosis and other conditions in man and animals.[222]

The familial periodic paralyses have been divided into three types: hypo, normal, and hyperkalemic.[223-229] The hyperkalemic form is also known as adynamia episodica hereditaria. In these conditions, abnormalities of membrane transport are apparent and it has been suggested that the endoplasmic reticulum is abnormal.[217]

In human progressive muscular dystrophy of the Duchenne type, a change was reported[230] in phospholipid (lecithin) fatty acid composition with increase of oleic acid and decrease of linoleic acid. Such changes may not be specific because a similar pattern was observed in hyperthyroid rats.[231]

d. Glycogen Deposition Diseases. The nervous system is involved prominently in one of the glycogen deposition disorders (Type 2 of Cori, idiopathic generalized glycogenosis, Pompe's disease, neuromuscular form of generalized glycogenosis, etc.). Prominent changes in lipid metabolism are noted in several glycogen deposition diseases,[232] but the reasons for the changes have not been fully determined. Of general interest are the reports of deficiencies of more than one enzyme and induction of enzyme activity.[233]

e. Disorders in Experimental Animals. In general, hereditary neuropathological states in experimental animals have received little attention. Outstanding exceptions are the studies of scrapie, a viral disorder in sheep, and the production of ceroid and similar lipid pigments in several species. Many conditions are found in experimental animals that are clearly similar to conditions in humans.[234-236] Of interest is the metachromatic leukodystrophy in mink in which both brain cerebroside and sulfatide fail to increase normally during development[237] and the counterpart in mink of the Chediak–Higashi syndrome.[238] In dogs, both the Krabbe type (globoid cell) leukodystrophy[239] and amaurotic familial idiocy[240] are found. A large number of neurological mutants of mice are also available, although possible change in lipid metabolism has not been sought. It is apparent that hereditary disorders in experimental animals should be studied more extensively.

VI. ACKNOWLEDGMENTS

The participation of Drs. Claude Baxter, Gene Kritchevsky, and Gerald Simon in work on the comparative aspects of brain lipid composition and Dr. A. N. Siakotos on brain subcellular particles is gratefully acknowledged, as are the helpful comments and criticisms of Dr. George Edgar. This work was supported in part by U.S. Public Health Service Grants NB-01847-10 and NB-06237-12 from the National Institute of Neurological Diseases and Blindness, and Contract DA-18-035-AMC-335 (A) from the U.S. Army Edgewood Arsenal, Maryland.

VII. REFERENCES

1. R. Paoletti, R. Fumagalli, E. Grossi, and P. Paoletti, Studies on brain sterols in normal and pathological conditions, *J. Am. Oil Chemists' Soc.* **42**:400–404 (1965).
2. G. Rouser, G. Kritchevsky, and A. Yamamoto, Column chromatographic and associated procedures for separation and determination of phosphatides and glycolipids, *in Lipid Chromatographic Analysis* (G. V. Marinetti, ed.), Vol. I, pp. 99–162, Marcel Dekker, New York (1967).
3. G. Rouser, G. Kritchevsky, A. N. Siakotos, and A. Yamamoto, Determination of lipid composition of brain and its subcellular structures, *in An Introduction to Neuropathology: Method and Diagnosis* (C. G. Tedeschi, ed.), Little, Brown, Boston (in press).
4. H. Debuch and G. Wendt, Über eine neue Gruppe von colaminhaltigen Glycerinphosphatiden aus Gehirn, *Hoppe–Seyler's Z. Physiol. Chem.* **348**:471–474 (1967).
5. G. Rouser, G. Kritchevsky, A. Yamamoto, A. G. Knudson, Jr., and G. Simon, Accumulation of a glycerolphospholipid in classical Niemann–Pick disease, *Lipids* **3**:287–290 (1968).
6. M. A. Wells and J. C. Dittmer, The identification of glycerophosphorylglycerol phosphate as the deacylation product of a new brain lipid, *J. Biol. Chem.* **241**:2103–2105 (1966).
7. E. Klenk and M. Doss, Über das Verkommen von Estercerebrosiden im Gehirn, *Hoppe-Seyler's Z. Physiol. Chem.* **346**:296–298 (1966).
8. Y. Kishimoto, M. Wajda, and N. S. Radin, 6-Acyl galactosyl ceramides of pig brain: Structure and fatty acid composition, *J. Lipid Res.* **9**:27–33 (1968).
9. A. L. Prensky, The relationship of biochemical and morphological information in the central nervous system: The problem of sampling, *J. Am. Oil Chemists' Soc.* **44**:667–679 (1967).
10. G. Rouser, C. Galli, and G. Kritchevsky, Lipid class composition of normal human brain and variations in metachromatic leucodystrophy, Tay–Sachs, Niemann–Pick, chronic Gaucher's and Alzheimer's diseases, *J. Am. Oil Chemists' Soc.* **42**:404–410 (1965).
11. J. Folch, M. Lees, and G. H. Sloane–Stanley, A simple method for the isolation and purification of total lipides from animal tissues, *J. Biol. Chem.* **226**:497–509 (1957).
12. G. Rouser, A. J. Bauman, G. Kritchevsky, D. Heller, and J. O'Brien, Quantitative chromatographic fractionation of complex lipid mixtures: Brain lipids, *J. Am. Oil Chemists' Soc.* **38**:544–555 (1961).
13. G. Rouser, G. Kritchevsky, D. Heller, and E. Lieber, Lipid composition of beef brain, beef liver, and the sea anemone: Two approaches to quantitative fractionation of complex lipid mixtures, *J. Am. Oil Chemists' Soc.* **40**:425–454 (1963).

14. G. Rouser, G. Kritchevsky, C. Galli, A. Yamamoto, and A. Knudson, Variation in lipid composition of human brain during development and in sphingolipidoses: Use of two-dimensional thin layer chromatography, *in Inborn Disorders of Sphingolipid Metabolism* (S. M. Aronson and B. W. Volk, eds.), pp. 303–316, Pergamon Press, New York (1967).

15. A. N. Siakotos and G. Rouser, Analytical separation of nonlipid water soluble substances and gangliosides from other lipids by dextran gel chromatography, *J. Am. Oil Chemists' Soc.* **42**:913–919 (1965).

16. O. H. Lowry, N. J. Rosebrough, A. L. Farr, and R. J. Randall, Protein measurement with the Folin phenol reagent, *J. Biol. Chem.* **193**:265–275 (1951).

17. G. Rouser, G. Kritchevsky, G. Simon, and G. Nelson, Quantitative analysis of brain and spinach leaf lipids employing silicic acid column chromatography and acetone for elution of glycolipids, *Lipids* **2**:37–40 (1967).

18. G. L. Feldman and G. Rouser, Ultramicro fatty acid analysis of polar lipids: Gas-liquid chromatography after column and thin layer chromatographic separation, *J. Am. Oil Chemists' Soc.* **42**:290–293 (1965).

19. M. Bürger, Die chemische biomorphose des menschlichen Zentralnervensystems, *Medizinische* **1**:561–567 (1956).

20. G. Rouser and A. Yamamoto, Curvilinear regression course of human brain lipid composition changes with age, *Lipids* **3**:284–287 (1968).

21. G. Rouser and A. Yamamoto, Changes in human brain during development and aging: Total lipid and lipid class composition (in preparation).

22. A. Yamamoto and G. Rouser, Changes in human brain during development and aging: Fatty acids of phospholipids and glycolipids (in preparation).

23. E. Marshall, R. Fumagalli, R. Niemiro, and R. Paoletti, The change in fatty acid composition of rat brain phospholipids during development, *J. Neurochem.* **13**:857–862 (1966).

24. S. Fleischer and G. Rouser, Lipids of subcellular particles (A review with new data), *J. Am. Oil Chemists' Soc.* **42**:588–607 (1966).

25. G. Rouser, G. J. Nelson, S. Fleischer, and G. Simon, Lipid composition of animal cell membranes, organelles and organs, *in Biological Membranes* (D. Chapman, ed.), Academic Press, New York, pp. 5–69 (1968).

26. S. Fleischer, G. Rouser, B. Fleischer, A. Casu, and G. Kritchevsky, Lipid composition of mitochondria from bovine heart, liver, and kidney, *J. Lipid Res.* **8**:170–180 (1967).

27. L. F. Eng and E. P. Noble, The maturation of rat brain myelin, *Lipids* **3**:157–162 (1968).

28. P. Parsons and R. E. Basford, Brain mitochondria. VI. The composition of bovine brain mitochondria, *J. Neurochem.* **14**:823–840 (1967).

29. M. L. Cuzner, A. N. Davison, and N. A. Gregson, Turnover of brain mitochondrial membrane lipids, *Biochem. J.* **101**:618–626 (1966).

30. K. H. Slotta, Mitochondrial phospholipids in rats of various ages, *J. Gerontol.* **18**:326–330 (1963).

31. S. Lovtrup and L. Svennerholm, Chemical properties of brain mitochondria, *Exptl. Cell Res.* **29**:298–313 (1963).

32. D. M. Derry and L. S. Wolfe, Gangliosides in isolated neurones and glial cells, *Science* **158**:1450–1452 (1967).

33. T. Tsumita and M. Iwanaga, Chemical composition of isolated nuclei from bovine brain cortex and white matter, *Japan. J. Exptl. Med.* **35**:11–21 (1965).

34. J. L. Nussbaum, R. Bieth, and P. Mandel, Phosphatides in myelin sheaths and repartition of sphingomyelin in the brain, *Nature* **198**:586–587 (1963).

35. L. M. Seminario, N. Hren, and C. J. Gomez, Lipid distribution in subcellular fractions of the rat brain, *J. Neurochem.* **11**:197–207 (1964).

36. E. M. Kreps, K. G. Manukyan, M. V. Patrikeeva, A. A. Smirnov, N. Y. Chenykaeva, and E. V. Chirkovskaya, Phospholipids in subcellular fractions of chicken brain during ontogenesis, *Zhurnal Evolyutsionnoi Bipkh. i Fiziol.* 1:16 (1965) (*Fed. Proc. Translation Supplement* 25:T277–282 (1966)).

37. M. L. Cuzner, A. N. Davison, and N. A. Gregson, The chemical composition of vertebrate myelin and microsomes, *J. Neurochem.* 12:469–481 (1965).

38. E. F. Soto, L. S. de Bohner, and M. D. C. Calvino, Chemical composition of myelin and other subcellular fractions isolated from bovine white matter, *J. Neurochem.* 13:989–998 (1966).

39. J. N. Cumings, E. J. Thompson, and H. Goodwin, Sphingolipids and phospholipids in microsomes and myelin from normal and pathological brains, *J. Neurochem.* 15:243–248 (1968).

40. J. Eichberg, Jr., V. P. Whittaker, and R. M. C. Dawson, Distribution of lipids in subcellular particles of guinea-pig brain, *Biochem. J.* 92: 91–100 (1964).

41. J. S. O'Brien and E. L. Sampson, Lipid composition of the normal human brain: gray matter, white matter, and myelin, *J. Lipid Res.* 6:537–544 (1965).

42. L. A. Autilio, W. T. Norton, and R. D. Terry, The preparation and some properties of purified myelin from the central nervous system, *J. Neurochem.* 11:17–27 (1964).

43. M. L. Cuzner, A. N. Davison, and N. A. Gregson, Chemical and metabolic studies of rat myelin of the central nervous system, *Ann. N.Y. Acad. Sci.* 122:86–94 (1965).

44. W. T. Norton and L. A. Autilio, Chemical composition of bovine CNS myelin, *Ann. N.Y. Acad. Sci.* 122:77–85 (1965).

45. F. Wolfgram, Macromolecular constituents of myelin, *Ann. N.Y. Acad. Sci.* 122:104–115 (1965).

46. E. B. Thompson and M. W. Kies, Current studies on the lipids and proteins of myelin, *Ann. N.Y. Acad. Sci.* 122:129–147 (1965).

47. H. Pilz and E. Mehl, Untersuchungen zur Lipoidzusammensetzung des menschlichen Myelins, *Z. Physiol. Chem.* 346:306–309 (1966).

48. J. S. O'Brien, E. L. Sampson, and M. B. Stern, Lipid composition of myelin from the peripheral nervous system. Intradural spinal roots, *J. Neurochem.* 14:357–365 (1967).

49. N. L. Banik and A. N. Davison, Desmosterol in rat brain myelin, *J. Neurochem.* 14:594–596 (1967).

50. M. E. Smith, R. Fumagalli, and R. Paoletti, The occurrence of desmosterol in myelin of developing rats, *Life Sci.* 6:1085–1091 (1967).

51. F. Young and F. H. Hulcher, Cholesterol esters in myelin and the component fatty acids, *Proc. Soc. Exptl. Biol. Med.* 123:385–387 (1966).

52. J. Eichberg and R. M. C. Dawson, Polyphosphoinositides in myelin, *Biochem. J.* 96:644–650 (1965).

53. R. H. Michell and J. N. Hawthorne, The site of diphosphoinositide synthesis in rat liver, *Biochem. Biophys. Res. Commun.* 21:333–338 (1965).

54. G. C. Quarton, T. Melnechrik, and F. O. Schmitt, eds., *The Neurosciences*, Rockefeller University Press, New York (1967).

55. C. W. M. Adams, ed., *Neurohistochemistry*, Elsevier, New York (1965).

56. R. L. Friede, *Topographic Brain Chemistry*, Academic Press, New York (1966).

57. S. M. Blinkov and I. I. Glezer, *The Human Brain in Figures and Tables* (*A Quantitative Handbook*), Basic Books and Plenum Press, New York (1968).

58. J. B. Stanbury, J. B. Wyngaarden, and D. S. Fredrickson, eds., *The Metabolic Basis of Inherited Disease*, 2nd ed., Blakiston Div.-McGraw Hill, New York (1966).

59. T. R. Harrison, R. B. Adams, I. L. Bennett, W. H. Resnik, G. W. Thorn, and M. M. Wintrobe, eds., *Principles of Internal Medicine*, Blakiston Div.-McGraw Hill, New York (1966).

60. R. T. C. Pratt, *The Genetics of Neurological Disorders*, Oxford University Press, New York (1967).
61. W. Blackwood, W. H. McMenemey, A. Meyer, R. N. Norman, and D. S. Russel, eds., *Greenfield's Neuropathology*, Williams and Wilkins, Baltimore (1963).
62. J. B. Sidbury, E. K. Smith, and W. Harlan, An inborn error of short-chain fatty acid metabolism, *J. Pediat.* **70**:8–15 (1967).
63. E. Klenk and W. Kahlke, Über das Vorkommen der 3,7,11,15-Tetramethyl-hexadecansäure in den Cholesterinestern und anderen Lipoidfraktionenen der Organe bei einem Krankheitsfall unbekannter Genese (Berdacht auf Heredopathia atactica polyneuritiformis) (Refsum-Syndrom), *Hoppe–Seyler's Z. Physiol. Chem.* **333**:133–139 (1963).
64. D. Steinberg, F. Q. Vroom, W. K. Engel, J. Cammermeyer, C. E. Mize, and J. Avigan, Refsum's disease—a recently characterized lipidosis involving the nervous system: combined clinical staff conference at the National Institutes of Health, *Ann. Int. Med.* **66**:365–395 (1967).
65. E. H. Kolodny, W. K. Hass, B. Lane, and W. D. Drucker, Refsum's syndrome, *Arch. Neurol.* **12**:583–596 (1965).
66. D. S. Fredrickson, Cerebroside lipidosis: Gaucher's disease, *in The Metabolic Basis of Inherited Disease* (J. B. Stanbury, J. B. Wyngaarden, and D. S. Fredrickson, eds.), Ch. 27, pp. 565–585, McGraw-Hill, New York (1966).
67. R. O. Brady, J. N. Kanfer, R. M. Bradley, and D. Shapiro, Demonstration of a deficiency of glucocerebroside-cleaving enzyme in Gaucher's disease, *J. Clin. Invest.* **45**:1112–1115 (1966).
68. J. Austin, D. McAfee, D. Armstrong, M. O'Rourke, L. Shearer, and B. Bachhawat, Abnormal sulphatase activities in two human diseases (metachromatic leukodystrophy and gargoylism), *Biochem. J.* **93**:15C–17C (1964).
69. B. K. Bachhawat, J. Austin, and D. Armstrong, A cerebroside sulphotransferase deficiency in a human disorder of myelin, *Biochem. J.* **104**:15C–17C (1967).
70. C. C. Sweeley and B. Klionsky, Glycolipid lipidosis: Fabry's disease, *in The Metabolic Basis of Inherited Disease* (J. B. Stanbury, J. B. Wyngaarden, and D. S. Fredrickson, eds.), Ch. 29, pp. 618–632, McGraw-Hill, New York (1966).
71. R. O. Brady, A. E. Gal, R. M. Bradley, E. Martensson, A. L. Warshaw, and L. Laster, Enzymatic defect in Fabry's disease, *N. Engl. J. Med.* **276**:1163–1167 (1967).
72. G. J. Kremer and R. Denk, Angiokeratoma corposis diffusum (Fabry) Lipoidchemische Untersuchungen des Harnsedimenst, *Klin. Wschr.* **46**:24–26 (1968).
73. D. S. Fredrickson and E. G. Trams, Ganglioside lipidosis: Tay–Sachs disease, *in The Metabolic Basis of Inherited Disease* (J. B. Stanbury, J. B. Wyngaarden, and D. S. Fredrickson, eds.), Ch. 25, pp. 523–538, McGraw-Hill, New York (1966).
74. K. Sandhoff, U. Andreae, and H. Jatzkewitz, Deficient hexosaminidase in an exceptional case of Tay–Sachs disease with additional storage of kidney globoside in visceral organs, *Life Sci.* **7**:283–288 (1968).
75. D. S. Fredrickson, Sphingomyelin lipidosis: Niemann-Pick disease, *in The Metabolic Basis of Inherited Disease* (J. B. Stanbury, J. B. Wyngaarden, and D. S. Fredrickson, eds.), Ch. 28, pp. 586–617, McGraw-Hill, New York (1966).
76. R. O. Brady, J. N. Kanfer, M. N. Mock, and D. S. Fredrickson, The metabolism of sphingomyelin. II. Evidence of an enzymatic deficiency in Niemann-Pick disease, *Proc. Natl. Acad. Sci. U.S.* **55**:366–369 (1966).
77. P. B. Schneider and E. P. Kennedy, Sphingomyelinase in normal human spleens and in spleens from subjects with Niemann–Pick disease, *J.Lipid Res.* **8**:202–209 (1967).
78. L. Schiff, W. K. Schubert, A. J. McAdams, E. L. Spiegel, and J. F. O'Donnell, Hepatic cholesterol ester storage disease, a familial disorder, *Am. J. Med.* **44**:538–546 (1968).

79. J. W. Farquhar and P. Ways, Abetalipoproteinemia, *in The Metabolic Basis of Inherited Disease* (J. B. Stanbury, J. B. Wyngaarden, and D. S. Fredrickson, eds.), Ch. 24, pp. 509–522, McGraw-Hill, New York (1966).

80. R. S. Lees, Immunological evidence for the presence of B protein (apoprotein of α-lipoprotein) in normal and abetalipoproteinemic plasma, *J. Lipid Res.* **8**:396–405 (1967).

81. G. B. Phillips and J. T. Dodge, Phospholipid and phospholipid fatty acid and aldehyde composition of red cells of patients with abetalipoproteinemia (acanthocytosis). Evidence for essential fatty acid deficiency in man, *J. Lab. Clin. Med.* **71**:629–653 (1968).

82. H. N. Hoffman II and D. S. Fredrickson, Tangier disease (familial high density lipoprotein deficiency). Clinical and genetic features in two adults, *Am. J. Med.* **39**:582–593 (1965).

83. D. S. Fredrickson, Familial high density lipoprotein deficiency: Tangier disease, *in The Metabolic Basis of Inherited Disease* (J. B. Stanbury, J. B. Wyngaarden, and D. S. Fredrickson, eds.), Ch. 23, pp. 486–508, McGraw-Hill, New York (1966).

84. K. R. Norum and E. Gjone, Familial plasma lecithin: cholesterol acyltransferase deficiency. Biochemical study of a new inborn error of metabolism, *Scand. J. Clin. Lab. Invest.* **20**:231–243 (1967).

85. J. A. Glomset, The plasma lecithin:cholesterol acyltransferase reaction, *J. Lipid Res.* **9**:155–167 (1968).

86. A. F. Liber and H. G. Rose, Saturated hydrocarbons in follicular lipidosis of the spleen, *Arch. Path.* **83**:116–122 (1967).

87. H. Bernheimer and F. Seitelberger, Über das Verhalten der Ganglioside im Gehirn bei 2 Fällen von spätinfantiler amaurotischer Idiotie, *Wien Klin. Wschr.* **80**:163–164 (1968).

88. E. Sluga and T. Majdetzki, Zur ultrastruktur des Speichermaterials von spätinfantiler amaurotischer Idiotie, *Acta Neuropath.* **9**:254–272 (1967).

89. F. Seitelberger, H. Jacob, and R. Schnabel, The myoclonic variant of cerebral lipidosis, *in Inborn Disorders of Sphingolipid Metabolism* (S. M. Aronson and B. W. Volk, eds.), pp. 43–74, Pergamon Press, London, (1967).

90. L. Klinken–Rasmussen and H. Dyggve, A case of late infantile amaurotic idiocy of the myoclonus type, *Am. J. Obstet. Gynecol.* **91**:172–186 (1965).

91. D. M. Derry, J. S. Fawcett, F. Andermann, and L. S. Wolfe, Late infantile systemic lipidosis. Major monosialogangliosidosis. Delineation of two types, *Neurology* **18**:340–348 (1968).

92. B. H. Landing, F. N. Silverman, J. M. Craig, J. M. Jacoby, M. E. Lahey, and D. L. Chadwick, Familial neurovisceral lipidosis, *Am. J. Dis. Child.* **108**:503–522 (1964).

93. J. S. O'Brien, M. B. Stern, B. H. Landing, J. K. O'Brien, and G. N. Donnell, Generalized gangliosidosis, *Am. J. Dis. Child.* **109**:338–346 (1965).

94. R. M. Norman, H. Urich, A. H. Tingey, and R. A. Goodbody, Tay–Sachs' disease with visceral involvement and its relation to Niemann–Pick disease, *J. Path. Bact.* **78**:409–421 (1959).

95. C. Davison and S. A. Jacobson, Generalized lipidosis in a case of amaurotic familial idiocy, *Am. J. Dis. Child.* **52**:345–360 (1936).

96. K. Suzuki, Cerebral G_{M1}-gangliosidosis: chemical pathology of visceral organs, *Science* **159**:1471–1472 (1968).

97. N. Gonatas and J. Gonatas, Ultrastructural and biochemical observations on a case of infantile lipidosis and its relationship to Tay–Sachs' disease and gargoylism, *J. Neuropath. Exp. Neurol.* **24**:318–340 (1965).

98. K. Suzuki, K. Suzuki, and G. C. Chen, Morphological, histochemical and biochemical studies on a case of systemic late infantile lipidosis (generalized gangliosidosis), *J. Neuropath. Exp. Neurol.* **27**:15–38 (1968).

99. R. M. Norman, A. H. Tingey, C. G. H. Newman, and S. P. Ward, Tay–Sachs' disease with visceral involvement and its relationship to gargoylism, *Arch. Dis. Child.* **39**:634–640 (1964).

100. L. Schneck, B. J. Wallace, A. Saifer, and B. W. Volk, A clinical, biochemical and electron microscopic study of late infantile amaurotic family idiocy, *Am. J. Med.* **39**:285–295 (1965).

101. S. P. Bessman and R. Baldwin, Imidazole aminoaciduria in cerebromacular degeneration, *Science* **135**:789–791 (1962).

102. H. Kufs, Über eine Spätform der amaurotischen Idiotie und ihre heredofamiliaren Grundlagen, *Z. Ges. Neurol. Psychiat.* **95**:169–188 (1925).

103. R. A. Kritzler, J. Y. Terner, J. Lindenbaum, J. Magidson, R. Williams, R. Preisig, and G. B. Phillips, Chediak–Higashi syndrome. Cytologic and serum lipid observations in a case and family, *Am. J. Med.* **36**:583–594 (1964).

104. J. H. Kanfer, R. Richards, J. P. Kampine, S. Handmaker, and R. A. Yankee, Alteration of the sphingolipid content in leucocytes from patients with Chediak–Higashi syndrome, *Life Sci.* **6**:2661–2664 (1967).

105. A. C. Woods and J. R. Duke, Coats' disease. I. Review of the literature, diagnostic criteria, clinical findings and plasma lipid studies, *Brit. J. Ophthalmol.* **47**:385–412 (1963).

106. J. R. Duke and A. C. Wood, Coats' disease. II. Studies on the identity of the lipids concerned, and the probable role of mucopolysaccharides in its pathogenesis, *Brit. J. Ophthalmol.* **47**:413–434 (1963).

107. W. H. McMenemey, The dementias and progressive diseases of basal ganglia, in *Greenfield's Neuropathology* (W. Blackwood, W. H. McMenemey, A. Myer, R. M. Norman, and D. S. Russell, eds.), Ch. 9, pp. 520, Williams and Wilkins, Baltimore (1963).

108. G. B. Solitare and J. B. Lamarche, Alzheimer's disease and senile dementia as seen in mongoloids: neuropathological observations, *Am. J. Mental Defic.* **70**:840–848 (1966).

109. G. Rouser, C. Galli, and G. Kritchevsky, Lipid class composition of normal human brain and variations in metachromatic leukodystrophy, Tay–Sachs, Niemann–Pick, chronic Gaucher's, and Alzheimer's diseases, *J. Am. Oil Chemists' Soc.* **42**:404–410 (1965).

110. G. Rouser, G. Feldman, and C. Galli, Fatty acid composition of human brain lecithin and sphingomyelin normal individuals, senile cerebral cortical atrophy, Alzheimer's disease, metachromatic leukodystrophy, Tay–Sachs, and Niemann–Pick diseases, *J. Am. Oil Chemists' Soc.* **42**:411–412 (1965).

111. K. Suzuki, R. Katzman, and S. R. Korey, Chemical studies on Alzheimer's disease, *J. Neuropathol. Exptl. Neurol.* **24**:211–224 (1965).

112. G. D. Cherayil and A. E. Cyrus, Jr., The quantitative estimation of glycolipids in Alzheimer's disease, *J. Neurochem.* **13**:579–590 (1966).

113. G. D. Cherayil, Fatty acid composition of brain glycolipids in Alzheimer's disease, senile dementia, and cerebrocortical atrophy, *J. Lipid Res.* **9**:207–214 (1968).

114. S. R. Korey, R. Katzman, and J. Orloff, A case of Jakob–Creutzfeldt disease. 2. Analysis of some constituents of the brain of a patient with Jakob–Creutzfeldt disease, *J. Neuropathol. Exptl. Neurol.* **20**:95–104 (1961).

115. H. Simpson, Encephalopathy and fatty degeneration of the viscera. Acid-base observations, *Lancet* **2**:1274–1277 (1966).

116. J. Peremans, P. J. DeGraef, G. Strubbe, and G. DeBlock, Familial metabolic disorder with fatty metamorphosis of the viscera. *J. Pediat.* **69**:1108–1112 (1966).

117. G. S. Golden and D. Duffell, Encephalopathy and fatty degeneration of the viscera associated with chickenpox, *Pediatrics* **36**:67–74 (1967).

118. R. Jenkins, A. Dvorak, and J. Patrick, Encephalopathy and fatty degeneration of the viscera associated with chickenpox, *Pediatrics* **39**:769–771 (1967).

119. A. S. Dekaban, Plasma lipids in epileptic children treated with the high fat diet, *Arch. Neurol.* **15**:177–184 (1966).

120. C. Hooft, P. de Laey, J. Herpol, F. de Loore, and J. Verbeeck, Familial hypolipidemia and retarded development without steatorrhea. Another inborn error of metabolism?, *Helv. Paediat. Acta* **17**:1–23 (1962).

121. S. Farber, J. Cohen, and L. Uzman, Lipogranulomatosis. A new lipoglyco-protein storage disease, *J. Mt. Sinai Hosp.* **24**:816–837 (1957).

122. S. M. Bierman, T. Edgington, V. D. Newcomer, and C. M. Pearson, Farber's disease. A disorder of mucopolysaccharide metabolism with articular, respiratory and neurologic manifestations, *Arthrit. Rheum.* **9**:620–630 (1966).

123. A. L. Prensky, G. Ferreria, S. Carr, and H. W. Moser, Ceramide and ganglioside accumulation in Farber's lipogranulomatosis, *Proc. Soc. Exptl. Biol. Med.* **126**:725–728 (1967).

124. C. A. Hellwig and P. N. Wilkinson, Lipid studies in Hashimoto disease, *Growth* **26**:297–307 (1962).

125. S. R. Korey and H. Winograd, Biochemical alterations in a case of Heller's disease, *Am. J. Dis. Child.* **97**:668–675 (1959).

126. J. Bell, Huntington's chorea, *in The Treasury of Human Inheritance* (R. A. Fisher and L. S. Penrose, eds.), Vol. IV, pp. 1–67, Cambridge University Press, London (1948).

127. G. J. M. Hooghwinkel, P. F. Borri, and G. W. Pruyn, Biochemical studies in Huntington's chorea. II. Composition of blood lipids, *Acta Neurol. Scand.* **42**:213–220 (1966).

128. P. F. Borri, W. M. Op Den Velde, G. J. M. Hooghwinkel, and G. W. Bruyn, Biochemical studies in Huntington's Chorea. VI. Composition of striatal neutral lipids, phospholipids, glycolipids, fatty acids, and amino acids, *Neurology* **17**:172–178 (1967).

129. I. S. Friedman, H. Cohn, M. Zymaris, and M. G. Goldner, Hypocholesteremia in idiopathic steatorrhea, *Arch. Int. Med.* **105**:112–120 (1960).

130. J. H. Menkes, M. Alter, G. K. Steigleder, D. R. Weakley, and J. H. Sung, A sex-linked recessive disorder with retardation of growth, peculiar hair, and focal cerebral and cerebellar degeneration, *Pediatrics* **29**:764–779 (1962).

131. J. S. O'Brien and E. L. Sampson, Kinky hair disease. II. Biochemical studies, *J. Neuropathol. Exptl. Neurol.* **25**:523–530 (1966).

132. R. D. Adams and E. P. Richardson, Jr., The demyelinative diseases of the human nervous system. A classification; A review of salient neuropathologic findings; Comments on recent biochemical studies, *in Chemical Pathology of the Nervous System* (J. Folch–Pi, ed.), pp. 162–194, Pergamon Press, New York (1966).

133. W. Zeman, W. Demyer, and H. F. Falls, Pelizaeus–Merzbacher disease, *J. Neuropath. Exptl. Neurol.* **23**:334–354 (1964).

134. H. Jatzkewitz and E. Mehl, The fate of C_{24}-fatty acids during lipophilic myelin breakdown in the central nervous system. I. C_{24}-fatty acid deficiency in the lipophilic decomposition and breakdown products, *Z. Physiol. Chem.* **329**:264–277 (1962).

135. R. M. Norman and A. H. Tingey, Sudanophil leucodystrophy and Pelizaeus–Merzbacher disease, *in Brain Lipids and Lipoproteins and the Leucodystrophies* (J. Folch–Pi and H. J. Baur, eds.), Elsevier, New York (1963).

136. B. Gerstl, L. J. Rubinstein, L. F. Eng, and M. Tavaststjerna, A neurochemical study of a case of sudanophilic leukodystrophy, *Arch. Neurol.* **15**:603–614 (1966).

137. R. M. Norman, A. H. Tingey, P. W. Harvey, and A. M. Gregory, Pelizaeus–Merzbacher disease: a form of sudanophil leucodystrophy, *J. Neurol. Neurosurg. Psychiat.* **29**:521–529 (1966).

138. V. W. D. Schenk, F. C. Stam, and A. M. Batenburg–Plenter, A family with sudanophilic leucodystrophy, *Acta Neuropathol.* **9**:233–243 (1967).

139. R. M. Norman, A. H. Tingey, J. C. Valentine, and H. J. Hislop, Sudanophil leucody-strophy: a study of intersib variation in the form taken by the demyelinating process, *J. Neurol. Neurosurg. Psychiat.* **30**:75–82 (1967).
140. A. H. Tingey and G. W. F. Edgar, The chemistry of the leucodystrophies, *J. Neurochem.* **10**:817–823 (1963).
141. F. Lindlar, K. Nagal, and A. Vogel, Veränderungen an den Hirnlipoiden bei einem Fall von diffuser Sklerose, *Z. Physiol. Chem.* **347**:1–6 (1966).
142. M. Adachi and S. M. Aronson, Studies on spongy degeneration of the central nervous system (Van Bogert–Bertrand type) *in Inborn Disorders of Sphingolipid Metabolism* (S. M. Aronson and B. W. Volk, eds.), pp. 129–147, Pergamon Press, London (1967).
143. G. A. Jervis, R. C. Harris, and J. H. Menkes, Cerebral lipidosis of unclear nature, *in Cerebral Sphingolipidoses* (S. M. Aronson and B. W. Volk, eds.), pp. 101–118, Academic Press, New York (1962).
144. A. Rosenberg, The sphingolipids from the spleen of a case of lipidosis, *in Cerebral Sphingolipidoses* (S. M. Aronson and B. W. Volk, eds.), pp. 119–123, Academic Press, New York (1962).
145. A. Sawitsky, G. A. Human, and J. B. Hyman, An unidentified reticuloendothelial cell in bone marrow and spleen, *Blood* **9**:977–985 (1954).
146. I. L. Thompson and W. C. Moloney, Lipid storage disease, *Blood* **27**:49–56 (1966).
147. T. I. Malinin, Unidentified reticuloendothelial cell storage disease, *Blood* **17**:675–686 (1961).
148. D. G. Cogan and D. D. Federman, Retinal involvement with reticuloendotheliosis of unclassified type, *Arch. Ophthalmol.* **71**:489–491 (1964).
149. M. N. Silverstein, D. G. Young, W. H. ReMine, and G. L. Pease, Splenomegaly with rare morphologically distinct histiocyte, *Arch. Int. Med.* **114**:251–257 (1964).
150. B. M. Rifkind and J. A. Boyle, Blood lipid levels, thyroid status, and glucose tolerance in progressive partial lipodystrophy, *J. Clin. Pathol.* **20**:52–55 (1967).
151. B. G. P. Shafiroff and Q. Y. Kau, Lipodystrophy of adipose areolar tissue in the retroperitoneal space, *Surgery* **59**:696–702 (1966).
152. P. Clarkson, Lipodystrophies, *Plast. Reconstruct. Surg.* **37**:499–503 (1966).
153. D. S. Fredrickson, Lipidosis and xanthomatosis, *in Principles of Internal Medicine* (T. R. Harrison, R. D. Adams, I. L. Bennett, Jr., W. H. Resnik, G. W. Thorn, and M. M. Wintrobe, eds.), Ch. 106, pp. 579, McGraw-Hill, New York (1966).
154. W. Orkin, R. W. Goltz, R. A. Good, A. Michael, and J. Fisher, A study of multicentric reticulohistiocytosis, *Arch. Dermatol. Syph.* **89**:640–654 (1964).
155. M. V. Barrow, F. W. Sunderman, Jr., R. L. Hackett, and W. S. Colvin, Identification of tissue lipids in lipoid dermatoarthritis (multicentric reticulohistocytosis), *Am. J. Clin. Pathol.* **47**:312–325 (1967).
156. A. N. Davison and M. Wajda, Cerebral lipids in multiple sclerosis, *J. Neurochem.* **9**:427–432 (1962).
157. G. W. F. Edgar, Anatomo-chemical research in demyelinating conditions and inborn errors of metabolism, *in Proc. Vth Int. Cong. Neuropath.*, Zurich, Sept. 1965 (Excerpta Med. Intl. Cong. Series No. 100, pp. 350–363).
158. B. Gerstl, M. G. Tavaststkerna, R. B. Hayman, L. F. Eng, and J. K. Smith, Alterations in myelin fatty acids and plasmalogens in multiple sclerosis, *Ann. N.Y. Acad. Sci.* **122**:405–415 (1965).
159. C. B. Courville, Multiple sclerosis as an incidental complication of a disorder of lipid metabolism. I. Close resemblance of the lesions resulting from fat embolism to the plaques of multiple sclerosis, *Bull. Los Angeles Neurol. Soc.* **24**:60–76 (1959).

160. C. B. Courville. Multiple sclerosis as an incidental complication of a disorder of lipid metabolism. II. A survey of the geographical, clinical and biochemical evidence; the significance of endogenous fat embolism, *Bull. Los. Angeles Neurol. Soc.* **24**:77–88 (1959).

161. C. B. Courville, Multiple sclerosis as an incidental complication of a disorder of lipid metabolism. III. Treatment with heparin of acute exacerbations of the disease, *Bull. Los Angeles Neurol. Soc.* **24**:89–105 (1959).

162. J. Bernsohn, and L. M. Stephanides, Aetiology of multiple sclerosis, *Nature* **215**:821–823 (1967).

163. H. R. Gray, J. H. Graham, and W. C. Johnson, Necrobiosis lipoidica: a histopathological and histochemical study, *J. Invest. Dermatol.* **44**:369–380 (1965).

164. D. Cowen and E. V. Olmstead, Infantile neuroaxonal dystrophy, *J. Neuropathol. Exptl. Neurol.* **22**:175–236 (1963).

165. W. S. Hartroft and E. A. Porta, Ceroid, *Am. J. Med. Sci.* **250**:324–345 (1965).

166. N. R. Di Luzio and F. Costales, Inhibition of the ethanol and carbon tetrachloride induced fatty liver by antioxidants, *Exptl. Mol. Pathol.* **4**:141–154 (1965).

167. M. Comporti, A. Hartman, and N. R. Di Luzio, Effect of in vivo and in vitro ethanol administration on liver peroxidation, *Lab. Invest.* **16**:616–624 (1967).

168. E. H. Oppenheimer and E. C. Andrews, Ceroid storage disease in children (2 cases), *Pediatrics* **23**:1091–1102 (1959).

169. C. A. Pallis, S. Duckett, and A. G. E. Pearse, Diffuse lipofuscinosis of the central nervous system, *Neurology* **17**:381–394 (1967).

170. R. Hunt, Dyssynergia cerebellaris myoclonica-primary atrophy of the dentate system: a contribution to the pathology and symptomatology of the cerebellum, *Brain* **44**:490–538 (1921).

171. B. H. Landing and H. S. Shirkey, A syndrome of recurrent infection and infiltration of viscera by pigmented lipid histiocytes, *Pediatrics* **20**:431–438 (1957).

172. M. J. Carson, D. L. Chadwick, C. A. Brubaker, R. S. Cleland, and B. H. Landing, Thirteen boys with progressive septic granulomatosis, *Pediatrics* **35**:405–412 (1965).

173. J. Bartman, R. L. Van de Velde, and F. Friedman, Pigmented lipid histiocytosis and susceptibility to infection: ultrastructure of splenic histiocytes, *Pediatrics* **40**:1000–1002 (1967).

174. B. Holmes, P. G. Quie, D. Windhorst, and R. A. Good, Fatal granulomatous disease of childhood, *Lancet* **1**:1225–1228 (1966).

175. J. Hallervorden and H. Spatz, Eigenartige Erkrankung im extrapyramidalen System mit besonderer Beteiligung des Globus pallidus und der Substantia nigra, *Z. Ges. Neurol. Psychiat.* **79**:254–302 (1922).

176. E. P. Richardson, Jr., A. Torvik, and R. D. Adams, Degenerative diseases of the nervous system, *in Principles of Internal Medicine* (T. R. Harrison, R. D. Adams, I. L. Bennett, W. H. Resnik, G. W. Thorn, and M. M. Wintrobe, eds.), Ch. 211, pp. 1234–1249, McGraw-Hill, New York (1966).

177. P. B. Diezel, J. A. Rossner, N. Koppang, P. Ritzhaupt, and D. Bartling, Juvenile form of amaurotic family idiocy. A contribution to the morphological, histochemical and electromicroscopic aspects, *in Inborn Disorders of Sphingolipid Metabolism* (S. M. Aronson and B. W. Volk, eds.), pp. 23–42, Pergamon Press, New York (1967).

178. S. Donahue, W. Zeman, and I. Watanabe, Electron microscopic observations in Batten's disease, *in Inborn Disorders of Sphingolipid Metabolism* (S. M. Aronson and B. W. Volk, eds.), pp. 3–22, Pergamon Press, New York (1967).

179. F. Seitelberger and K. Simma, On the pigment variant of amaurotic idiocy, *in Cerebral Sphingolipidoses* (S. M. Aronson and B. W. Volk, eds.), pp. 29–47, Academic Press, New York (1962).

180. P. J. Moe and A. E. Hansen, Reticuloendothelial granuloma, *Am. J. Dis. Child.* **99**:175–184 (1960).

181. J. M. Hill, R. J. Speer, and H. Gedikoglu, Secondary lipidosis of spleen associated with thrombocytopenia and other blood dyscrasias treated with steroids, *Am. J. Clin. Pathol.* **39**:607–615 (1963).

182. S. L. Saltzstein, Phospholipid accumulation in histiocytes of splenic pulp associated with thrombocytopenic purpura, *Blood* **18**:73–88 (1961).

183. B. H. Landing, L. Strauss, A. C. Crocker, H. Braunstein, W. L. Henley, J. R. Will, and M. Sanders, Thrombocytopenic purpura with histiocytosis of the spleen, *New Engl. J. Med.* **265**:572–577 (1961).

184. C. O. Burdick, Lipidosis of the spleen, *Arch. Pathol.* **79**:583–587 (1965).

185. S. Warren and H. F. Root, Lipoid containing cells in the spleen in diabetes with lipemia, *Am. J. Pathol.* **2**:69–80 (1926).

186. I. Tellez–Nagel and D. H. Harter, Subacute sclerosing leukoencephalitis: Ultrastructure of intranuclear and intracytoplasmic inclusions, *Science* **154**:899–901 (1966).

187. W. T. Norton, S. E. Poduslo, and K. Suzuki, Subacute sclerosing leukoencephalitis. II. Chemical studies including abnormal myelin and an abnormal ganglioside pattern, *J. Neuropathol. Exptl. Neurol.* **25**:582–597 (1966).

188. R. Ledeen, K. Salsman, and M. Cabrera, Gangliosides in subacute sclerosing leukoencephalitis: isolation and fatty acid composition of nine fractions, *J. Lipid Res.* **9**:129–136 (1968).

189. F. M. Enzinger and E. B. Helwig, Whipple's disease: A review of the literature and report of fifteen patients, *Arch. Pathol. Anat. Physiol.* **336**:238–269 (1963).

190. J. Caroli, C. Julien, and B. Bonneville, Recent advances in Whipple's disease (intestinal lipodystrophy), *Rev. Franc. Etudes Clin. Biol.* **10**:362–380 (1965).

191. M. J. Phillips and J. M. Finlay, Bacilli-lipid associations in Whipple's disease, *J. Pathol. Bact.* **94**:131–137 (1967).

192. M. Wolman, V. V. Sterk, S. Gatt, and M. Frenkel, Primary familial xanthomatosis with involvement and calcification of the adrenals, *Pediatrics* **28**:742–757 (1961).

193. A. C. Crocker, G. F. Vawter, E. B. D. Neuhauser, and A. Rosowsky, Wolman's disease: three new patients with a recently described lipidosis, *Pediatrics* **35**:627–640 (1965).

194. A. Dorfman, Hereditary diseases of connective tissues: The Hurler syndrome, *in The Metabolic Basis of Inherited Disease* (J. B. Stanbury, J. B. Wyngaarden, and D. S. Fredrickson, eds.), 2nd ed., pp. 963–994, McGraw-Hill, New York (1966).

195. V. A. McKusick, D. Kaplan, D. Wise, W. B. Hanley, S. B. Suddarth, M. E. Sevick, and A. E. Maumanee, The genetic mucopolysaccharidoses, *Medicine* **44**:445–483 (1965).

196. H. E. Williams, Heritable disorders of mucopolysaccharide metabolism, *Calif. Med.* **106**:306–311 (1967).

197. L. E. Glynn, Diseases of collagen and related tissues, *in Intern. Rev. Connective Tissue Res.* (D. A. Hall, ed.), Vol. 2, pp. 213–241, Academic Press, New York (1964).

198. R. Ledeen, K. Salsman, J. Gonatas, and A. Taghavy, Structure comparison of the major monosialogangliosides from brains of normal human, gargoylism, and late infantile systemic lipidosis. Part I, *J. Neuropathol. Exptl. Neurol.* **24**:341–351 (1965).

199. T. Taketomi and T. Yamakawa, Glycolipids of the brain in gargoylism, *Japan. J. Exp. Med.* **37**:11–21 (1967).

200. R. Brunish and B. Sørensen, Urinary excretion of acid mucopolysaccharides and hydroxyproline in psoriasis, *Dermatologica* **130**:165–172 (1965).

201. E. J. Moynahan, Hyalinosis cutis et mucosae (Lipoid Proteinosis). Demonstration of a new disorder of mucopolysaccharide metabolism, *Proc. Roy. Soc. Med.* **59**: 1125–1126 (1966).

202. L. N. Green, I. Herzog, and D. Aberfeld, A case of hypertrophic interstitial neuritis coexisting with dementia and cerebellar degeneration, *J. Neuropathol. Exptl. Neurol.* **24**: 682–688 (1965).

203. E. Boder and R. P. Sedgwick, Ataxia-telangiectasia: A familial syndrome of progressive cerebellar ataxia, oculocutaneous telangiectasia and frequent pulmonary infection, *Univ. So. Calif. Med. Bull.* **9**:15–27 (1957).

204. E. Boder and R. P. Sedgwick, Ataxia-telangiectasia: A familial syndrome of progressive cerebellar ataxia, oculocutaneous telangiectasia and frequent pulmonary infection, *Pediatrics* **21**:526–554 (1958).

205. E. Boder and R. P. Sedgwick, Ataxia-telangiectasia: A review of 101 cases, *Little Club Clinics Develop. Med.* **8**:110–118 (1963).

206. P. Fireman, M. Boesman, and D. Gitlin, Ataxia-telangiectasia, a dysgamma-globulin-aemia with deficient $\gamma_1 A(\beta_2 A)$-globulin, *Lancet* **1**:1193–1195 (1964).

207. R. D. A. Peterson, W. D. Kelly, and R. A. Good, Ataxia-telangiectasia: its association with a defective thymus, immunological-deficiency disease and malignancy, *Lancet* **1**:1189–1193 (1964).

208. R. K. Young, K. F. Austen, and H. W. Mosher, Abnormalities of serum gamma IA globulin and ataxia-telangiectasia, *Medicine* **43**:423–432 (1964).

209. L. N. Green, I. Herzog, and D. Aberfeld, A case of hypertrophic interstitial neuritis coexisting with dementia and cerebellar degeneration, *J. Neuropathol. Exptl. Neurol.* **24**:682–689 (1965).

210. J. H. Austin, Observations on the syndtome of hypertrophic neuritis (the hypertrophic interstitial radiculo-neuropathies), *Medicine* **35**:187–237 (1956).

211. P. D. Bedford and F. E. James, Family with progressive hypertrophic polyneuritis of Dejerine and Sottas, *J. Neurol. Neurosurg. Psychiat.* **9**:46–51 (1956).

212. F. Anderman, D. Lloyd-Smith, H. Mavor, and G. Mathieson, Observations on hyper-trophic neuropathy of Dejerine and Sottas, *Neurology* **12**:712–724 (1962).

213. D. C. Gajdusek, Kuru, *Trans. Roy. Soc. Trop. Med. Hyg.* **57**:151–169 (1963).

214. C. Kidson, Kuru as a model of hormonally programmed central nervous degeneration, *Lancet* **2**:830–831 (1965).

215. N. K. Gonatas, Axonic and synaptic lesions in neuropsychiatric disorders, *Nature* **214**:352–354 (1967).

216. Parts III (pp. 212–420) and XI (1179–1320), *in The Metabolic Basis of Inherited Disease* (J. B. Stanbury, J. B. Wyngaarden, and D. S. Fredrickson, eds.), 2nd ed., McGraw-Hill, New York (1966).

217. G. M. Shy, Chemical and morphological abnormalities in muscle disease, *Ann. N. Y. Acad. Sci.* **138**:232–245 (1966).

218. G. M. Shy, D. H. Silberberg, S. H. Appel, M. M. Mishkin, and E. H. Godfrey, A genera-lized disorder of nervous system, skeletal muscle and heart resembling Refsum's disease and Hurler's syndrome, I. Clinical, pathologic and biochemical characteristics, *Am. J. Med.* **42**:163–168 (1967).

219. N. K. Gonatas, I. Evanelista, and J. Martin, A generalized disorder of nervous system, skeletal muscle and heart resembling Refsum's disease and Hurler's syndrome. II. Ultra-structure, *Am. J. Med.* **42**:169–178 (1967).

220. R. F. Coleman, A. W. Nienhuis, W. J. Brown, T. L. Munsat, and C. M. Pearson, New myopathy with mitochondrial enzyme hyperactivity. *J. Am. Med. Assoc.* **199**:624–630 (1967).

221. L. Vitali–Mazza, P. Anversa, and O. Visioli, Ultrastrukturelle Veränderungen des Herz-muskels bei Ratten unter polyensäurearmer Diät, *Virchow's Arch. Path. Anat.* **342**:38–39 (1967).

222. M. D. Haust, Crystalloid structures of hepatic mitochondria in children with heparitin sulfate mucopolysaccharidosis (Sanfillipo type), *Exptl. Mol. Pathol.* **8**:123–134 (1968).

223. G. M. Shy, T. Wanko, P. T. Rowley, and A. G. Engel, Studies in familial periodic paralysis, *Exptl. Neurol.* **3**:53–121 (1961).

224. C. M. Pearson, The periodic paralyses: differential features and pathological observations in permanent myopathic weakness, *Brain* **87**:341–354 (1964).
225. D. C. Poskanzer and D. N. S. Kerr, A third type of periodic paralysis with normokalemia and favorable response to sodium chloride, *Am. J. Med.* **31**:328–342 (1961).
226. W. J. Jaffurs, R. H. Herman, M. K. McDowell, and J. M. Blumberg, Hyperkalemic paralysis (Adynamia episodica hereditaria): Ultrastructural studies of muscle in two cases, *Metabolism* **12**:740–750 (1963).
227. A. G. Engel, E. H. Lambert, J. W. Rosevear, and W. N. Tauxe, Clinical and electromyographic studies in a patient with primary hypokalemic periodic paralysis, *Am. J. Med.* **38**:626–640 (1965).
228. E. L. Howes, Jr., H. M. Price, C. M. Pearson, and J. B. Blumberg, Hypokalemic periodic paralysis. Electronmicroscopic changes in the sarcoplasm, *Neurology* **16**:242–256 (1966).
229. D. L. Odor, A. N. Patel, and L. A. Pearce, Familial hypokalemic periodic paralysis with permanent myopathy, *J. Neuropath. Exptl. Neurol.* **26**:98–114 (1967).
230. A. Takagi, Y. Muto, Y. Takahashi, and K. Nakao, Fatty acid composition of lecithin from muscles in human progressive muscular dystrophy, *Clin. Chim. Acta* **20**:41–42 (1968).
231. J. J. Peifer, Disproportionately higher levels of myocardial docosahexaenoate and elevated levels of plasma and liver arachidonate in hyperthyroid rats, *J. Lipid Res.* **9**:193–199 (1968).
232. R. A. Field, Glycogen deposition diseases, *in The Metabolic Basis of Inherited Disease* (J. B. Stanbury, J. B. Wyngaarden, and D. S. Frederickson, eds.), 2nd ed., pp. 141–177, McGraw-Hill, New York (1966).
233. S. W. Moses, S. Levin, R. Chayota, and K. Steinitz, Enzyme induction in a case of glycogen storage disease, *Pediatrics* **38**:111–121 (1966).
234. J. R. M. Innes and L. Z. Saunders, Diseases of the central nervous system of domesticated animals and comparisons with human neuropathology, *in Advances in Veterinary Science* (C. A. Brandly and E. L. Jungherr, eds.), Vol. 3, pp. 33–196, Academic Press, New York (1957).
235. M. W. Fox, Diseases of possible hereditary origin in the dog: A bibliographic review, *J. Hered.* **56**:169–76 (1965).
236. D. F. Patterson and W. Medway, Hereditary diseases of the dog, *J. Am. Veterin. Med. Assoc.* **149**:1741–1754 (1966).
237. H. A. Anderson, Leucodystrophy in mink: A biochemical study, *Acta Neuropathol.* **7**:297–304 (1967).
238. D. B. Windhorst, Studies on a hereditary defect involving lysosomal structure, *Fed. Proc.* **25**:358 (Abstr. #954) (1966).
239. T. F. Fletcher, J. H. Kurtz, and D. G. Low, Globoid cell leukodystrophy (Krabbe type) in the dog, *J. Am. Veterin. Med. Assoc.* **149**:165–172 (1966).
240. E. Karbe and B. Schiefer, Familial amaurotic idiocy in male German shorthair pointers, *Pathol. Veterin.* **4**:223–232 (1967).

Chapter 9

MYELIN

Lewis C. Mokrasch

Research Laboratory
McLean Hospital
Belmont, Massachusetts

I. INTRODUCTION

The examination of myelin as an entity isolated in milligram quantities dates to 1961 when three groups of investigators isolated and identified myelin[1-3] by modifications of older methods for the separation of subcellular particles. These centrifugation techniques were then modified further to permit relatively quick and reliable isolations based on centrifugation in density-gradient media of sucrose.[4-10] Myelin may also be obtained by methods based on hypotonic treatments of brain subcellular particles[11] and by centrifugation on density gradients of Ficoll.*[12]

Whether myelin, as isolated by the methods cited above, truly represents myelin as it is observed in white matter is an important question and there is reason to suspect that the two are not identical.[13] Only one example need be cited here. Laatsch,[14] among others, reports that the myelin he isolated was completely soluble in chloroform–methanol mixtures; therefore, the myelin protein is proteolipid.[15] However, Lees has shown that a removal of ions, which is accomplished by the density gradient techniques, results in an increase in the apparent amount of proteolipid, as defined by the solubility criterion, in rat brain preparations.[16] Thus it can be argued that residual axoplasmic proteins in the isolated preparations can be artifactitiously solubilized in chloroform–methanol mixtures or that some proteins which are truly myelin proteins may not be soluble in the chloroform–methanol extracts of white matter.[17] Absence of normal cytoplasmic constituents from purified myelin is not consistent with the notion that the Schwann cell or oligodendrocyte cytoplasm intercalates the myelin winding and that even inclusions such as mitochondria are observed in electron micrographs.[18] Finally, the use of proteolipid protein as a "marker" for myelin is of limited value, since proteolipid protein is found in other organelles of the nerve tissue[19-21] and of other tissues.[15,21,22] In fact, the reservations expressed

* © Pharmacia, Uppsala, Sweden.

here have been raised about all the other subcellular structures which have been isolated from broken cell preparations. Even though valid questions exist about the integrity of these subcellular fractions, the isolated structures provide excellent insight into the chemistry of the whole cell.

Without indulging in an extended discussion of the composition of white matter, which will be covered elsewhere in this handbook, some comparisons of white matter and isolated myelin would be of interest.

Based on the proteolipid content of white matter and isolated myelin, Norton and Autilio calculate that white matter averages 40%[23] or 50%[24] myelin, and Brante[25] estimated that 50% of the white matter is myelin. Cuzner, Davison, and Gregson[26] quote values from 36 to 53% for the myelin contents of mammalian white matter.

A classic assumption has been that the nonmyelin portion of white matter is nearly the same in composition as cortical tissue. From the data presented in the tables which follow, it can be calculated that about 50% of the white matter lipids are nonmyelin in origin. The phospholipids are 50–55% extramyelin and the glycolipids and cholesterol are 35–45% extramyelin. About 75% of white matter protein is nonmyelin in origin. No single myelin constituent is exclusively localized in myelin. Since the composition of white matter and gray matter will be discussed in detail in another section of the Handbook, it would be repetitious to compare white matter and myelin in detail here. However, several of the references cited deal specifically with this comparison.[5,23,25,31]

II. CHEMICAL COMPOSITION OF MYELIN

A. Gross Composition

The data are presented in this chapter in terms of the dry weight of myelin wherever possible. Expressing the results on the basis of the fresh weight of myelin would be preferable from a physiological viewpoint, but no method of myelin preparation permits the isolation of a product with unchanged hydration because myelin is osmotically sensitive.[34,35] The water content of tightly wound myelin has been estimated as approximately 40%, based on X-ray diffraction studies[36] and on a model lipid–protein unit with the composition approximating that of myelin.[37] This figure is probably a minimum value, excluding interlaminar gaps containing Schwann cell or oligodendrocyte cytoplasm and higher degrees of hydration which may occur normally, especially in myelin with a higher protein content. Nevertheless, the water content of myelin from various sources is probably less variable than the content of any other constituent.

Referring to Table I, the eight animal species represented do not permit taxonomic deductions about gross myelin composition. There is poor agreement among different investigators on the content of some myelin constituents for the same species. Whether the disagreement derives from a lack of

TABLE I

Gross Composition of Myelin

Source	Constituent, mg/g of dry myelin				Yield, mg of fresh weight
	Protein	Glycolipid	Phospholipid	Sterol	
Human white matter[32]	218	196	371	190	—
Human white matter[26]	456	136 ± 8	238 ± 6	178 ± 7	125, 155
Bovine white matter[5]	250 ± 27	222 ± 9	321 ± 12	197 ± 10	45
Bovine white matter[24]	196 ± 15	236 ± 19	361 ± 17	174 ± 9	—
Bovine white matter[21]	295 ± 25	266	319	170	115 ± 15
Bovine dorsal roots[30]	(241)[a]	126	444	188	—
Ovine white matter[31]	298	(232)[a]	255	215	—
Rabbit whole brain[26]	380	171	310	160	62.7
Rat whole brain[8]	479 ± 39	113 ± 11	290 ± 30	122 ± 15	38
Rat whole brain[33]	253 ± 17	(307)[a]	309 ± 5	130 ± 13	—
Pigeon whole brain[26]	474	152	199	154	81
Dogfish whole brain[b][26]	560 ± 90	72	260	96	41 ± 10
Frog whole brain[c][26]	390	69	364	156	14

[a] Estimated by difference.
[b] *Scyllium canicula.*
[c] *Rana temporaria.*

morphological purity of the preparations or from alterations in composition induced by the isolation procedure cannot be decided until many more analyses are reported and new isolation procedures are designed. Since "heavy" and "light" myelin, where the two forms are isolated,[5,38] do not differ significantly in their lipid analyses, the analytical values for the two forms are combined as an average in this review, much as some investigators combine the two fractions themselves prior to further experimental treatment.

The protein and phospholipid contents of myelin from the central nervous system each approximate one third of the total dry weight. Glycolipid and sterol contents are each about one sixth of the total dry weight.

Myelin isolated from the peripheral nervous system of the cow,[30] rat,[39] guinea pig,[39] and squirrel monkey[40] generally is richer in phospholipid and has less glycolipid than that of the corresponding central nervous system myelin. Unfortunately the data for the latter three animals are not presented in terms which are convertible to milligrams per gram of dry myelin.

B. Glyceryl Lipids in Myelin

1. Phospholipids by Class

There are a few consistencies evident in the contents of glyceryl lipids in myelin from various sources which permit a few generalizations to be made.

TABLE II
Glyceryl Lipids[a]

	Phosphatidyl ethanolamine	Phosphatidal ethanolamine	Phosphatidyl serine	Phosphatidal serine	Phosphatidyl choline	Phosphatidal choline	Phosphatidyl inositol	Phosphatidic acid[b]	Total plasmalogen
Human white matter[29]	...128 ± 6...		...78 ± 9[c]...		...85 ± 3...		—	—	—
Human white matter[26]	...104 ± 10...		...43...		...48 ± 2...		4.2	1.1	90.0 ± 7.5
Bovine white matter[24]	(35 ± 7.5)[d]	96 ± 7	(47 ± 11)[d]	1.9 ± 0.4	(80 ± 9)[d]	2.5 ± 2	7.5 ± 2	—	100 ± 9
Bovine white matter[26]	...104 ± 15...		...59 ± 3...		...61 ± 5...		11	—	86
Bovine spinal root[30]	...121, 109...		...66, 57...		...94, 100...		—	—	—
Rabbit brain[26]	...98...		...44...		...56...		9.3	21	—
Rat brain[26]	(18)[d]	114	...53...		(64)[d]	27	18	9	172
Pigeon brain[26]	...83...		...18...		...51...		5.8	16	93
Dogfish brain[26]	...107...		...21...		...64...		7.1	15	122
Frog brain[26]	...134...		...27...		...143...		9.2	6.1	—

[a] Mg/g dry myelin.
[b] Includes cardiolipin.
[c] Includes inositol phosphatides.
[d] Estimated by difference.

The ethanolamine phospholipids amount to approximately 10 % of the weight of myelin solids (Table II). The ethanolamine phospholipid content is virtually the same for central and peripheral nervous system myelins.[30,39,40] Most of the ethanolamine phospholipid is phosphatidal ethanolamine; in one of the studies on human myelin the authors report that the ethanolamine plasmalogen content of a sample of ethanolamine phospholipids was 100 %.[26]

The serine, choline, and inositol glyceryl lipids are much more variable in their occurrence in myelin than are the ethanolamine lipids. The plasmalogen contents of these classes decrease in this order of their "bases": serine > choline > inositol. The plasmalogen content of the inositol lipids is usually close to zero, and the plasmalogen content of the serine phospholipids is usually less than 10 % of the total serine phospholipid.

The polyphosphoinositide content of guinea pig myelin has been reported by Eichberg and Dawson.[41] The polyphosphoinositide phosphorus content is 5–6 % of the total lipid phosphorus and about three quarters of this is triphosphoinositide phosphorus. The authors suggest that the triphosphoinositide content of guinea pig brain is predominantly localized in the myelin.

2. Aliphatic Acid Content of Phospholipids

Undoubtedly, more information on the acyl contents of myelin lipids will become available through the use of gas–liquid chromatography, but present data are the product of a single group.[27,30] Generally, oleic acid is the predominant fatty acid in all classes of glyceryl phospholipids in the samples examined (Table III). Except for a slight decrease of saturated acids and increase in olefinic acids, the acyl contents of the glyceryl lipids are not greatly changed with age.[27] The content of C_{16} acyl groups in the choline lipids is nearly as great as that of the C_{18} groups, whereas in the serine and ethanolamine lipids the C_{16} content is much smaller.

To some degree, the fatty acid composition depends on diet and can be modified by it.[42] A slight increase in polyunsaturated acids can be induced as a result of feeding rats a diet rich in sunflower seed oil.

Among lower animals, no work is yet reported on the fatty acid content of fractions of isolated myelin. It has been reported that the fatty acid pattern of myelinated nerves of the garfish resembles that of the ethanolamine and serine lipids of human myelin.[43]

3. Aldehyde Content of Phospholipids

The aldehyde pattern of the phospholipids does not closely resemble the aliphatic acid pattern. Branched-chain aldehydes are present in significant amounts. The C_{16} aldehydes are present in larger proportion than the C_{16} acids. There is a noticeable trend with increasing age to diminution of the C_{16} content, increase in the C_{18} content and decrease in saturation of the aldehydes.[27] The aldehyde contents of samples derived from bovine spinal roots

TABLE III

Fatty Acid Content of Myelin Glycerophosphatides[a]

Acid class	Bovine spinal roots[30]			Human white matter[27]						
	Ethanolamine glycerophosphatide	Serine glycerophosphatide	Choline glycerophosphatide	Ethanolamine glycerophosphatide		Serine glycerophosphatide		Choline glycerophosphatide	
				10 month[b]	55 year[b]	10 month	55 year	10 month	55 year
14:0[c]	1.2	1.4	2.9	0.7	0.4	0.5	0.3	2.1	0.4
16:0	21.0	7.2	40.0	6.0	6.5	1.4	2.6	35.1	40.1
16:1	1.8	1.3	tr	1.3	0.4	0.6	0.6	2.5	0.8
17:0	0.3	1.0	tr	—	—	—	—	—	—
18:0	5.0	17.6	11.0	28.2	7.7	47.7	40.0	14.6	6.1
18:1	42.0	63.5	36.0	38.1	72.5	30.5	43.3	40.1	51.6
18:2	1.1	0.6	5.2	1.4	tr	0.4	tr	0.3	0.6
18:3	0.4	0.8	tr	—	—	—	—	—	—
20:0	1.1	1.1	0.3	—	—	—	—	—	—
20:1	4.0	2.7	2.0	2.9	3.9	0.2	3.6	0.9	—
20:2	1.3	—	—	—	2.0	tr	1.4	tr	—
20:3	1.7	—	1.0	3.5	0.8	1.7	tr	3.0	—
20:4	2.0	tr	0.7	3.1	1.6	2.2	tr	tr	0.4
22:4	1.8	0.6	—	—	tr	tr	tr	tr	—
22:5	9.5	1.2	—	4.7	5.1	6.1	2.3	tr	—
22:6	—	—	—	4.5	0.6	4.3	2.3	tr	tr

[a] In mole percent, each component, of total acyl groups of each lipid.
[b] Age of donor.
[c] Chain length: number of olefin bonds.

TABLE IV

Aldehyde Content of Myelin Plasmalogens[a]

Aldehyde class	Bovine spinal roots[30]		Human white matter[27]			
	Phosphatidal ethanolamine	Phosphatidal serine	Phosphatidal ethanolamine		Phosphatidal serine	
			10 month[b]	55 year[b]	10 month[b]	55 year[b]
14:0[c]	0.5	0.8	tr	1.4	1.8	0.6
15:0	—	—	tr	tr	—	tr
16:0	19.0	25.3	45.2	29.0	42.5	28.5
16:0 (branched)	3.6	6.5	0.9	3.4	tr	tr
16:1	—	—	tr	tr	tr	2.1
17:0	4.2	4.8	2.2	2.2	2.4	3.2
18:0	34.2	35.9	27.5	21.0	29.2	17.5
18:0 (branched)	3.4	4.9	0.5	1.3	0.8	tr
18:1	34.1	17.1	23.7	41.7	20.7	43.0
18:2	1.0	4.7	—	—	—	—

[a] In percent, each component, of total aldehydes of each lipid.
[b] Age of donor.
[c] Chain length:number of olefin bonds.

are otherwise fairly similar to those of samples derived from human white matter (Table IV).

C. Sphingolipids in Myelin

1. Sphingolipid Content by Class

Sphingomyelin is one of the cerebral constituents which is relatively concentrated in myelin; 38% of rat brain sphingomyelin[33] and 70% of bovine white matter sphingomyelin is in the myelin.[24] The sphingomyelin content of peripheral myelin is about double that of central myelin,[30,39,40] which usually contains 40–50 mg/g dry myelin (Table V).

Central myelin contains two to three times as much cerebroside by weight as sphingomyelin. The cerebroside content of peripheral myelin is lower than that of central myelin,[30,39,40] approximating the sphingomyelin content. An anomaly exists in the fact that myelin is much richer in cerebroside than the glial membranes from which it is formed.[47] The probable explanation is that some of the cerebrosides are deposited after myelination has begun. Unlike sphingomyelin, cerebrosides are not concentrated in myelin.[46]

The amount of sulfatide in myelin is of the order of 20–30 mg/g dry myelin, or approximately one sixth of the total glycolipid. In human brain, approximately 60% of the sulfatide belongs to the kerasin class (saturated

TABLE V

Sphingolipids in Myelin[a]

	Sphingomyelin	Cerebroside	Sulfatide	Ganglioside[f]
Human brain[29]	46.4 ± 4.8	...170 $\pm$ 17...		—
Human brain[23]				
10 month[b]	46	137	54	—
55 year[c]	44	160	34	—
Human brain[44]	58.5 ± 5.4	46 ± 9^d	31 ± 5^d	—
	46 (C_{24})	85 ± 10^e	22 ± 5^e	—
	29 (C_{18})	—	—	—
Human brain[26]	47, 38	136 ± 85	22	—
Bovine brain[26]	52 ± 12	170 ± 9	—	—
Bovine brain[48]	—	—	—	2.4, 2.5
Bovine brain[24]	—	...222 $\pm$ 4, 211 $\pm$ 4...		0.14, 0.39
Bovine spinal roots[30]	121, 123	110, 104	19,19	—
Rabbit brain[26]	53	110	26	—
Guinea pig brain[38]	32, 51	...358...		5, 2
Rat brain[45]	—	—	—	0.47 ± 0.08
Rat brain[46]	—	—	—	0.58
Rat brain[26]	33 ± 9	140 ± 25	28	—
Pigeon brain[26]	17	131	20	—
Frog[26]	36	69	—	—
Dogfish[26]	40	45	28	—

[a] Mg component/g dry wt of myelin; means $\pm$ deviation, where n = 3 to 9.
[b] Ceramide content: 12 mg/g.
[c] Ceramide content: 7 mg/g.
[d] Kerasin type.
[e] Cerebron type.
[f] As mg N-acetyl neuraminic acid/g dry wt of myelin.

fatty amide), whereas 60% of the cerebroside belongs to the cerebron class (hydroxy fatty amide).[44]

The ganglioside content of myelin exhibits more variation in reported values than any other component. The ganglioside concentration in myelin changes little during myelination,[45] the total amount exactly paralleling the amount of myelin. Approximately 90% of the ganglioside in mature rat myelin is the normal monosialoganglioside, rising from a proportion of about 50% for myelin from 15-day-old rats.[45] Trisialoganglioside comprises about 2% of the total in mature myelin and about 12% in myelin from 15-day-old rats. The two disialogangliosides are present in equal amounts throughout myelination and together roughly represent 9% of the total gangliosides in the mature myelin. The ganglioside pattern of normal and pathological human myelin seems to be more complicated than that of rat myelin due to the presence of uncharacterized components.[49]

TABLE VI

Unsubstituted Fatty Acid Content of Myelin Sphingolipids[a]

Class	Ceramide	Sphingomyelin		Cerebroside		Sulfatide	
	9 year	10 month	55 year	10 month	55 year	10 month	55 year
14:0	tr	0.3	0.4	1.9	0.4	0.9	0.3
16:0	12.8	9.0	5.4	9.3	2.0	4.0	8.5
18:0	23.7	62.8	33.6	14.6	7.8	10.0	3.9
18:1	5.8	2.3	0.4	6.6	6.2	2.0	1.3
20:0	1.3	0.8	0.5	1.2	1.1	1.7	1.5
22:0	2.2	1.3	0.8	4.3	1.4	5.7	1.4
22:1	1.3	0.5	0.2	1.0	0.2	2.3	tr
23:0	3.7	0.5	1.4	1.2	2.9	1.0	1.2
23:1	1.4	0.4	0.6	0.6	0.3	0.3	tr
24:0	11.9	3.7	8.0	15.2	14.2	24.0	14.7
24:1	25.8	10.0	40.0	27.1	38.8	31.0	42.7
25:0	1.6	0.5	1.8	2.0	2.9	1.8	6.1
25:1	3.1	1.6	3.6	3.6	12.9	2.6	7.3
26:0	tr	1.0	tr	1.6	0.9	1.7	2.0
26:1	2.8	2.6	2.4	7.3	5.4	8.5	8.3

[a] Percent, each component, of total unsubstituted fatty acids in each sphingolipid class.[29]

2. Aliphatic Acid Content of Sphingolipids

The unsubstituted fatty acid content of myelin (Table VI) differs from that of the glyceryl lipids in having an abundance of longer chain acids. The pattern of changes with age is similar, however. Except for the high stearic acid content of the sphingomyelin in the myelin of the 10-month-old child, the principal unsubstituted acid in all the sphingolipids is nervonic acid, which increases in amount with age at the expense of lignoceric acid, the second most abundant acid.

The acid patterns of cerebrosides and sulfatides in myelin are very similar, reflecting their close metabolic relationship. The comparatively high stearic acid content of sphingomyelin from myelin is the only similarity to the fatty acid profiles of other phospholipids.

The hydroxy acid patterns of the glycolipids (Table VII) differ some-what from the unsubstituted acid patterns. Here there is a predominance of the saturated acid, cerebronic acid, and a lesser proportion of hydroxy-nervonic acid. The content of C_{18} hydroxy acid is also lower than that of the C_{18} unsubstituted acid in the compounds isolated from central myelin,[27,29] although it is higher in the glycolipids obtained from peripheral myelin.[30]

TABLE VII

Hydroxy Fatty Acids in Myelin Sphingolipids[a]

| | Human white matter | | | | | Bovine spinal roots[30] | |
| | Cerebroside[27] | | Sulfatide[27] | | Whole myelin[29] | Cerebroside | Sulfatide |
	10 month[b]	55 year	10 month[b]	55 year			
14:0	1.1	0.8	0.2	2.1	—	2.2	1.0
16:0	0.8	0.3	3.1	6.3	0.8 ± 0.6	2.3	1.6
17:0	—	—	—	—	—	tr	tr
18:0	1.3	0.6	2.4	3.4	1.1 ± 0.6	21.2	13.8
18:1	tr	tr	1.1	0.3	—	—	—
19:0	—	—	—	—	—	tr	tr
20:0	4.5	1.2	3.6	1.0	0.5 ± 0.1	1.4	1.8
21:0	—	—	—	—	0.2 ± 0.1	—	—
22:0	9.5	12.0	1.1	13.3	7.5 ± 1.3	13.2	14.1
22:1	—	—	—	—	0.3 ± 0.1	—	—
23:0	9.2	12.8	7.4	10.0	16.3 ± 2.9	8.0	5.4
23:1	—	—	—	—	1.1 ± 0.7	—	—
24:0	49.2	32.2	42.7	40.0	38.8 ± 3.5	39.3	43.6
24:1	12.3	24.0	19.5	9.6	20.8 ± 5.5	8.0	12.9
25:0	4.4	4.0	3.6	3.5	5.2 ± 0.5	3.8	4.8
25:1	1.9	tr	2.8	5.0	2.6 ± 0.8	—	—
26:0	1.2	2.1	3.2	1.7	1.2 ± 0.1	—	—
26:1	4.6	7.0	8.0	2.3	3.7 ± 0.2	—	—

[a] Mole percent each acid of total hydroxy acids of each sphingolipid.
[b] Age of donor.

The hydroxy acid content of cerebrosides rises from 38% of the total cerebroside fatty acids in the myelin from a 6-year-old human to 82% of that from a 55-year-old human. The hydroxy acid content of sulfatide, however, remains relatively constant at about 25% of the total sulfatide fatty acid, without large variation with age.[27]

3. Sphingolipid Bases

The examination of the bases of sphingolipids isolated from myelin has been largely neglected except for the report of Pilz and Mehl,[44] who find the ratio of C_{24} to C_{18} bases to be 5:3. Some inferences can be drawn from the base composition of white matter sphingomyelin and from the assumption that most of the sphingomyelin is located in the myelin. Thus, 85% of the bases of rabbit brain sphingomyelin is the normal sphingosine,[50] and in the rabbit spinal cord, 70%. Dihydrosphingosine is 2% of the rabbit brain sphingomyelin base and 22% of that in the rabbit spinal cord. Minor amounts of C_{16}-dihydrosphingosine and C_{20}-sphingosine were also detected.

D. Sterols in Myelin

Cholesterol and its related compounds, like ethanolamine phospholipids and sphingomyelin, are highly concentrated in the myelin of nervous tissue. Calculations for bovine myelin[24] indicate that almost four fifths of white matter cholesterol is in myelin. For practical purposes, one may assume that the sterol content of mature myelin (Table I) is almost entirely cholesterol. Other sterols and derivatives do exist in myelin, however. Cholesterol esters have been analyzed in bovine myelin[51] and desmosterol has been detected in rat myelin.[52,53]

Occurrence of cholesterol esters in elevated quantity is characteristic of either pathological or immature myelin; however, small amounts persist in mature myelin.[51] The fatty acid composition of these esters has oleic acid as the most abundant acid, followed by stearic and palmitic acids. The aliphatic acid pattern does not closely resemble that of any other central myelin ester, but is curiously similar to the pattern of peripheral myelin serine phospholipid.[30]

Desmosterol, which has been known to exist in white matter, is now unequivocally shown to be a myelin constituent.[52] In young rats (12 days), desmosterol amounts to about 10% of the sterols from brain myelin and between $1-3\%$ of the sterols from spinal cord myelin. The proportion of brain myelin desmosterol decreases quickly with age, reaching values close to zero at twenty days. Desmosterol persists in the spinal cord myelin, remaining near 1% of the total sterol thereof. The desmosterol content of brain myelin and its persistence can be increased by treatment with drugs which inhibit cholesterol synthesis.[53]

E. Molar Ratios of Lipids

The concept of a unitary structure for biological membranes[55] implies a unitary molecular substructure. A unitary molecular structure has been assumed to imply a unitary chemical composition. This notion has been abetted by the ambiguous affinities of the OsO_4 and $KMnO_4$ used for electron microscopic visualization. If there is a constancy in chemical composition in membranes, there would be a constancy in the molar ratios of the constituents of membranes.

It is clear, however, that stoichiometric ratios are not appropriate for metabolic or equilibrium systems, even if some of the components exhibit "metabolic stability."[26,56] Moreover, since the hypothesis of the chemical similarity of membranes is not valid, the validity of the underlying related concept is questionable.

When the molecular ratios of the components of myelin from various sources as reported by various authors are compared (Table VIII) it is clear that the data display a comparatively high degree of dispersion, although there are some values for which the agreement is close. Nevertheless, the use of molar ratios is an alternative way to express differences between

TABLE VIII
Molar Ratios (Cholesterol = 1)

	Human	Squirrel monkey[40]		Bovine		Guinea pig		Rat	
	CNS[27,49]	CNS	PNS	CNS[24,39]	PNS[30]	CNS[39,38]a	PNS[39]	CNS[8,33,39]b	PNS[39]
Glycolipid	0.59, 0.44	0.50	0.43	0.48, 0.47	0.28	0.42 ± 0.05	1.8	0.47 ± 0.14	0.4
Total phospholipid	1.01, 0.75	0.77	0.93	0.75, 0.76	1.16	1.09 ± 0.17	1.56	1.14 ± 0.19	0.21
Choline phosphoglyceride	0.26, 0.20	0.14	0.13	0.18, 0.18	0.28	0.31 ± 0.08	0.33	0.28 ± 0.06	0.29
Ethanolamine phosphoglyceride	0.38, 0.29	0.33	0.36	0.58, 0.33	0.30	0.55 ± 0.09	0.07	0.46 ± 0.07	0.45
Serine phosphoglyceride	0.11, 0.08	0.13c	0.17c	0.05, 0.12c	0.30	0.22 ± 0.03c	—	0.20 ± 0.12c	0.23c
Inositol phosphoglyceride	— 0.01	—	—	0.02	—	—	0.53d	—	—
Sphingomyelin	0.19, 0.14	0.14	0.26	0.11, 0.11	0.31	0.11 ± 0.02	—	0.09 ± 0.03	0.24
Plasmalogen	0.31, 0.23	0.27	0.30	0.27, 0.26	0.29	—	—	—	—

a Mean of 3 reported values ± mean deviation.
b Mean of 5 or 6 reported values ± mean deviation.
c Sum of serine phospholipid and inositol phospholipid.
d Sum of serine phospholipid, inositol phospholipid, and sphingomyelin.

membranes from difference cellular sites,[56] between morphologically similar membrane systems, such as central and peripheral myelin in the same species, and between membranes of the same site but different species.

F. Myelin Proteins

1. General Discussion

The examination of proteins from isolated myelin has lagged somewhat behind the study of other myelin constituents. Probably this is due in part to the difficulties in handling the proteins, and in purifying and characterizing them. In two cases, proteins were isolated from white matter and were correctly assumed to be myelin proteins. These proteins are proteolipid protein[57] and an allergic encephalitic antigen.[58,59] There are other myelin proteins, but the number is small.[60,61]

Loosely defined, all the myelin proteins are proteolipid protein by virtue of their solubility in chloroform–methanol mixtures. If one adds the requirement that the proteolipid protein remain in solution with the lipids in the chloroform-rich phase of a chloroform–methanol–water two phase system,[62] it seems probable that only one of the proteins is the classical proteolipid protein.

2. Proteolipid and Related Proteins

Since this important group will be reviewed separately in another chapter, detail will be spared. Histological evidence[63] verified that proteolipid protein was a myelin constituent of both central and peripheral myelin and that it was fully extractable by the proper solvent mixture. Only recently has it been demonstrated that the classical proteolipid protein amounts to about 60% of total bovine myelin protein,[60] and that this protein can be isolated in a pure state, undenatured and freed of adventitious lipids.[64] Studies on the rat brain have shown that about 50% of the myelin protein is proteolipid protein[21] and that it occurs in other portions of the brain.[19,21]

Several other proteinaceous components of the nerve, believed to be myelin components, have occupied investigators and may be derivatives of proteolipid protein. They are neurokeratin,[65] the trypsin-resistant protein residue,[66] and neurosclerin.[67] Each of these is conspicuous by its insolubility. This property and the amino acid analyses[68,70] are consistent with the hypothesis that these materials are mixtures of denatured proteolipid and other proteins.[68] Whether a proteolipid characteristic of immature brain[71] and an acid-extractable proteolipid[72] are myelin constituents remains to be demonstrated.

3. Basic Proteins

By means of immunofluorescent techniques, the encephalitogenic component of white matter was shown to be an authentic myelin constituent.[73]

Extracts of white matter with acidified solvents permit purification of this material. The high molecular weight "Component IV" of Martenson and LeBaron[58] has been shown to be the same as the antigenic protein of Kies et al.[59,74] and was isolated in yields of 10 mg/g fresh bovine white matter. Another protein has been isolated from purified myelin by Lowden, Moscarello, and Morecki,[75] the amino acid composition of which (Table IX) is similar to that of a protein fraction isolated by Martenson and LeBaron ("acetic acid-eluted protein") and different from their "Component IV." This is also an encephalitogenic antigen and is isolated in yields of 10–20 mg/g of dry myelin. The molecular weight is quoted to be about 10,000.

4. Other Proteins and Peptides

Margolis has provided evidence of the existence of glycoproteins in isolated myelin.[76] The water-soluble brain protein of Moore[77] seems not to be a myelin component. Lowden, Wood, and Moscarello, have described the extraction of some low molecular weight peptides from myelin with mercaptoethanol which are encephalitogenic.[78]

The phosphatidopeptides of LeBaron[79] have been reexamined and shown to be concentrated in myelin.[80] These have been separated into 28

TABLE IX

Amino Acid Composition of Some Myelin Antigens[a]

	"Component IV"[58]	Acid-extractable protein[75]
Aspartic	7.0	8
Threonine	4.2	3
Serine	9.7	8
Glutamic	6.4	10
Proline	7.4	8
Glycine	15.1	10
Alanine	8.8	8
Cysteine	0.0	0.4–0.8
Valine	1.5	4
Methionine	1.2	0.8–1.2
Isoleucine	1.8	4
Leucine	6.4	6
Tyrosine	2.5	2
Phenylalanine	4.8	4
Lysine	7.6	6
Histidine	5.7	4
Arginine	10.1	6

[a] Mole percent of all amino acids except tryptophan.

fractions which yield phosphates of glycerol, inositol, and serine and a large amount of glycine, among other amino acids, when hydrolyzed.

G. Common Cellular Components Absent from Myelin or Present in Traces

1. Nucleic Acids

Although a low nucleic acid content has been taken as a criterion for the purity of isolated myelin,[24] it seems implausible that nucleic acids should be completely absent from myelin or any other cellular structure. In fact, small amounts of nucleic acids have been detected in the purest myelin and in a number of animals (Table X). The higher reported values may indeed represent contamination with other cellular structures, however.

Reported base compositions for myelin RNA (Table XI) show no close or consistent similarities to axoplasmic RNA[82] or that of other subcellular particles,[83] and may vary as a result of neural activity.[84]

2. Amino Acids

The only reported examination of myelin for free amino acids[85] shows a relative concentration profile (Table XII) which is not greatly different from that of whole brain. The absolute concentrations in myelin are much less than in whole brain.

3. Inorganic Material

The inorganic cation content of myelin is very low, even when recalculated in terms of myelin water rather than myelin solids (Table XIII). The

TABLE X

Nucleic Acid Content of Myelin[a]

Source	DNA	RNA
Bovine white matter[5]		
"Crude"	...(0.45 $\pm$ 0.15)[c]...	
"Light"	...(0.07–0.15)[c]...	
Guinea pig brain[65]d	—	3.9
Rat brain[2]	2.6	0.8–1.6
Rat brain[81]b	0.18 $\pm$ 0.02	0.1 $\pm$ 0.02
Rat brain[57]	...(0.25 $\pm$ 0.09)[c]...	
Goldfish Mauthner neuron[82]	—	0.6–1.0

[a] Mg nucleic acid/g dry myelin.
[b] Recalculated, assuming a myelin content of 38 mg dry myelin/g whole brain[8]; $N = 6$.
[c] DNA and RNA combined; $N = 7$.
[d] Recalculated, assuming 300 mg protein/g dry myelin.

TABLE XI

Base Composition of Myelin RNA[a]

	Adenine	Guanine	Cytosine	Uracil
Goldfish Mauthner myelin[82]b	20.9 ± 0.8	37.6 ± 1.7	23.4 ± 1.5	18.3 ± 1.2
Guinea pig myelin[83]c	21.9 ± 0.8	31.5 ± 0.4	26.9 ± 0.4	19.8 ± 0.5

[a] Mole percent, each base, of total base.
[b] Mean ± average of standard deviations; N = 5–9 for each of three sets of values.
[c] Mean ± standard deviation; N = 6.

relative concentration of the monovalent and divalent cations is very different from that found in whole brain, cytoplasm, cerebrospinal fluid, or blood plasma. It probably reflects the affinity of acidic lipids for these ions.[41,86]

4. Miscellaneous

Enzyme activities will be neglected here, to be covered in another chapter devoted to myelin metabolism. Among the coenzymes, however, guinea pig myelin may be calculated to have the following amounts of the pyridine nucleotides (in $m\mu$moles/g dry myelin): NAD, 130; NADH, 0; NADP, 21; NADPH, 15.[87]

By electron-microscopic autoradiography, norepinephrine and 5-hydroxytryptamine have been shown to be absent or present in vanishingly

TABLE XII

Free Amino Acids in Guinea Pig Myelin[85]a

Alanine	1.9
γ-Amino butyric acid	2.9
Aspartic acid	6.9
Glutamic acid	16.1
Glutamine	2.7
Glycine	2.7
Serine	2.4
Threonine	0.2
Unidentified	3.4

[a] Micromoles/g dry myelin, uncorrected for nonquantitative recovery. Recalculated, assuming a myelin content of 33 mg dry wt/g whole brain.

TABLE XIII

Metal Content of Myelin

Source	Metal	μg-atom/g dry wt
Human[29]a	Ca	4.0 ± 1.6
	Mg	13.5 ± 3.4
	Zn	0.45 ± 0.19
	Cu	0.20 ± 0.13
	Mn	0.055 ± 0.042
Guinea pig[85]b	K	1.8
Guinea pig[41]b	Na	3.6, 3.6
	K	1.7, 1.7
	Ca + Mg	11.8, 11.4

a Mean $\pm$ mean deviation; $N = 5$.
b Recalculated, assuming 33 mg myelin dry wt/g whole brain.[50]

small amounts in myelin.[88,89] DeRobertis *et al.* found about 5 mμmoles acetylcholine/g myelin.[4]

The lipoprotein complex which accumulates in aging neurons, lipofuscin, has been shown to be neither derived from myelin nor to be associated with it.[90]

III. STRUCTURE–COMPOSITION CORRELATIONS

A. Ultrastructure

It is beyond the scope of this review to deal *in extenso* on ultrastructural studies. The reader is instead referred to a number of excellent reviews and symposia.[91–93]

B. Lipid–Protein Complexes

The genealogy of some current hypothetical models for the molecular structure of myelin[95,96] includes the unit membrane hypothesis[55] and the classical model of Danielli.[97] In its simplest form, the myelin membrane consists of a double sheet of protein (osmiophilic), enclosing a matrix of lipids (non-osmiophilic) spirally wound so that the protein layers are tightly contiguous with the protein layers of the adjacent lamellae.

Briefly, Vandenheuval's model, carefully constructed with attention to probable molecular dimensions, requires that the hydrophobic groups on the protein layers are oriented externally to the lipid core and promote the stability of the myelin windings by forming a continuous, relatively hydrophobic milieu between juxtaposed protein layers. The more polar or ionic protein groups are oriented toward the polar or ionic groups of the lipid core. The lipids are arranged with the relatively nonpolar fatty side chains of the lipids directed toward the opposite protein layer so that the longer fatty acyl groups of the two lipid layers may meet and interdigitate in the center of the lamella.

This model is consistent with the basic assumptions of the electron microscopist that the protein is the osmiophilic portion of myelin and that the lipids are not stained well by OsO_4. The assumption about the reactivity of myelin proteins toward OsO_4 has been demonstrated to be literally correct.[98] The assumption about nonreactivity of the lipid moiety has been challenged by evidence that the unsaturated lipids are indeed osmiophilic.[56]

Thus, there are two sets of inconsistencies in the model under discussion: (1) the hypothetical hydrophobic interlammelar orientation of myelin proteins and the osmotic responsiveness of the myelin winding,[34,35] and (2) the orientation of the unsaturated acids (osmiophilic) and the lack of osmiophilic staining at the lamellar center.

Alternative suggestions have been advanced for a hydrophobic, non-ionic binding of lipids and proteins,[56,99] which would account for the observed osmiophilic and osmotic properties of myelin: the osmiophilic side chains of the lipids would be bound to the osmiophilic proteins and would probably be seen as a single-electron dense line in micrographs; if the hydrophobic groups of the protein were turned inward to bind the lipids, the hydrophilic groups would presumably be turned out, to account for the osmotic reactivity of myelin.

C. Myelin Membranes and Ions

The peculiar, special relationship of calcium to cellular membranes[100] appears to obtain in the case of myelin. The polyphosphoinositides[41] are presumably associated with the myelin calcium content and with the ion-exchange properties of extracts derived from myelin.[86] The view that calcium is essential to the morphological integrity of myelin is strongly supported by the work of Wolman,[101] who showed that in addition to the familiar effects due to variation in the tonicity of the medium, the structure of excised myelinated nerves could be destroyed or stabilized by alterations in the sodium–calcium ratio of the perfusing medium. Similarly, it was shown that the chemical composition of myelin was preserved in aqueous systems containing calcium, but not in others.[102] Not all of the lipids in isolated myelin are extractable in aqueous systems by cations other than calcium ion. Some of the cation-insensitive lipids exhibit extractability in the presence of certain anions.[103]

The function, if any, of myelin in maintenance of the electrolyte economy of the axon is not clear. The old concept of myelin functioning principally as an insulator is probably inaccurate [see Schmitt[91]]. It is clear that cationic substances can permeate myelin,[104] but whether the transport is inter-laminar or translaminar is not known.

The failure of the metal content of myelin (Table XIII) to reflect the composition of other biological fluids cannot be cited as evidence that myelin is not involved in electrolyte balance. Every preparation procedure for myelin is designed to remove contaminating cytoplasm and probably succeeds in removing that part of the Schwann cell or oligodendrocyte cytoplasm which should be accounted as part of myelin.

The review of literature pertinent to isolated myelin was concluded in December 1967.

IV. REFERENCES

1. J. D. Patterson and J. B. Finean, Ultracentrifugal fractionation of nerve tissue, *J. Neurochem.* **7**:251–258 (1961).
2. P. Mandel, T. Borkowski, S. Harth, and R. Mardell, Incorporation of ^{32}P in ribonucleic acid of subcellular fractions of various regions of the rat central nervous system, *J. Neurochem.* **8**:126–138 (1961).
3. C. August, A. N. Davison, and F. Maurice-Williams, Phospholipid metabolism in nervous tissue. 4. Incorporation of ^{32}P into the lipids of subcellular fractions of the brain, *Biochem. J.* **81**:8–12 (1961).
4. E. deRobertis, A. Pellegrino de Iraldi, G. Rodriguez de Lores Arnaiz, and L. Salganicoff, Cholinergic and non-cholinergic nerve endings in rat brain. I. Isolation and subcellular distribution of acetylcholine and acetylcholinesterase, *J. Neurochem.* **9**:23–35 (1962).
5. L. A. Autilio, W. T. Norton, and R. D. Terry, Preparation and some properties of purified myelin from the central nervous system, *J. Neurochem.* **11**:17–27 (1964).
6. R. H. Laatsch, M. W. Kies, S. Gordon, and E. C. Alvord, Jr., The encephalomyelitic activity of myelin isolated by ultracentrifugation, *J. Exptl. Med.* **115**:777–788 (1962).
7. E. G. Gray and V. P. Whittaker, The isolation of nerve endings from brain; an electron microscopic study of cell fragments derived by homogenization and centrifugation, *J. Anat. (London)* **96**:79–88 (1962).
8. L. M. Seminario, N. Hren, and C. J. Gomez, Lipid distribution in subcellular fractions of the rat brain, *J. Neurochem.* **11**:197–207 (1964).
9. F. H. Hulcher, E. V. Spudis, and M. G. Netsky, Encephalomyelitis induced by a white matter fraction, *Arch. Neurol.* **8**:1–23 (1963).
10. J. L. Nussbaum and P. Mandel, Distribution of phosphatides in myelin, mitochondria and microsomes of rat brain, *Bull. Soc. Chem. Biol.* **47**:395–408 (1965).
11. F. H. Hulcher, Physical and chemical properties of myelin, *Arch. Biochem. Biophys.* **100**:237–244 (1963).
12. R. Tanaka and L. G. Abood, Isolation from rat brain of mitochondria devoid of glycolytic activity, *J. Neurochem.* **10**:571–576 (1963).
13. F. Wolfgram, Macromolecular constituents of myelin, *Ann. N.Y. Acad. Sci.* **122**:104–115 (1965).
14. R. H. Laatsch, Fractionation of myelin constituents by a two phase system, *Federation Proc.* **22**:316 (1963).
15. J. Folch and M. Lees, Proteolipids, a new type of tissue lipoproteins. Their isolation from brain, *J. Biol. Chem.* **191**:807–817 (1951).

16. M. B. Lees, Effect of salts on the solubility of myelin, *Federation Proc.* **25**:767 (1966).
17. F. Wolfgram and A. S. Rose, A study of some component proteins of central and peripheral nerve myelin, *J. Neurochem.* **8**:161–168 (1961).
18. J. D. Robertson, The unit membrane of cells and mechanisms of myelin formation, *in Ultrastructure and Metabolism of the Nervous System*, Assoc. Res. Nerv. Ment. Dis., Vol. XL, pp. 94–158 (1962).
19. L. C. Mokrasch, Incorporation of ^{14}C-amino acids into the proteolipid of subcellular preparations of rat brain *in vitro*, *J. Neurochem.* **13**:49–58 (1966).
20. N. Robinson, Proteolipids and proteins in subcellular particles of human brain, *Clin. Chim. Acta* **13**:541–545 (1966).
21. A. J. Tolani and L. C. Mokrasch, Incorporation of ^{14}C-amino acids into proteolipid protein of subcellular fractions from rat brain, heart and liver, *Life Sci.* **6**:1771–1774 (1967).
22. M. Murokami, Y. Ozawa, and S. Funahashi, Proteolipid from beef heart muscle, II. Chemical composition and subcellular distribution, *J. Biochem. (Tokyo)* **54**:166–172 (1963).
23. W. Norton and L. A. Autilio, The chemical composition of bovine CNS myelin, *Ann. N.Y. Acad. Sci.* **122**:77–85 (1965).
24. W. T. Norton and L. A. Autilio, The lipid composition of purified bovine brain myelin, *J. Neurochem.* **13**:213–222 (1966).
25. G. Brante, Studies on lipids in the nervous system with special reference to quantitative chemical determination and topical distribution, *Acta Physiol. Scand.* **18**:Suppl. 63 (1949).
26. M. L. Cuzner, A. N. Davison, and N. A. Gregson, The chemical composition of vertebrate myelin and microsomes, *J. Neurochem.* **12**:469–481 (1965).
27. J. S. O'Brien and E. L. Sampson, Lipid composition of the normal human brain: Gray matter, white matter and myelin, *J. Lipid Res.* **6**:537–544 (1965).
28. J. S. O'Brien and E. L. Sampson, Fatty acid and fatty aldehyde composition of the major brain lipids in normal human gray matter, white matter and myelin, *J. Lipid Res.* **6**:545–551 (1965).
29. B. Gerstl, L. F. Eng, R. B. Hayman, M. G. Tavaststjerna, and P. R. Bond, On the composition of human myelin, *J. Neurochem.* **14**:661–670 (1967).
30. J. S. O'Brien, E. L. Sampson, and M. B. Stern, Lipid composition of myelin from the peripheral nervous system, *J. Neurochem.* **14**:357–365 (1967).
31. S. R. Korey, M. Orchen, and M. Brotz, Studies of white matter. I. Chemical composition and respiration of neuroglial and myelin enriched fractions of white matter, *J. Neuropath. Exptl. Neurol.* **17**:430–438 (1958).
32. J. S. O'Brien and E. L. Sampson, Myelin membrane: A molecular abnormality, *Science* **150**:1613–1614 (1965).
33. J. L. Nussbaum, R. Bieth, and P. Mandel, Phosphatides in myelin sheaths and repartition of sphingomyelin in the brain, *Nature* **198**:586–587 (1963).
34. J. D. Robertson, Structural alterations in nerve fibers produced by hypotonic and hypertonic solutions, *J. Biophys. Biochem. Cytol.* **4**:349–364 (1958).
35. J. B. Finean and P. F. Millington, Effects of ionic strength of immersion medium on the structure of peripheral nerve myelin, *J. Biophys. Biochem. Cytol.* **3**:89–94 (1957).
36. J. B. Finean, Electron microscope and electron diffraction studies of the effects of dehydration on the structure of nerve myelin, *J. Biophys. Biochem. Cytol.* **8**:13–29 (1960).
37. F. A. Vandenheuval, Structural studies on biological membranes: The structure of myelin, *Ann. N.Y. Acad. Sci.* **122**:57–76 (1965).
38. J. Eichberg, V. P. Whittaker, and R. M. Dawson, Distribution of lipids in subcellular particles of guinea pig brain, *Biochem. J.* **92**:91–100 (1964).
39. M. J. Evans and J. B. Finean, The lipid composition of myelin from brain and peripheral nerve, *J. Neurochem.* **12**:729–734 (1965).

40. L. A. Horrocks, Composition of myelin from peripheral and central nervous systems of the squirrel monkey, *J. Lipid Res.* **8**:569–576 (1967).
41. J. Eichberg and R. M. C. Dawson, Polyphosphoinositides in myelin, *Biochem. J.* **96**:644–650 (1965).
42. L. Rathbone, Effect of diet on the fatty acid compositions of serum, brain, brain mitochondria and myelin in the rat, *Biochem. J.* **97**:620–628 (1965).
43. R. J. Light and O. M. Easton, Saponifiable fatty acids of the myelinated and unmyelinated nerve fibers of the garfish, *J. Neurochem.* **14**:141–142 (1967).
44. H. Pilz and E. Mehl, Studies on the lipid composition of human myelin, *Z. Physiol. Chem.* **346**:306–309 (1966).
45. K. Suzuki, S. E. Poduslo, and W. T. Norton, Gangliosides in the myelin fraction of developing rats, *Biochim. Biophys. Acta* **144**:375–381 (1967).
46. M. W. Spence and L. S. Wolfe, Gangliosides in the developing brain. Isolation and composition of subcellular membranes enriched in gangliosides, *Can. J. Biochem.* **45**:671–688 (1967).
47. A. N. Davison, M. L. Cuzner, N. L. Banick, and J. Oxbury, Myelinogensis in the rat brain, *Nature* **212**:1373–1374 (1966).
48. H. Weigandt, The subcellular localisation of gangliosides in brain, *J. Neurochem.* **14**:671–674 (1967).
49. W. T. Norton, S. E. Poduslo, and K. Suzuki, Subacute sclerosing leukoencephalitis, II. Chemical studies including an abnormal ganglioside pattern, *J. Neuropathol. Exptl. Neurol.* **25**:582–597 (1966).
50. H. P. Schwarz, I. Kostyk, A. Marmolejo, and C. Sarappa, Long chain bases of brain and spinal cord of rabbits, *J. Neurochem.* **14**:91–97 (1967).
51. F. Young and F. H. Hulcher, Cholesterol esters in myelin and the component fatty acids, *Proc. Soc. Exptl. Biol. Med.* **123**:385–387 (1966).
52. M. E. Smith, R. Fumagilli, and R. Paoletti, The occurrence of desmosterol in myelin of developing rats, *Life Sci.* **6**:1085–1091 (1967).
53. N. L. Banik and A. N. Davison, Desmosterol in rat brain lipids, *J. Neurochem.* **14**:594–596 (1967).
54. L. F. Eng and M. E. Smith, The cholesterol complex in the myelin membrane, *Lipids* **1**:296 (1966).
55. J. D. Robertson, The unit membrane, *in Electron Microscopy in Anatomy* (J. D. Boyd, F. R. Johnson, and J. D. Lever, eds.), pp. 74–99, Edward Arnold, Ltd., London (1961).
56. E. D. Korn, Structure of biological membranes, *Science* **153**:1491–1498 (1966).
57. J. Folch-Pi, Composition of the brain in relation to maturation, *in Biochemistry of the Developing Nervous System* (H. Waelsch, ed.), pp. 121–136, Academic Press, New York (1955).
58. R. E. Martenson and F. N. LeBaron, Studies on the acid-extractable proteins of bovine brain white matter, *J. Neurochem.* **13**:1469–1479 (1966).
59. M. W. Kies, E. B. Thompson, and E. C. Alvord, The relation of myelin proteins to experimental allergic encephalomyelitis, *Ann. N.Y. Acad. Sci.* **122**:148–160 (1965).
60. L. Autilio, Fractionation of myelin proteins, *Federation Proc.* **25**:764 (1966).
61. C. W. Cotman and H. R. Mahler, Resolution of insoluble proteins in rat brain subcellular fractions, *Arch. Biochem. Biophys.* **120**:384–396 (1967).
62. J. Folch-Pi, Recent studies on the chemistry of the brain and their relation to the structure of the myelin sheath, *Exposes Ann. Biochim. Med.* **21**:81–95 (1959).
63. F. Wolfgram and A. S. Rose, A study of some component proteins of central and peripheral nerve myelin, *J. Neurochem.* **8**:161–168 (1961).
64. L. C. Mokrasch, A rapid purification of proteolipid protein adaptable to large quantities, *Life Sci.* **6**:1905–1909 (1967).

65. A. Ewald and W. Kuhne, Concerning a new constituent of the nervous system, *Verhandl. Naturhist. Med. Verins Heidelberg* 1:457–489 (1877).

66. F. N. LeBaron and J. Folch, Isolation from brain tissue of a trypsin–resistant protein fraction containing combined inositol and its relation to neurokeratin, *J. Neurochem.* 1:101–108 (1956).

67. L. L. Uzman, Lipophilic peptides and proteins of brain. I. Their relation to development of the brain and myelin formation, *Arch. Biochem. Biophys.* 76:474–489 (1958).

68. F. N. LeBaron, The relation of certain lipid-protein complexes of nervous tissue to neurokeratin, *in Brain Lipids and Lipoproteins and the Leucodystrophies* (J. Folch and H. Bauer, eds.), pp. 31–41, Elsevier Publishing Co., Amsterdam (1963).

69. L. L. Uzman and H. Rosen, Lipophilic peptides and proteins of brain. II. Composition of the neurosclerin fraction, *Arch. Biochem. Biophys.* 76:490–495 (1958).

70. F. Wolfgram and A. S. Rose, The amino acid composition of central and peripheral nerve neurokeratin, *J. Neurochem.* 9:623–627 (1962).

71. A. L. Prensky and H. W. Moser, Changes in the amino acid composition of proteolipids of white matter during maturation of the human nervous system, *J. Neurochem.* 14:117–121 (1967).

72. F. Wolfgram, A new proteolipid fraction of the nervous system, *J. Neurochem.* 13:461–470 (1966).

73. H. C. Rauch and S. Raffel, Immunofluorescent localisation of encephalitogenic protein in myelin, *J. Immunol.* 92:452–455 (1964).

74. M. W. Kies, E. C. Alvord, R. E. Martenson, and F. N. LeBaron, Encephalitogenic activity of bovine basic proteins, *Science* 151:821–822 (1966).

75. J. A. Lowden, M. A. Moscarello, and R. Morecki, The isolation and characterisation of an acid-soluble protein from myelin, *Can. J. Biochem.* 44:567–577 (1966).

76. R. U. Margolis, Acid mucopolysaccharides and proteins of bovine white matter and myelin, *Biochim. Biophys. Acta* 141:91–102 (1967).

77. B. W. Moore, A soluble protein characteristic of the nervous system. *Biochem. Biophys. Res. Commun.* 19:739–744 (1965).

78. J. A. Lowden, D. D. Wood, and M. A. Moscarello, Encephalitogenic activity in mercaptoethanol extracts of myelin, *Can. J. Biochem.* 45:148–151 (1967).

79. F. N. LeBaron and E. E. Rothleder, Brain lipid–peptide complexes extracted with acidified solvents, *in Biochemistry of Lipids* (G. Popjak, ed.), pp. 1–7, Pergamon Press, New York (1960).

80. P. Mandel and M. Ledig, Amidated phosphopeptides in nervous tissue, *Biochem. Biophys. Res. Commun.* 24:275–279 (1966).

81. R. Balazs and W. A. Cocks, RNA metabolism in subcellular fractions of brain tissue, *J. Neurochem.* 14:1035–1055 (1967).

82. A. Edstrom, The ribonucleic acid in the mauthner neuron of the goldfish, *J. Neurochem.* 11:309–314 (1966).

83. S. Yamagami and Y. Kawokita, Base composition of ribonucleic acid of subcellular fractions from guinea pig brain, *J. Neurochem.* 11:899–900 (1964).

84. B. Jakonbek and J. E. Edstrom, Ribonucleic acid changes in the mauthner axon and myelin sheath after increased functional activity, *J. Neurochem.* 12:845–849 (1965).

85. J. L. Mangan and V. P. Whittaker, The distribution of free amino acids in subcellular fractions from guinea pig brain, *Biochem. J.* 98:128–137 (1966).

86. G. Leitch, Some ion exchange properties of a myelin extract from bovine optic nerve, *Proc. Soc. Exptl. Biol. Med.* 121:1253–1256 (1966).

87. A. Lindall and J. D. Frantz, Synaptosome pyridine nucleotide content, *J. Neurochem.* 14:771–774 (1967).

88. G. K. Aghajanian and F. E. Bloom, Electron microscopic autoradiography of rat hypo-thalamus after intraventricular ^{3}H-norepinephrine, *Science* **153**:308–310 (1966).

89. G. K. Aghajanian and F. E. Bloom, Localisation of tritiated serotonin in rat brain by electron microscopic radioautography, *J. Pharmacol. Exptl. Therap.* **156**:23–30 (1966).

90. T. Samorajski, J. R. Keefe, and J. M. Ordy, Intracellular localization of lipofuscin age pigments in the nervous system, *J. Gerontol.* **19**:262–276 (1964).

91. S. R. Korey (ed.), *The Biology of Myelin*, Hoeber Harper, New York (1959).

92. S. R. Korey, A. Pope, and E. Robins (eds.), *Ultrastructure and Metabolism of the Nervous System*, Williams and Wilkins Co., Baltimore (1962).

93. H. E. Whipple (ed.), Research in demyelinating diseases, *Ann. N.Y. Acad. Sci.* **122**:Article 1 (1965).

94. A. Peters, Further observations on the structure of myelin sheaths in the central nervous system, *J. Cell Biol.* **20**:281–296 (1964).

95. F. A. Vandenheuval, Structural studies of biological membranes: The structure of myelin, *Ann. N.Y. Acad. Sci.* **122**:57–76 (1965).

96. F. A. Vandenheuval, Lipid–protein interactions and cohesional forces in the lipoprotein systems of membranes, *J. Am. Oil Chem. Soc.* **43**:258–264 (1966).

97. H. Davson and J. F. Danielli, *The Permeability of Natural Membranes*, Cambridge University Press, Cambridge (1952).

98. L. Napolitano, F. LeBaron, and J. Scaletti, Preservation of myelin lamellar structure in the absence of lipid, *J. Cell Biol.* **34**:817–826 (1967).

99. D. F. Hoelzl Wallach and P. H. Zahler, Protein conformations in cellular membranes, *Proc. Natl. Acad. Sci.* **56**:1552–1559 (1966).

100. J. F. Manery, Effects of Ca ions on membranes, *Federation Proc.* **25**:1804–1810 (1966).

101. M. Wolman, Myelin breakdown *in vitro*, *Biochim. Biophys. Acta* **102**:261–268 (1965).

102. M. Wolman and H. Wiener, Structure of the myelin sheath as a function of concentration of ions, *Biochim. Biophys. Acta* **102**:269–279 (1965).

103. M. Wolman, H. Wiener, and J. J. Bubis, On the nature of cation-insensitive bonds of myelin, *Israel J. Chem.* **4**:53–58 (1966).

104. M. Singer and M. M. Solpeter, Transport of tritium-labelled L histidine through the Schwann and myelin sheaths into the ion of peripheral nerves, *Nature* **210**:1225–1227 (1966).

Chapter 10

STEROLS

R. Paoletti, E. Grossi-Paoletti, and R. Fumagalli

Institutes of Pharmacology
Universities of Milan and Cagliari
Italy

I. INTRODUCTION

The high concentration of cholesterol in the nervous tissues and its localization in cell membranes, notably myelin, indicate an important role for this sterol and related compounds in growth, maturation, and metabolism of the brain. However, very little is known, not only about sterol function, but, until recently, about sterol composition and pathways of biosynthesis and metabolism in the brain, especially when compared with the liver. This lack of information is due to the great difficulty in isolating and identifying other sterols in the presence of large amounts of cholesterol and by the fact that age, anatomical areas, routes of administration greatly modify the results obtained in studying brain cholesterol biosynthesis and turnover.

The recent introduction of drugs inhibiting specific steps of cholesterol biosynthesis in the nervous system has made possible a closer approach to solving these complex problems by inducing selective accumulation of precursors which are easier to separate and identify. These methods and the introduction of sensitive analytical procedures are the bases for the rapid progress made recently in this area.

II. ISOLATION AND IDENTIFICATION OF BRAIN STEROLS

A. Sterol Extraction

Brain sterols may be isolated either from total lipid extracts or from the non-saponifiable fraction. Total lipids from brain are extracted using chloroform/methanol 2:1 according to the method of Folch *et al.* and its modifications.[1,2] Total lipids may then be purified from water-soluble non-lipid contaminants using Sephadex columns.[3] Neutral lipids containing free sterols and sterol esters are then separated from the polar lipids using silicic acid[4] or diethylaminoethylcellulose (DEAE) column chromatography.[5,6] Silicic acid columns may also allow the separation of neutral lipids into their major classes: hydrocarbons, free sterols, triglycerides, and sterol esters.[7] Column fractionation on triethylaminoethylcellulose (TEAE) has been introduced recently for the same purpose with more reproducible results.[8]

When total sterols are considered, brain saponification in alcoholic KOH can be used and sterols are separated from other components of the non-saponifiable fraction (squalene and polyprenols) using alumina[9,10] or silicic acid column chromatography[11]; alternatively, sterols containing a hydroxyl group in 3-β position.are purified by digitonin[12] or tomatin[13] precipitation. Using digitonin precipitation, some selective loss of sterols may occur during washing of the precipitated digitonides.[14]

B. Spectrophotometric Procedures

Cholesterol concentrations in brain are easily measurable in terms of colorimetric reactions: the Lieberman–Burchard[15] and Zlatkis[16] procedures are widely used. To evaluate total cholesterol with the former color reaction, it is necessary to hydrolyze the cholesterol esters[17]; this step is avoided with the Zlatkis procedure. However, cholesterol esters are not present in mature adult brain. Differential spectrophotometric techniques are required for sterols other than cholesterol.[18,19] However, quantitative results are obtained only when considerable concentrations of the other sterols are present. Sterols containing conjugated diene systems, such as 7-dehydrocholesterol, can be measured readily by UV spectrophotometry.[20]

A complete evaluation of the sterol patterns, however, requires the use of chromatographic techniques. A difficulty is due to the presence of large amounts of some sterols, primarily cholesterol, together with, sometimes, extremely low concentrations of others. The structural differences among the sterols to be examined should also be considered. Brain sterols, with a few exceptions,[21] have only one alcoholic function in 3-β position and differ mainly in molecular size as well as in the degree of unsaturation. Therefore, the chromatographic separations should be based on these structural differences.

C. Chromatographic Methods

1. Thin-Layer Chromatography (TLC)

Fractionation of sterol mixtures of biological origin using TLC has recently been reviewed by Copius-Peereboom and Beekes.[22] Although the resolution of structurally related sterols on silica gel G or aluminum oxide layers is not easily achieved, some separations have been reported.[23–25] The use of reverse phase TLC improves the separation of sterols. For instance, cholesterol and desmosterol are separated, as acetyl-derivatives, using the reverse phase system: undecane/acetic acid–acetonitrile $(1:3)$[22] which, on the other hand, does not resolve either cholesterol from lanosterol and from Δ^7-cholestenol, or desmosterol from zymosterol. The use of brominated derivatives improves the separation of some critical pairs of sterols,[26] but sterols containing diene conjugated systems or devoid of the double bond in position 5-6 may undergo complete decomposition.[22] Sterols differing in the presence, number, and position of double bonds are more easily separated by means of thin-layer plates impregnated with silver nitrate. This argentation TLC displays to a great extent structural differences based on different degrees of unsaturation. The separation of cholesterol and related sterols, in the free form or as the acetyl derivatives, has been obtained.[22,27–31]

2. Column Chromatography

Column chromatography is, in comparison to TLC, a more time-consuming procedure, but with some definite advantages.

In TLC, one or at most two solvent systems can be applied and this may not always be sufficient for a complete resolution of complex sterol mixtures of biological origin.

In column chromatography, solvent systems of varying degree of polarity are used and the resolution of structurally related sterols is therefore improved. Another favorable feature is the possibility of processing larger amounts of sterols, so that the detection of minor components ($< 0.5\%$ of total sterols) is possible. Since the polarity of the 3-β hydroxyl group, common to all brain sterols, tends to overshadow other less polar functional groups, such as double bonds, it is preferable to convert sterols to derivatives with blocked 3-β alcoholic function. Sterol derivatives, such as azoyl esters,[32] iodobenzoates,[33] and acetates,[34,35] have been used. Sterol separation can be accomplished on alumina or silicic acid columns.[36–40] However, argentation column chromatography is preferred for its property to separate sterols according to the presence, number, and position of the double bonds. This procedure is now widely used[31,35,41,42] because it allows quantitative separations.[35] A review of silver nitrate chromatography in sterol and lipid fractionation has appeared recently.[43] In this laboratory, the elution of cholesterol and its metabolically related brain sterols, C27 saturated, C29 Δ^8, C28 Δ^8, C27 Δ^8, cholesterol, C29 $\Delta^{8,24}$, C28 $\Delta^{8,24}$, C27 $\Delta^{8,24}$, C27 $\Delta^{5,24}$, has been achieved using silver nitrate chromatography.[44]

3. Gas–Liquid Chromatography (GLC)

This highly sensitive method leads to a rapid fractionation of complex mixtures of sterols and provides qualitative and quantitative data, as well as useful structural information, on their components. In analytical separation of sterols, thin-film packed columns are used.[45] The separation of mixtures of sterols biologically related to cholesterol needs the choice of liquid phases showing specific retention effects for the structural differences involved.

While nonpolar liquid phases such as methylpolysiloxanes are preferred when sterols differ mainly in molecular size and shape, liquid phases with specific retention effects for carbon–carbon unsaturation should be used when dealing with sterols differing for presence, number, and position of double bonds. Among them, PhSi 191–43 and OV–17 (phenylmethylpoly-siloxanes), EGSS-X (methylpolysiloxane–ethylene glycol succinate copo-lymer), CHDMS (cyclohexanedimethanol succinate), CNSi (a cyanoethyl-methylpolysiloxane), and linear polyesters (neopentylglycol succinate, and diethyleneglycol succinate) have been used. For quantitative estimation of sterols by GLC, proper deactivation of the solid support[46] and the use of sterol derivatives are needed to avoid a partial loss of the compounds during the analyses. Sterol derivatives have to be volatile, thermostable, and less polar than the free sterols: acetyl esters,[47] trifluoroacetylesters,[48] methyl ethers,[49] trimethylsilyl ethers,[50] chloromethyldimethylsilyl ethers,[51] and dimethylsilyl ethers[52] have been used.

The behavior of sterols in GLC, as in other systems of partition chroma-tography, provides information regarding their structure. Among others, Horning *et al.* have proposed the use of a "Steroid Number" for the charac-terization of steroids on nonpolar[53] and polar stationary phases.[54] However, it should be stressed that these methods lead to tentative identifi-cation only and that independent evidence is required.

Problems involved in GLC of steroids are discussed in a number of reviews.[55–58]

D. Sterol Identification

The chromatographic investigation of sterols does not provide final information as to their structure. However, its principal advantage is the exclusion of many of the possible structures hypothesized. Definite proof of structure should be based on a combination of various techniques: melting point of sterols after crystallization, spectrophotometric techniques, and optical rotation, which is valuable for the location of the nuclear double bonds and presence of methyl groups in the ring system.[59] Conjugated diene systems in the steroidal nucleus are detected by means of UV absorption[60] and the location of double bonds is facilitated by the use of IR spectro-metry[37,61] and nuclear magnetic resonance.[61] The combination of IR spectrometry with GLC has considerable potentialities in structural studies of micro amounts of sterols.

A major advance in structural investigations of sterols is the combination of gas chromatography with mass spectrometry.[62–65] This technique com-bines the advantages of GLC (high resolution) and mass spectrometry (characteristic fragmentation patterns), thus providing an excellent approach to structural studies of micro amounts of sterols present in complex mixtures of biological origin. The information obtained refers mainly to the molecular weight and to the number and position of double bonds in the ring as well as in the side chain. The fragmentation patterns of cholesterol and other meta-bolically related sterols have been reported recently.[66–69]

E. Radioactivity Determination

The radioactivity evaluation of biosynthetically labeled sterols is based on their isolation in pure form, obtained by the chromatographic techniques. To avoid possible contaminants, the isolated compound should be crystallized as such, or, preferably, after preparation of proper derivatives until specific radioactivity remains constant.[70,71] Since trace amounts of sterols are often dealt with, authentic standard carrier, when available, is added before crystallization. The possibility that sterols can form mixed crystals should also be taken into account. For determination of radioactivity, liquid scintillation techniques allowing differential isotopic counting are generally employed.

The simultaneous determination of sterol amount and radioactivity using radiogaschromatography[72] is of considerable help. This procedure is based on continuous counting either of the intact sterols in the effluent gas or of their labeled combustion products (carbon dioxide and water). General problems of radiogaschromatography have been reviewed recently.[73]

III. BRAIN STEROL COMPOSITION

A. Sterols in Adult Brain

Cholesterol is the only major sterol present in mature nervous tissue and its concentration is exceptionally high, about 10 % of the dry weight. In most mammalian species, including man, it represents about 20 mg/g fresh weight.

Cholesterol accumulation in brain is associated with the formation of myelin and, in myelinated brain, its turnover rate is very slow, in contrast with the immature brain. Other sterols have been identified in brain extracts: cholestanol and Δ^7 cholestenol are present in commercial preparations of cholesterol from spinal cord and brain of cattle,[74] and Δ^7 sterols are also present in human brain fixed in formalin.[75] Cholestanol, a catabolite of cholesterol,[76] has been detected in adult brain[77,78] and sterols with conjugated double bonds were also found.[79,80] 24-Hydroxycholesterol,[81] 7-β- and 7-α-hydroxycholesterol[21] were identified and Nicholas has shown the presence of an acidic fraction of cholesterol catabolites in brain.[82] Desmosterol, which is an important component of brain sterols in immature brain, is barely detectable in adult brain by gas chromatographic or colorimetric techniques. More recent investigations in this laboratory,[44] using larger samples of brain sterols, have led to the detection and quantitation in adult rat brain of many more sterols, biosynthetically related to cholesterol, in addition to cholesterol and desmosterol.

In the brain of three-month-old rats, desmosterol represents 0.8 % of the total sterols, followed by cholestanol (0.12 %), and by other sterols with 27, 28, and 29 carbon atoms and one or two double bonds, in the nucleus (8-9 position or 7-8) and in the lateral chain (in position 24) as shown in Table I.

TABLE I

Sterols Detected in Brain of Adult Rat[a]

Sterols	Percent of total sterols
C29 Δ^8	0.007
C28 Δ^8	0.011
C27 Δ^8	0.155
C29 $\Delta^{8,24}$	0.082
C28 $\Delta^{8,24}$	0.050
C27 $\Delta^{8,24}$	0.047
C27 $\Delta^{7,24}$	0.070
C27 $\Delta^{5,24}$	0.797
C27 Δ^5	98.000
C27 Δ^0	0.122
Unidentified	0.699

[a] Reported from Weiss et al.[44]

B. Sterols in Brain Development

During the maturation of the nervous tissue, an increase in the concentration of the various lipids has been demonstrated by many investigators.[83–86]

The concentration of cholesterol increases sharply during brain development. Folch-Pi[87] found a manifold increase of cholesterol in mouse brain from birth to 90 days, and Mandel and Bieth[88] demonstrated an increase of cholesterol concentration during the development of the rat brain from the newborn to the adult. The same is also true for chick embryo brain.[89] Actually the concentration of cholesterol increases further in the brain of adult rats[85] and mice[84] in contrast to the histological findings, which indicates a complete deposition of myelin. This fact agrees well with the recent demonstration given by Norton that in rat brain the total amounts of myelin increase linearly up to the 450th day of age.[90] During development, brain tissue actively synthesizes cholesterol to meet the requirements of myelin deposition. It is possible that some precursors of cholesterol might be detected in the embryonic tissue in larger amounts than in mature brain. In fact, it was demonstrated that the developing brain of chick embryo,[91] rat,[92,93] and man[93] contains desmosterol in considerable amounts. Desmosterol–cholesterol ratios decrease regularly in correlation with brain maturation, particularly with myelin deposition, as shown in several animal species: rat, mouse, guinea pig, chick embryo, and man (Table II).[94]

Fish, Boyd, and Stokes[91] established a precursor product relationship between desmosterol and cholesterol in chick embryo brain which seems to be valid also for other animal species because larger quantities of desmosterol are found in the brain of rat, mouse, guinea pig, chick embryo, and man when the biosynthetic activity is high. The reason for desmosterol accumulation

TABLE II

Desmosterol in Brain of Various Animal Species at Different Periods of Life[a]

Species	Before birth	Desmosterol as percent of total sterols	At birth	Desmosterol as percent of total sterols	Adult	Desmosterol as percent of total sterols
Man	6-month gestation	5.6		1.6		n.d.[b]
Rabbit	—	—		18.0		n.d.
Guinea pig	1-month gestation	18.1		traces		n.d.
Rat	2-week gestation	19.6		26.2 / 23.7[c]		0.8
Mouse	—	—		22.0		n.d.
Chicken	14-day egg incubation	10.7 / 11.2[d]		1.4 / 1.2[d]		n.d.

[a] Data from author's laboratory unless otherwise stated.

[b] n.d. denotes nondetectable.

[c] Calculated from the value of desmosterol/cholesterol ratio at birth, reported by Kritchevsky et al.[96]

[d] Fish et al.[91]

TABLE III

Sterols Detected in Brain of Chick Embryo (As Percent of Total Sterols)[a]

Sterols	11-Day incubation	18-Day incubation
C29 Δ^8	0.033	0.027
C28 Δ^8	0.015	0.255
C27 Δ^8	0.016	n.d.
C29 $\Delta^{8,24}$	0.154	0.314
C28 $\Delta^{8,24}$	0.140	0.198
C27 $\Delta^{8,24}$	0.066	0.105
C27 $\Delta^{7,24}$	0.013	0.125
C27 $\Delta^{5,24}$	7.574	6.865
C27 Δ^5	89.330	90.590
C27 Δ^0	0.450	0.426
Unidentified	2.208	1.092

[a] Reported from Weiss et al.[44]

in brain during maturation is still uncertain; a high control mechanism might be involved, linked to a limited NADPH supply. Other sterols were detected in the brain of chick embryos and immature rats: Holstein, Fish, and Stokes detected zymosterol[95] and Kritchevsky et al.[96] the presence of another sterol, tentatively called 7-24 cholestadien-3β-ol, in the brain of immature rats.

In our laboratory,[44] a considerable number of sterols have been detected and identified in the unsaponifiable fraction of chick embryo brain lipids (Table III). The sterols are the same detected in adult rat brain, but the concentrations are higher, and, comparing chick embryos of two different ages, 11 and 18 days of incubation, the absolute amounts of these sterols are higher at 18 days than at 11.

C. Distribution of Sterols and Sterol Esters

Cholesterol is present in mature brain in the free form only[97,98] and LeBaron and Folch[99] postulated that it is combined with water-soluble proteins in the myelin sheath. However, esterified cholesterol is present in the developing brain. Cholesterol esters were detected in the brain[100] and in the cord[101] of the chick embryo, and are also present in fetal human brain[101] and in newborn human brain,[83,102] while Cumings[97] found trace amounts of cholesterol esters in adult white and infant gray matter. More recently, Tichy detected cholesterol esters and measured their fatty acid composition in human white matter,[103] showing that the fatty acid pattern differs from plasma cholesterol esters. The amounts detected, however, vary considerably, probably due to methodological differences, and it seems difficult to correlate the concentration changes with brain maturation.

The brain of the developing rabbit from the 17th to the 33rd day of age contains minute amounts of esterified cholesterol, never exceeding 0.7% of the total, which cannot all be accounted for by plasma contamination.[104] Grafnetter, Grossi, and Morganti[105] measured the esterified sterols in the brain of the developing chick embryo from the 8th to the 36th day from the beginning of incubation. The concentration of the sterol esters decrease regularly during brain development from 3.0 to 0.3% of the total sterols. The sterol and fatty acid composition of brain sterol esters were also examined[106] and it was demonstrated that the sterol esterified is mostly cholesterol while desmosterol, even if present in considerable amounts in immature brain, is almost completely in free form.

Up to the present, it seems difficult to establish the role of esterified sterols in developing brain. Cholesterol distribution in the various areas of the brain is uneven. White matter contains three to four times more sterols than gray.[83,107] The highly myelinated areas of human brain contain up to 42 mg/g (wet.wt.) of cholesterol, while other areas have as low as 7 mg of sterol per gram[108,109] Histochemical data confirm that cholesterol distribution varies from area to area of the brain.[110]

More recently, the availability of methods for the separation and purification of brain subcellular fractions gave new impulse to the studies of lipid distribution in subcellular compartments of the brain, with special regard to purified myelin. For this purpose, ultracentrifugation techniques were applied, using sucrose gradient,[111,112] as recently reviewed in detail by Smith.[113] The lipid composition of the various subcellular fractions was investigated by many authors; however, there are some differences in the results obtained, probably due to different methods of subfractionation and lipid analysis.

Cholesterol is always the most abundant lipid component in purified myelin. The amounts detected in myelin from various animal species range from 0.5 to 0.7 μmoles/mg lipid with consistent results for several animal species, such as rat,[113–115] guinea pig,[114,116] ox,[115,117] and man.[117,118] It is now accepted that myelin lipid composition changes with brain maturation, in support of which it was shown that desmosterol is present also in myelin sterols,[119–121] but in concentrations lower than in the total brain homogenate, and its concentration decreases regularly with brain maturation in the same sense.[119,120] Myelin cholesterol accounts for approximately 78% of brain cholesterol[117] and the exact location of the remaining fraction is difficult to establish because of lack of sufficient purity of many of the subcellular fractions isolated from the brain.

IV. BIOSYNTHESIS OF BRAIN STEROLS

Cholesterol is synthesized in brain during growth and in connection with myelin deposition, and the measurement of breakdown rate in brain and plasma of cholesterol biosynthesized from labeled precursors indicates that labeled sterols found in the central nervous system are largely the result of endogenous synthesis.[122] In adult brain, the rate of incorporation of labeled precursors into sterols is generally very low and the question of whether the synthesis of new sterols occurs is still open; the present status is summarized later (**IV. C**). The pathways leading to cholesterol from simple precursors (acetate, mevalonic acid) have been investigated only partially in brain, while such steps are well known in the mammalian liver. The pathways of sterol biosynthesis in the liver were reviewed recently by Olson.[123]

A. From Acetate to Squalene

Acetate, leucine, glucose, and mevalonate are incorporated in brain cholesterol, glucose being the most efficient *in vivo* precursor[122] in contrast with the more efficient incorporation of mevalonic acid into liver cholesterol. In brain, glucose has been found to be a better precursor than acetate for sterols,[124] while acetate and glucose are incorporated into cholesterol at similar rates when rat brain stem and spinal cord preparations are used *in vitro*.[125]

In vitro results obtained in this laboratory using brain slices and comparing acetate, butyrate, mevalonate, and glucose, regard acetate as a very efficient precursor when using tissue of 10- and, to a lesser degree, 60-day-old rats.[126]

The observation that mevalonate, while very efficiently incorporated in liver cholesterol,[127] is much less utilized in brain,[128] is of considerable interest. Potassium mevalonate is utilized far better than mevalonolactone by brain from suckling rats, thus suggesting that mevalonic lactonase is lacking in the brain, in comparison with the liver, in the same animal.[129] A proper evaluation of these results should take into account several factors besides the chemical form of the precursor. Mevalonate is more efficiently utilized than acetate for sterol biosynthesis by brain slices when the incubation time is prolonged,[130] while the opposite is observed in the brain from 12-day-old animals.[131]

The route of administration is also critical: glucose is more efficiently incorporated into brain sterols than mevalonate if given intraperitoneally, while the opposite is observed after intracerebral inoculation.[96] Mevalonic acid-2-C[14] is more easily incorporated in brain stem and spinal cord than in the cortex (gray matter) of growing rats, which implies differential utilization according to the anatomical areas.[132,133] Kabara suggests that the observed differences may be related to the presence of multiple pools for cholesterol precursors in the nervous tissues, including a slowly metabolizing fraction containing "structural" sterols and precursors and a more rapidly metabolized "functional" fraction.[134]

The presence of labeled squalene in brain was recently demonstrated by Nicholas and Thomas[135] after intracerebral injection of labeled mevalonic acid in adult female rats. Independently, Agranoff[136] found squalene and the enzyme isopentenylpyrophosphate isomerase in adult central nervous tissue, which suggests that the pathways from the isopentenyl unit to squalene may be similar to that known for mammalian liver (Fig. 1). However, in unpublished data Grossi showed that the feedback mechanism active in the liver cannot be shown in the brain.[137]

B. From Squalene to Cholesterol

In mammalian tissues, squalene is transformed into lanosterol according to a concerted mechanism which starts with an OH^+ group in position 2 of squalene, involves a series of hydride and methyl shifts, and ends with the final expulsion of a proton in position 9 of lanosterol.[138] It has also been proposed that the cyclization of squalene to cholesterol starts from 2-3 mono-epoxy squalene, and the convertibility of radioactive 2-3 epoxy squalene to lanosterol in rat liver homogenates in anaerobic conditions has been shown.[139,140] Lanosterol, a sterol with 30 carbon atoms, is considered to be

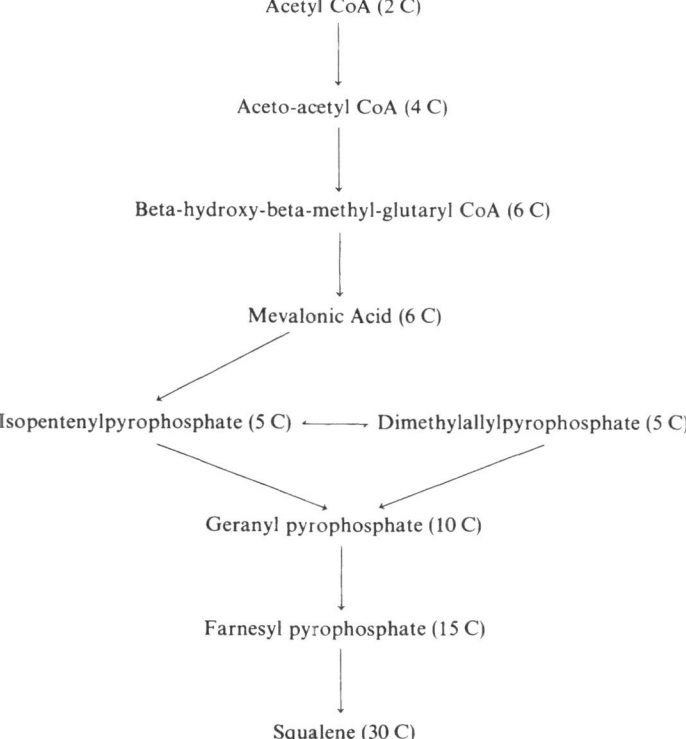

Fig. 1. Pathway of cholesterol biosynthesis from acetyl CoA to squalene.

the main precursor of cholesterol. In rat tissues, the following steps of trans-formation have been described: (1) loss of the methyl group in 14 and, later, of the two methyl groups in 4 (with formation of C29, C28, and C27 sterols); (2) isomerization of the double bond from 8 to 7; (3) introduction of the double bond in 5; (4) reduction of the double bond in 7; (5) reduction of the double bond in 24.

Cholesterol precursors with 30, 29, and 28 carbon atoms in animal tissues have a double bond in position 7 or 8 and the demethylation takes place both in compounds with a saturated and an unsaturated lateral chain.

The isomerization from Δ^8 to Δ^7 is catalyzed by a specific enzyme[61] with an irreversible reaction, which does not require oxygen.[141]

Isomerization does not take place in the presence of a methyl group in position 14-α or of either one of the two methyl groups in position 4. The pro-cess does not involve a molecular shift of hydride from position 7 to position 9-α, as shown in this laboratory,[142] but it may involve the uptake of an atom of solvent hydrogen.[143,144]

The mechanism of introduction of a double bond in Δ^5 has not yet been clarified; it requires oxygen, but probably not NADPH.[145] The saturation of the double bond in 7 in liver homogenates and in intestinal tissues requires NADPH, but not oxygen,[146] and it is irreversible.[60]

The reduction of the double bond in 24 may be specifically inhibited by the drug triparanol[147] with an accumulation of desmosterol in liver, blood, and in growing nervous tissues. This indicates that, in mammalian tissues, cholesterol biosynthesis may take place through precursors with an unsatu-rated lateral chain.[148] However, the reduction of the double bond is possible at any level between lanosterol and cholesterol, and precursors with a satu-rated lateral chain may also be required.[149] A direct demonstration that C29, C28, and C27 precursors (Fig. 2) are necessary for brain cholesterol biosynthesis was made recently in this laboratory.[44] Mevalonate injected in 11- and 18-day-old chick embryos is largely incorporated in brain desmos-terol and cholesterol. High specific activities have been detected in the C29 and C28 precursors, which for the most part have been identified (see Sec. III). It is interesting to note that in chick embryo and adult rat brain the specific activity of cholesterol is lower than in many of the precursors. The fractions corresponding to the $\Delta^{8,24}$ sterols have a very high activity, indicating that these sterols are important intermediates in brain cholesterol synthesis. In the same experiments, the Δ^8 sterols have a lower activity compared to the $\Delta^{8,24}$ intermediates.[44]

C. An Appraisal of the *in Vitro* Versus *in Vivo* Experiments

Cholesterol formation in mature nervous tissue and the relative import-ance of locally biosynthesized sterols in comparison with the uptake of peri-pheral and blood sterols at different ages have been a matter of dispute for the past twenty years. The early work by Waelsch, Sperry, and Stoyanoff[150,151] led to the conclusion that deuterium incorporation into nonsaponifiable

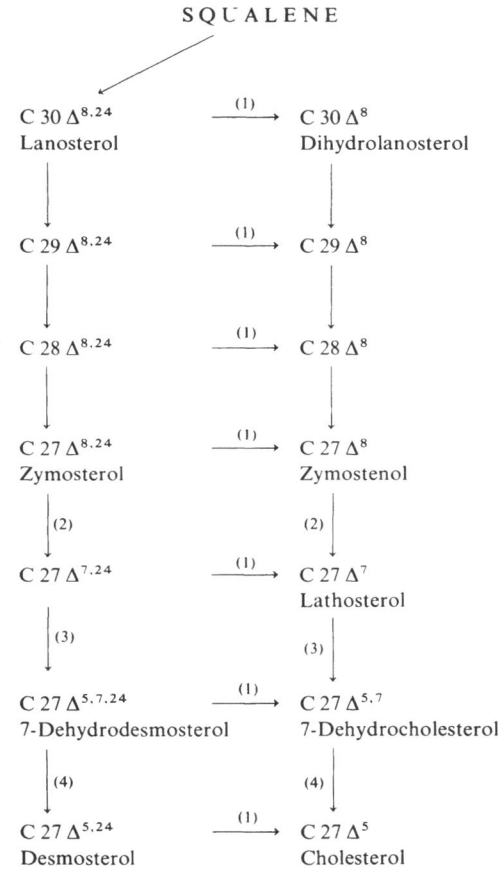

SQUALENE

C 30 $\Delta^{8,24}$ $\xrightarrow{(1)}$ C 30 Δ^8
Lanosterol   Dihydrolanosterol

C 29 $\Delta^{8,24}$ $\xrightarrow{(1)}$ C 29 Δ^8

C 28 $\Delta^{8,24}$ $\xrightarrow{(1)}$ C 28 Δ^8

C 27 $\Delta^{8,24}$ $\xrightarrow{(1)}$ C 27 Δ^8
Zymosterol   Zymostenol

(2)   (2)

C 27 $\Delta^{7,24}$ $\xrightarrow{(1)}$ C 27 Δ^7
                           Lathosterol

(3)   (3)

C 27 $\Delta^{5,7,24}$ $\xrightarrow{(1)}$ C 27 $\Delta^{5,7}$
7-Dehydrodesmosterol   7-Dehydrocholesterol

(4)   (4)

C 27 $\Delta^{5,24}$ $\xrightarrow{(1)}$ C 27 Δ^5
Desmosterol   Cholesterol

(1) Δ^{24}—reductase
(2) Δ^8—Δ^7 isomerase
(3) Δ^5—Δ^6 dehydrogenase
(4) Δ^7—reductase

Fig. 2. Suggested pathways from squalene to cholesterol in brain.

lipids of adult rat brain was negligible in comparison with growing animals.

Later results by Srere et al.[152] and Rossiter[153] confirmed these data and Azarnoff[154] was able to show a lack of acetate incorporation in human brain cholesterol *in vitro*, while the incorporation did occur in human brain tumour preparations. A consistent incorporation of labeled acetate and pyruvate has, however, been observed in rats after intracisternal injection[155] or intracerebral inoculation.[156]

This difference is not due to lack of penetration across the blood–brain barrier of labeled acetate,[130] but probably, as pointed out by Dobbing,[157] a direct injection into the brain raises the amount of precursor available to brain for biosynthesis. On the other hand, a considerable incorporation of labeled mevalonic acid has been shown recently by Nicholas and Aexel in cell-free extracts of adult rat brain incubated with glucose and cofactors.[158] These results underline the importance of the experimental conditions for evaluating brain sterol synthesis and the need to correlate the *in vitro* results with *in vivo* administration of the radioactive precursor.

V. DRUGS ACTING ON BRAIN STEROLS

The possibility that some drugs might act on brain cholesterol has been investigated. It might be expected that adult brain reacts differently from immature brain for the following reasons: in adult brain the turnover of cholesterol is slower and the penetration of drugs is conditioned by the presence of a fully developed blood–brain barrier. In studying the effect of certain psychotropic drugs, such as chlorpromazine[159] and imipramine,[160] on brain lipid synthesis in our laboratory, we showed that, while the phospholipid synthesis is stimulated by low doses of drugs (10^{-5}M *in vitro* and 3.5 mg/kg *in vivo*), cholesterol biosynthesis is affected only by higher doses of the drugs (10^{-3}M *in vitro*), which also affect the respiratory rate and show other side effects. Only in this case there is a marked reduction of incorporation into cholesterol of precursors, such as labeled acetate, mevalonate, or glucose. Methylphenidate (Ritalin), which was claimed to lower cholesterol levels in adult brain,[134] has an inhibitory effect on brain cholesterol synthesis from acetate[161] and glucose.[162] Metrazol also inhibits the incorporation of labeled acetate and mevalonate without any effect on squalene and lanosterol utilization for cholesterol synthesis.[163,164]

Some experimental conditions, however, may induce increased brain cholesterol synthesis. Insulin stimulates cholesterol biosynthesis *in vitro* in presence of glucose but has no effect on the incorporation of labeled mevalonate and glucose.[165] This stimulating effect is demonstrable both in adult and in developing rat brain and agrees very well with the effect on lipid synthesis obtained in peripheral nerves.[131] A treatment of growing rats with Triton WR 1339, which stimulates acetate incorporation into liver cholesterol, also has a stimulating effect in the brain.[166]

The radioprotective agent, cysteamine, is able to protect against blockade of cholesterol biosynthesis in the brain[167] induced by ionizing radiation, while a radiosensitizer, Synkavit, potentiates this radiation effect.[167] All the drugs described above affect the first steps of cholesterol biosynthesis (Fig. 1) interfering with the incorporation of the simplest precursors of cholesterol.

However, another important group of drugs acts on cholesterol biosynthesis at the last stages (Fig. 2) and induces accumulation of sterol precursors of cholesterol in liver, blood, and other tissues. Under certain conditions, these drugs induce remarkable changes in brain sterol composition. The effect is quantitatively different in developing and adult brain in relation to the different biosynthetic rates. Triparanol, [1-(p-β-dimethylamino-ethoxy)-phenyl]1-(p-tolyl)-2p-chloro-phenyl-ethanol, which specifically blocks the Δ^{24} reductase, thus inducing an accumulation of desmosterol in plasma and peripheral tissues,[19] has no effect after chronic treatment on human brain sterols,[168] but it may induce, after a 10-day treatment, an easily detectable accumulation of desmosterol (4% of total sterols) in rat brain.[169] After 15 days of alternative treatment, desmosterol (1.6%) is detectable also in brain of adult mice.[170] In developing mice, when desmosterol is normally present in brain, Triparanol treatment induces an increase of desmosterol content up to 50% of the total sterols.[171,172] Other drugs acting on cholesterol synthesis at the same level as Triparanol, such as 20,25-diazacholesterol, induce similar accumulation of desmosterol in rat brain,[169] but the appearance of desmosterol is more rapid, probably due to the easier penetration of diazacholesterol derivatives than Triparanol across the blood–brain barrier. AY-9944 [(trans-1,4 bis(-2 dichlorobenzyl-amino-ethyl-cyclohexane-dihydrochloride], a drug inhibiting sterol Δ^7-reductase and inducing accumulation of 7-dehydrocholesterol in tissues,[173] induces a deposition of the same compound in the brain of immature rats (up to 37% of the total sterols after a 6-day treatment), but only traces of it in adult brain.[174] The incorporation of radioactive precursors into cholesterol was inhibited with concomitant appearance of sterol precursors.[175]

It is interesting to note that AY-9944, given to pregnant rats for six days before delivery, induces accumulation of two other sterols, the major one being 5,7,24 cholestatrienol, which in the liver is present only after a double treatment with AY-9944 and Triparanol,[174] suggesting that the Δ^{24}-reductase may represent a limiting rate in brain. The reduction of desmosterol in the nervous system concomitant with brain maturation can be influenced by thyroid inhibition: propylthiouracil, given in the chick embryo at the eighth day of incubation, induces thyroid suppression, decrease of the body weight gain, and prolongation of incubation time. These effects are accompanied by a slower curve of disappearance of desmosterol. Thyroxine administration under the same conditions increases the rate of desmosterol disappearance from the brain.[176]

VI. BRAIN STEROLS IN PATHOLOGICAL CONDITIONS

A. Nontumoral Diseases

Sterol concentrations in brain are affected by a number of pathological conditions, mainly fasting, malnutrition, Wallerian degeneration, demyelinating diseases, and lipidosis.

1. Fasting and Malnutrition

Cholesterol metabolism in brain has been found sensitive to fasting and malnutrition, particularly during the developmental period. The rate of cholesterol deposition in brain of suckling rats is significantly reduced by undernourishment[177]; dogs with deficient protein caloric intake during the suckling period show lower cholesterol concentrations in brain.[178] Short periods of fasting affect the incorporation of labeled precursors into cholesterol in young and adult animals and the data reported, although with some disagreements,[179,180] show the sensitivity of brain cholesterol metabolism to these experimental conditions.

2. Wallerian Degeneration

In the degenerating nerve, along with a reduction of free cholesterol, there is the appearance of esterified cholesterol, which is normally absent from the intact peripheral nerves.[181,182] The possibility that the esters originate from cholesterol present in the nerve itself is clearly demonstrated.[183] In the period of regeneration following, cholesterol is actually synthesized *in situ* for the newly forming structures.[184]

3. Demyelinating Diseases

In multiple sclerosis, along with modifications in other lipid classes, a loss of free cholesterol and the appearance of large amounts of cholesterol esters occur, particularly in the plaques of sclerosis.[108,185–187] In inflammatory demyelinating affections, such as Schiller's disease, acute disseminated encephalomyelitis, and subacute sclerosing leukoencephalitis, a loss of cholesterol and the presence of variable amounts of its esters is a common feature,[188–190] while in metachromatic leukoencephalopathy the loss of brain cholesterol is not accompanied by the appearance of sterol esters.[191–193] In this disease, the presence of an unidentified sterol has been reported.[194] Esterified cholesterol in brain also was observed in Pelizaeus–Merbacher's disease.[192] In experimental rabbit encephalomyelitis, cholesterol biosynthesis is not affected in brain[125] and no sterols, other than cholesterol, have been detected by GLC in the brain of guinea pigs[195] with the same disease.

4. Brain Lipidosis

In Niemann–Pick's disease, brain cholesterol content is increased and small amounts of cholesterol esters are also present[196]; in the infantile form, the increase of cholesterol occurs in the cortex.[197,198] Cholesterol esters are relevant components of brain lipids in the brain localization of Hand–Schuller–Christian's disease,[199] while in Tay–Sachs' infantile amaurotic idiocy, brain cholesterol levels are generally not affected.[200] In this disease, the incorporation of tritiated water into brain cholesterol suggests that some synthesis takes place.[201]

B. Brain Tumors

As in many nontumoral diseases, cholesterol esters, absent from normal adult brain, occur in brain tumors. Their presence has been described in human gliomas, meningiomas, and neurinomas.[83,202,203] The finding of desmosterol, a steroidal precursor of cholesterol, in normal developing brain and in some spontaneous or transplantable brain tumors, suggests possible correlations between embryonic and tumoral growth. The presence of desmosterol in developing brain is considered the result of the slow rate of the desmosterol reductase, which is a limiting step in brain tissue when a high rate of cholesterol biosynthesis takes place.[95] Since in brain tumors this synthesis is high, in contrast with normal adult brain,[204] the presence of low concentrations of desmosterol in some brain tumors is not surprising.

Desmosterol has been detected in both transplantable and spontaneous brain tumors. Among twelve transplantable murine brain tumors, this sterol has been detected in three cases: two ependymomas and one spongioblastic glioblastoma. In spontaneous brain tumors, desmosterol was present in one case of a dog glioblastoma. The experimental findings correlate well with the detection of desmosterol in human glioblastomas. The relative amounts of desmosterol in these tumors depend on two factors: the purity of the tumors from surrounding brain tissue and malignancy of the tumor itself. Sometimes other sterols may also be detected.[94,205,206]

The use of drugs selectively interfering at one or more steps of the cholesterol pathway after squalene cyclization leads to modifications of the sterol patterns in brain tumors. Mice bearing transplantable brain tumors when treated with Triparanol or with 20,25-diazacholesterol accumulate desmosterol in various peripheral tissues as well as in the tumors.[170] Instead, using AY-9944, an accumulation of 7-dehydrocholesterol in experimental tumors is observed.

An interesting observation is that the two above-mentioned cholesterol precursors accumulate in the experimental brain tumors but not in the normal brain (Table IV). Human subjects bearing various types of brain tumors and treated with Triparanol for different periods of time before undergoing surgical operation accumulate desmosterol in blood as well as in the brain tumor. In the normal surrounding brain, when available, desmosterol was not detectable.[170]

The accumulation of cholesterol precursors in brain tumors and their absence in normal brain could be due to various factors: differential permeability to the drugs used, different levels of sterol biosynthesis, or different penetration through the blood–brain barrier of sterols synthesized by peripheral tissues. The findings that labeled precursors are actively incorporated into cholesterol in brain tumors[204,207] could play in favor of the importance of local synthesis.

It is interesting to observe that the concentrations of desmosterol in the cerebrospinal fluid of patients treated with Triparanol are significantly higher

TABLE IV

Effect of Triparanol and AY-9944 in Mice on Accumulation of Cholesterol Precursors in Normal Brain and in a Subcutaneously Transplantable Ependymoma

Tissue[a]	Triparanol[b]		AY-9944[c]	
	Total sterols, mg/g f.w. ± SE	Desmosterol, % ± SE	Total sterols, mg/g f.w. ± SE	7-Dehydrocholesterol, % ± SE
Brain	(8)[d] 14.89 ± 0.76	tr	(5) 14.14 ± 0.62	tr
Ependymoma ZE	(8) 1.57 ± 0.17	23.3 ± 2.41	(5) 1.84 ± 0.07	30.4 ± 1.75

[a] Obtained from the same animals.
[b] 30 mg/kg i.p. for 3 days.
[c] 20 μmoles/kg s.c. for 5 days.
[d] The number of cases is shown in parentheses.

in the presence of brain tumors. This finding can be of practical importance for the detection of human brain tumors.[208]

VII. ACKNOWLEDGMENTS

The original investigations reported in this review have been partially supported by the Association for the Aid of Crippled Children and by the National Institute for Neurological Diseases and Blindness, N.I.H. Grant H.E. 09970/03.

VIII. REFERENCES

1. J. Folch, M. Lees, and G. H. Sloane-Stanley, A simple method for the isolation and purification of total lipids from animal tissues, *J. Biol. Chem.* **226**:497–509 (1957).
2. G. Rouser, G. Kritchevsky, C. Galli, A. Yamamoto, and A. G. Knudson, Jr., in *Inborn Disorders of Sphingolipids Metabolism* (S. M. Aronson and B. W. Volk, eds.), pp. 303–316, Pergamon Press, Oxford/New York (1966).
3. A. N. Siakotos and G. Rouser, Analytical separation of nonlipid water soluble substances and gangliosides from other lipids by dextran gel column chromatography, *J. Am. Oil Chem. Soc.* **42**:913–919 (1965).
4. B. Borgström, Investigation on lipid separation methods: Separation of phospholipids from neutral fat and fatty acids, *Acta Physiol. Scand.* **25**:101–110 (1952).
5. G. Rouser, A. J. Bauman, G. Kritchevsky, D. Heller, and J. O'Brien, Quantitative chromatographic fractionation of complex lipid mixtures: Brain lipids, *J. Am. Oil Chem. Soc.* **38**:544–555 (1961).
6. G. Rouser, C. Galli, and G. Kritchevsky, Lipid class composition of normal human brain and variations in metachromatic leucodistrophy, Tay–Sachs, Niemann–Pick, chronic Gaucher's and Alzheimer's diseases, *J. Am. Oil Chem. Soc.* **42**:404–410 (1965).
7. M. G. Horning, E. A. Williams and E. C. Horning, Separation of tissue cholesterol esters and triglycerides by silicic acid chromatography, *J. Lipid Res.* **1**:482–485 (1960).
8. G. Rouser, personal communication.
9. E. Heftmann, B. E. Wright, and G. U. Liddel, The isolation of Δ^{22}-stigmastane-3β-ol from *Dictiostelium discoideum*, *Arch. Biochem. Biophys.* **91**:266–270 (1960).
10. V. C. Joshi, J. Jayaraman, and T. Ramasarma, Incorporation of mevalonic acid-2-C^{14} into Ubichromenol and coenzyme Q in rat, *Indian J. Exptl. Biol.* **1**:113–123 (1963).
11. G. J. Schroepfer, Jr., and I. Y. Gore, Chromatographic separation of allylic alcohols on silicic acid columns: analysis of the nonsaponifiable lipids on an ascites tumor derived from a benzpyrene-induced sarcoma, *J. Lipid Res.* **4**:266–269 (1963).
12. W. M. Sperry and M. Webb, A revision of the Schoenheimer–Sperry method for cholesterol determination, *J. Biol. Chem.* **187**:97–110 (1950).
13. J. J. Kabara, J. T. McLaughlin, and C. A. Riegel, Quantitative microdetermination of cholesterol using tomatine as precipitating agent, *Anal. Chem.* **33**:305–307 (1961).
14. L. J. Goad, and T. W. Goodwin, The biosynthesis of sterols in higher plants, *Biochem. J.* **99**:735–746 (1966).
15. D. Kritchevsky, Analysis of cholesterol in *Cholesterol*, pp.232–255, John Wiley & Sons Inc., New York, and Chapman & Hall Ltd., London (1958).
16. A. Zlatkis, B. Zak, and A. J. Boyle, A new method for the direct determination of serum cholesterol, *J. Lab. Clin. Med.* **41**:486–492 (1953).

17. L. L. Abell, B. B. Levy, B. B. Brodie, and F. E. Kendall, A simplified method for the estimation of total cholesterol in serum and demonstration of its specificity, *J. Biol. Chem.* **195**:357–366 (1952).
18. J. Glover, Determination of cholesterol and 7-dehydrosterols, in monograph No. 2 *The Determination of Sterols*, Society for Analytical Chemistry, London, pp. 10–17 (1964).
19. J. Avigan, D. Steinberg, H. E. Vroman, M. J. Thompson, and E. Mosettig, Studies of cholesterol biosynthesis: I. The identification of desmosterol in serum and tissues of animals and man treated with Mer-29, *J. Biol. Chem.* **235**:3123–3126 (1960).
20. M. Glover, J. Glover, and R. A. Morton, Provitamin D_3 in tissues and the conversion of cholesterol to 7-dehydro-cholesterol *in vivo*, *Biochem. J.*, **51**:1–9, (1952).
21. K. Schubert, G. Rose, and M. Bürger, Über das Vorkommen von dihydroxysterinen im Menschlichen Gehirn, *Hoppe-Seylers Z. physiol. Chem.* **326**:235–241 (1961).
22. J. W. Copius-Peereboom and H. W. Beekes, The analysis of mixtures of animal and vegetable fats, V. Separation of sterol acetates by thin-layer chromatography in reversed-phase systems and on silica gel G-silver nitrate layers, *J. Chromat.* **17**:99–113 (1965).
23. J. W. Copius-Peereboom and H. W. Beekes, The analysis of mixtures of animal and vegetable fats, III. Separation of some sterols and sterol acetates by thin-layer chromatography, *J. Chromat.* **9**:316–320 (1962).
24. R. D. Bennett and E. Heftmann, Thin-layer chromatography of sterols, *J. Chromat.* **9**:359–362 (1962).
25. J. Avigan, D. S. Goodman, and D. Steinberg, Thin-layer chromatography of sterols and steroids, *J. Lipid Res.* **4**:100–101 (1963).
26. S. Fabro, Paper chromatography of cholesterol and desmosterol after bromination, *J. Lipid Res.* **3**:481–483 (1962).
27. J. R. Claude, Séparation du cholestérol, du desmostérol et du 5-dihydrocholestérol par chromatographie en couche mince après propionylation, *J. Chromat.* **17**:596–599 (1965).
28. A. S. Truswell and W. D. Mitchell, Separation of cholesterol from its companions, cholestanol and Δ^7-cholestenol, by thin-layer chromatography, *J. Lipid Res.* **6**:438–441 (1965).
29. P. D. Klein, J. C. Knight, and P. A. Szczepanik, The behavior of sterols on silica surfaces and at other interfaces, *J. Am. Oil Chem. Soc.* **43**:275–280 (1966).
30. N. W. Di Tullio, C. S. Jacob, and W. L. Holmes, Thin layer chromatography and identification of free sterols, *J. Chromat.* **20**:354–357 (1965).
31. H. E. Vroman and C. F. Cohen, Separation of sterol acetates by column and thin-layer argentation chromatography, *J. Lipid Res.* **8**:150–152 (1967).
32. D. R. Idler and C. A. Baumann, Skin sterols. II. Isolation of Δ^7-cholestenol, *J. Biol. Chem.* **195**:623–628 (1952).
33. W. M. Stokes, F. C. Hickey, and W. A. Fish, Chromatography of I^{131}-labeled esters, *J. Am. Chem. Soc.* **76**:5174–5175 (1954).
34. P. D. Klein and P. A. Szczepanik, The differential migration of sterol acetates on silica gels and its application to the fractionation of sterol mixtures, *J. Lipid Res.* **3**:460–466 (1962).
35. G. Galli and E. Grossi-Paoletti, Quantitative separation of C-27 sterol precursors of cholesterol, *Lipids* **2**:84–85 (1967).
36. F. Gautschi and K. Bloch, On the structure of an intermediate in the biological demethylation of lanosterol, *J. Am. Chem. Soc.* **79**:684–689 (1957).
37. W. M. Stokes, W. A. Fish, and F. C. Hickey, Metabolism of cholesterol in the chick embryo. II. Isolation and chemical nature of two companion sterols, *J. Biol. Chem.* **220**:415–430 (1956).
38. A. A. Kandutsch and A. E. Russell, Preputial gland tumor sterols. I. The occurrence of 24,25-dihydrolanosterol and a comparison with liver and the normal gland, *J. Biol. Chem.* **234**:2037–2042 (1959).

39. I. D. Frantz, Jr., Chromatography of unesterified sterols on silicic acid-super-cel, *J. Lipid Res.* **4**:176–178 (1963).
40. R. B. Clayton, A. N. Nelson, and I. D. Frantz, Jr., The skin sterols of normal and Triparanol treated rats, *J. Lipid Res.* **4**:166–176 (1963).
41. G. Galli and E. Grossi-Paoletti, Separation of cholesterol–desmosterol acetates by thin-layer and column chromatography on silica gel G-silver nitrate, *Lipids,* **2**:72–75 (1967).
42. S. Shefer, S. Hauser, and E. H. Mosbach, Biosynthesis of cholestanol: 5α-cholestan-3-one reductase of rat liver, *J. Lipid Res.* **7**:763–769 (1966).
43. L. J. Morris, Separations of lipids by silver ion chromatography, *J. Lipid Res.* **7**:717–732 (1966).
44. J. F. Weiss, G. Galli, and E. Grossi-Paoletti, Sterols with 29, 28 and 27 carbon atoms metabolically related to cholesterol, occurring in developing and mature brain, *J. Neurochem.* **15**:563–575 (1968).
45. W. J. A. VandenHeuvel, C. C. Sweeley, and E. C. Horning, Separation of steroids by gas-chromatography, *J. Am. Chem. Soc.,* **82**:3481–3482 (1960).
46. E. C. Horning, K. C. Maddock, K. V. Anthony, and W. J. A. VandenHeuvel, Quantitative aspects of gas chromatographic separations in biological studies, *Anal. Chem.* **35**:526–532 (1963).
47. H. H. Wotiz and H. F. Martin, Studies in steroid metabolism: X. Gas-chromatographic analysis of estrogens, *J. Biol. Chem.* **236**:1312–1316 (1961).
48. W. J. A. VandenHeuvel, J. Sjövall, and E. C. Horning, Gas-chromatographic behaviour of trifluoroacetoxy steroids, *Biochim. Biophys. Acta* **48**:596–599 (1961).
49. R. B. Clayton, Gas liquid chromatography of sterol methyl ethers, *Nature* **190**:1071–1072 (1961).
50. T. Luukkainen, W. J. A. VandenHeuvel, E. O. A. Haahti, and E. C. Horning, Gas-chromatographic behaviour of trimethylsilyl ethers of steroids, *Biochim. Biophys. Acta* **52**:599–601 (1961).
51. B. S. Thomas, C. Earborn, and D. R. M. Walton, Preparation and gas chromatography of steroid chloromethyldimethylsilyl ethers, *Chemical Commun.* 408 (1966).
52. W. R. Supina, R. F. Kruppa, and R. S. Henly, Use of dimethylsilyl ether derivatives in gas chromatography, *J. Am. Oil Chem. Soc.* **44**:74–76 (1967).
53. W. J. A. VandenHeuvel and E. C. Horning, A study of retention-time relationships in gas-chromatography in terms of the structure of steroids, *Biochim. Biophys. Acta* **64**:416–429, (1962).
54. R. J. Hamilton, W. J. A. VandenHeuvel, and E. C. Horning, An extension of the steroid number concept to relationships between the structure of steroids and their gas-chromatographic retention times observed with selective phases, *Biochim. Biophys. Acta* **70**:679–687 (1963).
55. E. C. Horning, W. J. A. VandenHeuvel, and B. G. Creech, Separation and determination of steroids by gas chromatography, *in Methods of Biochemical Analysis* (D. Glick, ed.), Vol. XI, pp. 69–147, Interscience, New York (1963).
56. H. H. Wotiz and S. J. Clark, *Gas Chromatography in the Analysis of Steroid Hormones,* Plenum Press, New York (1966).
57. *Gas Liquid Chromatography of Steroids.* Proceedings of a Symposium held at the University of Glasgow, April 4–6, 1966 (J. K. Grant, ed.) Cambridge University Press (1967).
58. R. Fumagalli, *Analytical Gas Chromatography of Cholesterol and Sterol Precursors,* in press.
59. D. H. R. Barton and J. D. Cox, The application of the method of molecular rotation differences in steroids, Part VII. Olefinic unsaturation at the 8(9) position, *J. Chemical Soc.* **1949**:214–219.

60. I. D. Frantz, Jr., A. T. Sanghvi, and R. B. Clayton, Detection of a sterol with the probable structure $\Delta^{5,7,24}$ cholestatrien-3β-ol in the intestinal wall of guinea pigs treated with triparanol, *J. Biol. Chem.* **237**:3381–3383 (1962).

61. I. D. Frantz, Jr., T. J. Scallen, A. N. Nelson, and G. J. Schroepfer, $\Delta^{7,24}$-cholestadien-3β-ol, a probable intermediate in cholesterol synthesis, *J. Biol. Chem.* **241**:3818–3821 (1966).

62. R. Ryhage, Use of a mass spectrometer as a detector and analyzer for effluents emerging from high temperature gas liquid chromatography columns, *Anal. Chem.* **36**:759–764 (1964).

63. R. Ryhage and J. Sjövall, Direct mass-spectrometry of steroids in gas chromatography, *Biochem. J.*, **92**:2P–3P (1964).

64. E. Stenhagen, II. Massenspektrometrie. Jetziger Stand der Massenspektrometrie in der organischen Analyse, *Z. Anal. Chem.* **205**:109–124 (1964).

65. J. T. Watson and K. Biemann, High resolution mass spectra of compounds emerging from a gas chromatograph, *Anal. Chem.* **36**:1135–1137 (1964).

66. S. S. Friedland, G. H. Lane, Jr., R. T. Longman, K. E. Train, and M. J. O'Neal, Jr., Mass spectra of steroids, *Anal. Chem.* **31**:169–174 (1959).

67. H. Budzikiewicz, C. Djerassi, and D. H. Williams, *Structure Elucidation of Natural Products by Mass Spectrometry*, Vol. II, Holden-Day, Inc., San Francisco (1964).

68. C. J. W. Brooks, W. A. Harland, and G. Steel, Squalene, 26-hydroxycholesterol and 7-ketocholesterol in human atheromatous plaques, *Biochim. Biophys Acta* **125**:620–622 (1966).

69. G. Galli and S. Maroni, Mass spectrometric investigations of some unsaturated sterols biosynthetically related to cholesterol, *Steroids* **10**:189–197 (1967).

70. R. B. Clayton and K. Bloch, Biological synthesis of lanosterol and agnosterol, *J. Biol. Chem.* **218**:305–318 (1956).

71. F. Snyder and C. Piantadosi, Labeling and radiopurity of lipids, *Advan. Lipid Res.* **4**:257–283 (1966).

72. L. Swell, Simultaneous determination of mass and radioactivity of labeled sterols and steroids by radiogaschromatography, *Anal. Biochem.* **16**:70–83, (1966).

73. A. Karmen, Measurement of carbon-14 and tritium in the effluent of a gas chromatography column, *J. Am. Oil. Chem. Soc.* **44**:18–25 (1967).

74. L. F. Fieser, Cholesterol and companions. III. Cholestanol, lathosterol and Chetone 104, *J. Am. Chem. Soc.* **75**:4395–4403 (1953).

75. K. Nakanishi, B. K. Bhattacharyya, and L. F. Fieser, Cholesterols and companions. V. Microdetermination of Δ^7 stenols, *J. Am. Chem. Soc.* **75**:4415–4417 (1953).

76. H. Werbin, I. L. Chaikoff, and M. R. Imada, 5α-cholestan-3β-ol: its distribution in tissues and its synthesis from cholesterol in the guinea pig, *J. Biol. Chem.* **237**:2072–2077 (1962).

77. I. H. Page and E. Mueller, Notiz über das Vorkommen von Dihydrocholesterin in menschlichen Gehirn, *Z. Physiol. Chem.* **204**:13–14 (1932).

78. R. Shoenheimer, H. Van Behring, and R. Hummel, Mitteilung: Untersuchung der Sterine aus verschiedenen Organen auf ihren Gehalt an gesättigten Sterinen, *Z. Physiol. Chem.* **192**:93–96, (1930).

79. I. H. Page and W. Menschick, Über das Vorkommen von Ergosterin im menschlichen Gehirn, *Biochem. Z.* **231**:446–459 (1931).

80. E. M. Koch and F. C. Koch, Provitamin D potency of some sterol derivatives, *J. Biol. Chem.* **116**:757–768 (1936).

81. A. Ercoli and P. De Ruggieri, The constitution of cerebrosterol: a hydroxycholesterol isolated from horse brain, *J. Am. Chem. Soc.* **75**:3284 (1953).

82. H. J. Nicholas, Cholesterol turnover in the central nervous system, *J. Am. Oil Chem. Soc.* **42**:1008–1012 (1965).

83. G. Brante, Studies on the lipids in the nervous system, with special reference to quantitative chemical determination and topical distribution, *Acta Physiol. Scand.* **18**, Suppl. 63:1–189 (1949).

84. J. Folch, J. Casals, A. Pope, J. A. Meath, F. N. LeBaron, and M. Lees, Chemistry of myelin development *in The Biology of Myelin*, (S. R. Korey, ed.), pp. 122–137, Harper & Bros., New York (1959).

85. Y. Kishimoto, W. E. Davies, and N. S. Radin, Developing rat brain: changes in cholesterol, galactolipids and the individual fatty acids of gangliosides and glycerophosphatides, *J. Lipid Res.* **6**:532–535 (1965).

86. C. Galli and D. Re Cecconi, Lipid changes in rat brain during maturation, *Lipids* **2**:76–82 (1967).

87. J. Folch-Pi, *in Biochemistry of the Developing Nervous System* (H. Waelsch, ed.), pp. 121–132, Academic Press, Inc., New York (1955).

88. P. Mandel and R. Bieth, La répartition des diverses fractions lipidiques au cours du développement du cerveau chez le rat, *Bull. Soc. Chim. Biol.* **33**:973–981 (1951).

89. P. Mandel, R. Bieth, and R. Stoll, La repartition des diverses fractions lipidiques dans le cerveau de l'embryon de poulet durant la seconde partie de l'incubation, *Compt. Rend. Soc. Biol.* **143**:1224–1226 (1949).

90. W. T. Norton, personal communication.

91. W. A. Fish, J. E. Boyd, and W. M. Stokes, Metabolism of cholesterol in the chick embryo. III. Localization and turnover of desmosterol (24-dehydrocholesterol), *J. Biol. Chem.* **237**:334–337 (1962).

92. D. Kritchevsky and W. L. Holmes, Occurrence of desmosterol in developing rat brain, *Biochem. Biophys. Res. Comm.* **7**:128–131 (1962).

93. R. Fumagalli and R. Paoletti, The identification and significance of desmosterol in developing human and animal brain, *Life Sci.* **2**:291–295 (1963).

94. R. Paoletti, R. Fumagalli, E. Grossi-Paoletti, and P. Paoletti, Studies on brain sterols in normal and pathological conditions, *J. Am. Oil Chem. Soc.* **42**:400–404 (1965).

95. T. J. Holstein, W. A. Fish, and W. M. Stokes, Pathway of cholesterol biosynthesis in the brain of the neonatal rat, *J. Lipid Res.* **7**:634–638 (1966).

96. D. Kritchevsky, S. A. Tepper, N. W. Di Tullio and W. L. Holmes, Desmosterol in developing rat brain, *J. Am. Oil Chem. Soc.* **42**:1024–1028 (1965).

97. J. N. Cumings, *in Cerebral Lipidoses* (J. N. Cumings and A. Lowenthal, eds.), pp. 112–121, Blackwell, Oxford (1957).

98. C. W. M. Adams and A. N. Davison, The form in which cholesterol occurs in the adult C.N.S., *J. Neurochem.* **5**:293–296 (1960).

99. F. N. LeBaron and J. Folch, The effect of *p*H and salt concentration on aqueous extraction of brain proteins and lipoproteins, *J. Neurochem.* **4**:1–8 (1959).

100. P. Mandel and R. Bieth, Development biochimique du cerveau de l'embryon de poulet. II. les lipids, *Bull. Soc. Chim. Biol.* **32**:109–115 (1950).

101. C. W. M. Adams and A. N. Davison, The occurrence of esterified cholesterol in the developing nervous systems, *J. Neurochem.* **4**:282–289 (1959).

102. A. C. Johnson, A. R. McNabb, and R. J. Rossiter, Lipids of normal brain, *Biochem. J.* **43**:573–577 (1948).

103. J. Tichy, Cholesterol esters in the white matter of adult human brain serum and cerebrospinal fluid, *J. Neurochem.* **14**:555–559 (1967).

104. R. Clarenburg, I. L. Chaikoff, and M. D. Morris, Incorporation of injected cholesterol into the myelinating brain of the 17-day-old rabbit, *J. Neurochem.* **10**:135–143 (1963).

105. D. Grafnetter, E. Grossi, and P. Morganti, Occurrence of sterol esters in the chicken brain during prenatal and postnatal development, *J. Neurochem.* **12**:145–149 (1965).

106. R. Fumagalli, D. Grafnetter, E. Grossi, and P. Morganti, Steroli liberi ed esterificati nel tessuto nervoso dell'embrione di pollo durante lo sviluppo con particolare riguardo al desmosterolo, *Atti Accad. Med. Lomb.* **18**:535–540 (1963).
107. A. C. Johnson, A. R. McNabb, and R. J. Rossiter, Concentration of lipids in the brain of infants and adults, *Biochem. J.* **44**:494–498 (1949).
108. C. M. Plum and S. E. Hansen, Studies on multiple sclerosis, *Acta Psych. Neurol. Scand.* **35**, Suppl. 141:184–302 (1960).
109. E. Robins, K. M. Edik, and D. E. Smith, Distribution of lipids in the cerebellar cortex and its subjacent white matter, *J. Biol. Chem.* **220**:677–682 (1956).
110. C. W. M. Adams, *in Neurochemistry* (K. A. Elliot, I. H. Page, and J. H. Quastel, eds.), pp. 85–112, Chas. C. Thomas, Springfield, Ill. (1962).
111. V. P. Whittaker, The isolation and characterization of acetylcholine containing particles from brain, *Biochem. J.* **72**:694–706 (1959).
112. L. A. Autilio, W. T. Norton, and R. D. Terry, The preparation and some properties of purified myelin from the central nervous system, *J. Neurochem.* **11**:17–27 (1964).
113. M. E. Smith, *in Advances in Lipid Research* (R. Paoletti and D. Kritchevsky, eds.), Vol. 5, pp. 241–278, Academic Press, New York (1967).
114. M. J. Evans and J. B. Finean, The lipid composition of myelin from brain and peripheral nerve, *J. Neurochem.* **12**:729–734 (1965).
115. M. L. Curner, A. N. Davison, and N. A. Gregson, The chemical composition of vertebrate myelin and microsomes, *J. Neurochem.* **12**:469–481 (1965).
116. J. Eichberg, V. P. Whittaker, and R. M. C. Dawson, Distribution of lipids in subcellular particles of guinea pig brain, *Biochem. J.* **92**:91–100 (1964).
117. W. T. Norton and L. A. Autilio, The lipid composition of purified bovine brain myelin, *J. Neurochem.* **13**:213–222 (1966).
118. J. S. O'Brien and E. L. Sampson, Lipid composition of the normal human brain: gray matter, white matter and myelin, *J. Lipid Res.* **6**:537–544 (1965).
119. M. E. Smith, R. Fumagalli, and R. Paoletti, The occurrence of desmosterol in myelin of developing rats, *Life Sci.* **6**:1085–1091 (1967).
120. W. T. Norton, personal communication.
121. N. L. Banik and A. N. Davison, Desmosterol in rat brain myelin, *J. Neurochem.* **14**:594–596 (1967).
122. J. J. Kabara and G. T. Okita, Brain cholesterol: biosynthesis with selected precursors *in vivo*, *J. Neurochem.* **7**:298–304 (1961).
123. J. A. Olson, The biosynthesis of cholesterol, *Ergeb. Phys. Chem. Exptl. Pharmak.* **56**:173–214 (1965).
124. H. Moser and M. L. Karnovsky, Studies on the biosynthesis of glycolipides and other lipides of the brain, *J. Biol. Chem.* **234**:1990–1997 (1959).
125. M. E. Smith, Lipid biosynthesis in the central nervous system in experimental allergic encephalomyelitis, *J. Neurochem.* **11**:29–37 (1964).
126. E. Grossi, P. Paoletti, and R. Paoletti, An analysis of brain cholesterol and fatty acid biosynthesis, *Arch. Int. de Physiol. et de Biochimie* **66**:564–572 (1965).
127. P. Tavormina, M. A. Gibbs, and J. L. Huff, The utilization of β-hydroxy-β-methyl-δ-valerolactone in cholesterol biosynthesis, *J. Am. Chem. Soc.* **78**:4489–4499 (1956).
128. S. Garattini, R. Paoletti, and P. Paoletti, Lipid biosynthesis *in vivo* from acetate-1-C^{14} and 2-C^{14} and mevalonic 2-C^{14} acid, *Arch. Biochem. Biophys.* **84**:254–258 (1959).
129. R. Fumagalli, E. Grossi, M. Poggi, P. Paoletti, and S. Garattini, Cholesterol synthesis in rat brain: differential incorporation of mevalonolactone-2-C^{14}, *Arch. Biochem. Biophys.* **99**:529–533 (1962).

130. H. J. Nicholas, Cholesterol: The metabolism of cholesterol in the central nervous system, *J. Kansas Med. Soc.* **62**:358–361 (1961).
131. E. Grossi, P. Paoletti, and M. Poggi, The effect of insulin on brain cholesterol and fatty acid biosynthesis, *World Neurol.* **3**:209–215 (1962).
132. S. R. Korey, ed., *The Biology of Myelin*, Harper and Row (Hoeber), New York (1959).
133. S. R. Korey and A. Stein, *in Regional Neurochemistry* (S. Kety and J. Elkes, eds.), pp. 175–189, Pergamon Press, New York (1961).
134. J. J. Kabara *in Advances in Lipid Research* (R. Paoletti and D. Kritchevsky, eds.), Vol. V, pp. 279–327, Academic Press, New York (1967).
135. H. J. Nicholas and B. E. Thomas, The metabolism of cholesterol and fatty acids in the central nervous system, *J. Neurochem.* **4**:42–49 (1959).
136. B. W. Agranoff, H. Eggerer, V. Henning, and F. Lynen, Biosynthesis of terpenes. VII. Isopentenyl pyrophosphate isomerase, *J. Biol. Chem.* **235**:326–332 (1960).
137. E. Grossi-Paoletti, unpublished observations.
138. A. Eschenmoser, L. Ruzicka, O. Jeger, and D. Arigoni, Zur Kenntnis der Triterpene. Eine stereochemische Interpretation der biogenetischen Isoprenregel bei den Triterpenen, *Helv. Chim. Acta* **38**:1890–1904 (1955).
139. E. J. Corey, W. E. Russey, and P. Ortiz de Montellano, 2,3-oxidosqualene, an intermediate in the biological synthesis of sterols from squalene, *J. Am. Chem. Soc.* **88**:4750–4751 (1966).
140. E. E. Van Tamelen, J. P. Willet, R. B. Clayton, and K. E. Lord, Enzymic conversion of squalene 2,3-oxide to lanosterol and cholesterol, *J. Am. Chem. Soc.* **88**:4752–4754 (1966).
141. J. L. Gaylor, C. V. Delwicke, and A. C. Swindell, Enzymatic isomerization ($\Delta^7-\Delta^8$) of intermediates of sterol biosynthesis, *Steroids* **8**:353–363 (1966).
142. L. Canonica, A. Fiecchi, M. Galli Kienle, A. Scala, G. Galli, E. Grossi-Paoletti, and R. Paoletti, The biological conversion of 5α-cholest-8-en-3β-ol to 5α-cholest-7-en-3β-ol in the biosynthesis of cholesterol, *Steroids* **11**:287–298 (1968).
143. M. Akhtar, D. C. Wilton, and K. H. Munday, The obligatory intermediary of cholest-5,7-diene system in the hepatic biosynthesis of cholesterol, *Biochem. J.* **101**:23c (1966).
144. G. J. Schroepfer, Jr., Lee Wen-Lui, and R. Kammereck, *in Proceedings of the 154th Meeting of the American Chemical Society*, Chicago, Ill., Sept. 1967, 140C.
145. M. E. Dempsey, J. D. Seaton, G. J. Schroepfer, Jr., and R. W. Trockman, The intermediary role of $\Delta^{5,7}$-cholestadien-3β-ol in cholesterol biosynthesis, *J. Biol. Chem.* **239**:1381–1387 (1964).
146. G. J. Schroepfer, Jr., and I. D. Frantz, Jr., Conversion of Δ^7-cholestenol-4-C^{14} and 7-dehydrocholesterol-4-C^{14} to cholesterol, *J. Biol. Chem.* **236**:3137–3140 (1961).
147. J. Avigan, D. Steinberg, M. J. Thompson, and E. Mosettig, The mechanism of action of Mer-29, *Progr. Cardiovascular Diseases* **2**:525–530 (1960).
148. J. D. Johnston and K. Bloch, *In vitro* conversion of zymosterol and dihydrozymosterol to cholesterol, *J. Am. Chem. Soc.* **79**:1145–1149 (1957).
149. D. S. Goodman, J. Avigan, and D. Steinberg, Studies of cholesterol biosynthesis. V. The time course and pathway of the later stages of cholesterol biosynthesis in the liver of intact rats, *J. Biol. Chem.* **238**:1287–1293 (1963).
150. H. Waelsch, W. H. Sperry, and V. A. Stoyanoff, Lipid metabolism in brain during myelination, *J. Biol. Chem.* **135**:297–302 (1940).
151. H. Waelsch, W. H. Sperry, and V. A. Stoyanoff, The influence of growth and myelination on the disposition and metabolism of lipids in the brain, *J. Biol. Chem.* **140**:885–897 (1950).
152. P. A. Srere, I. L. Chaikoff, S. S. Treitman, and L. S. Burstein, The extrahepatic synthesis of cholesterol, *J. Biol. Chem.* **182**:629–632 (1950).
153. R. Rossiter *in Metabolism of the Nervous System* (D. Richter, ed.), pp. 355–380, Pergamon Press, New York (1956).

154. D. L. Azarnoff, G. L. Curran, and W. P. Williamson, Incorporation of acetate-1-C^{14} into cholesterol by human brain tumors, *Fed. Proceedings* **16**:148 (1957).
155. P. McMillan, G. L. Douglas, and R. A. Mortensen, Incorporation of acetate 1-C^{14} and pyruvate-2-C^{14} into brain cholesterol in the intact rat, *Proc. Soc. Exptl. Biol. Med.* **96**:738–740 (1957).
156. H. J. Nicholas, Biosynthesis of cholesterol in the central nervous system, *Fed. Proceedings* **16**:324 (1957).
157. J. Dobbing, The blood–brain barrier, *Physiol. Rev.* **41**:130–188 (1961).
158. H. J. Nicholas and R. T. Aexel, Biosynthesis of cholesterol in cell-free extracts of adult rat brain, *Fed. Proceedings* **26**:342 (1967).
159. E. Grossi, P. Paoletti, and R. Paoletti, The *in vitro* and *in vivo* effects of Chlorpromazine on brain lipid synthesis, *J. Neurochem.* **6**:73–78 (1960).
160. R. Fumagalli, E. Grossi, and P. Paoletti, The effect of imipramine and desmethylimipramine on lipid biosynthesis in brain and liver, *J. Neurochem.* **10**:213–217 (1963).
161. J. J. Kabara, Brain cholesterol. VIII. Effect of methylphenidate (Ritalin) on the incorporation of specifically labeled acetate, *Proc. Soc. Exptl. Biol. Med.* **118**:905–908 (1965).
162. J. J. Kabara and C. A. Riegel, Brain cholesterol. IX. Effect of methylphenidate on the incorporation of specifically labeled glucose, *Biochem. Pharmacol.* **14**:1928–1930 (1965).
163. G. T. Alexander and R. B. Alexander, Inhibition of cholesterol synthesis *in vivo* by a convulsant, Metrazol, *Proc. Soc. Exptl. Biol. Med.* **115**:229 (1964).
164. G. T. Alexander and R. B. Alexander, Effect of Metrazol on isolated mammalian cells. II. Inhibition of synthesis of cholesterol, *Biochemistry* **1**:783–788 (1962).
165. R. A. Field and L. C. Adams, Insulin response of peripheral nerve. II. Effects on lipid metabolism, *Biochim. Biophys. Acta* **106**:474–479 (1965).
166. E. Grossi-Paoletti and R. Fumagalli, Effetto del Triton sulla sintesi del colesterolo e degli acidi grassi del tessuto cerebrale di ratto durante lo sviluppo, *Atti Accad. Med. Lomb.* **19**:377–380 (1964).
167. E. Grossi and P. Paoletti, Effetto della radiazioni e di farmaci radioprotettori e radiosensibilizzanti sul metabolismo lipidico, *Giorn. It. Chemiot.* **6–9**:225–229 (1962).
168. A. V. Chobanian and W. Hollander, Tissue distribution of cholesterol and 24-dehydrocholesterol during chronic triparanol therapy, *J. Lipid Res.* **6**:37–42 (1965).
169. R. Fumagalli and R. Niemiro, Effect of 20–25 diazacholesterol and triparanol on sterols particularly desmosterol in rat brain and peripheral tissues, *Life Sci.* **3**:555–561 (1964).
170. R. Fumagalli, E. Grossi-Paoletti, P. Paoletti, and R. Paoletti, Lipids in brain tumors. II. Effect of triparanol and 20–25 diazacholesterol on sterol composition in experimental and human brain tumors, *J. Neurochem.* **13**:1005–1010 (1966).
171. T. J. Scallen, R. M. Candie, and G. J. Schroepfer, Jr., Inhibition by Triparanol of cholesterol formation in the brain of the newborn mouse, *J. Neurochem.* **9**:99–103 (1961).
172. T. G. Scott and V. C. Barber, An enzyme histochemical and biochemical study on the effect of an inhibitor of cholesterol synthesis on myelinating mouse brain, *J. Neurochem.* **11**:423–429 (1964).
173. D. Dvornik, M. Kraml, J. Dubuc, M. Givner, and R. Gaudry, A novel mode of inhibition of cholesterol biosynthesis, *J. Am. Chem. Soc.* **85**:3309 (1963).
174. R. Fumagalli, R. Niemiro, and R. Paoletti, Investigation on the biogenetic reaction sequence of cholesterol in rat tissues through inhibition with AY-9944, *J. Am. Oil Chem. Soc.* **42**:1018–1023 (1965).
175. M. L. Givner and D. Dvornik, Agents affecting lipid metabolism. XV. Biochemical studies with the cholesterol biosynthesis inhibitor AY-9944 in young and mature rats, *Biochem. Pharmacol.* **14**:611–619 (1965).
176. R. Fumagalli, E. Grossi-Paoletti, and D. Grafnetter, unpublished observations.

177. J. Dobbing and J. B. Kersley, The influence of early nutrition on brain cholesterol accumulation during growth, *J. Physiol. (London)* **166**:34P (1963).
178. B. S. Platt, R. J. C. Steward, and S. N. Payne, Protein caloric deficiency and the nervous system in *Proc. Intern. Neurochem. Conf.* (G. B. Ansell, ed.), pp. 91–92, Oxford (1965).
179. J. J. Kabara, Brain cholesterol. XI. A review of biosynthesis in adult mice. *J. Am. Oil Chem. Soc.* **42**:1003–1008 (1965).
180. M. E. Smith, The effect of fasting on lipid metabolism of the central nervous system of the rat, *J. Neurochem.* **10**:531–536 (1963).
181. A. C. Johnson, A. R. McNabb, and R. J. Rossiter, Chemical studies of peripheral nerve during Wallerian degeneration, *Biochem. J.* **45**:500–505 (1949).
182. D. Kline, W. L. Magee, E. T. Pritchard, and R. J. Rossiter, Chemical studies of peripheral nerve during Wallerian degeneration. VII. Labeling of phospholipid and cholesterol from carboxy-^{14}C acetate, *J. Neurochem.* **3**:52–58 (1958).
183. G. Simon, Cholesterol ester in degenerating nerve: origin of cholesterol moiety, *Lipids* **1**:369–370 (1966).
184. W. L. Magee, J. F. Berry, M. Magee, and R. J. Rossiter, Chemical studies of peripheral nerve during Wallerian degeneration X. *In vitro* incorporation of radioactive inorganic phosphate into phosphatides and acid soluble phosphorus compounds, *J. Neurochem.* **3**:333–340 (1958).
185. J. N. Cumings, Lipid chemistry of the brain in demyelinating diseases, *Brain* **78**:554–563 (1955).
186. A. N. Davison and M. Wajda, Cerebral lipids in multiple sclerosis, *J. Neurochem.* **9**:427–432 (1962).
187. C. G. Honegger, Über die dünnschichtchromotographie von Lipiden. 1. Mitteilung. Untersuchungen von Gehirngewebe Multiple-Sklerose-Kranker und Normaler, *Helv. Chim. Acta* **45**:281–289 (1962).
188. C. W. M. Adams in *Neurohistochemistry*, Elsevier, Amsterdam, pp. 442–444 (1965).
189. J. N. Cumings in *Modern Scientific Aspects of Neurology* (J. N. Cumings, ed.), Arnold, London, p. 330 (1960).
190. W. T. Norton and S. Poduslo, Metachromatic leucodistrophy: Chemically abnormal myelin and cerebral biopsy studies of three siblings, *Abstracts Intern. Neurochem. Conf.*, Oxford, England, p. 82 (1965).
191. J. N. Cumings, Metachromatic leucodistrophy: some biochemical observations, *Proc. London Conf. Sci. Study of Mental Deficiency*, Vol. 2, pp. 449–453, May and Baker, London (1962).
192. J. N. Cumings in *Mechanism of Demyelination* (A. S. Rose and C. M. Pearson, eds.), pp. 58–71, McGraw-Hill, New York (1963).
193. M. Mossakowski, G. Mathieson, and J. N. Cumings, On the relation of metachromatic leucodystrophy and amaurotic idiocy, *Brain* **84**:585–604 (1961).
194. W. T. Norton, S. Poduslo, and K. Suzuki, Chemical findings including abnormal myelin and an abnormal ganglioside pattern in a case of subacute sclerosing leucoencephalitis, *Abstracts Intern. Neurochem. Conf.*, Oxford, England, p. 83 (1965).
195. A. Allegranza, R. Fumagalli, and P. Paoletti, unpublished data.
196. R. M. Norman, H. Urich, A. H. Tingey, and R. A. Goodboy, Tay-Sachs' disease with visceral involvement and its relationship to Niemann-Pick's disease, *J. Path. Bact.* **78**:409–421 (1959).
197. R. M. Norman, A. H. Tingey, and M. C. Fowler, The subacute form of Niemann–Pick's disease, *Proc. Vth Intern. Congr. Neuropath.* (P. Luethy, ed.), pp. 143–148, Excerpta Medica, Amsterdam (1965).
198. R. M. Norman, R. M. Forrester and A. H. Tingey, The juvenile form of Niemann–Pick's disease, *Arch. Dis. Childhood* **42**:91–96 (1967).

199. S. J. Thannhauser, The lipidoses: diseases of the cellular lipid metabolism, *Oxford Medicine*, Oxford, (1949).
200. R. Ohman, Chemical pathology of congenital amaurotic idiocy, *Abstracts Intern. Neurochem. Conf.*, Oxford, p. 84 (1965).
201. S. Gatt and E. R. Berman, Studies on brain lipids in Tay-Sachs' disease. III. Incorporation tritiated water into brain lipids, *J. Neurochem.* **10**:73–77 (1963).
202. K. Gopal, E. Grossi, P. Paoletti, and M. Usardi, Lipid composition of human intracranial tumors: a biochemical study, *Acta Neurochir.* **11**:333–347 (1963).
203. D. E. Slagel, J. C. Dittmer, and C. B. Wilson, Lipid composition of human glial tumor and adjacent brain, *J. Neurochem.* **14**:789–798 (1967).
204. P. Paoletti, A. H. Soloway, B. Whitman, and J. R. Messer, Lipid biosynthesis from labeled precursors in an experimental brain tumor bearing mice, *Neurochir.* **9**:12–18 (1966).
205. R. Fumagalli, E. Grossi, P. Paoletti, and R. Paoletti, Studies on lipids in brain tumours. I. Occurrence and significance of sterol precursors of cholesterol in human brain tumours, *J. Neurochem.* **11**:561–565 (1964).
206. R. Fumagalli, R. Paoletti, A. Allegranza, and P. Paoletti, Sterol composition of human and animal spontaneous and experimental brain tumours, *Proc. Vth Intern. Congr. Neuropath.* (F. Luthy and A. Bischoff, eds.), pp. 455–458, Excerpta Medica, Amsterdam (1965).
207. D. L. Azarnoff, G. L. Curran, and W. P. Williamson, Incorporation of acetate-1-C^{14} into cholesterol by human intracranial tumor *in vitro*, *J. Nat. Cancer Inst.* **21**:1109–1115 (1958).
208. F. A. Vandenheuvel, R. Fumagalli, R. Paoletti, and P. Paoletti, A possible biochemical procedure for the diagnosis of human brain tumours, *Life Sci.* **6**:439–444 (1967).

Chapter 11

GLYCOPROTEINS

Eric G. Brunngraber

Illinois State Psychiatric Institute
Chicago, Illinois

I. PROTEIN-BOUND HEXOSAMINE AND *N*-ACETYLNEUR-AMINIC ACID IN NERVOUS TISSUE

The presence of glycoproteins in the α-globulin fraction obtained by paper electrophoresis of soluble extracts prepared from nervous tissue was demonstrated by Hoffman and Schinko.[1] A considerable amount of periodic acid-Schiff (PAS)-positive protein also remained at the origin. A brain-specific, PAS-positive, *N*-acetylneuraminic acid (NANA)-containing protein, present in the α_2-globulin fraction of soluble extracts prepared from brain, has been studied by immunoelectrophoresis and immunoprecipitation by Warecka and Bauer.[2] Investigations that involve soluble extracts prepared from brain tissue are limited, since it is not possible to extract more than 15–30 % of the total protein content of brain tissue with aqueous solvents. The majority of the tissue protein remains associated with the particulate fractions that are sedimented by high speed centrifugation. Techniques that have been applied to the whole tissue have revealed that neural tissue contains polysaccharides that are usually associated with glycoproteins. Ejima[3] isolated material that contained hexosamine, galactose, mannose, and fucose from human cerebrum. Brante[4] found that 3.5, 3.5, and 22 % of the total brain hexosamine was extractable with acetone, ether, and chloroform–methanol (1:2, v/v), respectively. Hot water extracted an additional 8 %, but the insoluble defatted protein residue that remained contained 63 % of the total hexosamine present in brain. He was able to solubilize 70 % of this protein-bound hexosamine by the proteolytic action of trypsin. A third of the solubilized hexosamine was found to be dialyzable. Glegg and Pearce[5] reported that alkaline extraction of white matter yielded material that was PAS-positive and that yielded fucose, galactose, and mannose upon hydrolysis. Dische, Danilczenko, and Zelmenis[6] reported the presence in nervous tissue of a neutral polysaccharide that contained galactose, mannose, fucose, and hexosamine.

Svennerholm[7] noted that evaporation of chloroform–methanol
(1:2, v/v) extracts prepared from brain tissue yielded a residue that failed to
redissolve in this solvent. The amount of this insoluble material was decreased
in amount if brain were first extracted with acetone. He noted that the defatted brain tissue contained more NANA than the extracted lipid. Svennerholm suggested that brain tissue contains mucopolysaccharides, a portion of
which may be extracted with chloroform–methanol and would subsequently
interfere in the determination of gangliosides. These observations received
confirmation when Kuhn and Mueldner[8] and Gielen[9] isolated a glycoprotein from brain tissue by extraction with 90% acetone. The glycoprotein
was obtained in water-soluble form, fractionated and purified by ammonium
sulfate precipitation and anion-exchange chromatography on DEAE-
cellulose columns. The material had a molecular weight of 30,000 and contained two polysaccharide chains. One chain contained equimolar amounts
of glucose, galactose, NANA, and N-acetylglucosamine; the second chain
contained glucose, galactose, N-acetylgalactosamine, and NANA in a molar
ratio of 1:2:1:1. The quantitative contribution of this glycoprotein to the
total amount of NANA present in the tissue was not reported, but in view of
the observations of Brante that acetone extracts only 3.5% of the total brain
hexosamine, it appears that this glycoprotein fails to account for the bulk of
the brain hexosamine. Immunoelectrophoretic analysis indicated that this
glycoprotein was brain-specific.

Crevier[10] reported that alkaline extracts of defatted brain tissue contain a PAS-positive substance which upon chromatography and electrophoresis behaved like a non-sulfated mucoprotein.

Brunngraber and Brown[11–15] reported the isolation of glycopeptides
that accounted for most of the non-gangliosidic NANA present in rat brain.
The tissue[15] is successively extracted with chloroform–methanol (2:1, v/v)
and chloroform–methanol (1:2, v/v) in order to remove the gangliosides
quantitatively. A single extraction with chloroform–methanol (2:1, v/v) is
not effective in accomplishing this purpose, and several reports in the literature that equate the amount of NANA remaining in the defatted protein
residue after a single extraction give high values for "protein-bound" NANA.
The defatted residue after complete extraction of the gangliosides was subjected to the proteolytic action of papain for 24 hr at 60°C. The mixture was
centrifuged and the residue was again subjected to the proteolytic action of
fresh papain for an additional 24 hr. The precipitate obtained after centrifugation (the "papain-resistant" fraction) was analyzed for hexosamine and
NANA. The combined supernatants obtained as a consequence of the double
digestion procedure were dialyzed to yield the diffusable glycopeptides and
the nondiffusable sialomucopolysaccharides. The latter substance was
further purified by treatment with cetylpyridinium chloride in order to
remove by precipitation the nucleic acids and the glucuronic acid-containing
mucopolysaccharides. A balance sheet of the contribution of each of the
fractions obtained (Table I) to the total NANA and hexosamine present in
brain revealed that the gangliosides and the sialomucopolysaccharides res-

TABLE I

NANA- and Hexosamine-Containing Constituents of Rat Whole Brain[a]

	NANA		Hexosamine	
Fraction	μmoles/g	%	μmoles/g	%
Chloroform–methanol (2 : 1,v/v) extracts	1.73	52 ⎫		
Chloroform–methanol (1 : 2,v/v) extracts	0.49	15 ⎭	0.96	22
Sialomucopolysaccharides	0.61	18	0.89	20
Diffusable glycopeptides	0.39	12	1.63	37
Papain-resistant fraction	0.08	3	0.32	7
Uronic acid-containing mucopolysaccharides	0	0	0.55	13
Totals	3.30		4.35	

[a] NANA was determined by the thiobarbituric acid method of Warren[18] and corrected for destruction of NANA that occurs during the acid hydrolysis in 0.1 N sulfuric acid at 80°C required for this assay.[14]

pectively account for 68 and 18 % of the NANA present in brain tissue. The diffusable glycopeptides (12%) and the papain-resistant fraction (3%) account for the remainder. The sialomucopolysaccharides and the diffusable glycopeptides, both of which are derived from glycoproteins, account for over half of the total hexosamine present in brain.

Gangliosides are removed from the tissue by the two extractions with chloroform–methanol, although some glycoprotein[7,9] may be extracted. Fucose was absent in such extracts from rat brain; all of the hexosamine was present in the form of galactosamine. The sialomucopolysaccharides contain NANA, glucosamine, mannose, galactose, and fucose. The hexosamine present in the papain-resistant fraction was galactosamine, and this material probably also contains fucose, glucose, and other hexoses. The papain-resistant fraction fails to yield additional soluble NANA-containing substances even after five successive digestions with papain. Failure to completely extract the gangliosides in the isolation procedure described causes these substances to appear in the papain-resistant fraction; papain is relatively ineffective in solubilizing the gangliosides.[15]

The double digestion procedure with papain is not essential in order to obtain good recoveries of sialomucopolysaccharides. A single digestion generally releases 95 % of the sialomucopolysaccharides and diffusable glycopeptides present in brain tissue.

TABLE II

Molar Ratios of Carbohydrate Constituents of the Sialomucopolysaccharide Fractions Obtained by Anion Exchange Chromatography, Gel Filtration, and Electrophoresis[a]

Experiment	Fraction	NANA/fucose	Hexosamine/fucose	Hexose/fucose	Hexose/hexosamine
1. DEAE-Sephadex	I	0.80	2.48	4.11	1.65
	II	1.54	2.96	4.72	1.59
	III	2.44	4.83	5.33	1.09
	IV	4.34	5.77	8.00	1.38
2. Sephadex G-200	I	2.77	3.57	6.54	1.63
	II	1.34	3.05	5.00	1.82
3. Electrophoresis	I	5.11	4.39	5.91	1.35
	II	5.26	4.72	6.17	1.30
	III	3.70	3.97	5.41	1.37
	IV	3.18	3.67	4.65	1.28
	V	2.11	2.51	4.63	1.89
	VI	0.90	2.40	3.28	1.38

[a] Analytical data obtained for fractions prepared by anion exchange chromatography on DEAE-Sephadex [Fig. 1(a)], for gel filtration on Sephadex G-200 [Fig. 1(b)], and column electrophoresis [Fig. 1(c)].

II. GLYCOPROTEINS AND THE PROTEIN-LINKED SIALOMUCOPOLYSACCHARIDES

A. Properties of the Protein-Linked Sialomucopolysaccha-rides Isolated from Whole Rat Brain

The isolation procedure of the sialomucopolysaccharides has indicated that these substances are nondialyzable and not precipitated by cetylpyridinium chloride. Upon subjection to gel filtration on Sephadex G-200, the sialomucopolysaccharides are largely excluded from the gel matrix and thus appear to have a molecular weight in excess of 100,000 (Fig. 1(b)]. This result is obtained if the elution is carried out with distilled water. If elution is carried out with a solution containing 0.05 M tris hydrochloride buffer, pH 7.5, and 0.1 M potassium chloride, the sialomucopolysaccharides are retarded upon gel filtration.[14,16] Their elution volume is that expected for a substance with a molecular weight of ten to thirteen thousand. Analysis of the first ("high" molecular weight) and second ("low" molecular weight) portion of the sialomucopolysaccharide material to emerge from Sephadex G-200 columns using water as eluant [Fig. 1(b) and Table II] revealed that the retarded material has a lower content of NANA. The greater the NANA content of a sialomucopolysaccharide fraction, the greater is its apparent

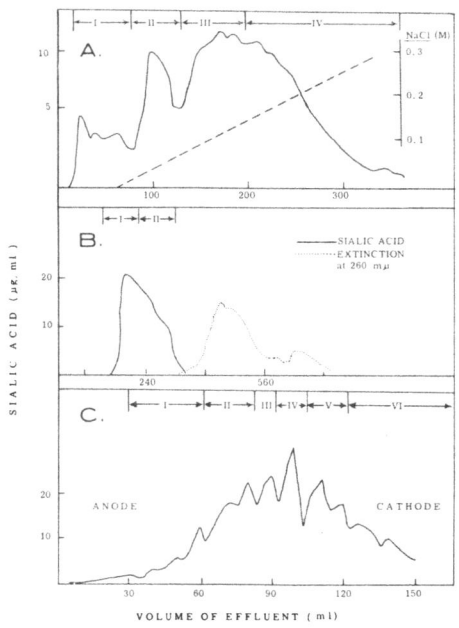

Fig. 1. Three methods that demonstrate the heterogeneity of the sialomucopolysaccharides isolated from brain glycoproteins. (a) Anion-exchange chromatography[12,14] on DEAE-Sephadex A-50 columns (1 × 22 cm). Sialomucopolysaccharides are adsorbed to the anion exchanger and subsequently eluted by a linear gradient of sodium chloride concentration at pH 7.0. Contents of test tubes containing fractions I, II, III, and IV (top of graph) were combined and analyzed (Table II). (b) Gel filtration on a Sephadex 200 column (34 × 4 cm) using water as eluting agent.[14,16] This chromatogram demonstrates the separation of an impurity that adsorbs light at 260 mμ from the sialomucopolysaccharides. Contents of test tubes containing fractions I and II (top of graph) were combined and analyzed (Table II). (c) Column electrophoresis on formalated cellulose[17] in glycine–sodium hydroxide buffer, pH 10.3, using LKB 3400 column electrophoresis. Contents of test tubes containing fractions I, II, III, IV, V, and VI (top of graph) were combined and analyzed (Table II).

molecular weight. It appears that the degree of aggregation of a sialmucopoly-saccharide fraction is dependent upon the number of NANA residues present in the molecule.

The heterogeneity of the sialomucopolysaccharides is also revealed by subjecting the substances to anion exchange chromatography on columns of Dowex 1-chloride[13] or DEAE-Sephadex[12,14] [Fig. 1(a) and Table II] and column electrophoresis[17] [Fig. 1(c) and Table II]. All fractions obtained contain NANA, glucosamine, mannose, galactose, and fucose. The molar ratios of these constituents vary. The NANA content of the sialomucopoly-saccharide fraction increases with the increase in negative charge or the affinity to the anion exchanger. The increase in NANA appears to be cor-related with a decrease in fucose content.[14] On the other hand, the ratio of hexose/hexosamine does not appear to vary beyond that expected by the limitations of the analytical methods used. It is believed[14] that all sialo-mucopolysaccharide molecules possess the same repeating unit consisting of hexose and hexosamine. The molecules differ from each other in the number of fucose and NANA residues attached to the basic repeating unit. Some galactosamine appears to be present[14] but this appears to be less than 20% of the total hexosamine. In addition to fucose, sialomucopolysaccharides from dog brain contain rhamnose as their methylpentose constituent. Sialomucopolysaccharides have been isolated from human brain by Heijlman and Roukema.[19]

If the thiobarbituric acid method of Warren[18] is used to determine NANA, the maximal values for NANA are obtained after hydrolysis for 1 hr in 0.1 N sulfuric acid at 80°C. The value obtained is multiplied by 1.09 to correct for degradation of NANA during acid hydrolysis.[14] Prolonged treatment of sialomucopolysaccharides with *Vibrio cholerae* neuraminidase liberated all of the NANA content; it appears, however, that all of the NANA residues are not equally susceptible to the cleaving action of this enzyme. Sixty percent of the NANA was released within 40 min of incuba-tion; prolonged incubation was required to liberate the rest.

B. Dialyzable Hexosamine Liberated by Papain Digestion of Brain Glycoproteins

The dialyzable hexosamine released by digestion of defatted brain tissue by the proteolytic action of papain has not received extensive study, but it appears to be present in a glycopeptide that contains considerable amounts of hexose. There is considerable circumstantial evidence[20] that this material and the sialomucopolysaccharides are part of the same macromolecular complex. Dilute aqueous buffers and solutions containing the nonionic detergent Triton X-100 solubilize the same percentage of the total sialo-mucopolysaccharides and dialyzable hexosamine-containing glycopeptides present in the treated tissue sample. The subcellular distribution of the glycopeptides resembles that of the sialomucopolysaccharides.[21] All attempts to purify the glycoproteins that are solubilized by means of the detergent Triton X-100 by column chromatography on calcium hydroxylapatite yield fractions that contain both the sialomucopolysaccharides and dialyzable hexosamine-containing glycopeptides.[20] It has not been possible to obtain a fraction enriched in one constituent relative to the other.

It is probable that the parent glycoprotein molecule contains more than one kind of polysaccharide chain. For example, Spiro[22,23] has found two distinct types of carbohydrate units attached to the glycoprotein constituent of the basement membrane of the bovine renal cortex glomerulus. One is a disaccharide unit containing glucose and galactose. The other is a sialomucopolysaccharide that contains variable amounts of NANA, fucose, galactose, mannose, and glucosamine. The thyroglobulin molecule[24] contains nine chains of a polysaccharide of molecular weight 1050 containing five residues of mannose and one residue of glucosamine. In addition, it contains 14 chains of a polysaccharide of molecular weight 3200 and that contain NANA, fucose, galactose, mannose, and glucosamine. A glycoprotein isolated from bovine aorta[25] contains two polysaccharide chains, one containing glucosamine and galactose and the other containing NANA, fucose, mannose, galactose, and glucosamine.

C. Solubility of the Glycoproteins

Particulate fractions prepared from rat brain were successively extracted with 0.02 M tris hydrochloride buffer, pH 8.0, and water.[20] The combined supernatants obtained after high-speed centrifugation contain only 10–15% of the protein-linked sialomucopolysaccharides and dialyzable hexosamine-containing glycopeptides present in the original material. Extraction of the residue obtained by this procedure with solutions containing 0.5% Triton X-100 solubilizes an additional 53–61% of the sialomucopolysaccharides and dialyzable hexosamine-containing glycopeptides. There is no significant difference in the molar ratios of the sugars present in the sialomucopolysaccharides solubilized by aqueous buffers or detergent, compared to the ratios of the sialomucopolysaccharides remaining in the sediment after the extraction. This indicates that the extraction methods are not specific in the type of sialomucopolysaccharide molecule that is extracted; the possibility remains, however, that the protein portion of the glycoprotein molecule determines its extractability. Treatment of particulate fractions from rat brain with butanol or laurylsulfate was not successful in solubilizing more than a small part of the total glycoprotein present[26] in the tissue sample. Preliminary experiments[20] with phospholipase indicate that at least 15% of the glycoproteins that contain sialomucopolysaccharides and diffusable hexosamine-containing glycopeptides can be solubilized. Presumably, at least a portion of the glycoproteins are rendered insoluble by association with phospholipids. D'Monte and Talwar[27] reported that gentle shaking of excised cortical areas from monkey cerebral cortex with 1.5 M urea at 4°C released some protein and nucleic acid, but no hexosamine and NANA. Glycoproteins are apparantly not linked by weak interactions, such as hydrogen bonds.

Attempts to fractionate the glycoproteins extracted with solutions containing Triton X-100 by means of column chromatography on calcium hydroxylapatite have failed to separate glycoproteins from proteins that are free of carbohydrate constituents[20,26]; some fractions obtained show an increase in the value of micrograms NANA per milligram protein compared to that of the original extract. The procedure used did achieve considerable purification of several enzymes and isozymes present in such extracts.[28] Rechromatography of protein fractions obtained by gradient or stepwise elution techniques[26] provided chromatograms that suggested the presence of at least three different glycoproteins. More recent work[20] has cast doubt upon this interpretation, since it has been found that a portion of the Triton-solubilized glycoproteins that were adsorbed to the gel upon chromatography failed to be retained upon rechromatography. The NANA content of this fraction is higher than that of the original extract. Higher values for micrograms NANA per milligram protein can also be obtained if Triton extracts are prepared using as starting material purified subcellular fractions obtained from the crude mitochondrial fraction by density-gradient centrifugation.

Soluble extracts prepared from the combined mitochondrial and microsomal fraction from rat brain by means of Triton X-100 contain about $8-10\ \mu$g protein-bound NANA/mg protein. It can be calculated that the carbohydrate content of the protein in these extracts is approximately 3.2%. Glycoproteins from other sources that have been well characterized generally contain approximately 10–33% carbohydrate, suggesting that a three- to tenfold purification of the glycoproteins present in the Triton X-100 extract may be sufficient to obtain pure brain glycoprotein. However, it must be borne in mind that a few glycoproteins, such as the human blood group substances, have a carbohydrate content that accounts for 80–90% of the molecule.

The anomalous behavior of the Triton X-100-solubilized glycoprotein on calcium hydroxylapatite may be due to aggregation and depolymerization of the glycoprotein molecules. Evidence for such a physical interaction has been demonstrated for the red blood cell membrane,[29] and for the urinary T and H glycoprotein.[30] The urinary glycoprotein is composed of an aggregation of smaller glycoprotein units of molecular weight of 28,000, and it has been suggested that this substance initially forms a part of the external glycoprotein layer of cell membranes present in the kidney.

The effectiveness of Triton X-100 in solubilizing glycoproteins is of some interest in view of the report that this detergent selectively solubilizes the synaptic membranes of the synaptosomes,[31] leaving the synaptic junctional complex intact. Triton X-100, however, does solubilize and fragment other structures as well.[28]

The gangliosides and the papain-resistant fraction (Table I) do not share the solubility characteristics of the glycoproteins. These substances are not extracted with aqueous solvents; Triton X-100 solubilizes 30 and 50% of the gangliosides and the papain-resistant fraction, respectively.[20]

The finding that 10–15% of the sialomucopolysaccharide-containing glycoprotein is solubilized by aqueous buffer is of interest in view of the demonstration of the presence of glycoproteins in the α-globulin fraction obtained by electrophoresis of soluble brain extracts. The brain-specific glycoprotein demonstrated by Warecka and Bauer[2] may represent the sialomucopolysaccharide-containing glycoprotein. On the other hand, the glycoprotein demonstrated by these workers may be identical with the brain-specific glycoprotein isolated by extraction with acetone by Gielen.[9] It is not clear whether Gielen's glycoprotein is extracted by the use of dilute aqueous buffers.

For additional information regarding glycoproteins in brain tissue, the reader is referred to Bogoch (Chapter 5 of this volume). Bogoch and co-workers have solubilized glycoproteins from brain tissue by means of solvents of low ionic strength and subsequently subjected the extracted protein to anion-exchange chromatography and electrophoresis in order to obtain several glycoprotein species. The relationship of these glycoproteins to the sialomucopolysaccharide-containing protein or Gielen's[9] glycoprotein, or both, remains to be established.

III. SUBCELLULAR LOCALIZATION OF GLYCOPROTEINS

The crude mitochondrial and microsomal fractions obtained from rat brain contain most of the gangliosides and glycoproteins present in the tissue.[13,21] The concentration of gangliosidic NANA and protein-bound NANA (μg NANA/mg protein) in the microsomal fraction is approximately double that found in the crude mitochondrial fraction. Separation of the crude mitochondrial fraction into subcellular fractions that are enriched in myelin, synaptosomes, and mitochondria by utilization of differential centrifugation through a discontinuous sucrose density gradient revealed that the largest percentage of the gangliosides and glycoproteins in the crude fraction are present in the synaptosomes.[21] Comparison of the subfractions obtained from crude mitochondrial and microsomal preparations, however, revealed an increase in the ratio of glycoprotein NANA to gangliosidic NANA present in particles of increasing density. The sedimentation characteristics of the diffusable hexosamine-containing glycopeptides was similar to that of the sialomucopolysaccharides. Furthermore, in several experiments in which particles subjected to osmotic shock prior to differential centrifugation were studied,[21,32] a shift of glycoproteins from more dense fractions to lighter fractions was observed. The data suggested that the gangliosides and the glycoproteins may not be localized in the same subcellular structure.

TABLE III

Subcellular Distribution of Gangliosidic NANA and Glycoprotein NANA

Experiment[b]	Fraction	% Distribution		RSA[a]		Ratio glycoprotein NANA/gangliosidic NANA
		Gangliosidic NANA	Glycoprotein NANA	Gangliosidic NANA	Glycoprotein NANA	
1	O	1.0	9.5	0.31	0.37	1.19
	D	8.3	5.6	1.04	0.70	0.67
	E	21.7	15.9	3.62	2.65	0.73
	F	30.1	17.2	2.51	1.43	0.57
	G	19.8	27.8	1.53	2.14	1.40
	H	12.1	12.7	0.60	0.63	1.05
	I	7.0	11.3	0.44	1.03	2.34
2	P2A	15.8	7.7	0.72	0.35	0.36
	P2B	22.9	12.2	2.36	1.25	0.36
	P2C	31.4	30.7	1.54	1.51	0.70
	P2D	17.0	19.5	0.92	1.05	0.83
	P2E	13.0	30.0	0.44	1.01	1.77
3	P2B-a	21.5	9.8	1.72	0.78	0.45
	P2B:b	17.6	12.6	1.54	1.10	0.71
	P2B-c	20.3	16.4	1.21	0.98	0.81
	P2B-d	18.5	20.8	0.88	0.98	1.11
	P2B-e	21.8	40.2	0.57	1.06	1.86

[a] RSA denotes percent NANA recovered in fraction/percent protein recovered in fraction.

[b] Experiment 1: Water-treated (osmotically shocked) mitochondrial fraction from rat whole brain (P2) was centrifuged at $10,000 \times g$ for 20 min to remove the larger mitochondria and myelin fragments. The cloudy supernatant was subjected to centrifugation through a discontinuous sucrose density gradient (0, 0.4, 0.6, 0.8, 1.0, and 1.2 M sucrose). The procedure is that of Whittaker, Michaels, and Kirkland.[37] Experiment 2: The crude mitochondrial fraction (P2) from cat cerebellum was subjected to centrifugation through a discontinuous sucrose density gradient (0.32, 0.8, 1.0, 1.2, and 1.3 M sucrose) as described by De Robertis et al.[38] Experiment 3: A nerve ending preparation from rat whole brain (P2B) was subjected to centrifugation through a discontinuous sucrose density gradient (0.32, 0.9, 1.0, 1.1, and 1.2 M sucrose) by the procedure of Dekirmenjian and Brunngraber.[33]

Dekirmenjian and Brunngraber[33] utilized more sophisticated centrifugation procedures (Table III) to demonstrate that the gangliosides and the glycoproteins differ in their distribution patterns. Neither substance appears to be an important constituent of the synaptic vesicles, lysosomes, nuclei, ribosomes, mitochondria, or myelin. Both are enriched in those fractions that contain membranal fragments: the membranal fragments of the microsomal fraction and the membranes of the synaptosomes. On the other hand, the ratio of glycoprotein NANA to gangliosidic NANA in any subfraction obtained by three different procedures appeared to be a function of the particle density. Cell membranes present in the denser fractions are associated with synaptosomes (Table III, Exp. 1, fractions F, G, and H; Exp. 2, fractions P2C and P2D) and cell membranes that are present in the lighter fractions originate from the membranes of the cell body, dendrites, and axons (Table III, Exp. 1, fractions D and E; Exp. 2, fractions P2A and P2B). Support for the hypothesis that axonal membranes are present in lighter fractions was obtained by Johnston and Larramendi.[34] Electron microscopic examination of the P2B fraction (Table III, Exp. 2) derived from the cerebellum revealed that this fraction is rich in axons, the morphological features of which appear to be similar to those expected for parallel fibers (Fig. 2). In the same experiment, subfractions P2C and P2D were enriched in synaptosomes. The data of Table III suggest that the membranes associated with the synaptosomes contain a higher concentration of glycoprotein relative to that of gangliosides than do other portions of the neuronal cell surface.[62]

An alternate hypothesis can be offered. If different neurons differ in the amount of glycoprotein and ganglioside that they contain, one might expect that membranes derived from the glycoprotein-rich neurons have higher densities than membranes derived from ganglioside-rich neurons.

Two other findings related to the data of Table III are of considerable interest. The distribution of acetylcholinesterase closely parallels the distribution of the gangliosides. Furthermore, the glycoproteins[33] and the sodium and potassium-activated ATPase[35] are enriched in membranes derived from the synaptosomes. This finding is reinforced by the observations that Triton X-100, which is effective in solubilizing the presynaptic membrane of the synaptosomes,[31] is also quite effective in solubilizing glycoproteins and ATPase. The histochemical localization of ATPase and glycoprotein is also similar.[36]

Areas of the brain in which cell bodies predominate (cerebral gray, basal ganglia, thalamus, and cerebellum) contain a higher concentration of gangliosides and glycoproteins[15] than do areas that are predominantly myelinated fiber tracts (cerebral white, corpus callosum, pons, and medulla). On the other hand, the ratio of glycoprotein/gangliosides is generally greater in white matter than it is in gray matter, indicating a higher concentration of glycoproteins relative to that of gangliosides in areas that consist predominantly of myelinated fiber tracts and glia. To reconcile these findings with the observations recorded above, we can take note of the work of Lumsden[47] and Singh,[46] who suggested that periodic acid-Schiff positive substances are

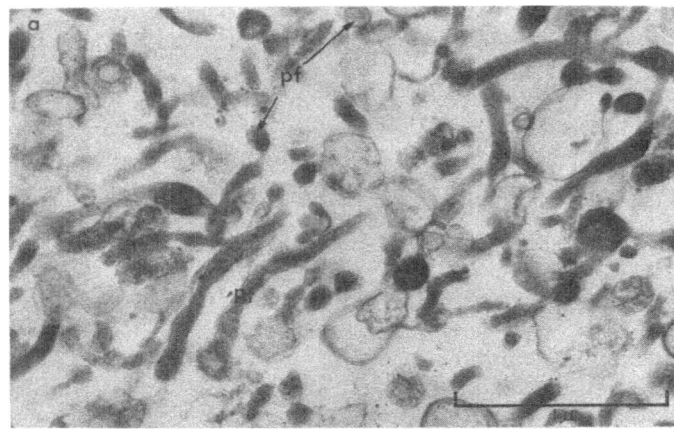

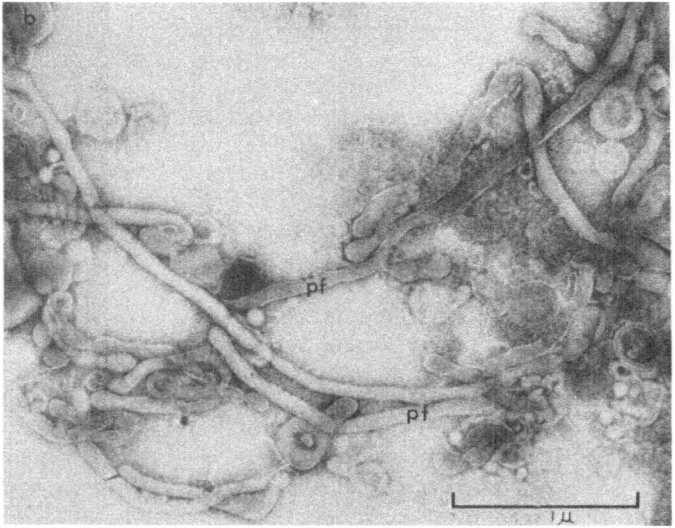

Fig. 2. Electron micrographs of subcellular fractions enriched in axons prepared from cat cerebellum by Johnston and Larramendi.[34] (a) Thin section of P2B layer obtained from subfractionated crude mitochondrial pellet of the cat cerebellum. This field is dominated by longitudinal sections and cross sections of tubular elements (pf) believed to represent the non-myelinated axons of the parallel fibers. (b) Negatively stained suspension from P2B layer obtained from sucrose density gradient centrifugation experiments. This fraction abounds in tubular elements such as shown here (pf). The tubules are interpreted as being intersynaptic segments of nonmyelinated axons.

present in glial processes. One may postulate that homogenization of the tissue causes fragments of the oligodendroglial processes to sediment along with the nerve endings. As will be noted in the next section, Rambourg and Leblond[36] find that all cell types possess a carbohydrate-rich coat. If glial membranes contain a higher concentration of glycoprotein relative to that of gangliosides, one might expect this to be reflected in those subcellular fractions that are enriched in membranes that originate from the glia.

IV. HISTOCHEMISTRY AND GLYCOPROTEINS

Protein-bound sialomucopolysaccharides give a positive reaction in the periodic acid-Schiff test.[39] Application of this histochemical technique to nervous tissue has generally revealed that the neuronal cell body and processes give a weak or negative reaction.[40,42] However, Sulkin[43] reported that the mucoproteins are present in granular form in the cytoplasm of all nerve cells and suggested that these may be elaborated to form a ground substance. Hess[40,41] reported the presence of a PAS-positive substance between axons and dendrites; the material was believed to be a neutral polysaccharide that formed an intercellular ground substance. The findings were difficult to reconcile with the discovery that there is very little intercellular space in brain tissue.[44,45] Singh[46] and Lumsden[47] produced evidence in support of the claim that PAS-positive substances accumulate in glia and are present in glial processes that make up a great deal of the interneuronal space. Wolman[48] found that during the first stages of myelin degeneration there is an increase in the amount of demonstrable polysaccharides along the degenerating nerve fiber. Shanklin and Azzam[49] reported that the incisures of Schmidt–Lantermann along axons have a central core rich in proteins and carbohydrates.

Rambourg and Leblond[36] provided evidence that all cell types possess a carbohydrate-rich coat. The literature is extensively reviewed and these authors present their own findings based on light- and electron-microscopic investigations that indicate that nerve cells are outlined by a discontinuous reactive line that can be traced along dendrites and axons. The cell processes are surrounded by a layer, the density of which increases in the synaptic clefts (Fig. 3). Bondareff[50,51] also provided evidence that mucopolysaccharides are present in intercellular gaps and especially in the synaptic cleft. Pease[52] found such material to be present in synaptic clefts and neuropil and suggested that the substance provides an extracellular compartment in which ions may accumulate and move. Barer[53] found mucopolysaccharides between nerve cell swellings and blood or effector cells and thought that these may play a role in storage and release of active substances.

Histochemical and electron microscopic data have generally been considered to provide evidence of the presence of an intercellular substance, although Rambourg and Leblond consider the material to be part of the cell coat. Robertson's[54] view that the plasma membrane consists of a lipid

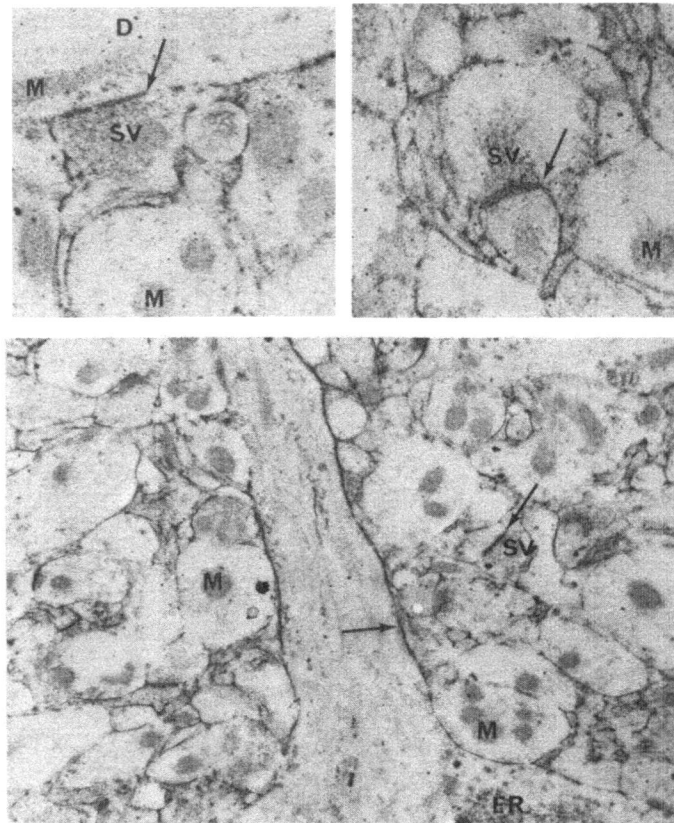

Fig. 3. Electron micrographs demonstrating the application of the periodic acid–silver methenamine technique to the cerebral cortex of the rat by Rambourg and Leblond.[36] Upper left: Neuropil. The staining of the intercellular space separating the dendrite (D) in the upper part of the picture from other processes is sharply increased in the region of the synaptic cleft (arrow). The matrix of the mitochondria (M) and synaptic vesicles (SV) also take up some stain. × 42,000. Upper right: Synapse. The synaptic cleft is indicated by an arrow. × 42,000. Bottom: Nerve cell. The long dendrite of the nerve cell whose perikaryon is visible at the base is sharply outlined by cell coat material (arrow). In the neuropil, nerve and glial processes are separated from each other by stained material. A synaptic cleft is indicated by an oblique arrow. M, mitochondrion; SV, synaptic vesicles; ER, ergastoplasm. × 20,000.

bileaflet bounded by an outer surface containing glycoproteins and an inner surface consisting of protein has gained general acceptance and appears to be consistent with the data available. The fact that the brain glycoproteins are largely insoluble and appear to be associated with phospholipids suggests that the bond between the external glycoproteins and the underlying lipid is not readily ruptured. It may be noted additionally that the red blood cell membrane appears to contain an insoluble glycoprotein that contains NANA, hexosamine, and hexose and that can be solubilized by butanol treatment of red blood cell ghosts.[55] Butanol is an effective agent that acts by disrupting phospholipid–protein complexes. It is now known that a variety of cell types possess a negative surface charge that is due to the presence of NANA.

V. METABOLISM OF THE GLYCOPROTEINS

Autoradiographic techniques have indicated that the Golgi region is the site of synthesis of the glycoproteins.[56] Lawford and Schachter[57] showed that, in the liver, hexosamine is incorporated into newly synthesized polypeptides while these are still attached to ribosomes: additional hexosamine units are incorporated after detachment from the ribosomes as the polypeptide traverses the channel of the rough- and smooth-surfaced endoplasmic reticulum on its way out of the cell. N-acetylneuraminic acid, the terminal group in the glycoproteins, is incorporated primarily during the final stages of this process, while the glycoprotein is within the smooth-surfaced endoplasmic reticulum. Barondes[88] has found that there is a lapse of time before intracisternally injected labeled amino acids appear in proteins present in synaptosomes, which is due presumably to the time of transit required between synthesis in the cell body and appearance in the nerve endings. Incorporation of glucosamine into proteins of the synaptosomes is more immediate, suggesting that the polysaccharide chains are built up, in part at least, at the site of the synapse. Bogoch et al.[89] studied the incorporation of labeled glucose into glycoproteins isolated from brain and tumor tissues. Robinson, Molnar, and Winzler[53] injected glucosamine-1-^{14}C intravenously and measured its incorporation into the proteins of various rat tissues. The amount of label incorporated into one gram of brain tissue was approximately 20% that of the liver. Brain was considered to have relatively low activity. However, it should be remembered that the contribution of the nerve cell body to the total weight of a given sample of nervous tissue is small. Thus, Oehlert, Schultze, and Maurer[59] reported that the protein synthesis in brain tissue occurs predominantly in the ganglion cell. Activity was poor in glia and white matter. Synthesis of glycoprotein may therefore proceed quite rapidly in the nerve cell body. McGuire et al.[60] incubated particulates with UDP-galactose-^{14}C and found that such preparations transferred galactose from UDP-galactose to N-acetylglucosamine or to orosmucoid that had been pretreated with sialidase and β-galactosidase. Particulate fractions from brain tissue had appreciable activity, being approximately 20% that found using liver microsomes. The rate of transfer of galactose to N-acetylglucosamine was similar to that observed with liver.

Catabolic studies have been few in number. Weil-Malherbe and Drysdale[61] reported little change in the brain content of hexosamine after incubating slices aerobically or anaerobically. Brunngraber and Brown[15] have found that subjection of brain tissue to autolysis by incubating the tissue for 1 hr at 37°C had no effect on the amount of sialomucopolysaccharides isolated from the tissue, or on the molar ratio of its sugar constituents.

The carbohydrate units become attached to the peptide chains of the protein molecule by means of their nucleoside diphosphoglycosyl derivatives. Several of the enzymes that metabolize these derivatives have been partially purified from brain. Shoyab, Pattabiramen, and Bachhawat[63] reported the purification of the enzyme that synthesizes CMP-NANA from CTP and NANA. Joseph and Bachhawat[64] reported the purification of the enzyme that synthesizes NANA from N-acetylmannosamine 6-phosphate and phosphoenolpyruvate. Pattabiramen and Bachhawat[65] reported on the purification and anatomical localization of several enzymes that are involved in glucosamine metabolism. Degradative enzymes that may play a role in the metabolism of the glycoproteins have been partially purified. Gatt and Rapport[66] partially purified β-galactosidase; Frohwein and Gatt[67] reported the purification of β-N-acetylglucosaminase and β-N-acetylgalactosaminidase from calf brain.

One gram brain tissue is capable of synthesizing 0.1 μmole NANA/hr[64]; CMP-NANA is synthesized at a rate of 0.45 μmole/hr.[63] UDP-acetylglucosamine pyrophosphorylase, UDP-acetylglucosamine pyrophosphatase, N-acetylglucosamine kinase, and glucosamine 6-phosphate deaminase convert 1.2–7.8, 6–18, 1.2–3.4, and 3.7–4.9 μmoles of substrate per gram tissue per hour.[65] If these reactions proceed at these rates *in vivo*, metabolism of the nucleoside diphosphoglycosyl derivatives is such to suggest that the turnover of the glycoproteins is neither very rapid nor very slow.

VI. PERSPECTIVE AND SPECULATION

A great variety of sialoproteins have been isolated and characterized in recent years and an excellent review on glycoproteins has recently been published.[68] Most of these glycoproteins are secreted and are found in blood plasma, urine, and other biological fluids. Recently several publications have appeared in which sialoproteins derived from plasma membranes were described. The percentage of carbohydrate present in glycoproteins derived from the red blood cell membrane[55] and the kidney basement membrane[22,23] is about 8–10%. A glycoprotein extracted from bovine aorta is probably about 20–25% carbohydrate.[25] Gielen's glycoprotein[9] from brain contains approximately 33% carbohydrate. Among the large number of extracellular glycoproteins that have been studied are many that have been found to contain sialomucopolysaccharide units similar to the one isolated from brain. Recent work has shown that hog intrinsic factor,[69] human macroglobulin,[70] ceruloplasmin,[71] and thyroglobulin[24] contain 33, 10, 10, and 10 % carbohydrate, respectively. The determination of the percentage of carbohydrate in brain glycoprotein will require its isolation and characterization; however, the aforementioned observations suggest that this figure will turn out to fall between 10 and 33 % carbohydrate. Since one gram of brain tissue contains approximately 100 mg protein and 1.5 mg carbohydrate derived from glycoproteins, it can be calculated that approximately 4–15 % of the total protein content of brain is glycoprotein.

Eylar[72] has noted that most proteins that are secreted are glycoproteins, while most intracellular proteins are not. It appears that the brain glycoproteins are not an exception. Although synthesized in the perikaryon, presumably in the Golgi region, they migrate to the cell surface. Secretion appears to be limited to a process in which the glycoproteins become deposited upon the outer portion of the cell surface. There they form a cell coat[36] or the superficial protein layer of the plasma membrane.[54] An intriguing problem for future research is a consideration of their subsequent metabolism. The surface glycoprotein may return to the cell in order to be catabolized. Alternatively, it may be elaborated from the cell surface and enters the intercellular areas, eventually to reach the CSF by means of the perivascular spaces, or it may enter the blood supply. Lowenthal, Karcher, and Van Sande[73] have postulated that glycoproteins present in the CSF may originate within the CNS. The few metabolic and enzymatic studies on brain glycoproteins suggest that these substances may have a moderate turnover rate. It may be pointed out here that pathological conditions due to metabolic aberrations may occur because of errors in transport of cellular substances from one site of the cell to another, as well as aberrations in biochemical synthesis. The synthesis of a macromolecule can conceivably proceed normally, but failure to transport it properly may lead to storage problems within the cell.

The subcellular distribution studies of the glycoproteins locates these substances at points of intercellular contacts and at those portions of the cell involved in transmission of the nerve cell impulse. A functional role for the glycoproteins has been suggested by a few studies. Agostini and Schulman[74] have studied interfacially oriented layers consisting of phospholipoprotein-polysaccharide monolayers and suggest that these control the flux of water and ions. Flux may be triggered or enhanced by adding acetylcholine. Kramer[75] reported the possibility of conformational changes of NANA on the cell surface, since he noted a transient elevation of electrophoretic mobility of CHO cells prior to cell division. Wooley and Gommi[76] studied the effects of neuraminidase on the sensitivity of muscle and suggested that gangliosides may act as a receptor for serotonin. The implication of the gangliosides in transport phenomena by McIlwain[77] and receptor function by Wooley and Gommi[76] may realistically be extended to the glycoproteins.

Kelly and Greiff[78] reported that there was an increase in brain neuraminidase activity 36 hrs after the intracerebral injection of nonneurotropic influenza virus; convulsions and death were maximal at this time. Gangliosidic and lipid-free NANA were reduced 5.5 and 18.8% respectively. These workers believe that there was a causal relation between death and the level of neuraminidase in brain. Glick and Githens[79] reported that the cleavage of NANA present on the surface of the L1210 leukemic cell by the use of neuraminidase inhibited the transport of potassium, regardless of the direction of flow. On the other hand, Evans and McIlwain[80] found that exposure of guinea pig cerebral cortex slices to neuraminidase had no effect on their metabolic response to electric stimulation, although about 50% of the tissue NANA groups were removed. Gainer[81] noted no effect of neuraminidase on the resting potential of muscle. Muscle membrane preparations contain protein-bound sialomucopolysaccharides.[82]

Functional change in nerve cells is accompanied by changes in nucleic acid and protein metabolism.[83] Barondes[84] and Roberts[85] have suggested that the regulatory mechanism concerned with protein synthesis may be altered in such a way by the electrical activity of the nerve cell that morphological and other changes may be produced in synaptic connections. The work of Dische et al.[86,87] may be of interest in this connection. These authors found that the type of glycoprotein molecule secreted by the dog submaxillary gland was dependent on the nature or intensity of the stimulus. Variations in the dose of pilocarpine, for example, do not affect the sum of fucose and NANA of the secreted glycoproteins, but profoundly affect the ratio of NANA/fucose.

The inherent polymorphism of the glycoprotein molecule—the many possible permutations and combinations in the location, chemical structure, and configuration of the polysaccharide side chains—suggests that the glycoprotein is a reasonable candidate for future investigations designed to elucidate macromolecular alterations that are due to functional activity in the central nervous system.

VII. REFERENCES

1. G. Hoffmann and H. Schinko, Elektrophoretische Trennung von Hirngewebe, *Klin. Wochschr.* **34**:86–90 (1956).
2. K. Warecka and H. Bauer, Studies on 'brain specific' proteins in aqueous extracts of brain tissue, *J. Neurochem.* **14**:783–787 (1967).
3. T. Ejima, A mucopolysaccharide from human cerebral cortex, *Tohoku J. Exptl. Med.* **61**:163–169 (1955).
4. G. Brante, in *Metabolism of the Central Nervous System* (D. Richter, ed.), pp. 112–120, Pergamon Press, New York (1957).
5. R. E. Glegg and R. H. Pearce, Chemical extraction of metachromatic and periodic acid-Schiff positive carbohydrates from cerebral tissue, *J. Comp. Neurol.* **106**:291–297 (1956).

6. Z. Dische, A. Danilczenko, and G. Zelmenis, *in Chemistry and Biology of Mucopolysaccharides* (G. E. W. Wolstenholm and M. O'Connor, eds.), pp. 116–139, Little, Brown and Co., Boston (1958).
7. L. Svennerholm, On sialic acid in brain, *Acta Chem. Scand.* **10**:694–696 (1956).
8. R. Kuhn and H. Müldner, Über Glyko-lipo-sialo-proteide des Gehirns, *Naturwissenschaften* **51**:635–636 (1964).
9. W. Gielen, Über ein Hirnspezifische Glykoprotein, *Naturwissenschaften* **53**:504–5 (1966).
10. M. Crevier, A histochemical and biochemical study of the polysaccharide substances of the developing nervous system of the rat and the relation with the appearance of cholinesterase activity, *Can. J. Biochem. Physiol.* **36**:275–288 (1958).
11. E. G. Brunngraber and B. D. Brown, Preparation of sialomucopolysaccharides from brain mitochondrial fractions, *Biochim. Biophys. Acta* **69**:581–582 (1963).
12. E. G. Brunngraber and B. D. Brown, Heterogeneity of sialomucopolysaccharides prepared from whole rat brain, *Biochim. Biophys. Acta* **83**:357–360 (1964).
13. E. G. Brunngraber and B. D. Brown, Fractionation of brain macromolecules—II. Isolation of protein-linked sialomucopolysaccharides from subcellular, particulate fractions from rat brain, *J. Neurochem.* **11**:449–459 (1964).
14. E. G. Brunngraber and B. D. Brown, Preparation and properties of sialomucopolysaccharides obtained from rat brain, *Biochem. J.* **103**:65–72 (1967).
15. E. G. Brunngraber, B. D. Brown, and V. Aguilar, Isolation and determination of nondiffusible sialofucohexosaminoglycans derived from brain glycoproteins and their anatomical distribution in bovine brain, *J. Neurochem.* (in press).
16. E. G. Brunngraber and G. Whitney, Effect of ionic strength of eluting solutions on behavior of sialomucopolysaccharides from rat brain on Sephadex G-200, *J. Chromatog.* **32**:749–750 (1968).
17. C. Di Benedetta, E. G. Brunngraber, and G. Whitney, Heterogeneity of sialomucopolysaccharides isolated from rat brain, *Federation Proc.* **27**:399 (1968).
18. L. Warren, The thiobarbituric acid assay of sialic acids, *J. Biol. Chem.* **234**:1971–1975 (1959).
19. J. Heijlman and P. A. Roukema, Isolation of sialomucopeptides from human brain, *Biochim. Biophys. Acta* **127**:269–271 (1966).
20. E. G. Brunngraber, V. Aguilar, and A. Aro, Chromatography on calcium hydroxylapatite of glycoproteins solubilized by Triton X-100 from particulates prepared from rat brain, *Arch. Biochem. Biophys.* (in press).
21. E. G. Brunngraber, H. Dekirmenjian, and B. D. Brown, The distribution of protein-bound N-acetylneuraminic acid in subcellular fractions of rat brain, *Biochem. J.* **103**:73–78 (1967).
22. R. G. Spiro, Studies on the renal glomerular basement membrane. Preparation and chemical composition, *J. Biol. Chem.* **242**:1915–1922 (1967).
23. R. G. Spiro, Studies on the renal glomerular basement membrane. Nature of the carbohydrate units and their attachment to the peptide portion, *J. Biol. Chem.* **242**:1923–1932 (1967).
24. R. G. Spiro, The carbohydrate units of thyroglobulin, *J. Biol. Chem.* **240**:1603–1610 (1965).
25. B. Radhakrishnamurthy and G. S. Berenson, Glycopeptides from bovine aorta glycoprotein, *J. Biol. Chem.* **241**:2106–2112 (1966).
26. E. G. Brunngraber and E. A. Bejnarowicz, *in Protides of the Biological Fluids* (H. Peeters, ed.), Vol. 13, pp. 201–205, Elsevier Publishing Co., Amsterdam (1966).
27. B. D'Monte and G. P. Talwar, Chemical composition and immunological specificity of urea-extractable macromolecules from three regions of the monkey brain, *J. Neurochem.* **14**:743–754 (1967).
28. E. G. Brunngraber and V. Aguilar, Fractionation of brain macromolecules—I. Chromatography on hydroxylapatite of enzymes solubilized by Triton X-100 from the rat brain mitochondrial fraction, *J. Neurochem.* **9**:451–461 (1962).

29. S. Bakerman and G. Wasemiller, Studies on structural units of human erythrocyte membrane. I. Separation, isolation, and partial characterization, *Biochemistry* **6**:1100 (1967).
30. M. Maxfield, in *Glycoproteins* (A. Gottschalk, ed.), pp. 446–461, Elsevier Publishing Company, Amsterdam (1966).
31. E. De Robertis, Ultrastructure and cytochemistry of the synaptic region, *Science* **156**:907–914 (1967).
32. E. G. Brunngraber, V. A. Ziboh, and W. G. Occomy, in *Protides of the Biological Fluids* (H. Peeters, ed.), Vol. 13, pp. 207–210, Elsevier Publishing Co., Amsterdam (1966).
33. H. Dekirmenjian and E. G. Brunngraber, Distribution of protein-bound N-acetylneuraminic acid in subcellular particulate fractions prepared from rat whole brain, *Biochim. Biophys. Acta* (in press).
34. N. L. Johnston and L. M. H. Larramendi, The separation and identification of fractions of non-myelinated axons from the cerebellum of the cat, *Exptl. Brain Res.* **5**:326–340 (1968).
35. R. J. A. Hosie, The localization of adenosine triphosphatases in morphologically characterized subcellular fractions of guinea pig brain, *Biochem. J.* **96**:404–412 (1965).
36. A. Rambourg and C. P. Leblond, Electron microscopic observations on the carbohydrate-rich cell coat present at the surfaces of cells in the rat, *J. Cell Biol.* **32**:27–53 (1967).
37. V. P. Whittaker, I. A. Michaelson, and R. J. A. Kirkland, The separation of synaptic vesicles from nerve-ending particles (synaptosomes), *Biochem. J.* **90**:293–303 (1964).
38. E. De Robertis, A. Pellegrino De Iraldi, G. Rodriquez de Lores Arnaiz, and L. Salganicoff, Cholingic and non-cholinergic nerve endings in rat brain—I. Isolation and subcellular distribution of acetylcholine and acetylcholinesterase, *J. Neurochem.* **9**:23–35 (1962).
39. S. S. Spicer and L. Warren, The histochemistry of sialic acid containing mucoproteins, *J. Histochem. Cytochem.* **8**:135–137 (1960).
40. A. Hess, The ground substance of the central nervous system revealed by histochemical staining, *J. Comp. Neurol.* **98**:69–91 (1953).
41. A. Hess, Further histochemical studies on the presence and nature of the ground substance of the central nervous system, *J. Anat. (London)* **92**:298–303 (1958).
42. I. J. Young and L. G. Abood, Histological demonstration of hyaluronic acid in the central nervous system, *J. Neurochem.* **6**:89–94 (1960).
43. N. M. Sulkin, The distribution of mucopolysaccharides in the cytoplasm of vertebrate nerve cells, *J. Neurochem.* **5**:231–235 (1960).
44. E. W. Dempsey and G. B. Wislocki, An electron microscopic study of the blood–brain barrier in the rat, employing silver nitrate as a vital stain, *J. Biophys. Biochem. Cytol.* **1**:245–256 (1955).
45. R. W. G. Wyckoff and J. Z. Young, The motoneurone surface, *Proc. Roy. Soc. London* **B144**:440–450 (1955–1956).
46. R. Singh, On histochemical localization of the ground substance in the central nervous system of *Psittacula krameri*, *J. Histochem. Cytochem.* **12**:712–713 (1964).
47. C. E. Lumsden, in *Biology of Neuroglia* (W. F. Windle, ed.), pp. 141–161, Charles C. Thomas, Springfield, Ill. (1958).
48. M. Wolman, Histochemical study of changes occurring during degeneration of myelin, *J. Neurochem.* **1**:370–376 (1956–1957).
49. W. M. Shanklin and N. A. Azzam, Histological and histochemical studies on the incisures of Schmidt-Lantermann, *J. Comp. Neurol.* **123**:5–9 (1964).
50. W. Bondareff, An intercellular substance in rat cerebral cortex: Submicroscopic distribution of ruthenium red, *Anat. Rec.* **157**:527–535 (1967).
51. W. Bondareff, Demonstration of an intercellular substance in mouse cerebral cortex, *Z. Zellforsch. Mikroskop. Anat.* **81**:366–373 (1967).
52. D. C. Pease, Polysaccharides associated with the exterior surface of epithelial cells: Kidney, intestine, brain, *J. Ultrastruct. Res.* **15**:555–588 (1966).

53. R. Barer, Speculation on the storage and release of hormone and transmitter substances, *Bibliotheca Anat.* **8**:72–75 (1967).
54. J. D. Robertson, Ultrastructure of excitable membranes and the crayfish median-giant synapse, *Ann. N.Y. Acad. Sci.* **94**:339–389 (1961).
55. A. H. Maddy, The properties of the protein of the plasma membrane of ox erythrocytes, *Biochim. Biophys. Acta* **117**:193–200 (1966).
56. M. Peterson and C. P. Leblond, Synthesis of complex carbohydrates in the Golgi region, as shown by radioautography after injection of labelled glucose, *J. Cell Biol.* **21**:143–148 (1964).
57. G. R. Lawford and H. Schachter, Biosynthesis of glycoproteins by liver, *J. Biol. Chem.* **241**:5408–5418 (1966).
58. G. B. Robinson, J. Molnar, and R. J. Winzler, Biosynthesis of glycoproteins, *J. Biol. Chem.* **239**:1134–1141 (1964).
59. W. Oehlert, B. Schultze, and W. Maurer, Autoradiographische Untersuchung der Grösse des Eiweissstoffwechsels der verschiedenen Zellen des Zentralnervensystems, *Beitr. Pathol. Anat. Allgem. Pathol.* **119**:343–376 (1958).
60. E. J. McGuire, G. W. Jourdian, D. M. Carlson, and S. Roseman, Incorporation of D-galactose into glycoproteins, *J. Biol. Chem.* **240**:PC4112–PC4115 (1965).
61. H. Weil-Malherbe and A. C. Drysdale, Ammonia formation in brain–III. The role of the protein amide groups and of hexosamine, *J. Neurochem.* **1**:250–255 (1956–1957).
62. H. Dekirmenjian, E. G. Brunngraber, N. L. Johnston, and L. M. H. Larramendi, Distribution of gangliosides, glycoprotein-NANA, and acetylcholinesterase in axonal and synaptosomal fractions of cat cerebellum, *Exptl. Brain Res.* (in press).
63. M. Shoyab, T. N. Pattabiramen, and B. K. Bachhawat, Purification and properties of the CMP-N-acetylneuraminic acid synthesizing enzyme from sheep brain, *J. Neurochem.* **11**:639–646 (1964).
64. R. Joseph and B. K. Bacchawat, Purification and properties of N-acetylneuraminic acid-synthesizing enzyme from sheep brain, *J. Neurochem.* **11**:517–526 (1964).
65. T. N. Pattabiramen and B. K. Bachhawat, Regional distribution of four enzymes associated with aminosugar metabolism in sheep brain, *J. Neurochem.* **11**:55–60 (1964).
66. S. Gatt and M. M. Rapport, Isolation of β-galactosidase and β-glucosidase from brain, *Biochim. Biophys. Acta* **113**:567–576 (1966).
67. Y. Z. Frohwein and S. Gatt, Separation of β-N-acetylglucosaminidase and β-N-acetylgalactosaminidase from calf brain cytoplasm, *Biochim. Biophys. Acta* **128**:216–218 (1966).
68. A. Gottschalk (ed.), *Glycoproteins, Their Composition, Structure, and Function*, Elsevier Publishing Co., Amsterdam (1966).
69. D. R. Highley, M. C. Davies, and L. Ellenbogen, Hog intrinsic factor—II. Some physicochemical properties of vitamin B_{12}-binding fractions from hog, *J. Biol. Chem.* **242**:1010–1015 (1967).
70. F. Miller and H. Metzger, Characterization of a human macroglobulin—I. The molecular weight of its subunit, *J. Biol. Chem.* **240**:3325–3333 (1965).
71. G. A. Jamieson, Studies on glycoproteins—I. The carbohydrate portion of human ceruloplasmin, *J. Biol. Chem.* **240**:2019–2027 (1965).
72. E. D. Eylar, On the biological role of glycoproteins, *J. Theoret. Biol.* **10**:89–113 (1965).
73. A. Lowenthal, D. Karcher, and M. Van Sande, Electrophoretic studies of central nervous system proteins, *Exptl. Neurol.* **1**:233–247 (1959).
74. A. M. Agostini and J. H. Schulman, Polysaccharides and flux of water through liquid membranes, *Soc. Chem. Ind. (London) Monograph* **19**:37–55 (1965).
75. P. M. Kramer, Configurational change of surface sialic acid during mitosis, *J. Cell Biol.* **33**:197–199 (1967).
76. D. W. Wooley and B. W. Gommi, Serotonin receptors: V. Selective destruction by neuraminidase plus EDTA and reactivation with tissue lipids, *Nature* **202**:1075–1078 (1964).

77. H. McIlwain, *Chemical Exploration of the Brain*, Elsevier Publishing Company, Amsterdam (1963).

78. R. T. Kelly and D. Greiff, Neuraminidase and neuraminidase-labile substrates in experimental influenza virus encephalitis, *Biochim. Biophys. Acta* **110**:548–553 (1965).

79. J. L. Glick and S. Githens, III, Role of sialic acid in potassium transport of L1210 leukemia cells, *Nature* **208**:88 (1965).

80. W. H. Evans and H. McIlwain, Excitability and ion content of cerebral tissues treated with alkylating agents, tetanus toxin, or a neuraminidase, *J. Neurochem.* **14**:35–44 (1967).

81. H. Gainer, Plasma membrane structure: Effects of hydrolases on muscle resting potential, *Biochim. Biophys. Acta* **135**:570–576 (1967).

82. L. G. Abood, K. Kurahasi, E. G. Brunngraber, and K. Koketsu, Biochemical analysis of isolated bullfrog sarcolemma, *Biochim. Biophys. Acta* **112**:330–339 (1966).

83. H. Hyden, Protein metabolism in the nerve cell during growth and function, *Acta Physiol. Scand.* **6** (Suppl. 17):1–136 (1943).

84. S. H. Barondes, Relationship of biological regulatory mechanisms to learning and memory, *Nature* **205**:18–21 (1965).

85. E. Roberts, The synapse as a biochemical self-organizing micro-cybernetic unit, *Brain Res.* **2**:117–166 (1966).

86. Z. Dische, C. Pallavicini, H. Kavasaki, N. Smirnow, L. J. Cizek, and S. Chien, Influence of the nature of the secretory stimulus on the composition of the carbohydrate moiety of glycoproteins of the submaxillary saliva, *Arch. Biochem. Biophys.* **97**:459–469 (1962).

87. Z. Dische, *in Protides of the Biological Fluids* (H. Peeters, ed.), Vol. 13, pp. 1–20, Elsevier Publishing Company, Amsterdam (1966).

88. S. H. Barondes, Incorporation of radioactive glucosamine into macromolecules at nerve endings, *J. Neurochem.* **15**:699 (1968).

89. S. Bogoch, W. Sacks, G. Korsh, W. H. Sweet, and P. C. Belval, *in Protides of the Biological Fluids* (H. Peeters, ed.), Vol. 15, Elsevier Publishing Company, Amsterdam (1968).

Chapter 12

MUCOPOLYSACCHARIDES

Richard U. Margolis

Department of Pharmacology
New York University School of Medicine
New York, New York

I. CHEMISTRY OF NERVOUS TISSUE ACID MUCOPOLY-SACCHARIDES

Several excellent surveys of the chemistry and biochemistry of the acid mucopolysaccharides* and related compounds have recently been published.[1-4] However very little attention has been given to their presence and possible role in nervous tissue.

The basic structure of this group of heterologous polymers consists of alternating residues of a uronic acid and a hexosamine, the latter usually acetylated. An ester sulfate group may also be present. The only acid mucopolysaccharides whose presence in the nervous system has been firmly established are hyaluronic acid, chondroitin 4-sulfate, and chondroitin 6-sulfate.[5-8] The repeating unit of the nonsulfated mucopolysaccharide, hyaluronic acid, contains equimolar proportions of glucuronic acid and N-acetylglucosamine, whereas the disaccharide repeating unit of chondroitin 4-sulfate and chondroitin 6-sulfate contains equimolar proportions of glucuronic acid, N-acetylgalactosamine, and sulfate (Fig. 1). However, sulfated mucopolysaccharides with degrees of sulfation greater or less than the usual equimolar one have also been shown to occur, including a chondroitin sulfate from human brain with a sulfate to uronic acid ratio of $0.3:1$.[9]

---

* The acid mucopolysaccharides are distinguished from the "neutral mucopolysaccharides" (such as the blood group substances) and the glycoproteins in that the latter two classes of compounds contain neutral sugars (hexoses and pentoses) and often sialic acid in addition to hexosamines, but no uronic acid. They also lack a serially repeating unit and contain a relatively low number of sugar residues in the heterosaccharide. The mucopolysaccharides are also known as glycosaminoglycans or glycosaminoglycouronans in the nomenclature proposed by Jeanloz.[2] In this chapter the following terms will be used, according to the nomenclature of Jeanloz, in place of the older names of these polysaccharides, which are given in parentheses: chondroitin 4-sulfate (chondroitin sulfate A), chondroitin 6-sulfate (chondroitin sulfate C), dermatan sulfate (chondroitin sulfate B), heparan sulfate (heparitin sulfate), and keratan sulfate (keratosulfate).

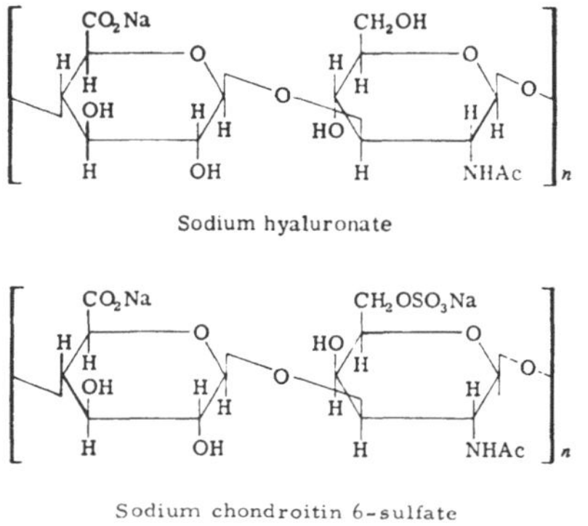

Sodium hyaluronate

Sodium chondroitin 6-sulfate

Fig. 1. Repeating disaccharide units of hyaluronic acid and chondroitin 6-sulfate. Chondroitin 4-sulfate is isomeric to chondroitin 6-sulfate with the ester sulfate on the 4-position of the galactosamine residue.

Mucopolysaccharides are usually isolated in combination with varying amounts of protein, depending upon the method of preparation. The mucopolysaccharide–protein linkages of chondroitin 4-sulfate[10] and of heparin[11] have both been demonstrated to consist of a glycosidic bond between xylose and the hydroxyl group of serine and to contain the sequence, glucuronosyl–galactosyl–galactosyl–xylosylserine, leading to the following proposed structure of the linkage region:

$$\text{-GlcUA-}(1 \rightarrow 3)\text{-}\underset{\underset{\text{SO}_4}{|}}{\text{GalNAc}}\text{-}(1 \rightarrow 4)\text{-GlcUA-}(1 \rightarrow 3)\text{-Gal-}(1 \rightarrow 3)\text{Gal-}(1 \rightarrow 4)\text{-Xyl-O-}\underset{\underset{\text{protein}}{|}}{\overset{\overset{\text{protein}}{|}}{\text{serine}}}$$

The amino acid composition of the chondroitin sulfate isolated from brain after extensive proteolytic digestion closely resembles that previously found in cartilage, with serine being the predominant amino acid in both.[8] Since galactose and xylose (in a 2:1 molar ratio) have also been found in a glycopeotide isolated from brain chondroitin sulfate,[12] it would appear that the chondroitin sulfate–protein linkage in brain is probably the same as that in cartilage.

The question of whether hyaluronic acid occurs covalently linked to protein has been a matter of controversy for some time, but recent evidence tends to support the existence of such a hyaluronic acid–protein complex.[13] Stary and co-workers[12,14] have reported the occurrence of arabinose in hyaluronic acid from human brain, and arabinose has been found in a concentration of somewhat less than 0.2 % in a preparation of highly purified hyaluronic acid from bovine brain.[8] This hyaluronic acid had a molecular weight of 140,000 and would therefore permit a maximum of one arabinose per molecule of hyaluronic acid, possibly as part of a hyaluronic acid–protein linkage. The amino acid composition of bovine brain hyaluronic acid shows that glutamic acid occurs in highest concentration, followed by slightly lower and approximately equimolar amounts of alanine and aspartic acid.[8] Similar results have been obtained by Rodén for hyaluronic acid from umbilical cord,[15] suggesting that one of these amino acids may play a role in a hyaluronic acid–protein linkage analogous to serine in chondroitin sulfate and heparin.

In addition to the mucopolysaccharides discussed above, there have also appeared reports of the presence in brain of dermatan sulfate,[16,17] heparan sulfate,[17,18] and a polysaccharide containing galacturonic acid.[19,20] Dermatan sulfate is a hybrid molecule resembling chondroitin 4-sulfate but containing both L-iduronic and D-glucuronic acid.[21] The structure of heparan sulfate is not yet well-defined, but recent studies have demonstrated that a majority of the heparan sulfate molecules of aorta consist predominantly of a rather large segment composed of repeating units of N-acetylglucosamine and glucuronic acid.[21a] Most of the N-acetylglucosamine residues in this segment (which is located near the protein linkage region) are free of ester sulfate, whereas other portions of the molecule contain N- and O-sulfated glucosamine.

II. DISTRIBUTION

A. Central Nervous System

The concentrations of hyaluronic acid and chondroitin sulfate in the central nervous system of a number of species are given in Tables I and II.

A study of the acid mucopolysaccharide concentration in various areas of sheep brain[6] did not reveal any striking differences, although the concentration tended to be highest in cerebrum and lowest in the brainstem and cerebellum. Analysis of bovine and human brain indicated a higher concentration of total mucopolysaccharides in gray matter than in white matter,[8,20] although this difference was not apparent in sheep brain.[6]

Over 25% of the total mucopolysaccharide in sheep brain choroid plexus was originally identified as heparin, with much lower concentrations in 14 other brain areas analyzed.[6] However, later work by these investigators indicated that this fraction represents dermatan sulfate rather than heparin,[17] and that dermatan sulfate was present in four whole human brains which were examined (fetal through adult), representing 2–5% of the total acid mucopolysaccharides. Choroid plexus was also found to have the highest concentration of chondroitin 4- and 6-sulfates, but contained no hyaluronic acid.[6]

B. Nervous Tissue Other Than CNS

A mucopolysaccharide, chemically identified as hyaluronic acid, has been histochemically demonstrated in the axoplasm and neurolemma sheath of amphibian and mammalian peripheral nerve.[22] Hyaluronic acid, chondroitin 4-sulfate, chondroitin 6-sulfate, heparan sulfate, and dermatan sulfate have also all been reported to be present in the peripheral nerve of monkey.[9]

Retina (from rat and cattle) has also been shown to contain mucopolysaccharides. A chemical study of the mucopolysaccharides from cattle retina revealed the presence of glucosamine, galactosamine, uronic acid,

TABLE I

Concentration of Acid Mucopolysaccharides in the Central Nervous System[a]

Species	Percent in fresh tissue	Percent of lipid-free protein residue	Reference
Human			
Gray matter	0.12–0.17	—	20
White matter	0.1	—	20
Cerebrum	—	0.41	58
Whole brain	0.08–0.1	0.9	14, 20
Whole brain	—	0.36	17
Fetus	—	0.32	17
Neonatal	—	0.63	17
One year	—	0.64	17
Seven months	—	0.3	18
Cow			
Whole brain	0.03	0.28	5
Whole brain	0.027	0.27	8
White matter	0.015	0.27	8
Spinal cord	0.01	0.20	5
Sheep[b]	0.01–0.023	—	6
Rat	0.014	—	6
Cow, pig, rabbit, monkey[c]	0.07–0.22	—	7

[a] Concentration in adult whole brain, if not otherwise indicated. Where necessary, concentrations have been calculated from data in the original papers on the basis of yield of acid mucopolysaccharide and uronic acid content.
[b] Range for various brain areas; whole brain was not examined.
[c] Data not given for individual species.

ester sulfate, and neutral sugars.[23] Although these compounds do not fit easily into the presently known categories of mucopolysaccharides, the analytical findings suggest that the major (galactosamine-containing) component from retina is either chondroitin or an undersulfated chondroitin sulfate, while the minor (glucosamine-containing) fraction may be a keratan sulfate-like substance.

It is noteworthy that in the Hurler syndrome, although mucopolysaccharides infiltrate most tissues, including brain and meninges (see below), and are found in high concentrations in urine, they do not pass into the cerebrospinal fluid. In a study of three patients with Hurler's syndrome, no mucopolysaccharide was found in the cerebrospinal fluid,[24] although less than 3 μg/ml would have been detectable by the methods used. This is surprising in view of the relatively low molecular weight (averaging 15,000 in the case of dermatan sulfate, and 2700 to 5500 for heparan sulfate) of the mucopolysaccharides isolated from urine and tissues in the Hurler syndrome. It would appear that the highly charged nature of these substances serves to exclude them from the CSF, since protein molecules of much greater molecular weight are capable of penetrating the blood–CSF barrier.

Hyaluronic acid has been reported to be present in the extracellular space of cockroach ganglia, forming a layer between the glial cells and separating the neuropil from the remainder of the ganglion.[25]

Hyaluronic acid and chondroitin 6-sulfate have also been isolated from the electric organ of the electric eel, in a yield of about 0.7 and 0.1 %, respectively, of the lipid-free dry weight.[26] Electric organs occur in several genera of elasmobranch and teleost fishes and are composed of a series of elements resembling motor endplates (electroplaques), each of which is embedded in a gelatinous connective tissue. Although the mucopolysaccharides appear to be present in the connective tissue portion of this neural structure, they have been shown to form complexes with curarizing quaternary ammonium compounds in the electric organ of *Electrophorus electricus*, and have therefore

TABLE II

Relative Amounts of Hyaluronic Acid and Chondroitin Sulfate in the Central Nervous System as Percent of Total Acid Mucopolysaccharide[a]

Species	Hyaluronic acid	Chondroitin sulfate		Reference
		4-Sulfate	6-Sulfate	
Human				
Fetal	35	41		17
Neonatal	54	19	12	17
One year	33	36	12	17
Adult	33	37	14	17
Seven months	60	(40 % heparan sulfate)		18
Cow				
Whole brain	44	56	—	5
Whole brain	50	44	6	8
Whole brain	19	61	20	7
White matter	52	29	19	8
Spinal cord	75	25	—	5
Sheep[b]	39–62	30–58		6
Rat	29	53		6
Monkey	78	—	23	7
Pig	—	100	—	7
Rabbit	—	—	100	7

[a] Whole brain, unless otherwise indicated. Where necessary, concentrations have been calculated from data in the original papers.
[b] Range for various brain areas; whole brain was not examined.

been proposed as a possible physiological "receptor" for acetylcholine and similar excitatory or inhibitory quaternary ammonium compounds. This topic is covered in more detail in the recent review by Szirmai[27] on the mucopolysaccharides of the electric organ and in the report by Trams and Lauter[28] on the isolation and characterization of a glycoprotein with ion-exchange capabilities from the electric organ of *Electrophorus electricus*.

III. ANATOMICAL LOCATION

A. Connective Tissue

Although the anatomical location of the nervous system acid mucopoly-saccharides has not yet been determined, an obvious location would be in connective tissue, in which they are found in other organs and tissues. Since most studies on nervous system acid mucopolysaccharides have been done after removal of meninges and superficial blood vessels, only the smaller blood vessels of the parenchyma and the "intravascular connective tissue strands" would remain as possible sources of connective tissue in the usual sense of the term. The intervascular connective tissue fibers have recently been extensively reinvestigated by Cammermeyer,[29,30] who found that they were present throughout the normal central nervous system, and especially in the gray matter. They vary in size and number according to region and species, being most numerous in the spinal cord and in the pigeon, and absent in the opossum.[29] Although their functional significance in normal brains is unclear, it appears that in addition to being a stabilizer for the vasculature these fine strands may be components of microcirculatory physiology, forming a pathway via which substances can be transported from one vascular territory to another without entering the nervous tissue itself.[30]

However, among the indications that brain is quite low in connective tissue content is the virtual absence of collagen in this organ.[8,31] The relatively greater collagen concentration reported for spinal cord[32] probably does not indicate any difference in the connective tissue concentration of brain and spinal cord parenchyma, but rather represents the collagenous epipial tissue known to cover the spinal cord but absent from brain.[33]

Elsewhere in the body where mucopolysaccharides occur as connective tissue components they are accompanied by considerably greater amounts of collagen. Therefore, if collagen concentration can be considered a valid index of the amount of connective tissue present in brain, it would appear unlikely that this can account for the mucopolysaccharide content of nervous tissue.

B. Myelin

A number of investigators have suggested the presence of acid muco-polysaccharides as components of the myelin sheath,[5,6,34–37] but in only one case[34] were these speculations based on actual experimental (histochemical) data concerning myelin. In a chemical study of purified brain myelin, no acid mucopolysaccharides were detected, although several myelin protein fractions contained approximately 1 % glucosamine, indicating the probable presence of glycoproteins.[8]

C. Localization on the Cellular Level

All reports bearing on the cellular localization of acid mucopolysaccharides in brain have unfortunately been based on histochemical techniques of questionable specificity.

Young and Abood,[38] using a modified Hale's stain, have reported the histochemical demonstration of mucopolysaccharides in neurons and their processes, with relatively little staining of the neuroglia. Most of the iron-reacting material was removed by hyaluronidase digestion.

Sulkin has reported the presence of cytoplasmic granules, which are revealed by the periodic acid–Schiff (PAS) technique or by metachromasia after sulfation and staining with toluidine blue, in the central and peripheral nervous system of a number of vertebrate species.[39] Although he claims that these granules represent mucopolysaccharides, recent work indicates that the PAS technique does not stain acid mucopolysaccharides (see below), and there is no evidence either that these substances are specifically stained by the sulfation–toluidine blue technique. This may represent another instance in which glycoproteins and other substances having a carbohydrate moiety are incorrectly referred to under the generic term of "mucopolysaccharides."

Hess originally reported finding a mucopolysaccharide ground substance in the extracellular space of brain, and related this to the presence of the blood–brain barrier.[40,41] This material, which was also demonstrated by periodic acid–Schiff staining, was not removed by hyaluronidase treatment and was considered by Hess to be a "neutral mucopolysaccharide." The reports of Hess could not be confirmed,[42,43] and it is unlikely that an extracellular ground substance, if present, would be capable of detection by light microscopy in the small 150–200 Å intercellular space of brain. Also arguing against the evidence of Hess is the fact that all of the mucopolysaccharides whose presence in brain has been well established are digestible by hyaluronidase,[5–8] although there is evidence for the presence of a small amount of hyaluronidase-resistant mucopolysaccharide in brain.[17] It has also been shown that the periodate–Schiff reaction gives no color production with pure acid mucopolysaccharides, except possibly for a very slight color yield with hyaluronic acid.[44] The PAS reaction is, however, known to stain glycoproteins,[19,45] which may be the compounds detected by Hess.

Recent electron-microscopic studies by Pease[46] have demonstrated a uniform layer of phosphotungstic acid-staining material separating the pre- and postsynaptic elements of cerebral cortical synapses. Cords and incomplete layers of this material, which he identifies as mucopolysaccharide, are also seen in the larger extracellular spaces of the neuropil. A similar "polysaccharide" coating has been found on the surface of epithelial cells in kidney and intestine as well as in brain.

The data of Bondareff,[47] obtained by use of different histochemical techniques, led him to essentially the same conclusions as those of Pease cited above. Bondareff examined by electron microscopy sections of rat cerebral cortex fixed with ruthenium red. The ruthenium red reaction was found to be uniformly and reproducibly distributed in the extracellular space, and Bondareff interprets his data as evidence for the presence of mucopolysaccharides in the extracellular space of brain. He has also demonstrated a similarly distributed material in cerebral cortex fixed by freezing and drying and subsequently stained with uranyl acetate.[48] The visualized density of the intercellular material was somewhat greater at pH 4.4 than at pH 2.0, supporting the concept that an anionic substance is present in the extracellular space of brain.

There is at present no evidence that the techniques of either Pease or Bondareff are specific for acid mucopolysaccharides, since they would also be expected to stain glycoproteins. These often also carry a net negative charge, especially when sialic acid residues are present, and glycoproteins have been reported to be present in brain in locations similar to those proposed for acid mucopolysaccharides.[49] However, the recent data of Pease and Bondareff, leading independently to similar conclusions using different techniques, have reopened the question of whether acid mucopolysaccharides may constitute the extracellular "ground substance" in the central nervous system.

IV. METABOLISM

Very little information is available on the metabolism of the acid mucopolysaccharides of brain.

Boström and Odeblad[50] noted that two days after intraperitoneal administration of ^{35}S-sulfate to adult rats there was a much greater incorporation of labeled sulfate into cerebral, cerebellar, and spinal cord gray matter than into white matter, as determined autoradiographically. Choroid plexus also showed a very high uptake of ^{35}S. Ringertz[51] found that the greatest incorporation of labeled sulfate into the gray matter of adult mouse brain (after ethanol–xylene extraction) occurred at 12–48 hr after injection.

Robinson and Green have studied the incorporation of ^{35}S-sulfate into various brain fractions in mature (150 g) rats.[52] The greatest incorporation into uncharacterized "sulfomucopolysaccharides" was observed two days after injection and declined progressively after that time. This crude ^{35}S-labeled extract was obtained after trypsin digestion of brain, followed by dialysis and precipitation with acetone. It was metachromatic with Azure A, nondialyzable, contained 19% ester sulfate, yielded glucosamine and galactosamine on hydrolysis, and was separable into two fractions by paper chromatography. Two days after injection of isotope, this extract incorporated twice as much ^{35}S as did sulfatide, with the reverse situation obtaining at 16 days after injection. However, at all times the incorporation into brain "sulfomucopolysaccharide" was considerably less than that observed in any of ten other organs and tissues examined. Subcellular fractionation experiments revealed that in brain most of the incorporation into "sulfo-mucopolysaccharides" occurred in the microsomal and supernatant fractions, whereas in kidney and liver one half of the total incorporation was found in the nuclei and cell debris.

Balasubramanian and Bachhawat have demonstrated the enzymic transfer of sulfate from 3'-phosphoadenosine 5'-phosphosulfate to muco-polysaccharides in rat brain.[53] The enzyme activity was highest with heparan sulfate or dermatan sulfate as acceptors, and less with chondroitin 4-sulfate or keratan sulfate. Chondroitin 6-sulfate and hyaluronic acid were inactive as acceptors. The enzyme activity was highest shortly after birth and decreased with increasing age of the animals.

Although arylsulfatase activity is known to be present in brain, a purified arylsulfatase from human brain (resembling arylsulfatase A from human liver) was found to be inactive in releasing ^{35}S-sulfate from labeled chondroitin sulfate.[54]

Distler and Jourdian have reported the isolation and characterization of a glucosyltransferase from chick embryo brain which catalyzes the transfer of glucose from UDP-^{14}C-glucose to D-xylosylserine, L-arabinosylserine, β-methylxyloside, galactose, or L-arabinose.[55] The enzyme was inactive with UDP-galactose, -acetylglucosamine, -acetylgalactosamine, or -glucuronic acid as donors, and towards α-methylxyloside, glucose, glucosamine, galacto-samine, serine, sphingosine, or glycogen as acceptors. Most of the activity is sedimented in a particulate fraction. It is present in a number of organs and tissues other than brain, being highest in epiphyseal cartilage and in skin. In view of recent reports on the presence of glucose and arabinose in hyaluronic acid from brain and of xylosylserine in chondroitin sulfate (see above), it would appear that this enzyme might be of significance in the formation of mucopolysaccharide–protein linkages in brain and other tissues.

V. PATHOLOGICAL ALTERATIONS

A. The Hurler Syndrome

This is an inherited disorder characterized clinically by short stature, mental retardation, unusual physiognomy, bony abnormalities, and often by corneal opacities. An intracellular storage and excessive urinary excretion of dermatan sulfate and heparan sulfate have been demonstrated in this disease, which has recently been comprehensively reviewed by Dorfman.[56]

Dermatan sulfate and heparan sulfate are usually found in very small amounts, if at all, in brain, but in patients with Hurler's syndrome they occur in relatively high concentration (up to 2% of the lipid-free dry weight) and often account for all of the acid mucopolysaccharide isolated from brain.[18, 57,58] Like liver, spleen, kidney, skin, and other organs and tissues in which these substances are known to accumulate, the relative proportions of the two vary widely in brain, ranging from 40 to 85% dermatan sulfate and 10 to 60% heparan sulfate.[18,57] As mentioned earlier, mucopolysaccharides are not found in the cerebrospinal fluid in this disease.[24]

Heparan sulfate isolated from the livers and urine of patients with Hurler's syndrome has been found to differ markedly from that obtained from normal human aorta. Two chemically and structurally distinct fractions are found, one of which is distinguished from the normal material by its low molecular weight (approximately one-fifth that of the aorta polysaccharide) and its markedly reduced amino acid content.[21a] The serine to xylose to galactose molar ratio of approximately 1:1:2 for this material indicates that protein binding is normal, albeit deficient, in the heparan sulfate from Hurler tissue. The other Hurler heparan sulfate fraction bears a close chemical relationship to heparin. It has been suggested that Hurler heparan sulfate may arise from partial hydrolysis of a parent molecule similar in structure to heparan sulfate from normal aorta. Separate studies of the dermatan sulfate of Hurler's tissues also indicate that the stored material is partially degraded.

Comparative studies by Matalon and Dorfman[59] on fibroblasts cultured from the skin of Hurler and normal patients have shown that the rates of synthesis and storage of dermatan sulfate and hyaluronic acid are increased within the Hurler cell. On the basis of these findings it has been postulated[21a] that both heparan sulfate and dermatan sulfate are partially degraded by intracellular (lysosomal) enzymes. Storage may result from the limited capacity to degrade these two polysaccharides.

B. Metachromatic Leukodystrophy

Metachromatic leukodystrophy (MLD) is an inherited disorder characterized clinically by progressive paralysis and dementia, which commonly becomes apparent during the second year of life and terminates fatally within 2 to 10 years. The disease is characterized pathologically by demyelination and an accumulation of metachromatic lipids (sulfatides) in the white matter of the central nervous system and in peripheral nerves, as well as in the kidney, liver, and certain other visceral organs. The sulfatide accumulation is thought to be the basis of all the manifestations of the disease.

Increased concentrations of acid mucopolysaccharides have been reported in the brain,* kidney, and urine of patients with metachromatic leukodystrophy.[58] The increased acid mucopolysaccharide levels in MLD appear to be accounted for almost entirely by an increase in heparan sulfate, with no excretion or tissue accumulation of dermatan sulfate as was found in cases of Hurler's syndrome by these authors and others cited above. The known deficiency of one or more sulfatases in MLD suggests that the

---

* Increased brain concentrations were found only in a variant form of MLD, in which cortical gray matter is involved and in which more than one sulfatase deficiency occurs.

observed increase in sulfated mucopolysaccharide levels may be related to an enzymatic defect of this type.

C. Alzheimer's and Jakob–Creutzfeldt Diseases

These two types of subacute, progressive presenile dementia are associated with neuronal loss and gliosis, and Jakob–Creutzfeldt disease is often also accompanied by various neurological abnormalities. They have been studied chemically by Suzuki and co-workers,[60,61] who found an average increase of 60 % in the mucopolysaccharide concentration of gray matter in three cases of Alzheimer's disease compared to five normal control specimens. There was no significant difference in acid mucopolysaccharide concentration in three cases of Jakob–Creutzfeldt disease. Only hyaluronic acid and chondroitin sulfate were found in all specimens examined, with no differences in the relative proportions of the two acid mucopolysaccharides. The authors suggested that the increased acid mucopolysaccharide found in Alzheimer's disease might represent the amyloid present in the core of the senile plaques which occur in this disease.

VI. EFFECTS OF HYALURONIDASE ON NERVOUS TISSUE

One indication that mucopolysaccharides may be important for the normal functioning of the central nervous system is that alterations in CNS activity have been reported following hyaluronidase* administration. Young[62] has demonstrated that the intraventricular administration to cats of 13,000 units of hyaluronidase daily for a period of 5 to 7 days results in marked neurological changes and seizure activity which are not seen after injection of an equivalent amount of cat serum protein. EEG abnormalities appeared 1 to 3 days after the beginning of hyaluronidase administration, and neurological changes, which were seen by the third or fourth day, were most pronounced after 5 to 6 days of enzyme administration. These effects were completely reversible and disappeared in two weeks after injections were stopped. Concomitant with the neurological and EEG abnormalities was a decrease in neuronal staining by colloidal iron using a modification of the Hale technique. No other histological changes were observed in the brains of hyaluronidase-treated animals, although multiple small hemorrhages were observed in spinal cord after microinjection of hyaluronidase into the gray matter.

* Throughout this chapter, hyaluronidase refers to the endohexosaminidase from bovine testis (EC 3.2.1.35) which catalyzes the cleavage of internal glycosidic bonds of hyaluronic acid, chondroitin 4-sulfate, and chondroitin 6-sulfate. Although dermatan sulfate has also been shown to be partially degraded by testicular hyaluronidase,[21] heparin, heparan sulfate, and keratan sulfate are not substrates for this enzyme. The animal mucopolysaccharidases have recently been reviewed by Mathews (*Methods in Enzymology*, Vol. 8, E. F. Neufeld and V. Ginsburg (eds.), pp. 654–662, Academic Press, New York, 1966), and mucopolysaccharidases of bacterial origin are discussed by Linker in the same volume (pp. 650–654).

Marcovici *et al.*[63] have also reported EEG seizure activity in dogs after application of hyaluronidase to the exposed cerebral cortex.

There are older studies which show an increased penetration of dyes[64,65] and of India ink particles[66] into brain after treatment with hyaluronidase. It has also been reported that intracisternal administration of hyaluronidase leads to an increase in CSF protein concentration up to ten times the normal value, as well as a shedding of epithelial cells from the choroid plexus and ependyma into the CSF.[67] Neurological effects were also observed, but no evidence of increased permeability of the blood–brain barrier as judged by penetration of intraperitoneally administered trypan blue into the central nervous system.

It is noteworthy that hyaluronidase activity could not be detected in brain, although present in 13 other organs tested except for pancreas.[68] Preliminary studies indicate that the presence of a hyaluronidase inhibitor in brain may account for the failure to detect enzyme activity in that organ.[68]

VII. CONCLUSIONS

It is evident from the preceding discussion that, although there is considerable information concerning the chemistry of nervous system mucopolysaccharides, knowledge of their metabolism and anatomical localization is still scanty. However, the chemical data currently available should provide a good foundation for further histochemical and metabolic studies.

There is some evidence to indicate that mucopolysaccharides may be present in the extracellular space in brain. This possibility, if proved correct, could be of considerable significance for our understanding of the blood–brain barrier phenomenon, since the intercellular localization of such highly charged polymers (especially in the case of the sulfated acid mucopolysaccharides) would be expected to influence the distribution of other ionized compounds and of high-molecular-weight substances in the central nervous system. Confirmation of such an extracellular localization of the acid mucopolysaccharides must await the availability of more specific histochemical techniques applicable on the electron-microscopic level.

VIII. REFERENCES

1. J. S. Brimacombe and J. M. Webber, *Mucopolysaccharides*, Elsevier, Amsterdam (1964).
2. R. W. Jeanloz and E. A. Balazs (eds.), *The Amino Sugars*, Academic Press, New York (1965–).
3. G. L. Dutton (ed.), *Glucuronic Acid, Free and Combined*, Academic Press, New York (1966).
4. S. Schiller, Connective and supporting tissues: Mucopolysaccharides of connective tissues, *Ann. Rev. Physiol.* **28**:137–158 (1966).
5. M. M. Szabo and E. Roboz-Einstein, Acidic polysaccharides in the central nervous system, *Arch. Biochem. Biophys.* **98**:406–412 (1962).
6. M. Singh and B. K. Bachhawat, The distribution and variation with age of different uronic acid-containing mucopolysaccharides in brain, *J. Neurochem.* **12**:519–525 (1965).

7. K. Onodera, S. Hirona, F. Horiuchi, and N. Kashimura, A comparative study of some animal brains with regard to content of acidic mucopolysaccharide, *Carbohydrate Res.* 3:234–238 (1966).

8. R. U. Margolis, Acid mucopolysaccharides and proteins of bovine whole brain, white matter and myelin, *Biochim. Biophys. Acta* 141:91–102 (1967).

9. B. K. Bachhawat and M. Singh, Isolation and characterization of mucopolysaccharides from nervous tissue, *Indian J. Biochem.* 4:27–28 (1967). [Suppl.: Abstracts of papers presented at the International Convention of Biochemists, Bangalore, 1967.]

10. L. Rodén and R. Smith, Structure of the neutral trisaccharide of the chondroitin 4-sulfate-protein linkage region, *J. Biol. Chem.* 241:5949–5954 (1966).

11. U. Lindahl, Further characterization of the heparin–protein linkage region, *Biochim. Biophys. Acta* 130:368–382 (1966).

12. A. H. Wardi, W. S. Allen, D. L. Turner, and Z. Stary, Isolation of arabinose-containing hyaluronate peptides and xylose-containing chondroitin sulfate peptides from protease-digested brain tissue, *Arch. Biochem. Biophys.* 117:44–53 (1966).

13. D. Hamerman, M. Rojkind, and J. Sandson, Protein bound to hyaluronate: chemical and immunological studies, *Federation Proc.* 25:1040–1045 (1966).

14. Z. Stary, A. H. Wardi, D. L. Turner, and W. S. Allen, Arabinose as a mucopolysaccharide component in human and animal brain tissue, *Arch. Biochem. Biophys.* 110:388–394 (1965).

15. L. Rodén, Structure of mucopolysaccharide–protein complexes, *Abstracts, Seventh International Congress of Biochemistry*, Tokyo, August 1967, Vol. 1, p. 69.

16. J. Clausen and A. Hansen, Acid mucopolysaccharides of human brain: Identification by means of infra-red analysis, *J. Neurochem.* 10:165–168 (1963).

17. M. Singh and B. K. Bachhawat, Isolation and characterization of glycosaminoglycans of human brain of different age groups, *J. Neurochem.* 15:249–258 (1968).

18. K. Meyer, P. Hoffman, A. Linker, M. M. Grumbach, and P. Sampson, Sulfated mucopolysaccharides of urine and organs in gargoylism (Hurler's syndrome). II. Additional studies. *Proc. Soc. Exptl. Biol. Med.* 102:587–590 (1959).

19. R. E. Glegg and R. H. Pearce, Chemical extraction of metachromatic and periodic acid–Schiff positive carbohydrates from cerebral tissue, *J. Comp. Neurol.* 106:291–297 (1956).

20. Z. Stary, A. Wardi, and D. Turner, Galacturonic acid in hydrolyzates of defatted human brain, *Biochim. Biophys. Acta* 83:242–244 (1964).

21. L.-Å. Fransson and L. Rodén, Structure of dermatan sulfate, I and II, *J. Biol. Chem.* 242:4161–4169, 4170–4175 (1967).

21a. J. Knecht, J. A. Cifonelli, and A. Dorfman, Structural studies on heparitin sulfate of normal and Hurler tissues, *J. Biol. Chem.* 242:4652–4661 (1967).

22. L. G. Abood and S. K. Abul-Haj, Histochemistry and characterization of hyaluronic acid in axons of peripheral nerve, *J. Neurochem.* 1:119–125 (1956).

23. E. R. Berman, Isolation of neutral sugar-containing mucopolysaccharides from cattle retina, *Biochim. Biophys. Acta* 101:358–360 (1965).

24. C. Friman, Hurler's syndrome. Lack of acid mucopolysaccharides in the cerebrospinal fluid, *Clin. Chim. Acta* 15:378–380 (1967).

25. D. E. Ashurst and N. G. Patel, Hyaluronic acid in cockroach ganglia, *Ann. Entomol. Soc. Am.* 56:182–184 (1963).

26. K. Meyer, E. Davidson, A. Linker, and P. Hoffman, The acid mucopolysaccharides of connective tissue, *Biochim. Biophys. Acta* 21:506–518 (1956).

27. J. A. Szirmai, in *The Amino Sugars* (R. W. Jeanloz and E. A. Balazs, eds.), Vol. 2A, pp. 251–256, Academic Press, New York (1965).

28. E. G. Trams and C. J. Lauter, Properties of electroplax protein, Part III, *Biochim. Biophys. Acta* 83:296–304 (1964).

29. J. Cammermeyer, A comparative study of intervascular connective tissue strands in the central nervous system, *J. Comp. Neurol.* **114**:189–200 (1960).

30. J. Cammermeyer, Cerebral intervascular strands of connective tissue as routes of transportation, *Anat. Rec.* **151**:251–260 (1965).

31. R. E. Neuman and M. A. Logan, The determination of collagen and elastin in tissues, *J. Biol. Chem.* **186**:549–556 (1950).

32. E. Roboz, N. Henderson, and M. W. Kies, A collagen-like compound isolated from bovine spinal cord—I, *J. Neurochem.* **2**:254–259 (1958).

33. J. W. Millen and D. H. M. Wollam, On the nature of the pia mater, *Brain* **84**:514–520 (1961).

34. M. Wolman, Histochemical studies of myelinization in the rat, *Bull. Res. Council Israel, Sect. E* **6**:163–167 (1957).

35. J. D. Robertson, Structural alterations in nerve fibers produced by hypotonic and hypertonic solutions, *J. Biophys. Biochem. Cytol.* **4**:349–364 (1958).

36. J. D. Robertson, in *Cellular Membranes in Development* (M. Locke, ed.), pp. 1–81, Academic Press, New York (1964).

37. G. Brante, in *Biochemistry of the Central Nervous System* (F. Brücke, ed.), Vol. 3, *Proc. 4th Intern. Congr. Biochem., Vienna*, 1958, pp. 291–300, Pergamon Press, New York (1959).

38. I. J. Young and L. G. Abood, Histological demonstration of hyaluronic acid in the central nervous system, *J. Neurochem.* **6**:89–94 (1960).

39. N. M. Sulkin, The distribution of mucopolysaccharides in the cytoplasm of vertebrate nerve cells, *J. Neurochem.* **5**:231–235 (1960).

40. A. Hess, The ground substance of the central nervous system revealed by histochemical staining, *J. Comp. Neurol.* **98**:69–92 (1953).

41. A. Hess, The ground substance of the central nervous system and its relation to the blood–brain barrier, *World Neurol.* **3**:118–124 (1962).

42. L. Ozzello, M. Lending, and F. D. Speer, The ground substance of the central nervous system in man, *Am. J. Pathol.* **34**:363–372 (1958).

43. R. Singh, On histochemical localisation of the ground substance in the central nervous tissue of *Psittacula krameri*, *J. Histochem. Cytochem.* **12**:712 (1964).

44. A. Dahlqvist, I. Olsson, and Å. Norden, The periodate–Schiff reaction: specificity, kinetics, and reaction products with pure substrates, *J. Histochem. Cytochem.* **13**:423–430 (1965).

45. A. G. E. Pearse, *Histochemistry*, 2nd ed., Little, Brown and Co., Boston (1960).

46. D. C. Pease, Polysaccharides associated with the exterior surface of the epithelial cells: kidney, intestine, brain, *J. Ultrastruct. Res.* **15**:555–588 (1966).

47. W. Bondareff, An intercellular substance in rat cerebral cortex: submicroscopic distribution of ruthenium red, *Anat. Rec.* **157**:527–535 (1967).

48. W. Bondareff, Demonstration of an intercellular substance in mouse cerebral cortex, *Z. Zellforsch.* **81**:366–373 (1967).

49. A. Rambourg and C. P. Leblond, Electron microscopic observations on the carbohydrate-rich coat present at the surface of cells in the rat, *J. Cell Biol.* **32**:27–53 (1967).

50. H. Boström and E. Odeblad, Autoradiographic observations on the uptake of S^{35}-labelled sodium sulphate in the nervous system of the adult rat, *Acta Psychiat. Neurol. Scand.* **28**:5–8 (1953).

51. N. R. Ringertz, On the sulphate metabolism of the mouse brain, *Exptl. Cell Res.* **10**:230–233 (1956).

52. J. D. Robinson, Jr., and J. P. Green, Sulfomucopolysaccharides in brain, *Yale J. Biol. Med.* **35**:248–256 (1962).

53. A. S. Balasubramanian and B. K. Bachhawat, Enzymic transfer of sulphate from 3′-phosphoadenosine 5′-phosphosulphate to mucopolysaccharides in rat brain, *J. Neurochem.* **11**:877–885 (1964).

54. A. S. Balasubramanian and B. K. Bachhawat, Purification and properties of an arylsulphatase from human brain, *J. Neurochem.* **10**:201–211 (1963).
55. J. Distler and G. W. Jourdian, Isolation and characterization of a novel glucosyltransferase from chick embryo brain, *Federation Proc.* **26**:345 (1967).
56. A. Dorfman, in *The Metabolic Basis of Inherited Disease* (J. B. Stanbury, J. B. Wyngaarden, and D. S. Fredrickson, eds.), 2nd. ed., pp. 963–994, McGraw-Hill Co., New York (1966).
57. K. Meyer and P. Hoffman, Hurler's syndrome, *Arthritis Rheumat.* **4**:552–560 (1961).
58. M. Bischel, J. Austin, and M. Kemeny, Metachromatic leukodystrophy (MLD). VII. Elevated sulfated acid mucopolysaccharide levels in urine and postmortem tissues, *Arch. Neurol.* **15**:13–28 (1966).
59. R. Matalon and A. Dorfman, Hurler's syndrome: Biosynthesis of acid mucopolysaccharides in tissue culture, *Proc. Natl. Acad. Sci.* **56**:1310–1316 (1966).
60. K. Suzuki, R. Katzman, and S. R. Korey, Chemical studies on Alzheimer's disease, *J. Neuropath. Exptl. Neurol.* **24**:211–224 (1965).
61. K. Suzuki and G. Chen, Chemical studies on Jakob–Creutzfeldt disease, *J. Neuropath. Exptl. Neurol.* **25**:396–408 (1966).
62. I. J. Young, Reversible seizures produced by neuronal hyaluronic acid depletion, *Exptl. Neurol.* **8**:195–202 (1963).
63. N. Marcovici, I. Stoica, A. Petrescu, and G. Marcovici, Effect of hyaluronidase on normal and experimentally damaged brain, *Rev. Roumaine Neurol.* **1**:37–47 (1964).
64. B. Freedman, Hyaluronidase effects on thionin-stained sections of brain, *Anat. Rec.* **115**:265–270 (1953).
65. J. L. Arteta, Effect of hyaluronidase on the cat's brain, *Proc. Soc. Exptl. Biol. Med.* **91**:440–442 (1956).
66. A. Bairati, Spreading factor and mucopolysaccharides in the central nervous system of vertebrates, *Experientia* **9**:461–462 (1953).
67. G. Owens and S. L. Clark, Some effects of injections of hyaluronidase into the subarachnoid space of experimental animals, *J. Neurosurg.* **8**:311–317 (1956).
68. D. Platt and R. Hartmann, Hyaluronidase, β-Glucuronidase- und β-Acetylglucosaminidase-Aktivität in epithelialen und mesenchymalen menschlichen Geweben, *Klin. Wochschr.* **45**:998–1004 (1967).

The following papers of interest have been published since the preparation of this chapter:

C. W. M. Adams and O. B. Bayliss, Histochemistry of myelin. VII. Analysis of lipid–protein relationships and absence of acid mucopolysaccharide, *J. Histochem. Cytochem.* **16**:119–127 (1968).

E. R. Berman and G. Bach, The acid mucopolysaccharides of cattle retina, *Biochem. J.* **108**:75–88 (1968).

F. E. Bloom and G. K. Aghajanian, Fine structural and cytochemical analysis of the staining of synaptic junctions with phosphotungstic acid, *J. Ultrastruct. Res.* **22**:361–375 (1968).

H. V. de Castejon, Demonstracion histoquimica de glucosaminoglucanos acidos (mucopolisacaridos) intraneuronales en el sistema nervioso central, *Acta Cien. Venezolana* **19**:13 (1968).

W. L. Cunningham and J. M. Goldberg, The determination of glycosaminoglycans present in various mammalian brains, *Biochem. J.* **110**:35P (1968).

J. T. Custod and I. J. Young, Cat brain mucopolysaccharides and their *in vivo* hyaluronidase digestion, *J. Neurochem.* **15**:809–813 (1968).

S. Ehrenpreis, Molecular aspects of cholinergic mechanisms, *in Drugs Affecting the Peripheral Nervous System* (A. Burger, ed.), pp. 1–78, Marcel Dekker, Inc., New York (1967).

J. C. Fratantoni, C. W. Hall, and E. F. Neufeld, The defect in Hurler's and Hunter's syndromes: Faulty degradation of mucopolysaccharides, *Proc. Natl. Acad. Sci.* **60**:699–706 (1968).

O. K. Langley and D. N. Landon, A light and electron histochemical approach to the node of Ranvier and myelin of peripheral nerve fibers, *J. Histochem. Cytochem.* **15**:722–731 (1967).

G. Quintarelli (ed.), *The Chemical Physiology of Mucopolysaccharides*, Little, Brown and Company, Boston (1968).

R. Singh, Some observations on the histochemistry of the neuropile tissue of the brain of the parrot (*Psittacula krameri*), *Acta Anat.* **68**:567–576 (1967).

G. M. Villegas and R. Villegas, Ultrastructural studies of the squid nerve fibers, *J. Gen. Physiol.* **51**:44S–60S (1968).

Chapter 13

IRON

Ole J. Rafaelsen and Bent Kofod

Psychochemical Laboratory
University Clinic of Psychiatry
Rigshospitalet
Copenhagen, Denmark

I. INTRODUCTION

From a chemical viewpoint, iron in the brain can be divided into two groups : heme compounds and nonheme compounds. Hemoglobin is not present in the brain outside the corpuscles in the vascular bed under normal conditions, but it appears under pathological conditions as a result of hemorrhage into the brain tissue or as a result of hemolysis. This hemoglobin is liberated, taken up by glia cells, and converted within days to hemosiderin. As in other tissues, the heme groups are then split off, and further decomposition of these prosthetic groups and of the globin residue takes place. Among the brain proteins containing iron, the cytochromes are of primary importance.

The nonheme compounds belong to a more heterogeneous group that is quantitatively larger; they also include subsequent products split from hemoglobin. The major part of iron in this group consists of ferritin, which contains ferrihydroxyl, and phosphate ions in micellar form. Ferritin represents a major part of the iron depot of the brain, if it is reasonable to use such a concept. Finally, it is probable that the brain contains small amounts of inorganic iron in a complex attachment to lipids. It is difficult to decide which of the above-mentioned iron-containing substances are observed histochemically, it is probably mostly ferritin and some of the other nonheme iron-containing compounds; perhaps a minor part of heme-bound iron can also be seen.

We know rather little about iron in the brain in health and disease. This is due to technical shortcomings of iron analyses used by many workers, which often lead to results that now must be considered erroneously high. More value can be attached to findings of relative distribution of iron in different parts of the brain, in different species, and in different developmental phases.

II. METHODS FOR IRON DETERMINATION IN CNS

Determination of the total iron content of brain will be described first, followed by a description of the determinations of the various iron-containing fractions or groups of substances.

Determination of total amount of iron in brain tissue necessitates a combustion of the sample, either by wet incineration with concentrated acids [mixture of nitric acid, sulfuric acid, and perchloric acid, possibly with the addition of a catalyst, such as vanadium[1]] or by dry incineration in a muffle furnace at 500–600°C, or *in vacuo* at 100–200°C, in an atmosphere of oxygen activated by high-frequency electromagnetic energy. Both types of dry incineration depend on prior removal of all water from the sample to avoid boiling when heating to the temperatures applied.

In these, and in all other methods for iron determination, it is of paramount importance to avoid contamination with iron from reagents, utensils, and blood. Reagents can, if necessary, be purified by distillation, extraction [with compounds forming extractable complexes with iron[1]], and recrystallization, etc. Utensils and instruments are most easily cleaned by a mixture of acid and detergent. By dry incineration in a muffle furnace, it is doubtful whether contamination can be avoided, as small amounts of iron will evaporate from heating coils and condense on the sample. Iron salts in the sample may also be volatile at the temperatures applied. Careful consideration of these problems is important before this method is chosen for iron determinations in brain tissue. As the iron content in the brain due to the blood is about 10 μg/g wet wt.,[2] it is very important to remove all blood from the vascular system before assay for iron.

Heme-bound iron can be extracted by acetone: hydrochloric acid and pyridine. The total concentration of heme can be determined by differential spectrometry of the extract.[3]

For nonheme iron, various methods have been suggested; in each one the iron is extracted either by warm acid[4] or by compounds forming complexes with iron [pyrophosphate,[5] 2,2'-dipyridyl,[6] etc.]. It is questionable, however, whether such methods are applicable. Warm acid decomposes hemoglobin, and most iron complex forming agents remove a small, but in the case of the brain still important, part of the heme-bound iron. Other methods depend on the fractionation of iron-containing compounds by extraction with various solvents,[7] but these methods may also be unsuitable, as good fractionation of the different groups of iron-containing compounds is difficult to obtain, the great amounts of brain lipids acting as detergents.

Many principles can be used for the final analysis of extracted iron: colorimetry, spectrography, neutron activation analysis [preferably combined with chromatography to separate the radioactive isotopes formed[8]], gas chromatography of volative iron complexes,[9] atomic absorption, and finally isotope dilution, which due to high sensitivity and simplicity of procedure seems to be the most suitable method.

Iron in CSF can in principle be determined in the same way as in serum (e.g., by simple colorimetry after splitting of protein-bound iron by acid).

As protein content in CSF under normal circumstances is very low, it is not necessary to precipitate proteins in methods with phenantrolines, in contrast to serum analysis, where protein precipitation is obligatory. Other methods are based on evaporation of the sample, followed by dry or wet incineration[8,10,11] and then colorimetric determination of the liberated iron. By incineration of samples, an increase in sensitivity can be obtained because of concentration of the sample.

Iron in CSF has also been determined by neutron activation analysis[8]; after activation, iron was separated by ion–resin exchange and measured by gamma counting.

In iron analysis of CSF, it is of the utmost importance to avoid contamination both at sampling of CSF and during analytical procedures; it will be shown later in the chapter that negligence of these factors has led to reports of too-high values for iron in CSF.[10,11]

Histochemical determination of brain iron starts by fixation, either by freezing or by chemical procedure (formaldehyde or ethanol). Formaldehyde fixation has the disadvantage that it leads to a leveling of differences in iron concentration in different parts of the sample; this phenomenon is also seen to some extent after treatment of the tissue by ethanol, but the changes here are smaller and more constant.[12] The slicing of the tissue by microtome also presents dangers for contamination from the steel knives.[13] Visualization macro- and microscopically is obtained by treating the tissue with ammonium sulfide, ferrocyanide plus acid,[12] or with other compounds forming colored iron complexes.[14]

III. EXPERIMENTAL ALTERATION OF IRON IN THE BRAIN

A condition resembling hemochromatosis was produced in rats by MacDonald and Pechet,[15] who administered a diet deficient in choline, but containing excess iron. After this diet, iron was demonstrable in most tissues, as seen in idiopathic hemochromatosis, but no increase was found by histochemical technique in brain or in bone marrow.

Tremorine and lysergic acid diethylamide (LSD) caused changes in rat brain iron in the experiments of Hadžović and co-workers.[16] Tremorine reduced brain iron 15 and 30 min after injection, but at 120 min an increase was seen. LSD led to a decrease in rat brain iron both 15 and 120 min after injection. The authors[17] later reported that tremorine caused a decrease in rabbit brain iron concentration in cortex and corpus striatum 30 min after injection. Also 2,2′,2″-tripyridine and 3-acetylpyridine reduced brain iron concentration.[18]

Rats dying after adrenalectomy have an increased brain iron.[19] When the animals after operation were treated with cortisone and hydrocortisone and survived for some days, the brain iron concentration was increased by 200%.

A condition phenomenologically resembling Hallervorden–Spatz's disease was produced in monkeys by Pentschew and Garro[20] by the intravenous administration of ferrous sulphate or manganous chloride. Histochemical changes were seen in various parts of the brain, and not predominantly in globus pallidus and substantia nigra, as is the case in Hallervorden–Spatz' disease.

Intravenous administration of iron salts to cats led to a decrease of niacin-containing coenzymes in various organs, including brain.[21] The concentration of free niacin remained constant. The administration of niacin led conversely to a decrease in the iron concentration of brain and other organs and an increase in serum iron concentration.

IV. IRON IN NORMAL BRAIN

A. Total Values and Regional Distribution

The first systematic investigations of iron in brain were undertaken by Guizzetti,[22] who found that different areas of the brain macroscopically displayed different intensities when stained for iron. Spatz[12] performed both macroscopical and microscopical investigations. Unfixed brain slices were placed in an ammonium sulfide solution, and the different areas of brain obtained precipitations of black ferrous sulfide according to their iron concentrations. The parts of brain were classified into four groups, according to the intensity of precipitation. Group I gave the strongest reaction, group IV gave no reaction at all. Group I consisted of globus pallidus and substantia nigra (especially zona reticulata). Group II: nucleus ruber, nucleus dentatus cerebelli, nucleus caudatus, putamen, and corpus subthalamicum Luysi. Group III: corpora mamillaria, parts of thalamus, and the whole cortex. Group IV: spinal cord and spinal ganglia. The neural part of the pituitary gave only a slight staining reaction, but locally intensely stained areas were often seen.

A parallelism between positive iron staining reaction and iron concentration determined by quantitative chemical technique was found by Wuth,[23] who reported iron concentration in different parts of brain from 140 to 664 μg/g dry wt. Widely different results for iron in brain have been obtained by various investigators. Tompsett[24] extracted iron with pyrophosphate and found an average value in twelve human brains of 9 μg/g wet wt. Cumings[25] found by dry incineration 333 μg/g dry wt. in white matter and 1129 μg/g dry wt. in globus pallidus. By spectroscopy, Alexander and Myerson[26] found 60 μg/g wet wt. in cortex.

A quantitative evaluation of histochemically demonstrable iron was performed by Wollemann,[27] who spectrophotometrically measured the color reaction after treatment of the tissue samples with potassium ferrous and potassium ferric cyanide. Standards were colored gelatine membranes of known thickness and with known iron content. Values from 30 μg/g dry wt. in spinal cord to 145 μg/g dry wt. in substantia nigra were obtained.

Bertha *et al.*[28] and Musil *et al.*[29,30] found between 25 and 1270 μg/g dry wt. These workers used dry incineration, followed by colorimetry or polarography. Dorfman[31] found in cortex 160 μg/g dry wt. and 550 μg iron/g dry wt. of pituitary gland. Hanig and Aprison[32] found about 20 μg/g wet wt. in rabbits, but they did not remove the blood from the brain before isolating the brain.

B. Iron-Containing Compounds

Ultrafiltrable (non-protein-bound) brain iron in dogs amounted to 16 % in gray matter and 25 % in white matter of total iron, which was 9.4 μg/g wet wt. in gray matter and 9.6 μg/g wet wt. in white matter;[33] these results were very much like those obtained by Tompsett[24] for inorganic iron.

A number of investigations quoted above regarding human and animal brain suggest that heme-bound iron constitutes only a small fraction of brain iron.[7] This is supported by the finding[2] in rat brain of nonheme iron practically identical to total iron, when all blood had been washed out of the vascular system by infusion of saline into the aorta in the living animal at sacrifice. Values of 10 μg/g wet wt. were obtained both by wet and dry incineration followed by colorimetry.

Ferritin in brain was isolated by Diezel,[34] and it was found that the amount of protein isolated corresponded reasonably well to the amount of iron determined histochemically. It was concluded that at least part of histochemically demonstrable iron exists as ferritin.

C. Iron in CNS in Relation to Age and Evolution

The iron content in the cerebrum of the human fetus in the third to the tenth month of pregnancy was found to increase in the course of growth.[35]

Iron in brain as a function of age was investigated by Guizzetti[22] and Spatz.[12] No part of embryonal brain gave a positive reaction, in contrast to embryonal liver, which already gave a strongly positive staining reaction for iron. In infants six months old, a positive staining for iron was seen in globus pallidus. Thereafter, substantia nigra and other parts of the brain followed in order of intensity of iron reaction as given above for adult human brain.

In mammals, iron staining reactions parallel the evolutionary stage of the animal according to Spatz[12] and Guizzetti.[22] Mouse brain gave a negative iron reaction, brain tissue from rabbit, cat, dog, and monkey gave a positive reaction of increasing intensity in globus pallidus and substantia nigra; in monkey there was also a positive reaction in nucleus dentatus, but not in nucleus ruber.

The influence of age on nonheme iron in human brain was studied by Hallgren and Sourander[4] by warm extraction in 5 N hydrochloric acid. The distribution of nonheme iron in brain followed the same pattern as given above for histochemical iron.[12] The amounts of nonheme iron in various parts of the brain were of the same order of magnitude, as found by various workers for total iron. Even in globus pallidus, low values of nonheme iron were found at birth. During childhood and adolescence, nonheme iron in globus pallidus increased very considerably, and after 30 years of age the amount was stable at around 200 μg/g wet wt. In thalamus, there was a similar increase from birth to the age of 30, after which the nonheme iron content declined. In putamen and nucleus ruber, a level was first attained around 50 to 60 years of age. In spinal cord the nonheme iron content was independent of age. One third of brain nonheme iron was extractable with buffer solutions; it was protein-bound and probably identical with ferritin.

The influence of age on total brain iron was investigated by Sundermann and Kempf[36] after dry incineration. In globus pallidus, putamen, nucleus caudatus, and nucleus ruber, iron content continued to increase during the whole life time, whereas in thalamus and cortex a maximum was reached at 30 to 40 years of age, iron content thereafter remaining at the same level or decreasing slowly in these parts of the brain.

Because our knowledge at the present time of normal values for iron in brain is insufficient, and because the results obtained by various workers are contradictory, new investigations need to be undertaken which avoid errors from iron contamination in the technical procedures and from blood contamination. At present, the results of investigations of brain iron in disease are difficult to interpret.

V. IRON IN BRAIN DISEASE

Technical shortcomings in brain iron analysis described above are unfortunately combined with clinical and histopathological lack of precision in many of the diseases in which abnormalities in brain iron have been claimed.

Hemochromatosis is a rare disorder of iron metabolism. It is characterized pathologically by excessive deposits of iron in the body, and clinically by liver cirrhosis, diabetes mellitus, skin pigmentation, and often cardiac affection. A typical case of hemochromatosis was described by Cammermeyer,[37] with iron deposits in most areas of the brain. The deposits were confined to phagocytic cells. Neumann[38] described another typical case of hemochromatosis with no iron pigment deposition in the brain at all. Sheldon[39] compared a typical case with two control patients. Histochemically, normal amounts of iron were found. Quantitative determination showed in several parts of the brain, however, values higher in hemochromatosis than in controls. Globus pallidus was contaminated by iron from an earlier hemorrhage, and similar complications make a number of other reports difficult to interpret.[38,40,41]

Herzenberg[42] described three patients with a cerebral type of iron deposition. As they had no diabetes, liver cirrhosis, or skin pigmentation, it seems confusing to use the term hemochromatosis. Their symptomatology consisted of ataxia, hearing disturbance, headache, and vomiting. In the most advanced of these cases, an intense histochemical reaction for iron was found in mid-brain, brain stem, and the pituitary gland.

Wilson's disease is a disturbance of copper metabolism. In three cases of this disease, Cumings[25] by dry incineration found increased total iron in gray matter, nucleus caudatus, and thalamus, but no change in white matter, globus pallidus, or putamen. Jakob and Pánczél[43] found, however, much lower cerebral and cerebellar cortical brain iron concentrations in a patient with Wilson's disease, but they had no control patients. By wet incineration and spectrography, Löwenthal[44] found increased iron values in white matter, globus pallidus, putamen, and brain stem, whereas there were normal or decreased values in cerebral and cerebellar cortical matter. It should be mentioned that Szanto and Gallyas[45] found abnormal general iron metabolism in a case of Wilson's disease.

In Hallervorden–Spatz' disease, Löwenthal[44] found no change in brain iron concentration in spite of pigment accumulation in glial cells, but Környey[46] found increased total iron in globus pallidus and thalamus in this disorder. Szanto and Gallyas[45] gave ^{59}Fe to patients with Hallervorden–Spatz' disease and found a normal general iron metabolism, but a slower turn-over of iron in the brains of these patients.

In Alzheimer's disease, Soniat[47] found, histochemically, increased iron reaction in nerve cells, in some oligodendroglial cells, and in the plaques characteristic of this disorder. When Hallgren and Sourander[48] investigated nonheme iron in brains from such patients, they found normal values; they concluded that the increased iron demonstrable by histochemical techniques was not caused by a true increase in brain iron, but was due to a redistribution of the normal iron content present. The investigations of Goodman[49] are in accordance with this view.

In Huntington's chorea, increased iron is often demonstrable in corpus striatum by histochemical technique.[50] Spectrographic iron examination in brains from such patients has given conflicting results.[51]

Kaschin–Beck's disease is supposed to be caused by increased iron in the diet, combined with vitamin deficiencies. Histochemically, brain iron has been found to be increased,[52] although the relative distribution followed normal patterns.[12] Most workers have found normal brain iron values in Sturge–Weber–Dimitri's disease.[53–55] A rusty coloring of globus pallidus and substantia nigra is seen in status pigmentatus, and microscopically the cells are full of pigment; staining for iron gives a strong positive reaction in the same areas.[56] Strong iron reaction is also obtained around old hemorrhages,[12] the iron being present as hemosiderin, especially in Hortega glial cells. Also, increase in total iron is often found in degenerated brain areas.[26]

Dementia paralytica is caused by spirochetal invasion of the brain tissue. Iron-containing pigment is demonstrable in glial cells, especially near blood vessels, in the cortex and in corpus striatum.[7,12,50,57] Similar results have been obtained by quantitative chemical determination of nonheme iron and total iron.[7]

Freeman[58] found less iron in brain cells of schizophrenic patients than in other psychotic patients both histochemically and chemically; nonpsychotic patients were not investigated as controls. Increased amounts of iron, demonstrated histochemically, has been reported by Strassmann[59] in globus pallidus of psychotic patients.

VI. IRON IN NORMAL AND ABNORMAL CEREBROSPINAL FLUID

Just as with brain tissue, various investigators have obtained very different results: most often an iron content about 30–40 μg/100 ml[9,10] of normal human cerebrospinal fluid. These investigators all used colorimetric methods. Kjellin[8] performed lumbar puncture with platinum needles, and, after wet incineration, determined iron by colorimetry and by neutron activation analysis. With both methods, spinal fluid iron was less than 1–2 μg/100 ml. Confirmatory evidence of cerebrospinal fluid iron much lower than previously assumed was obtained by Kofod and Rafaelsen,[2] who were unable to demonstrate iron in spinal fluid using a bathophenanthroline method with a sensitivity of 5 μg/100 ml. Besides these analytical problems, the sampling technique is also very important. Iron contamination from the lumbar puncture needle can be avoided as done by Kjellin[8] by using platinum, but modern disposable needles may be just as effective. A source of error is contamination with blood. One μl of blood in 10 ml spinal fluid is invisible (500 erythrocytes/μl spinal fluid), but gives a false addition to cerebrospinal fluid iron of 5 μg/100 ml. Counting of spinal fluid shortly after sampling and discarding, if a considerable amount of erythrocytes are found, will avoid this error.

Kjellin[8] assumed that all iron in cerebrospinal fluid is loosely bound to a protein, probably transferrin. If transferrin is present in spinal fluid protein in a similar ratio to total protein as it is in blood, it follows that it is most unlikely that iron should be present in spinal fluid in amounts as high as those reported in earlier investigations.

The findings reported in various pathological states can therefore be mentioned briefly. Lehmann and Kral[10,11] investigated psychotic patients and found lower values in patients with recent onset of schizophrenia than in chronic patients. Psychotic patients as a group did, however, not differ from nonpsychotic. In meningitis, encephalitis, poliomyelitis, and Guillain–Barré's syndrome, cerebrospinal fluid iron was increased.[60] New investigations of iron in cerebrospinal fluid is therefore urgently needed, using adequate sampling and radioactive methods: neutron activation analysis and probably isotope dilution.

VII. CONCLUSION

The study of iron in brain has been fraught with technical difficulties; this necessitates renewed investigation along several lines. Among the important research results which stand the test of time and re-examination are the mapping of brain centers according to relative distribution of iron and the demonstration of increasing iron in brain in the evolution of mammals and in the development of the human individual from embryo to adult.

The field of iron abnormalities in brain disease is, however, so vaguely illuminated by existing evidence that this area must be totally revised by new investigations in which radioactive techniques and perhaps immunological methods will be of paramount importance.

VIII. REFERENCES

1. W. A. Jones, Determination of traces of iron and copper in culture media prepared by enzymic digestion of muscle protein, *Biochem. J.* **43**:429–433 (1948).
2. B. Kofod and O. J. Rafaelsen (in press).
3. R. E. Basford, H. D. Tisdale, J. L. Glenn, and D. E. Green, Studies on the terminal electron transport system. VII. Further studies on the succinic dehydrogenase complex, *Biochim. Biophys. Acta* **24**:107–115 (1957).
4. B. Hallgren and P. Sourander, The effect of age on the nonhaemin iron in the human brain, *J. Neurochem.* **3**:41–51 (1958).
5. A. L. Foy, H. L. Williams, S. Cortell, and M. E. Conrad, A modified procedure for the determination of non-heme iron in tissue, *Anal. Biochem.* **18**:559–563 (1967).
6. R. Hill, A method for the estimation of iron in biological material, *Proc. Roy. Soc. (London)* **B107**:205–214 (1931).
7. A. H. Tingey, The iron content of the human brain, *J. Mental Sci.* **84**:980–984 (1938).
8. K. G. Kjellin, Determination of the iron content in the cerebrospinal fluid, *J. Neurochem.* **13**:413–421 (1966).
9. R. W. Moshier and J. E. Schwarberg, Determination of iron, copper and aluminium by gas–liquid chromatography, *Talanta* **13**:445–456 (1966).
10. H. E. Lehmann and V. A. Kral, Studies on the iron content of cerebrospinal fluid in different psychotic conditions, *A.M.A. Arch. Neurol. Psychiat.* **65**:326–336 (1951).
11. V. A. Kral and H. E. Lehmann, Further studies on the iron content of the cerebrospinal fluid in psychosis, *A.M.A. Arch. Neurol. Psychiat.* **68**:321–328 (1952).
12. H. Spatz, Über den Eisennachweis im Gehirn, besonders in Zentren des extrapyramidal-motorischen Systems, *Zentr. Gesamte Neurol. Psychiat.* **77**:261–390 (1922).
13. W. E. Lingren, G. Reeck, and R. Marson, Contamination of paper chromatograms by scissors, *J. Chromatog.* **24**:269 (1966).
14. P. B. Hukill and F. A. Putt, A specific stain for iron using 4,7-diphenyl-1,10-phenanthroline (bathophenanthroline), *J. Histochem. Cytochem.* **10**:490–494 (1962).
15. R. A. MacDonald and G. S. Pechet, Experimental hemochromatosis in rats, *Am. J. Pathol.* **46**:85–110 (1965).
16. S. Hadžović, B. Nikolin, and P. Stern, The effect of tremorine and lysergic acid diethylamide on the iron content of the rat brain, *J. Neurochem.* **12**:908–909 (1965).
17. P. Stern, S. Hadžović, and B. Nikolin, Bedeutung des Eisens im Corpus striatum für den Tremor, *Arch. Exptl. Pathol. Pharmakol.* **257**:67–68 (1967).
18. S. Hadžović, B. Nikolin, and P. Stern, The iron content of the brain of rats with choreiform movements, *Eur. J. Pharmacol.* **1**:15–17 (1967).
19. S. P. Voroshilovskaya and V. S. Raitses, Effect of adrenalectomy on the content of copper, zinc and iron in animal brain, *Patol. Fiziol. Eksperim. Terapiya* **10**:78–80 (1966). [*cf. Chem. Abstr.* **66**:62361 (Q) (1967).]
20. A. Pentschew and F. Garro, Experimental Hallervorden–Spatz's encephalopathy in monkeys, *J. Neuropathol. Exptl. Neurol.* **26**:146–148 (1967).

21. A. I. Gaidenko, Interrelation between iron and nicotinic acid metabolism, *Vopr. Pitaniya* **25**(5):37–41 (1966). [*cf. Chem. Abstr.* **66**:9123 (Q) (1967).]

22. P. Guizzetti, Principali risultati dell'applicazione grossolana a fresco delle reazioni isto-chimiche del ferro sul sistema nervosa centrale dell'uomo e di alcuni mammiferi domestici, *Riv. patol. nervosa e mentale* **20**:103–117 (1915).

23. O. Wuth, Über den Eisengehalt des Gehirns, *Zentr. Gesamte Neurol. Psychiat.* **84**:474–477 (1923).

24. S. L. Tompsett, The copper and "inorganic" iron content of human tissues, *Biochem. J.* **29**:480–486 (1935).

25. J. N. Cumings, The copper and iron content of brain and liver in the normal and in hepato-lenticular degeneration, *Brain* **71**:410–415 (1948).

26. L. Alexander and A. Myerson, Minerals in normal and in pathologic brain tissue, studied by micro-incineration and spectroscopy, *A.M.A. Arch. Neurol. Psychiat.* **39**:131–149 (1938).

27. M. Wollemann, A photometrical method for testing the presence of iron in the central nervous system, *Acta Morphol. Acad. Sci. Hung.* **1**:127–132 (1951).

28. H. Bertha, A. Musil, W. Haas, and O. Wawrschinek, Untersuchung über die regionale Kationenverteilung im menschlichen Gehirn, *Monatsh. Chem.* **93**:118–122 (1962).

29. A. Musil, H. Bertha, W. Haas, and O. Wawrschinek, Untersuchung über die regionale Kationenverteilung im menschlichen Gehirn, 2. Mitt.: Eisen, Kupfer, Calcium und Magnesium, *Monatsh. Chem.* **93**:536–540 (1962).

30. A. Musil, H. Lechner, O. Wawrschinek, W. Beyer, H. Wielinger, and H. H. Tagger, Untersuchungen über die regionale Ionenverteilung im menschlichen Gehirn, 4. Mitt.: Natrium, Kalium, Calcium, Magnesium, Eisen, Kupfer und Phosphor, *Monatsh. Chem.* **95**:1013–1016 (1964).

31. S. I. Dorfmann and S. A. Shipitsyn, The quantitative determination of some metals in the human brain, *Biokhimiya* **20**:136–139 (1955). [*cf. Chem. Abstr.* **49**:12651 (c) (1955).]

32. R. C. Hanig and M. H. Aprison, Determination of calcium, copper, iron, magnesium, manganese, potassium, sodium, zinc, and chloride concentrations in several brain areas, *Anal. Biochem.* **21**:169–177 (1967).

33. O. Ya. Duschechkina, Iron content of the brain tissue, *Fiziol. Zh. SSSR* **35**:284–292 (1949). [*cf. Chem. Abstr.* **44**:724 (c) (1950).]

34. P. B. Diezel, Iron in the brain, A chemical and histochemical examination, in *Biochemistry of the Developing Nervous System, Proc. First Intern. Neurochem. Symposium, Oxford*, 1954 (H. Waelsch, ed.), pp. 145–152, Academic Press, New York (1955).

35. Massato Kimitsuki, Inorganic substances in the cerebrum, *Fukuoka Igaku Zassi* **46**:998–1005 (1955). [*cf. Chem. Abstr.* **50**:12252 (e) (1956).]

36. A. Sundermann and G. Kempf, Über den physiologischen Eisengehalt einiger Stammhirnganglien und seine Abhängigkeit vom Lebensalter, *Z. Alternsforsch.* **15**:97–105 (1961).

37. J. Cammermeyer, Deposition of iron in paraventricular areas of the human brain in hemochromatosis, *J. Neuropathol. Exptl. Neurol.* **6**:111–127 (1947).

38. M. A. Neumann, Hemochromatotic pigmentation of the central nervous system, *A.M.A. Arch. Neurol. Psychiat.* **76**:355–368 (1956).

39. J. H. Sheldon, The iron content of the tissues in haemochromatosis, with special reference to the brain, *Quart. J. Med.* **21**:123–137 (1927).

40. F. H. Lewey and S. R. Govons, Hemochromatotic pigmentation of the central nervous system, *J. Neuropathol. Exptl. Neurol.* **1**:129–138 (1942).

41. M. A. Neumann, Hemochromatosis of the central nervous system, *J. Neuropathol. Exptl. Neurol.* **7**:19–34 (1948).

42. H. Herzenberg, Über Hämochromatose (Mit besonderer Berücksichtigung des Fe-Pigments im Gehirn), *Arch. Pathol. Anat. Physiol.* **260**:110–129 (1926).

43. I. Jakab and M. Pánczél, Contributions á la pathomorphologie et pathochimie de la maladie de Wilson–Westphal–Strümpell, *Acta Med. Acad. Sci. Hung.* **3**:341–346 (1952).

44. A. Löwenthal, Composition du cerveau en certains éléments minéraux et essai sur leur signification en pathologie, *Acta Neurol. Psychiat. Belg.* **59**:1161–1223 and 1227–1257 (1959).

45. J. Szanto and F. Gallyas, A study of iron metabolism in neuropsychiatric patients, *Arch. Neurol.* **14**:438–442 (1966).

46. St. Környey, Die Stoffwechselstörungen bei der Hallervorden–Spatzschen Krankheit, *Arch. Psychiat. Z. Gesamte Neurol.* **205**:178–191 (1964).

47. T. L. L. Soniat, Histogenesis of senile plaques, *A.M.A. Arch. Neurol. Psychiat.* **46**:101–114 (1941).

48. B. Hallgren and P. Sourander, The non-haemin iron in the cerebral cortex in Alzheimer's disease, *J. Neurochem.* **5**:307–310 (1960).

49. L. Goodman, Alzheimer's disease, a clinico-pathologic analysis of twenty-three cases with a theory on pathogenesis, *J. Nervous Mental Disease* **117**:97–130 (1953).

50. A. Metz, Die drei Gliazellarten und der Eisenstoffwechsel, *Zentr. Gesamte Neurol. Psychiat.* **100**:428–449 (1926).

51. C. B. Courville, R. E. Nusbaum, and E. M. Butt, Changes in trace metals in brain in Huntington's chorea, *Arch. Neurol.* **8**:481–489 (1963).

52. N. S. Hsiang, Brain pathology of Kaschin–Beck disease, *J. Oriental Med.* **33**:119–162 (1940). [*cf. Chem. Abstr.* **35**:511 (8) (1941).]

53. E. C. Eaves, A contribution to the study of deposits containing calcium and iron in the brain, *Brain* **49**:307–332 (1926).

54. A. H. Tingey, Iron and calcium in Sturge–Weber disease, *J. Mental Sci.* **102**:178–180 (1956).

55. N. Wachsmuth and A. Löwenthal, Détermination chimique d'éléments minéraux dans les calcifications intracérébrales de la maladie de Sturge–Weber, *Acta Neurol. Psychiat. Belg.* **50**:305–313 (1950).

56. M. Helfand, Status pigmentatus. Its pathology and its relation to Hallervorden–Spatz disease, *J. Nervous Mental Disease* **81**:662–675 (1935).

57. C. Trétiakoff and O. Caesar, Étude histochimique des composés du fer dans l'écorce cérébelleuse des aliénés, *Rev. Neurol.* **2**:220–242 (1926).

58. W. Freeman, Deficiency of catalytic iron in the brain in schizophrenia, *A.M.A. Arch. Neurol. Psychiat.* **24**:300–310 (1930).

59. G. Strassmann, Hemosiderin and tissue iron in the brain, its relationship, occurrence and importance, *J. Neuropathol. Exptl. Neurol.* **4**:393–401 (1945).

60. Nobumasa Kuki, Blood and spinal fluid ions in various children diseases, *Nippon Shonika Gakukai Zasshi* **63**:918–926 (1959). [*cf. Chem. Abstr.* **54**:17670 (a) (1960).]

Chapter 14

GRAY–WHITE MATTER DIFFERENCES

Jørgen Clausen

The Neurochemical Institute
Copenhagen, Denmark

I. INTRODUCTION—
Definition of Gray and White Matter: Cellular and Particulate Components

Gross anatomical inspection of the central nervous system (CNS) reveals that the shape taken by different parts is dependent upon accumulation of nerve cells referred to as nuclei, ganglia, or *gray matter* and bundles of nerve fibers referred to as tracts, funiculi, fasciculi, peduncles, commissures, or *white matter*. This old definition and differentiation between gray and white matter, based upon gross anatomical differences, still seems to be useful today, especially in approaching the physiological and pathological function of CNS. The gray and white matter will be discussed below in terms of cellular and chemical differences.

Light microscopic inspection of the cells present in gray and white matter reveals qualitatively the presence of the same cellular components in gray as in white matter.[1] Even the total amount of cells per weight unit seems comparable in these two compartments. Thus, as shown on Table I, the amounts of nuclei (DNA) per milligram brain tissue seem identical in the cerebral cortex and corpus callosum, representatives of gray and white matter, respectively. However, among the different parts of gray matter, there seem to be differences in the total amounts of cells (Table I); thus, the cerebellar cortex seems to contain about six times the amount of cells in the cerebral cortex. From the definition of gray and white matter it follows that the white matter is characterized by the presence of nerve fibers. However, these structures also occur in gray matter. The lamilar structure of the nerve sheath occurs not only in peripheral but also in the central nervous system. The special structure of the nerve sheath was first visualized by means of polarization studies,[3] later by means of X-ray diffraction studies,[4] and finally by means of electron microscopy.[5–9] Studies of the development of the nervous system have disclosed that the peripheral myelin sheath is

TABLE I

Estimates of Cell Density by DNA Determinations[a]

	DNA/nucleus, pg[b]	DNA/mg, μg^c	Nuclei/mg $\times 10^{3c}$
Cat			
Cerebral cortex	6.8 ; 7.9 ; 8.1	0.91	128
	Av, 7.1		
Cerebellar cortex	5.6 ; 6.5	5.73	806
Corpus callosum		0.96	135
Dog			
Cerebral cortex	5.5 ; 7.8	0.96	148
	Av, 6.5		
Cerebellar cortex	5.9 ; 6.6	3.68	566
Corpus callosum		0.98	151
Man			
Cerebral cortex	7.7 ; 8.1	0.93	131
	Av, 7.1		
Cerebellar cortex	5.3 ; 7.3	6.05	852
Primitive glial tumors[d]	10.1–15.7	1.3–11.7	99–1116
	Av, 12.2		
Astrocytomas	9.5–10.7	0.9–5.2	173–418
	Av, 10.0		
Oligodendroglioma	9.2	2.75	298

[a] Taken from Elliott.[2]
[b] 1 picogram (pg) = 10^{-12} g.
[c] Figures are averages.
[d] Medullublastoma, astroblastoma, glioblastoma, and unclassified.

derived from Schwann plasma membrane.[9–11] The peripheral axon is surrounded by a single sheet cell per internodal segment and is separated from its counterpart not only by the sheet cell cytoplasma but also by a basement membrane and an intercellular space. No such space is present in the central nervous system. Here the cells are closely packed with intricately interwoven cellular processes.[12]

The gray matter is characterized by the large amount of ganglion cells; the white matter, not only by the myelin sheath but also by the high proportion of glia cells.

The dendrites seem structurally related to the perikaryon; the axons seem more specialized.[1] The peripheral myelinated fibers may be covered with the so-called neurilemma or be without this substance. Also, the unmyelinated fibers may be covered with a neurilemma or may lack this substance. Myelin lacks the neurilemma sheath when derived from the oligodendroglia cell. The unmyelinated fibers are of two kinds: the naked axons and the fibers equipped with processes derived from cells outside the neuron, giving rise to the formation of a thin neurilemma. The naked axons are especially numerous in gray matter and some parts of the brain and spinal cord; every axon at its beginning from the nerve cell, as well as its terminal, is devoid of coverings. However, it seems possible to subdivide further the nerve fibers by electron microscopy of the glia cells attached to the fibers;[13] thus, studies of the optic nerve have revealed nonmyelinated nerve fibers associated with small (spider) astrocytes, finely myelinated fibers associated with oligodendrocytes.

The number of myelin lamellae in the sheath around the axons may show great variation. These variations seem related closely to the velocity of the impulses of conduction.[14–17] Several unclarified factors determine the extent to which the myelin sheath is developed. Among these factors seems to be the diameter of the axon.[18] However, Friede and Samorajski[19] also demonstrated a rectilinear relationship between the number of lamellae in the myelin sheath and the circumference of the axons for both the vagus and ischiaticus nerve.

The differentiation between gray and white matter cannot be completely clear but is mainly related to the presence of nerve cells and unmyelinated fibers in gray matter and the presence of mainly myelinated fibers in the white matter.

Throughout both gray and white matter, mesodermally derived connective tissue penetrates along with the blood vessels, but the main supporting tissue of CNS consists of the glia cells. Under this heading may be included ependyma cells and the proper neuroglia, including astrocytes, and oligodendroglia and mesoblastically derived microglia cells. The ependyma form a single layer of epithelial cells lining the ventricles and the central canal of the spinal cord. The microglia cells act as phagocytic cells in cases of pathologic changes. The neuroglia cells *sensu strictiori* include the protoplasmic astrocytes, fibroastrocytes, and oligodendroglia cells. The protoplasmic astrocytes are mainly localized in gray matter and are characterized by freely branching protoplasmic processes. The fibrous astrocytes are mainly localized in white matter. They appear with long unbranched fibers; both types of astrocytes are attached to blood vessels by means of the so-called foot processes. The oligodendroglia cells are smaller than the astrocytes. They are found in the white substance in rows along the nerve fibers, but they also penetrate into the gray matter as satellites of the nerve cells. The oligodendroglia may be of metabolic importance for the axon. The lack of intercellular space and lymph in CNS, the intimate contact between protoplasmic astrocytes and the neuron in the gray matter, and the intimate contact between oligodendroglia and the axon in the white matter, and finally the accumulation of mitochondria in the internodal glia cells, lacking a basement membrane towards the nodes, suggest a metabolic interaction between nerve cells and the glia as a basis for the function of the nerve cell.[13] Cytophysiological data[20,22] demonstrated that the activity of cytochrome oxidase and succino-oxidase in the spinal ganglion cells and their capsules is lower than in the glial cells. Lowry, Roberts, and Lewis[23,24] demonstrated activity of phosphoglucoisomerase, glutamic–aspartic transaminase, and glutamic dehydrogenase in the glial cells five times higher than that of the neurons. These experiments thus may indicate a higher respiratory rate and amino acid metabolism per unit volume of glia cells than nerve cells. Also, the respiratory enzyme activities of oligodendrocytes have been demonstrated to be greater than those nerve cells.[21,22] From these studies one can deduce that the glia satellite cells of the gray matter and the oligodendrocytes of the white matter are of importance not only for the support of the neuron and for development of the myelin sheath but also as sources of supply of enzymes and probably metabolic products formed during functional activity of the nerve cells and the axon. This is supported by the fact that, during stimulation of the vestibular nerve cells, the content of RNA, cytochrome oxidase, protein, and succinic oxidase increased in the nerve cell but decreased in the oligodendroglia cell.[21,22]

The gray–white matter differences do not exclusively originate from differences in composition and function of the neuron but mainly from differences in composition, function and distribution of the supporting connective tissue of CNS. The glial function is directly related to the metabolic activity during fetal development (myelogenesis) and to neural activity of the glial–neuronal system. The preparative isolation of nerve cells by freehand dissection suffers from the fact that plasma membranes of the cells are more or less damaged by the isolation procedure.

Until now it has not been possible to isolate and cultivate pure strains of different glial cells present in normal brain directly from either fetal or adult brains. Pure tissue or cell cultures of astrocytes, oligodendroglial cells, and ependyma cells can be obtained only by culture of brain tumors.[25] Chemical studies of pure brain tumor cell lines in tissue culture or direct chemical investigation of brain tumors have elucidated to some extent the individual differences between the constituents of the different glial cells. However, such studies cannot always be used as direct proof of the composition of the different glial cells, because changes may occur during malignant transformations. Malignant tumors may thus contain more esterified cholesterols and choline phosphatides but less ethanolamine phosphatides than normal tissue. Malignant tissue may also contain an abnormal lipid hapten, a dihexose-ceramide (cytolipin-H), and also more electrical charges on the surface of the plasma membrane than the normal plasma cell membrane.[28–40]

Gray and white matter differences in cellular particulate fractions have been elucidated mainly by means of gradient differential ultracentrifugation.[27,41,42]

By means of the method of Autilio, Norton, and Terry[43] it has been possible to isolate pure myelin, which by further gradient ultracentrifugation has been divided into the light and heavy myelin. Chemical analysis has verified that the chemical composition of myelin is similar to that of total white matter but different from that of total gray matter.

Nuclei, mitochondria, and synaptosomes are widely distributed in both gray and white matter, and the chemical composition of these structures cannot directly contribute to the elucidation of gray–white matter differences.

II. INORGANIC IONS AND WATER

Water is the major inorganic constituent of brain tissue, forming 80% of cerebral cortex and 70% of cerebral white matter. The highly active glial system, combined with a small extracellular space, causes rapid movement of water between brain tissue and the blood compartment.[44] The difference in water content and total amount of salts in gray and white matter originates from the enrichment of myelin in the white matter and causes the dry weight of white matter to be nearly twice as high as that of gray matter (Table II).

The distribution of electrolytes in brain tissue as in most tissue is characterized by an excess of inorganic cations over anions and is balanced by anions of large molecular weight. Thus, the contribution from lipids and proteins (80 μeq/g fresh wt.) exceeds that involving any other tissue.[46] The content of inorganic ions may be referred to wet or dry tissue weight. The first reference may conveniently be used in studies of ionic fluxes;[46] the latter expression, in other studies. The high content of K^+ in cerebral cortex (100 μeq/g wet wt.) causes K^+ to contribute two thirds of the cations and gives rise to an electrical potential (E) of 60 to 100 mV, partially opposed by an opposite Na^+ potential.[46]

Because of the correspondingly high content of polar lipids and cholesterol in the myelin membrane, the content of sodium and potassium ions (referred to brain tissue dry weight) is lower in white than in gray matter (Table III). However, this difference does not seem obvious when concentrations are referred to wet tissue. The glial cells exhibit a function similar to that of extracellular fluid in the extracerebral organs, thus the glial cells in particular have been considered high in electrolytes.[48,49]

TABLE II

Dry Weight and Sodium and Potassium Content of Brain Tissues of the Cat 24 hr After the Application of a Freezing Lesion[a]

	White				Gray			
	Dry weight	Na	K	Na/K	Dry weight	Na	K	Na/K
Normal[6]	32.0	58	82	0.74	18.2	68	90	0.76

[a] Freezing lesion = 80 mm^2, 20 sec; dry weight: milligrams per 100 mg fresh weight of tissue. Na and K: millimoles per kilogram fresh weight of tissue. From Pappius.[45]

TABLE III

Concentration of Electrolytes in Brain[a]

Constituent	Gray matter	White matter
	mmoles/kg dry	
Na$^+$	50–90	45–100
K$^+$	55–100	45–100
Ca$^+$	2.5–3.2	2.5–4.5
Mg$^+$	8.5–9.5	10.8–17.0
Cl$^-$	31–62	25–51
	mg P/100 g wet weight	
Total P	190–200	340–490
Inorganic P	29–64	11–60
Acid solution P	71–190	60–190
Lipid P	120–180	250–380
Protein P	22–29	25–42
RNA P	10–11	4–5
DNA P	4–9	5–9

[a] From Rossiter.[47]

The content of some inorganic anions other than sodium and potassium ions is indicated in Table III, from which it is obvious that the content of calcium, magnesium, and phosphorus is higher in white than in gray matter. This may be related to the enrichment of especially the myelin fraction in the white matter. Thus, the magnesium and calcium ions may occur bound to particulate anions. The higher content of phosphorus and sulfate in the white matter is also related to the myelin content; the high content of phosphatides and sulfatides in the myelin may explain these facts. The content of chlorine ions tends to be lower in white matter than in gray matter (Table III).

III. ORGANIC IONS

A. Amino Acids

As is the case with other metabolites, the content of free amino acids in gray and white matter is mainly determined by the uptake from or exit to the blood and spinal compartment, in some cases also by the synthesis in brain tissue and by the metabolic transformation of the amino acids (including protein synthesis).[50] Studies by Furst, Lajtha, and Waelsch[51] have demonstrated a preferentially higher uptake of C^{14}-labeled lysine in gray matter (cortex) than in white matter (corpus callosum). Similarly, a two-to-four times higher incorporation rate of S^{35} methionine has been demonstrated in gray matter than in white matter cells.[52] The incorporation seems dependent upon the exitable state of the tissue.[53]

The regional pattern of amino acids is related to differences in protein synthesis and to variable blood–brain and cerebrospinal fluid–brain barriers in the gray and white matter.[54] The content *in vivo* of lysine, GABA, glutamic acid, and aspartic acid has been determined to be low in cortex, of medium-amount in white matter, and highest in the spinal cord (Fig. 1).[55–65] The studies by Lajtha[63,64] seem to indicate a fast uptake of amino acids by white matter.

The increase in amino acid content in the rostral–caudal direction found by the above-mentioned authors may be related to differences in enzymic activities.[64] Thus, the increase in rostral–caudal direction of GABA-α-ketoglutarate-transaminase may be correlated to a higher content of glutamate–dehydrogenase activity in diencephalon than in more caudal regions.[58]

B. Ammonia

The conversion of ammonia in glutamine in the brain is regarded as the principal mechanism for protection against the toxic effect of the ammonium ion. Other methods for utilization of ammonia in the brain are

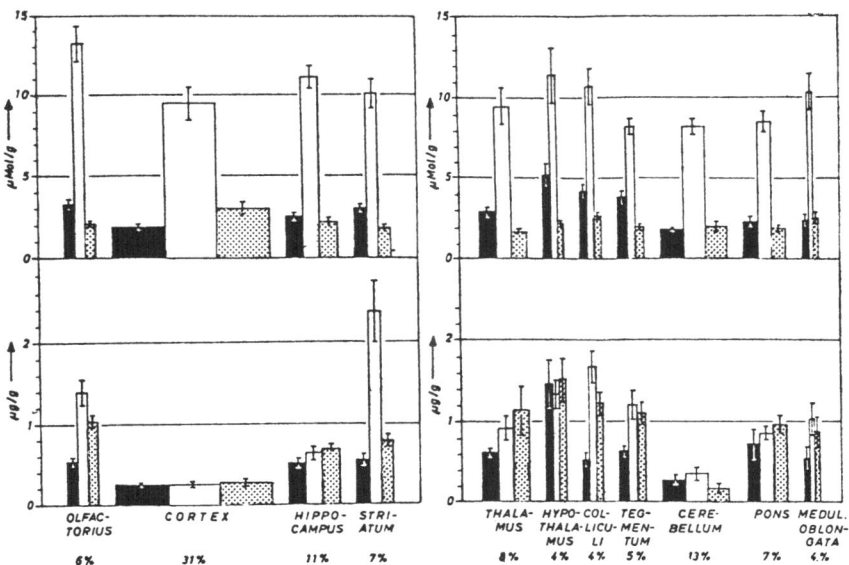

Fig. 1. Regional distribution of amino acids (above) and amines (below) in rat brains. Above: The black, white, and shaded areas denote γ-aminobutyric acid, glutamic acid, and aspartic acid, respectively. Below: The black, white, and shaded areas denote noradrenaline, dopamine, and serotonin, respectively. Taken from Popov et al.[65] [Courtesy of *Acta Biol. Med. Ger.*]

available. Thus, processes involving oxalo-acetic acid and pyruvic acid may contribute to the principal method for ammonia disposal.[66] The *in vivo* concentration of ammonia in different topographical areas of the brain is difficult to estimate, as a significant rise of the ammonia concentration in the brain occurs instantaneously upon death to values of about 57 ± 3.3 μmoles/100 g wet wt. of whole brain tissue.[67,68] Since the concentration of ammonia in the brain tissue is related to the availability of α-ketoglutaric acid as acceptor for the formation of glutamate and glutamine, the ammonia concentration in the brain tissue may be related directly to the concentration of glutamine and glutamate in the brain tissue of different topographic localizations. However, no studies seem to be available concerning the topographic *in vivo* variation in the concentration of free ammonia of brain tissue.

C. Transmitter Substances

Although some differences in the concentration of amino acids have been demonstrated between gray and white matter, the topographic differences in the content of amines are more pronounced.

By means of paper and gas chromatography and high-voltage electrophoresis,[69,70] it has been possible to demonstrate at least a dozen different organic amines of the central nervous system. Apart from the derivatives of catecholamines, serotonin, and GABA, the biological function of the amines is unknown.

The catecholamines and serotonin are compartmentalized by being localized in the peripheral parts of the neurons in the neighborhood of the synaptosomes, where they are synthesized, stored, and degraded.[71] However, the whole neuron and the neuropil also seem, as demonstrable by means of fluorescent microscopy, to contain the transmitter substances.[72,73] The storage of the transmitter substances in the synaptic vesicles of nerve endings causes the topographic localization of the transmitter substances to be closely related to both the density of synapses and the density of different chemically defined neurogenic pathways. Thus, in the central nervous system, both noradrenaline-containing and serotonin-containing pathways have been demonstrated by means of fluorescent microscopy technique.[73] Also histamine, acetylcholine, and GABA occur in highest concentration in the nerve endings.[74]

The highest content of organic amines is found in corpus striatum, hypothalamus, and coliculi; the lowest, in cortex, cerebellum, and white matter (see Fig. 1). Areas with high catecholamine content also show highest ratios of particulate-bound vs. free amines.[75] The differences between the topographic areas mentioned are highest concerning the content of dopamine but less concerning the noradrenaline and serotonin content. However, the topographic differences of CNS serotonin seem to be somewhat between that of noradrenaline and dopamine.[76–78] Also, the end product of dopamine, homovanillic acid, formed through the methyl-transferase pathway, also occurs in highest amounts in regions with high content of dopamine.[79]

Metabolic studies finally have shown the *in vivo* formation of noradrenaline and dopamine to be highest in the caudate nucleus but lowest in putamen, with the brainstem and spinal cord possessing intermediate values. The topographic localization of enzymes participating in the turnover of transmitter substances correspond well to that of the transmitter substances themselves. Thus, monoamine oxidase, catechol-*O*-methyl-transferase, and 5-hydroxy-tryptophan-decarboxylase-activities are nil or low in myelin (white matter) but high in mitochondria, nerve endings,[71] and midbrain nuclei.[80] Also, enzymes implicated in the chemical processes related to neurogenic transmission as ATPase, adenyl cyclase, *p*-nitrophenyl-phosphatase, and phosphodiesterase seem distributed as the transmitter substances.[73,81–84]

Acetylcholine concentration is highest in the brain stem and lowest in white matter.[85] Correspondingly, structures in brainstem and midbrain contain highest activities of enzymes implicated in the turnover of acetylcholine.

D. Organic Intermediate Metabolites

As already mentioned in the section concerning amino acids, the concentration of a certain metabolite in CNS is determined by (1) the exchange with blood and cerebrospinal fluid compartments and (2) the turnover rate. The gray–white matter differences concerning certain metabolites belonging to the different pathways of the intermediate metabolism are related to the cellular and particulate composition of these structures. Thus, glial elements, especially oligodendroglia, are low in content of hexokinase, phosphoglucoisomerase, lactic and malic dehydrogenase, glutamic dehydrogenase, and glutamic-aspartic transaminase but are relatively rich in glucose 6-phosphate and 6-phosphogluconate dehydrogenase, isocitric dehydrogenase, and succinic dehydrogenase; also, cytochrome oxidase and succinoxidase are present in higher levels in the glial cells than in the nerve cells.[9,21,22,85–89] The content of carboanhydrase is higher in the glial than in the neuronal elements.[86] These differences are of metabolic and functional importance; thus, it has been demonstrated by Hyden[21,22] that the high content of oxidative enzymes in the glial cells, especially the oligodendroglial cells, may cause many of these enzymes and certain substrates to be actively transported through the plasma membrane of the axon to the axonal compartment. This seems appropriate in order to compensate for the long distances of migration of proteins with enzymic activity synthesized in the RNA-rich compartment surrounding the nucleus of the nerve cell.

Even after strict indication of experimental conditions, the gray–white matter differences in content of substrates from main intermediate pathways seem uncertain because the cortical gray matter is situated outside the white matter. This difference causes slightly different times for deep-freezing (for instance, in freon-12). Differences in concentrations of metabolites may be induced hereby, since a delay in freezing causes a rapid fall in high-energy phosphates.[90]

For values of metabolites in outer, cortical layers and in total brains in normal and pathological conditions, the reader is referred to studies by Lowry and collaborators.[90–92]

IV. COLLOIDS

A. Proteins

1. Water-Soluble Proteins

The water-soluble proteins of CNS are derived either from the CNS itself or from the blood. The virtual lack of extracellular space in CNS[54] and the presence of glial footplates around CNS vessels, creating a barrier between the CNS and the blood compartment, the higher permeability of the CNS toward the cerebrospinal fluid compartment, and the vascular relationship between blood and spinal fluid causes the plasma proteins present in CNS either to be actively secreted into cerebrospinal fluid and from there transported to the CNS or to be actively taken up by the glial cells surrounding the capillaries of CNS. Only few studies of the similarities and differences between soluble proteins of brain tissue and plasma protein have been performed.[93–101] Both serum albumin and serum-β- and γ-globulins have been identified in the CNS.[93–96] The identification has been performed by immunoelectrophoresis of extracts from brain tissue, using antiserum from rabbits or horses against human serum.[93,96] The content of plasma proteins in different areas in CNS must be related to different blood–CNS barriers and different vascular permeabilities in different areas.[102] Thus, fluorescein-labeled albumin shows uptake in choroid plexus, area postrema, and tuber cinerum. Glial cells may show pinocytotic uptake of albumin but other areas of a normally functioning brain do not as a rule take up albumin. The edema caused by cortical injury is associated with penetration of plasma albumin into cortex and white matter.

Using specific antisera against brain tissue, one to five "specific" brain protein antigens have been found in immunoelectrophoresis.[98,100,101] These findings have been supported by chromatographic separation of brain tissue.[97,99] By means of tissue culture technique combined with immunoelectrophoresis of soluble proteins synthesized in tissue culture, it has been demonstrated that most plasma proteins present in CNS are not synthesized in the brain tissue[103] and, therefore, must originate from the blood or cerebrospinal fluid compartment.

The relative distribution of the different soluble brain tissue proteins separated by electrophoresis reveal that in all areas of the brain the β-globulin fraction is the predominant fraction, followed by α-globulin, γ-globulin, and albumin. When correlated to electrophoretic serum- and CSF-pattern possessing a preponderance of albumin fraction, therefore, the differences in relative distribution must be explained on the basis of the presence of proteins specific for the CNS. These proteins have electrophoretic mobilities corresponding to the β- and α-globulin areas.[100]

The topographic differences in the relative distribution of electrophoretically separated proteins of gray and white matter have been elucidated by Robertson[104] by means of paper electrophoresis. A significant relatively higher proportion of the β-globulin fraction in gray matter than in white matter was found associated with a significantly higher content of two slow-moving α-globulin fractions in white matter. No differences between the gray–white matter could be related to the fractions in γ-globulin, albumin, or prealbumin areas. This, however, seems to contradict the findings of Karcher, van Sande, and Löwenthal (1958; cited by Robertson[104]), who found differences in the quantitative distribution of two prealbumin fractions by means of agar–gel microelectrophoresis. More recent studies[99] using a two-dimensional combination of column chromatography on DEAE cellulose with starch–gel electrophoresis have revealed between 70 and 100 protein fractions. Quantitative correlation between these fractions and those of the liver have revealed a higher relative percentage of acidic proteins in brain tissue than in liver tissue, but the gray–white matter differences have not been elucidated.

Although the incorporation of labeled amino acids into proteins of the brain tissue has been studied both *in vitro* and *in vivo*, only scanty information on the turnover rate of the proteins in different areas of the brain has been given.[51] Studies by Piha *et al.*[105] demonstrated that the incorporation of labeled amino acids into soluble and insoluble proteins of the central nervous system is higher in the former than in the latter fraction and, furthermore, that the highest specific activities of soluble proteins were found in the brainstem and medulla and were secondary to those in the cerebellum. In these areas the maximum values for the incorporation were 30–50 % higher than in cerebral cortex and the basal ganglia. The rate of elimination of the proteins of the cerebral cortex was higher than in any other area of the brain, followed by the proteins of the basal ganglia, cerebellum, and pons. Also, a variation was found in the biological half-lives. By calculation of the successive slopes, at different times after the injection of radioactive amino acids, changes in the half-lives from the third to the twelfth day after injection were elucidated. Within the first 2–3 days the half-lives were found to be highest in cerebellum and lowest in cerebral cortex; the opposite was the case between the sixth and twelfth days. Similar results have been obtained by others.[106–111]

2. Structural Proteins

The lack of extracellular space—the high density of myelin in the CNS— causes this organ to be rich in particulate fractions. The particulate fractions are built up in agreement with the unit membrane structure,[112] and they contain lipoproteins, proteolipids, mucoproteins, glycoproteins, and particulate-bound enzymes more or less linked to the membrane structure. Therefore, they are characterized by a low solubility in water. Biomembrane protein, however, may be solubilized either by the deoxycholate method of Green[113] or by extraction with detergents, such as Triton-X-100.[114] The individual components of particulate fractions may be difficult to define, as every solubilization procedure causes disruption of the genuine membrane structure. Therefore, only the combination of physicochemical methods with chemical and morphological studies can characterize individual components of the membrane structures. Because of these difficulties, the individual components of CNS particulate fractions show variable composition in relation to the method used for isolation. Thus, for instance, the white matter proteolipid isolated by Folch and Lees[115] shows a variation in lipid content dependent on the method of isolation. The lipid content of the proteolipid contains 35 to 45 % (w/w) lipid when isolated from white matter by means of emulsion centrifugation but 20 % lipid isolated from white matter by dialysis against organic solvents[116] and finally down to 5 % lipid when prepared by silicium acid chromatography.[117] Further chemical treatment of the proteolipid described above has revealed it to be completely excluded by Sephadex G 100, and a major part comes out in the void volume of Sephadex G 200.[118] The main source of the proteolipid seems to be the myelin of the white matter.[119,120] However, in gray matter the content is ten times lower (Table IV).

It is also possible to isolate proteolipid by other than the chloroform method.[115] Gent et al.[121] showed it possible by means of lysolecithin to isolate a homogeneous lipid–protein–lysolecithin fraction from human

TABLE IV

Distribution of Proteolipids in the Human Nervous System[a]

Cerebral white matter		Cerebral cortex	
Corpus callosum	24.0	Frontal	2.6–3.3
Corona radiata		Sensory-motor	4.1–4.7
Anterior	24.0	Parietal, inferior	2.1–2.4
Posterior	24.0	Temporal	2.4–2.9
Cerebellar white matter	21.0	Occipital	5.4–6.0
Optic pathway, mainly chiasm	18.6	Cerebellar cortex	3.6–4.8
		Caudate	5.9
		Pulvinar	8.1
Spinal cord		Thalamus	11.6–14.1
Anterolateral columns	16.0		
Cervical	16.0	Mesencephalon	15.8
Thoracic	12.9–14.8	Pons	18.9
Lumbar and sacral	7.5–10.9	Medulla oblongata	16.5
Posterior columns		Vermis cerebelli	8.6
Cervical	15.3	Anterior spinal roots	3.3–4.5
		Posterior spinal roots	4.5
Thoracic	13.2	Sciatic nerve	1.1
Lumbar and sacral	14.5	Brachial plexus	0.3

[a] Results expressed as milligrams proteolipid protein per gram fresh tissue. From Amaducci.[120]

myelin and white matter. This complex gives rise to a typical turbidity curve as a function of the amount of lysolecithin and only one electrophoretic peak and a single symmetrical peak by ultracentrifugation. It is probable that the protein component solubilized by means of lysolecithin from myelin is identical with the proteolipid material isolated by Folch–Pi.[115] The neurosclerin isolated by Folch and Lees[122] is identical with the proteolipid.[119] A further argument for the presence of proteolipid in the myelin sheath is put forward by Klee and Sokoloff,[123] who demonstrated a high incorporation of amino acids in the myelin sheath and proteolipid during myelination. Although the studies mentioned cannot give any definite answer concerning the genuine composition of the biomembrane constituents, they show the presence in biomembranes of highly organized constituents and the possibility that the myelin fractions, like other particulate fractions of the brain, are composed of lipid–protein units that are removed as such in solubilization by detergents.

The distribution of other proteins characteristic of brain tissue between gray and white matter has been investigated only sparsely. Quamina and Bogoch[124] demonstrated, after extraction of brain matter with 0.32 M sucrose $+ 10^{-5}$ M calcium chloride by gradient ultracentrifugation, a higher content of the so-called cerebroproteins in gray matter nuclear fractions than in whole brain nuclear fractions, although the opposite was the case for the mitochondrial fractions. The identity of these fractions with those described above has not been proven. Several other antigenic constituents of the myelin sheath have been found.[125] Among these a basic protein possessing a molecular weight of 10,000 has been noted.[126,127] The similarities and differences between the gray and white matter content of particulate-bound enzymes seem directly related to the function of these two structures. As mentioned, the extracellular space of brain seems of minor importance and is substituted for by the glial system.[54] The particulate-bound enzymes of the plasma membrane of the glial cells seem thus to possess a biological function different from that normally encountered for surface membranes. The plasma membrane of the glial cells, i.e., the stabilized cells around the neurons, seems to act as a sort of mitochondrion possessing high cytochrome oxidase and monoamino oxidase activity.[128] Apart from the histochemical reviews by Colmant[129] and Friede,[130] at present no comprehensive description of the gray–white matter differences with regard to structural (and soluble) enzymes of the brain has been given. The reader is referred to the paragraphs covering the respective constituents of biomembranes. The quantitative topographic distribution of individual soluble and particulate enzymes in CNS has been elucidated by various authors.[131–134]

B. Lipids

In 1884, Thudicum[135] demonstrated that the brain tissue has a distribution of lipids different from all other organs. He isolated a series of new lipids from brain tissue and named them kephalin, phrenosin, sphingomyelin, etc. As a result of the development of the chromatographic methods of the

TABLE V

The Distribution of Cholesterol and Polar Lipids in Total Brain, Gray

Method	Tissue	Chol	Ce	Sulf	Pe	Pc
Microchemical	G	118.8	22.6		41.5	11.4
Determination	W	26.9	31.1		18.5	6.0
Determination	G	0.85	1.1		1.9	1.2
Determination	W	4.1	5.0		4.5	1.6
Determination	G	13.1	9.5		9.7	6.5
Determination	W	18.8	21.4		10.7	4.8
Determination	G	5.6	0.73	0.45	8.5	9.6
Determination	W	15.6	13.8	1.7	15.2	8.9
Chromatography + microchemical determination	T		16.8	3.5	17.0	11.0
Microchemical	G				136	7.9
Determination	W				178	4.1
Determination	W		19.4	5.3		
Chromatography + gravimetry	G		1.91	0.41	9.61	8.99
	W		10.67	3.95	12.11	8.84
Quantitative	G		12.6		43.1	24.9
TLC	W		33.1	5.5	24.9	14.7
Chromatography	G	14.7	8.7	2.9	9.3	8.7
+ microchemical determination	W	16.2	12.6	1.1	11.4	9.5
Quantitative						
TLC	Oligodendrogliomas		23.5		30.3	19.3
	Astrocytomas		12.6		36.1	24.8
	Glioblastomas		8.6		31.2	32.2
Chromatography + microchemical determination	LM	46.2	16.7	3.1	12.6	7.2
Chromatography + microchemical determination	W	41.6	19.0	3.9	10.9	8.7
Chromatography + microchemical determination	M	1.0		0.48	0.29	0.29
Chromatography + microchemical determination	M	21.4	13.5	5.0		
Chromatography + microchemical determination	W	20.0	13.6	5.5		
Chromatography + microchemical determination	M	1	0.31	0.10		
Chromatography + microchemical determination	W	1	0.33	0.11		
Chromatography + microchemical determination	M				0.42	0.29
Chromatography + microchemical determination	W				0.44	0.32
Chromatography + microchemical determination	M	1.3	0.73		0.432	0.187

[a] Abbreviations: G, gray matter; W, white matter; T, total brain; LM, light myelin; M, myelin; Chol, cholesterol; Ce, cerebroside; Sulf, sulfatide; Pe, phosphatidyl ethanolamine; Pc, phosphatidyl choline; Sph, sphingomyelin; Ps, phosphatidyl serin; and Pi, phosphatidyl inositides.

Matter, White Matter, and Myelin[a]

Sph	Ps	Pi	Unit	Reference	Notes
5.7			% lipid	Johnson[136]	
17.5			% lipid	adults	
0.35			% wet weight	Brante[28]	
1.2			% wet weight	adults	
5.5			% dry weight	Cumings[137]	
7.8			% dry weight	adults	
1.7			% dry weight	Svennerholm[139]	Age variations indicated
4.8			% dry weight	5 yr	
7.9	7.2	3.1	% total lipid	Rouser[138]	
25			% total lipid	adults	
52			mmoles/kg dry weight	Nayyar[29] (adults)	
10.4			% total lipid	Jatzkewitz[169]	
2.77	2.68		% dry weight	O'Brien[143]	
4.93	5.13		% dry weight	adults	
9.9	12.5		% total polar lipids	Christensen Lou[39]	
7.4	12.1		% total polar lipids	adults	
4.2	3.2		% dry weight	Fillerup and Mead[141]	Age variations indicated
6.8	2.2		28 yr	adults	
10.1	15.9		% total polar lipids	Christensen Lou and	
12.0	14.3		% total polar lipids	Clausen[39,40]	
6.2	6.7	0.6	Mole % total lipids	⌠Norton and	Bovine brain
6.1	8.3	0.8	Mole % total lipids	⌡ Autilio[142]	
0.11	0.11		Mole/mole cholesterol	O'Brien[143]	Human brain
6.2			% total lipid	Pilz and Mehl[144]	Human brain
7.0			Mole/mole cholesterol	Pilz and Mehl[144]	
0.15			Mole/mole cholesterol	Pilz and Mehl[144]	
0.17			Mole/mole cholesterol	Pilz and Mehl[144]	
0.25	0.29		mg P/mg dry weight	Soto[145]	Bovine brain
0.24	0.25		mg P/mg dry weight	Soto[145]	
0.172	0.176		Mole/mole lipid P	Horrocks[146]	Monkey brain

1930's, the differences of lipid compositions of gray and white matter were extensively investigated (Table V). The quantitative distribution of individual lipid components, through recent years, has been referred to on different bases: percentage of wet weight, percentage of dry weight, percentage of total lipid content, or the molar ratios, e.g., moles per mole cholesterol or per mole phospholipid. In brain tissue, white matter shows a relatively low content of choline phosphatides but a high content of cholesterol and ethanolamine phosphatides, while the lipid composition of myelin is nearly identical with that of white matter (Table V). The determination of ethanolamine phosphatides in gray matter seems to have given some different results for various investigators. The highest values have been obtained for these fractions in studies where the lipid extract was found by pure chemical procedures.[135, 137,29] However, by application of column chromatographic methods, lower values of phosphatidylethanolamine have been demonstrated.[140] The reasons for these discrepancies seem to originate from Rouser[147] describing a relatively unknown fraction being eluted just after phosphatidylethanolamine (also containing phosphatidylserine fraction). The unknown fraction seemed to show characteristics similar to those of phosphatidylethanolamine; thus, it was found to be ninhydrin positive. This fraction probably consists of oxidative decomposition products. The phosphatidylethanolamine fraction in the gray substance seems especially prone to such changes because of the high content of polyunsaturated fatty acids therein.[148,149]

Among the glycolipids, the gray matter is rich in gangliosides.[150,151] On the other hand, the cerebrosides and sulfatides seem predominant in white matter, especially in the myelin fraction (Table V).

The content of lipids in white matter increases significantly during fetal development. These changes are related to myelination. As myelin is rich in glycolipids and ethanolamine phosphatides, the constituents increase in amount.[139,152,153] However, age-dependent changes in adults[141] have not been found in either white or gray matter from the ages 26 to 80.

Topographic differences in the lipid composition between gray and white matter are related to differences in the distribution of axons and glial elements between the two substances. On the basis of studies of glial tumors, a relative preponderance of glycolipids has been found in oligodendroglia cells from oligodendroglioma.[39,40] The lipid composition of oligodendrogliomas seems nearly identical with the distribution found in the myelin fraction.[39, 40,142–146] The relatively high content of phosphatidyl inositides in white matter (Table III) must be related to the intimate linkage of this polar lipid to the structural proteins of the myelin (e.g. to proteolipid).[118,142,145,146]

The differences in the distribution of fatty acids in brain tissue are related to the differences in the distribution of the lipids. Thus, the gray matter contains more polyunsaturated fatty acids than white matter. On the other hand, the white matter and especially the myelin fraction contain a high percentage of long-chain fatty acids and hydroxy derivatives of fatty acids, explained by the relatively high percentage of cerebrosides and sulfatides in white matter myelin (Table VI).[139,148,149,154–168] The different chain length of the fatty acid causes the main polar lipid classes to appear as two or more fractions in thin-layer chromatography.[169] The prostaglandins, derived from ω-6-polyunsaturated C_{20}-fatty acids, also have been found in brain tissue; as they preferentially occur in synaptic vesicles, they may be implied in the transmitter function.[170]

C. Carbohydrates

The distribution of monosaccharides in brain tissue, especially that of glucose in different topographic areas of the brain, is difficult to estimate because of the rapid glycolysis occurring in the tissue. Thus, by deep freezing whole brain, King[90] found 2.88 mmoles glucose/g tissue in outer parts of brain, but 2.47 mmoles/kg in inner parts. Therefore, the differences in concentration of monosaccharides between gray and white matter cannot be deduced from such experiments (cf. p. 281).

The distribution of polysaccharides and acid mucopolysaccharides in brain matter has been studied in greater detail by several authors. Some of these substances show a lower turnover rate than glucose.

Only small amounts of glycogen have been demonstrated in brain tissue.[171] The turnover rate of this compound is high in brain tissue.

Until recently the presence in CNS of acid mucopolysaccharides [heparin, hyaluronic acid, chondroitin 4- and 6-sulfate (formerly called chondroitin sulfate A and C, respectively), dermatan sulfate (formerly called chondroitin sulfate B), and keratosulfate 1 and 2] has been doubted. However, in 1957 Brante[172] suggested the presence of acid mucopolysaccharides in the CNS. This was later supported by Young and Abood,[173] who by means of histochemical and paperchromatographic studies gave evidence for the presence of hyaluronic acid and chondroitin 4- and 6-sulfate. Chemical studies, using electrophoresis, IR spectroscopy, and quantitative estimation of acid mucopolysaccharides, added further evidence that acid mucopolysaccharides are present in brain tissue.[173–177]

Only a few attempts have been made to relate the presence of acid mucopolysaccharides to topographic differences. However, Abood and Abdul-Haj[174] demonstrated, by means of microscopic studies, the presence of hyaluronic acid in the peripheral nerve sheath. Indirectly, studies of Margolis[178] demonstrated the presence of acid mucopolysaccharides both in whole brain, in white matter, and in the myelin. Here both hyaluronic acid and chondroitin sulfate A and C were found. It was demonstrated that hyaluronic acid occurred in equal concentrations, in both whole brain and white matter. The total concentration of acid mucopolysaccharides of the whole brain was found to be about 0.03%, with one-half that amount in white matter.[178,179] However, on a lipid-free dry-weight basis the concentration is identical (0.27% in white matter and whole brain).

However, other polysaccharides also have been described in brain tissue. Brunngraber[114] described the presence of sialo-mucopolysaccharides of brain tissue, and Stary et al.[180] described the presence of mucopolysaccharide components of human brain-containing arabinose or xylose probably present in acid polysaccharides. The distribution of the arabinose and

TABLE VI

Fatty acids	Gray matter					White matter					Total brain			
	Serine glycerophosphatides[148]	Choline glycerophosphatides[148]	Ethanolamine glycerophosphatides[148]	Cerebroside sulfate[140]	Cerebroside[163]	Serine glycerophosphatides[148]	Choline glycerophosphatides[148]	Ethanolamine glycerophosphatides[148]	Cerebroside sulfate[140]	Cerebroside[163]	Triphosphoinositide[162]	Diphosphoinositide[162]	Phosphatidyl-myoinositol[162]	Normal sphingomyelin[139]
$C_{13:0}$	0.3^a	2.9^a	0.2^a	Tr	—	0.3^a	1.3^a	0.5^a	—	Tr				
$C_{14:0}$				1.6	0.5				2.4	6.1				
$C_{15:0}$				0.6	Tr				1.1	0.2				
$C_{15:1}$				Tr	Tr				1.1	Tr				
$C_{16:0}$	2.3	45.0	7.0	9.6	2.0	1.7	34.3	6.7	6.3	7.9	8.0	6.7	8.9	2.3
$C_{16:1}$	0.3	3.1	0.4	Tr	0.2	0.4	1.0	1.4	Tr	2.0	0.8	0.5	2.3	
$C_{17:0}$				1.6	0.4				Tr	1.8	1.3	1.2	2.6	0.2
$C_{18:0}$	25.4	9.3	27.7	3.6	1.7	35.8	13.4	9.0	2.3	2.1	39.8	35.9	47.7	38.8
$C_{18:1}$	21.5	31.4	12.3	3.8	0.3	37.7	45.2	42.4	1.1	Tr	18.1	14.7	8.4	
$C_{18:2}$	Tr	0.4	Tr			0.3	0.4	Tr						
$C_{18:3}$	0.6	—	—			0.5	—	Tr						
$C_{19:0}$				Tr	Tr				Tr	1.1				0.3
$C_{20:0}$	1.0	0.7	1.5	3.3	0.6	5.3	1.1	7.9	1.6	1.3	1.0	0.8	0.6	1.6
$C_{20:1}$	Tr	—	Tr			1.4	—	2.4				1.0	0.6	
$C_{20:2}$	0.7	Tr	0.5			0.6	—	1.6						
$C_{20:3}$	1.6	4.1	14.3			2.0	1.3	6.4			3.2	3.4	1.9	
$C_{20:4}$	—	—	—			0.7	—	0.4			21.9	26.4	21.8	
$C_{20:5}$														
$C_{21:0}$				1.6	Tr				Tr	Tr				0.3
$C_{21:1}$				0.5	—				Tr	Tr				

TABLE VI (cont.)

Fatty acids	Gray matter					White matter					Total brain			
	Serine glycerophosphatides[148]	Choline glycerophosphatides[148]	Ethanolamine glycerophosphatides[148]	Cerebroside-sulfate[140]	Cerebroside[163]	Serine glycerophosphatides[148]	Choline glycerophosphatides[148]	Ethanolamine glycerophosphatides[148]	Cerebroside sulfate[140]	Cerebroside[163]	Triphosphoinositide[162]	Diphosphoinositide[162]	Phosphatidyl-myoinositol[162]	Normal sphingomyelin[139]
$C_{22:0}$	5.0			12.3	8.5	0.6			13.2	8.9	3.6	5.0	2.1	1.6
C_{22} un [b]	3.3	—	2.4			4.2	—	1.3						
C_{22} un [c]		—	12.0				0.3	13.7						
$C_{22:1}$				0.5	Tr				0.2	Tr	2.3	4.2		0.6
$C_{22:4}$ [d]													0.8	
$C_{22:5}$ [d]													1.4	
$C_{22:5}$	—		—			0.9	—	0.5						
$C_{22:6}$	36.6	3.1	25.2	12.0	15.3	5.6	0.1	3.4	12.2	11.8				
$C_{23:0}$				0.3	Tr				Tr	Tr				2.1
$C_{23:1}$														1.0
$C_{24:0}$				21.7	27.3				41.2	29.9				4.3
$C_{24:1}$				15.0	26.4				8.4	15.8				31.7
$C_{25:0}$				3.1	5.1				2.0	4.3				3.7
$C_{25:1}$				4.6	6.8				5.4	3.2				6.4
$C_{26:0}$				Tr	Tr				Tr	Tr				0.2
$C_{26:1}$				3.4	6.0				2.0	4.0				4.9

[a] Analyses made in USC laboratory on one subject (55 yr).
[b] Possibly $C_{22:3}$, but identified by carbon number only.
[c] Possibly $C_{22:5}$ or $C_{22:6}$.
[d] Tentative assignment.

xylose-containing acid mucopolysaccharides seems to differ: They both contain a polypeptide group linked to the carbohydrate moiety, and they both contain more hexuronic acid than hexosamine. A detailed account for the distribution between gray and white matter has not been made.

D. Nucleic Acids and Nucleoproteins

As already mentioned, DNA occurs equally distributed between gray and white matter, thus indicating nearly an equal amount of cells in white and gray matter. However, Landolt, Hess, and Thalheimer[181] demonstrated by means of a microchemical technique[182] topographic differences within gray and white matter; thus, the lowest content of RNA (micrograms per cell unit) was found in the subcortical white matter, basis pedunculi, and corpus callosum, but the highest was found in the frontal cortex and subfornical organ. Similarly, the content of DNA was found to be slightly lower in gray matter (1.75 ± 0.1 μg/mg dry wt.) than in the frontal cortex (2.4 ± 0.11 μg/mg dry wt.). The proportion between RNA and DNA was found highest (2.17) in the frontal cortex and lowest in corpus callosum (white matter).

The nucleic acid occurs in high concentration in the Nissl substance and nucleolus of the nerve cell. The ergastroplasmic granules are considered to be ribonucleic protein particles in the cytoplasm of oligodendrocytes. The glia cells contain a high content of cytosine and the neurons, a relatively high percentage of guanine. Studies of the turnover rate of constituents of nucleic acids have revealed that pentose nucleic acid turns over rapidly in the nucleus of the neurons but at lower rates in the cytoplasm of the neurons. Furthermore, a rapid turnover of pentose nucleic acid has been found in oligodendrocytes, indicating a high uptake of precursors into pentose nucleic acid of white matter.[183]

V. REFERENCES

1. S. W. Ranson and S. L. Clark, *The Anatomy of the Nervous System. Its Development and Function*, Saunders, Philadelphia (1959).
2. K. A. C. Elliott, *in The Biology of Myelin* (S. R. Korey, ed.), pp. 230–236, Hoeber–Harper Book, New York (1959).
3. W. J. Schmidt, Doppelbrechung und Feinbau der Markscheide der Nervenfasern. *Z. Zellforsch. Mikroskop. Anat.* **23**:657 (1936).
4. F. O. Schmitt, R. S. Bear, and K. J. Palmer, X-ray diffraction studies on the structure of the nerve myelin sheath, *J. Cellular Comp. Physiol.* **18**:31–42 (1941).
5. H. Fernandez–Moran, Sheath and axon structures in the internode portion of vertebrate myelinated nerve fibres; electron microscope study of rat and frog sciatic nerves, *Exptl. Cell Res.* **1**:309–337 (1950).
6. H. Fernandez-Moran, *in Congreso Latino-Americano de Neurocirurgia VI*, Montevideo (1955); *Acta y Trabajos*, p. 599, Montevideo, Imprenta Rasqual H. Rosillo (1855).
7. B. B. Geren and J. Raskind, Development of the fine structure of the myelin sheath in sciatic nerves of chick embryos, *Proc. Natl. Acad. Sci.* **39**:880–884 (1953).

8. F. S. Sjöstrand, Lamellated structure of the nerve myelin sheath as revealed by high resolution electron microscopy, *Experientia* **9**:68 (1953).

9. S. A. Luse, Formation of myelin in the central nervous system of mice and rats as studied with the electron microscope, *J. Biophys. Biochem. Cytol.* **2**:777–784 (1956).

10. B. B. Geren, Formation from the Schwann cell of myelin in the peripheral nerves of chick embryos, *Exptl. Cell Res.* **7**:558–562 (1954).

11. E. DeRobertis, Morphogenesis of the retinal rods: An electron microscope study, *J. Biophys. Biochem. Cytol.* **2**, Suppl. 209 (1956).

12. S. A. Luse, in *The Biology of Myelin* (S. R. Korey, ed.), Hoeber–Harper Book, New York (1959).

13. C. P. Wendell-Smith, M. J. Blunt, F. Baldwin, and P. B. Paisley, Neurone-satellite cell relationship, *Nature* **205**:781–782 (1965).

14. H. S. Gasser and J. Erlanger, The role played by the size of the constituent fibers of a nerve trunk in determining the form of its action potential wave, *Am. J. Physiol.* **80**:522–547 (1927).

15. H. S. Gasser and J. Erlanger, Electrical signs of nervous action, University of Pennsylvania Press, Philadelphia (1937).

16. G. H. Bishop and P. Heinbecker, Differentiation of axon types in visceral nerve by means of the potential record, *Am. J. Physiol.* **94**:170–200 (1930).

17. F. Buchthal and A. Rosenfalck, Evoked action potential and conduction velocity in human sensory nerves, *Brain Res.* **3**:1–122 (1966).

18. D. Duncan, A relation between axon diameter and myelination determined by measurement of myelinate spinal root fibers, *J. Comp. Neurol.* **60**:437–472 (1934).

19. R. L. Friede and T. Samorajski, Relation between the number of myelin lamellae and axon circumference in fibers of vagus and sciatic nerves of mice, *J. Comp. Neurol.* **130**:223–232 (1967).

20. H. Hyden, S. Løvtrup, and A. Pigón, Cytochrome oxidase and succinoxidase activities in spinal ganglion cells and in glial capsule cells, *J. Neurochem.* **2**:304–311 (1958).

21. H. Hyden, in *Neurochemistry* (K. A. C. Elliott, J. H. Page, and J. H. Quastel, eds.), 2nd ed., C. C. Thomas, Springfield, Illinois (1962).

22. H. Hyden and P. W. Lange, A kinetic study of the neurone–glia relationship, *J. Cell Biol.* **13**:233–237 (1962).

23. O. H. Lowry, N. R. Roberts, and C. J. Lewis, The analysis of single cells, *J. Biol. Chem.* **222**:97–107 (1956).

24. O. H. Lowry, N. R. Roberts, and C. J. Lewis, Quantitative histochemistry of the retina, *J. Biol. Chem.* **220**:879–92, (1956).

25. C. E. Lumsden, in *Pathology of Tumors of the Nervous System* (D. S. Russell, L. J. Rubinstein, and C. E. Lumsden, eds.), p. 272, E. Arnold, London (1959).

26. S. P. R. Rose, Preparation of enriched fractions from cerebral cortex containing isolated metabolically active neuronal cells, *Nature* **206**:621–22 (1965).

27. S. P. R. Rose, Preparation of enriched fractions from cerebral cortex containing isolated, metabolically active neuronal and glial cells, *Biochem. J.* **102**:33–43 (1967).

28. G. Brante, Studies on lipids in the nervous system with special reference to quantitative chemical determination and topical distribution, *Acta Physiol. Scand.* (Suppl. 63) **18**:1–189 (1949).

29. S. N. Nayyar, R. E. McCaman, and R. F. Heimburger, Phospholipid composition of human brain tumors, *Federation Proc.* **19**:232 (1960).

30. K. Gopal, E. Grossi, P. Paoletti, and M. Usardi, Lipid composition of human intracranial tumors: A biochemical study, *Acta Neurochir. (Wien)* **11**:333–347 (1963).

31. H. M. Rapport, L. Graf, and J. Yariv, Immunochemical studies of organ and tumor lipids, *Arch. Biochem. Biophys.* **92**:438 (1961).

32. M. M. Rapport, L. Graf, and N. F. Alonzo, Comparison of human tumor and ox spleen cytosides, *Federation Proc.* **18**:307 (1959).
33. T. Kosaki, T. Ikoda, Y. Kotani, S. Nakagawa, and T. Saka, A new phospholipid, malignolipin in human malignant tumors, *Science* **127**:1176–1177 (1958).
34. D. P. Coman, Decreased mutual adhesiveness, a property of cells from squamous cell carcinomas, *Cancer Res.* **4**:625–629 (1944).
35. M. Abercrombie and E. J. Ambrose, Interference microscope studies of cell contacts in tissue culture, *Exptl. Cell Res.* **15**:332–345 (1958).
36. E. J. Ambrose, A. M. James, and J. H. B. Lowick, Differences between the electrical charge carried by normal and homologous tumor cells, *Nature (London)* **177**:576–577 (1956).
37. L. Purdom, E. J. Ambrose, and G. Klein, A correlation between electrical surface charge and some biological characteristics during the stepwise progression of a mouse sarcoma, *Nature (London)* **181**:1586–1587 (1958).
38. E. J. Ambrose, *Henry Ford International Symposium: Biological Interactions in Normal and Neoplastic Growth* (J. M. Brennan and W. L. Simpson, eds.), Churchill, London (1962).
39. H. O. Christensen Lou, J. Clausen, and F. Bierring, Phospholipids and glycolipids in the central nervous system, *J. Neurochem.* **12**:619–627 (1965).
40. H. O. Christensen Lou and J. Clausen, Polar lipids of oligodendrogliomas, *J. Neurochem.* **15**:263–264 (1968).
41. E. DeRobertis, A. Pellegrino de Iraldi, G. Rodriguez de Lores Arnaiz, and L. Salganicoff, Cholinergic and noncholinergic nerve endings in rat brain–I, *J. Neurochem.* **9**:23–35 (1962).
42. V. P. Whittaker, The isolation and characterization of acetylcholine containing particles from brain, *Biochem. J.* **72**:694–706 (1959).
43. L. A. Autilio, W. T. Norton, and R. D. Terry, The preparation and some properties of purified myelin from the central nervous system, *J. Neurochem.* **11**:17–27 (1964).
44. E. A. Bering, Water exchange of central nervous system and cerebrospinal fluid, *J. Neurosurg.* **9**:275 (1952).
45. H. M. Pappius, *in Biology of Neuroglia* (E. D. P. DeRobertis and R. Cawca, eds.), pp. 135–154, Elsevier, Amsterdam (1965).
46. H. McIlwain, *Chemical Exploration of the Brain*, Elsevier, Amsterdam (1963).
47. R. J. Rossiter, *in Neurochemistry* (K. A. C. Elliott, I. H. Page, and J. H. Quastel, eds.), pp. 40 and 42, C. C. Thomas, Springfield, Illinois (1962).
48. R. Katzman, Electrolyte distribution in mammalian central nervous system, *Neurology* **11**:27–36 (1961).
49. H. F. Bradford and S. P. R. Rose, Ionic accumulation and membrane properties of enriched preparation of neurons and glia from mammalian cerebral cortex, *J. Neurochem.* **14**:373–375 (1967).
50. G. Levi, R. Blasberg, and A. Lajtha, Substrate specificity of cerebral amino acid exit in vitro, *Arch. Biochem. Biophys.* **114**:339–351 (1966).
51. S. Furst, A. Lajtha, and H. Waelsch, Amino acid and protein metabolism of the brain— III. *J. Neurochem.* **2**:216–225 (1958).
52. F. T. Mérei and F. Gallyas, Quantitative determination of (^{35}S) methionine incorporated into proteins of cell groups or nuclei of the central nervous system, *J. Neurochem.* **11**:251–256 (1964).
53. J. Altman, Regional utilization of leucine-H^3 by normal rat brain, *J. Histochem. Cytochem.* **11**:741–750 (1963).
54. J. L. Fox, Development of recent thoughts on intracranial pressure and the blood-brain barrier, *J. Neurosurg.* **21**:909–967 (1964).

55. S. Berl and H. Waelsch, Determination of glutamic acid, glutamine, glutathione, and γ-aminobuturic acid and their distribution in brain tissue, *J. Neurochem.* 3:161–169 (1958).

56. I. P. Lowe, E. Robins, and G. S. Eyerman, The fluorimetric measurement of glutamic decarboxylase and its distribution in brain, *J. Neurochem.* 3:8–18 (1958).

57. R. A. Salvador and R. W. Albers, The distribution of glutamic-γ-aminobutyric transaminase in the nervous system of the rhesus monkey, *J. Biol. Chem.* 234:922–925 (1959).

58. R. W. Albers and R. O. Brady, The distribution of glutamic decarboxylase in the nervous system of the rhesus monkey, *J. Biol. Chem.* 234:926–928 (1959).

59. A. Lajtha and P. J. Mela, The brain barrier system—I. The exchange of free amino acids between plasma and brain, *J. Neurochem.* 7:210–217 (1961).

60. L. T. Graham, Jr., R. P. Shank, R. Werman, and M. H. Aprison, Distribution of some synaptic transmitter suspects in cat spinal cord: Glutamic acid, aspartic acid, γ-aminobutyric acid, glycine and glutamine, *J. Neurochem.* 14:465–472 (1967).

61. P. B. Müller and H. Langemann, Distribution of glutamic acid decarboxylase activity in human brain, *J. Neurochem.* 9:399–401 (1962).

62. L. Salganicoff and E. DeRobertis, Subcellular distribution of the enzymes of the glutamic acid, glutamine and γ-aminobutyric acid cycles in rat brain, *J. Neurochem.* 12:287–309 (1965).

63. G. Levi and A. Lajtha, The pattern of mammalian brain gangliosides—II, *J. Neurochem.* 12:629–638 (1965).

64. G. Levi, A. Cherayil, and A. Lajtha, Cerebral amino acid transport in vitro—III. *J. Neurochem.* 12:757–770 (1965).

65. N. Popov, W. Pohle, V. Rösler, and H. Matthies, Regionale Verteilung von γ-aminobuttersäure, Glutaminsäure, Asparaginsäure, Dopamin, Noradrenalin und Serotonin im Rattenhirn, *Acta Biol. Med. Ger.* 18:695–702 (1967).

66. R. J. Haslam and H. A. Krebs, The metabolism of glutamate in homogenates and slices of brain cortex, *Biochem. J.* 88:566–578 (1963).

67. D. Hathway, E. Mallinson, and D. A. Akintonwa, Effects of dieldrin, picrotoxin and telodrin on the metabolism of ammonia in brain, *Biochem. J.* 94:676–686 (1965).

68. H. Weil-Malherbe, in *Neurochemistry* (K. A. Elliot, I. H. Page, J. H. Quastel, eds.), pp. 321–330, C. C. Thomas, Springfield, Illinois (1962).

69. C. G. Honegger and R. Honegger, Votatile amines in brain, *Nature* 185:530–532 (1960).

70. T. L. Perry, S. Hansen, and L. MacDougal, Amines of human whole brain, *J. Neurochem.* 14:775–782 (1967).

71. E. DeRobertis, Adrenergic endings and vesicles isolated from brain, *Pharmacol. Rev.* 18:413–424 (1966).

72. A. Carlsson, B. Falch, and N. A. Hillarp, Cellular localization of brain monoamines, *Acta Physiol. Scand.* (*Suppl.* 196) 56:1–28 (1962).

73. A. Dahlström and K. Fuxe, Evidence for the existence of monoamine-containing neurons in the central nervous system, *Acta Physiol. Scand.* (Suppl. 232) 62:1–55 (1964).

74. E. DeRobertis, Ultrastructure and cytochemistry of the synaptic region, *Science* 156:907–914 (1967).

75. Y. Gutman and H. Weil-Malherbe, The intracellular distribution of brain catecholamines, *J. Neurochem.* 14:619–625 (1967).

76. J. Glowinski and L. L. Iversen, Regional studies of catecholamines in the rat brain—I. *J. Neurochem.* 13:655–669 (1966).

77. L. L. Iversen and J. Glowinski, Regional differences in the rate of turnover of norepinephrine in the rat brain, *Nature* 210:1006–1008 (1966).

78. L. L. Iversen and J. Glowinski, Regional studies of catecholamines in the rat brain—II, *J. Neurochem.* 13:671–82 (1966).

79. G. G. Gottfries, A. M. Rosengren, and E. Rosengren, The occurrence of homovanillic acid in human brain, *Acta Pharmacol. Toxicol.* **23**:36–40 (1965).
80. S. L. Manocha and G. H. Bourne, Histochemical mapping of monoamine oxidase and lactic acid dehydrogenase in the pons and mesencephalon of squirrel monkey, *J. Neurochem.* **13**:1047–1056 (1966).
81. W. Y. Cheung and L. Salganicoff, Cyclic 3′,5′-nucleotide phosphodiesterase: Localization and latent activity in rat brain, *Nature* **214**:90–91 (1967).
82. B. Formby and J. Clausen, Topographic studies of noradrenaline transmitter function, *Proc. 1st Intern. Congr. Neurochem., Strasbourg* (1967).
83. J. Clausen and B. Formby, Effect of noradrenaline on phosphatase activity in synaptic membrane of the rat brain, *Nature* **213**:389–390 (1867).
84. B. Formby and J. Clausen, Phosphatase activity related to synaptic transmitter function of noradrenaline in the central nervous system. In vitro studies, *Z. Physiol. Chem.* **349**:349–356 (1968); **349**:909–919 (1968).
85. J. H. Quastel, in *Neurochemistry* (K. A. C. Elliott, J. H. Page, and J. H. Quastel, eds.), C. C. Thomas, Springfield, Illinois (1962).
86. E. Giacobini, in *Morphological and Biochemical Correlates of Neural Activity* (M. M. Cohen and R. S. Snider, eds.), Hoeber–Harper Book, New York (1964).
87. O. H. Lowry, N. R. Roberts, K. Y. Lerner, M. L. Wu, and A. L. Farr, Quantitative histochemistry of brain: I. Chemical methods, *J. Biol. Chem.* **207**:1–17 (1959).
88. O. H. Lowry, N. R. Roberts, K. Y. Lerner, M. L. Wu, and A. L. Farr, Quantitative histochemistry of brain—III. Ammon's Horn, *J. Biol. Chem.* **207**:39–49 (1954).
89. O. H. Lowry, N. R. Roberts, and I. F. Kapphahn, Fluorimetric measurement of pyridine nucleotides, *J. Biol. Chem.* **224**:1047 (1957).
90. L. J. King, G. M. Schoepfle, O. H. Lowry, J. V. Passonneau, and S. Wilson, Effects of electrical stimulation on metabolites in brain of decapitated mice, *J. Neurochem.* **14**:613–618 (1967).
91. N. D. Goldberg, J. V. Passonneau, and O. H. Lowry, Effects of changes in brain metabolism in the levels of citric acid cycle intermediates, *J. Biol. Chem.* **241**:3997–4003 (1966).
92. L. J. King, O. H. Lowry, J. V. Passonneau, and V. Venson, Effects of convulsants on energy reserves in the cerebral cortex, *J. Neurochem.* **14**:599–611 (1967).
93. C. Chatagnon and P. Chatagnon, Étude des protéines cérébrales solubles de l'humain, *Ann. Biol. Clin.* **18**:427–437 (1960).
94. K. Kiyota, Electrophoretic protein fractions and the hydrophilic property of brain tissue—I, II, *J. Neurochem.* **4**:202–216 (1959).
95. D. Karcher, M. van Sande, and A. Lowenthal, Micro-electrophoresis in agar gel of proteins of the cerebrospinal fluid and central nervous system, *J. Neurochem.* **4**:135–140 (1959).
96. W. Gerhardt-Hansen and J. Clausen, Electrophoresis and immunoelectrophoresis of extractable proteins in brain tissue, *Danish Med. Bull.* **9**:9–13 (1962).
97. B. W. Moore, A Soluble protein characteristic of the nervous system, *Biochem. Biophys. Res. Commun.* **19**:739–744 (1965).
98. E. C. Laterre, J. F. Heremans, and A. Carbonara, Extracts from brain and kidney, *Clin. Chim. Acta* **10**:197–209 (1964).
99. B. W. Moore and D. McGregor, Chromatographic and electrophoretic fractionation of soluble proteins of brain and liver, *J. Biol. Chem.* **240**:1647–1653 (1965).
100. E. Kosinski and P. Grabar, Immunochemical studies of rat brain, *J. Neurochem.* **14**:273–281 (1967).
101. K. Warecka and H. Bauer, Studies on "brain specific" proteins in aqueous extracts of brain tissue, *J. Neurochem.* **14**:783–787 (1967).

102. J. Klatzo, J. Miguel, and R. Otenasek, The application of fluorescein labelled serum proteins (FLSP) to the study of vascular permeability in the brain, *Acta Neuropathol.* 2:144–160 (1962).
103. G. M. Hochwald, G. J. Thorbecke, and R. Asofsky, Sites of formation of immune globulins and of a component of C'3, *J. Exptl. Med.* 114:459–470 (1961).
104. D. M. Robertson, The paper electrophoretic distribution of soluble proteins in different regions of human brain, *J. Neurochem.* 5:145–149 (1960).
105. R. S. Piha, R. M. Bergstrøm, L. Bergstrøm, A. J. Uusitalo, and S. S. Oja, Studies in the metabolism of brain protein—I, *Ann. Med. Exptl. Biol. Fenniae* 41:485–497 (1963).
106. A. V. Palladin and N. Vertaimer, Protein renewal in the central nervous system in different functional states, *Dokl. Akad. Nauk SSSR* 102:319–321 (1955). [*cf. Chem. Abstr.* 49:14971 (1955).]
107. G. E. Vladimirov and A. P. Urinson, Glycine metabolism in the cerebral tissue of the rat in normal resting and in amytal-induced sleep, *Biochemistry* 22:665–707 (1957).
108. L. F. Pantchenko, *J. Physiol. USSR* 44:243–248 (1958).
109. D. H. Clouet and D. Richter, The incorporation of (^{35}S) labelled methionine into the proteins of the rat brain, *J. Neurochem.* 3:219–229 (1959).
110. D. Richter, M. K. Gaitonde, and P. Cohn, in *Structure and Function of the Cerebral Cortex* (D. B. Tower and J. P. Schadé, eds.), p. 340, Elsevier, Amsterdam (1960).
111. A. V. Palladin, in *Regional Neurochemistry* (S. S. Kety and J. Elkes, eds.), Pergamon Press, Oxford (1961).
112. J. D. Robertson, in *Cellular Membranes in Development* (M. Locke, ed.), pp. 1–81, Academic Press, New York (1964).
113. D. Green and S. Fleicher, The role of lipids in mitochondrial electron transfer and oxidative phosphorylation, *Biochim. Biophys. Acta* 70:554–582 (1963).
114. E. G. Brunngraber, in *Handbook of Neurochemistry* (A. Lajtha, ed.), Vol. I, pp. 223–244, Plenum Press, New York (1969).
115. J. Folch, J. Ascoli, M. Lees, J. A. Meath, and F. N. LeBaron, Preparation of lipid extracts from brain tissue, *J. Biol. Chem.* 191:833–841 (1951).
116. M. B. Lees, S. Carr, and J. Bolch, Purification of bovine brain white matter proteolipids by dialysis in organic solvents, *Biochim. Biophys. Acta* 84:464–466 (1964).
117. M. Matsumoto, R. Matsumoto, and J. Folch-Pi, The chromatographic fractionation of brain white matter proteolipids, *J. Neurochem.* 11:829–838 (1964).
118. D. Tenenbaum and J. Folch-Pi, The preparation and characterization of water-soluble proteolipid protein from bovine brain white matter, *Biochim. Biophys. Acta* 115:141–147 (1966).
119. J. Folch-Pi, in *Brain Lipids and Lipoproteins and Leucodystrophies* (J. Folch-Pi and H. Bauer, eds.), Elsevier, Amsterdam (1967).
120. L. Amaducci, in *Regional Neurochemistry* (S. S. Kety and J. Elkes, eds.), Pergamon Press, London (1961).
121. W. L. G. Gent, N. A. Gregson, D. B. Gammack, and J. H. Raper, The lipid protein unit in myelin, *Nature* 204:553–555 (1964).
122. J. Folch and M. Lees, Proteolipids, a new type of tissue lipoproteins, *J. Biol. Chem.* 191:807–817 (1951).
123. C. B. Klee and L. Sokoloff, Amino acid incorporation into proteolipid of myelin in vitro, *Proc. Natl. Acad. Sci.* 53:1014–1021 (1965).
124. A. Ouamina and S. Bogoch, Subcellular fractionation of glycoproteins and mucoids of human and rat brain, *Protides Biol. Fluids* 13:211–216 (1966).
125. P. J. van Alten and A. LaVelle, Antigenic changes in developing hamster brain using antisera to myelinated and unmyelinated brain, *Exptl. Neurol.* 14:115–133 (1966).
126. J. A. Lowden, M. A. Moscarello, and J. Morecki, *Can. J. Biochem.* 44:567 (1966).

127. S. E. Kornguth, J. W. Anderson, and G. Scott, Temporal relationship between myelino-genesis and the appearance of a basic protein in the spinal cord of the rat, *J. Comp. Neurol.* **127**:1–17 (1966).

128. E. Costa and B. B. Brodie, in *Progress in Brain Research*, Vol. 8, *Biogenic Amines* (H. E. Himwich and W. Himwich eds.), pp. 168–185, Elsevier, Amsterdam (1964).

129. H. J. Colmant, Ergebnisse der Enzymhistochemie am zentralen und peripheren Nerven-system, *Neurol. Psychiat.* **2**:61 (1961).

130. R. L. Friede, *Topographic Brain Chemistry*, Academic Press, New York (1966).

131. J. L. Strominger and O. H. Lowry, The quantitative histochemistry of brain, *J. Biol. Chem.* **213**:635–646 (1955).

132. E. Robins, N. R. Roberts, K. M. Eydt, O. H. Lowry, and D. E. Smith, Microdetermination of α-keto acids with special references to malic-lactic and glutamic dehydrogenases in brain, *J. Biol. Chem.* **218**:897–909 (1956).

133. D. B. McDougal, Jr., D. W. Schulz, J. V. Passonneau, J. R. Clark, M. A. Reynolds, and O. H. Lowry, Quantitative studies of white matter—I. Enzymes involved in glucose-6-phosphate metabolism, *J. Gen. Physiol.* **44**:487–498(1961).

134. D. B. McDougal, Jr., E. M. Jones, and U. I. Sila, Distribution of enzymes of the tricar-boxylic acid cycle in white matter, *Ultrastructure Metab. Nervous Sys.* **40**:182–188 (1962).

135. J. L. W. Thudichum, *A Treatise on the Chemical Constitution of Brain*, Bailliere, Tindall, and Cox, London (1884).

136. A. C. Johnson, A. R. McNabb, and R. J. Rossiter, Lipids of normal brain, *Biochem. J.* **43**:573–577 (1948).

137. J. N. Cumings, *Brain* **76**:553 (1953).

138. G. Rouser, G. Galley, and G. Kritchevsky, Lipid class composition of normal human brain and variations in metachromatic leucodystrophy (Tay-Sachs, Niemann-Pick, chronic Gaucher's and Alzheimers diseases), *J. Am. Oil Chemists' Soc.* **42**:404–410 (1965).

139. L. Svennerholm, in *Brain Lipids and Lipoproteins and the Leucodystrophies* (J. Folch-Pi and H. Bauer, eds.), Elsevier, Amsterdam (1963).

140. J. S. O'Brien, D. L. Fillerup, and J. F. Mead, Brain lipids. I. Quantification and fatty acid composition of cerebroside sulfate in human cerebral gray and white matter, *J. Lipid Res.* **5**:109–116 (1964).

141. D. L. Fillerup and J. F. Mead, The lipids of the ageing human brain, *Lipids* **2**:295–298 (1967).

142. W. T. Norton and L. A. Autilio, The chemical composition of bovine CNS myelin, *Ann. N.Y. Acad. Sci.* **122**:77–85 (1965).

143. J. S. O'Brien, Stability of the myelin membrane, *Science* **147**:1099–1107 (1965).

144. H. Pilz and E. Mehl, Untersuchungen zur Lipoidzusammensetzung der menschlichen myelins, *Z. Physiol. Chem.* **346**:306–309 (1966).

145. E. F. Soto, L. Seminario de Bohner, and M. C. Calvino, Chemical composition of myelin and other subcellular fractions isolated from bovine white matter, *J. Neurochem.* **13**:989–998 (1966).

146. L. A. Horrocks, Composition of myelin from peripheral and central nervous systems of the squirrel monkey, *J. Lipid Res.* **8**:569–576 (1967).

147. G. Rouser, A. J. Baumann, G. Kritchevsky, D. Heller, and J. S. O'Brien, Quantitative chromatographic fractionation of complex lipid mixtures, "brain lipids," *J. Am. Oil Chemists' Soc.* **38**:544 (1961).

148. J. S. O'Brien, D. L. Fillerup, and J. F. Mead, Brain lipids. I. Quantification and fatty acid and fatty aldehyde composition of ethanolamine, choline, and serine glycerophosphatides in human cerebral grey and white matter, *J. Lipid Res.* **5**:329–338 (1964).

149. Y. Kishimoto and N. S. Radin, Isolation and determination methods for brain cerebrosides, hydroxy fatty acids, and unsaturated and saturated fatty acids, *J. Lipid Res.* **1**:72–78 (1959).

150. R. Kuhn and H. Wiegandt, Über ein glucosaminhaltiges Gangliosid, *Z. Naturforsch.* **19b**:80–81 (1964).

151. L. Svennerholm, in *The Amino-Sugars*, Vol. IIA (E. A. Balazs and R. W. Jeanloz, eds.), pp. 381–400, Academic Press, New York (1965).

152. J. Clausen, H. O. Christensen Lou, and H. Andersen, Phospholipid and glycolipid patterns of infant and foetal brain. Thin-layer chromatographic studies, *J. Neurochem.* **12**:599–606 (1965).

153. C. Galli and D. ReCecconi, Lipid changes in rat brain during maturation, *Lipids* **2**:76–82 (1967).

154. G. Blix, Zur kenntnis der schwefelhaltigen Lipoidstoffe des Gehirns über Cerebronschwelferlsäure, *Z. Physiol. Chem.* **219**:82–98 (1933).

155. E. Klenk and P. Böhm, Zur kenntnis der Kaphalinfraktion des Gehirns, *Z. Physiol. Chem.* **288**:98–107 (1951).

156. H. Debuch, Beitrag zur chemischen konstitution der acetalphosphatide und zur Frage des Vorkommens des Colamin-Kephalins im Gehirn, *Z. Physiol. Chem.* **304**:109–137 (1956).

157. H. Jatzkewitz, Zwei typen von Cerebrosid-schwefelsäureestern als sog. "Prälipoide und Speicher Substanzen bei der Leukodystrophie Typ Scholz," *Z. Physiol. Chem.* **311**:279–282 (1958).

158. Y. Kishimoto and N. S. Radin, Structures of the normal unsaturated fatty acids of brain sphingolipids, *J. Lipid Res.* **4**:437–443 (1963).

159. H. Jatzkewitz, Cerebron- und Kerasin-schwefelsäureester als Speichersubstanzen bei der Leukodystrophie, typ Scholz, *Z. Physiol. Chem.* **320**:134–148 (1960).

160. Y. Kishimoto and N. S. Radin, Structures of the ester-linked mono- and diunsaturated fatty acids of pig brain, *J. Lipid Res.* **5**:98–102 (1964).

161. K. Bernhard and P. Lesch, Ein Beitrag zur Fettsäurezusammensetzung der Cerebroside, Sphingomyelin und Lecithine aus menschlichem Hirn, *Helv. Chim. Acta* **46**:1798–1801 (1963).

162. H. S. Hendrickson and C. E. Ballou, Ion exchange chromatography of intact brain phosphinositides on diethylaminoethyl cellulose by gradient salt elution in a mixed solvent system, *J. Biol. Chem.* **239**:1369–1373 (1964).

163. J. S. O'Brien and G. Rouser, The fatty acid composition of brain sphingolipids: sphingomyelin, ceramide, cerebroside and cerebroside sulfate, *J. Lipid Res.* **5**:339–343 (1964).

164. J. S. O'Brien and E. L. Sampson, Fatty acid and fatty aldehyde composition of the major brain lipids in normal human gray matter, white matter and myelin, *J. Lipid Res.* **6**:545–551 (1965).

165. L. F. Eng, B. Gerstl, R. B. Hayman, Y. L. Lee, R. W. Tietsort, and J. K. Smith, The 2-hydroxy fatty acids in white matter of infant and adult brains, *J. Lipid Res.* **6**:135–139 (1965).

166. S. G. Pakkala, D. L. Fillerup, and J. F. Mead, The very long chain fatty acids of human brain sphingolipids, *Lipids* **1**:449–450 (1966).

167. P. Lesch, S. Meier, and K. Bernhard, Die Neutrallipide aus Hirnen von Früh- und Neugeburten, *Helv. Chim. Acta* **49**:1215–1221 (1966).

168. P. Lesch, S. Meier, and K. Bernhard, Zur Kenntnis der Neutrallipide aus Säuglinshirnen, *Helv. Chim. Acta* **50**:207–212 (1966).

169. H. Jatzkewitz, Eine neue Methode zur Quantitativen Ultramikrobestimmung der Sphingolipoide aus Gehirn, *Z. Physiol. Chem.* **336**:25–39 (1964).

170. K. Kataoka, P. W. Ramwell, and S. Jessup, Prostaglandins: Localization in subcellular particles of rat cerebral cortex, *Science* **157**:1187–1189 (1967).

171. S. E. Kerr, C. W. Hampel, and J. J. Ghantus, The carbohydrate metabolism of brain—IV, *J. Biol. Chem.* **119**:405–421 (1937).

172. G. Brante, in *Metabolism of the Nervous System* (D. Richter, ed.), p. 112, Pergamon Press, London (1957).

173. I. J. Young and L. G. Abood, Histological demonstration of hyaluronic acid in the central nervous system, *J. Neurochem.* **6**:89–94 (1960).

174. L. G. Abood and S. K. Abdul-Haj, Histochemistry and characterization of hyaluronic acid in axons of peripheral nerve, *J. Neurochem.* **1**:119–125 (1956).

175. J. Clausen and P. Rosenkast, Isolation of acid mucopolysaccharides of human brain, *J. Neurochem.* **9**:393–398 (1962).

176. J. Clausen and A. Hansen, Acid mucopolysaccharides of human brain, *J. Neurochem.* **10**:165–168 (1963).

177. M. M. Szabo and E. Roboz-Einstein, Acidic polysaccharides in the central nervous system, *Arch. Biochem. Biophys.* **98**:406–412 (1962).

178. R. U. Margolis, Acid mucopolysaccharides and proteins of bovine whole brain, white matter and myelin, *Biochim. Biophys. Acta* **141**:91–102 (1967).

179. J. Clausen, H. V. Dyggve, J. C. Melchior, and H. O. Christensen Lou, Chemical studies in gargoylism, *Arch. Disease Childhood* **42**:62–69 (1967).

180. Z. Stary, A. H. Wardi, D. L. Turner, and W. S. Allen, Arabinose as a mucopolysaccharide component in human and animal brain tissue, *Arch. Biochem. Biophys.* **110**:388–394 (1965).

181. R. Landolt, H. H. Hess, and C. Thalheimer, Regional distribution of some chemical structural components of the human nervous system—I, *J. Neurochem.* **13**:1441–1452 (1966).

182. H. H. Hess and C. Thalheimer, Microassay of biochemical structural components in nervous tissues—I, *J. Neurochem.* **12**:193–204 (1965).

183. H. V. Koenig, An autoradiographic study of nucleic acid and protein turnover in the mammalian neuraxis, *J. Biophys. Biochem. Cytol.* **4**:785–792 (1958).

Chapter 15

CEREBRAL METABOLISM *IN VIVO*

William Sacks

Research Center
Rockland State Hospital
Orangeburg, New York

In reference to studies on the uptake of amino acids from blood, Waelsch[1] pointed out that "Experiments in which the amino acid concentration in blood is raised to an unphysiological level may elicit an aspect of the blood–brain barrier not operative under physiological conditions. For determining the uptake of amino acids of the brain under physiological conditions, accurate measurements of the arterio–venous differences would be required." Although brain slices *in vitro* and other brain preparations have proved of considerable value in biochemical research, it is true that, as Geiger[2] observed, "they obviously do not possess all the metabolic machinery which is involved in the physiological activity of the nerve cell."

In this chapter those studies which give some insight into the chemical environment of the intact brain and the effect of blood constituents on the physiological functioning of the brain will be described. Although there are many *in vivo* techniques in the literature, the investigations cited herein will be limited to those which use either the arterio–venous or brain perfusion methods.

I. ARTERIO–VENOUS METHODS

A. Cerebral Blood Flow Determination

In 1943, Dumke and Schmidt,[3] using the rhesus monkey, obtained the first quantitative data on a brain *in vivo* under conditions which might be considered as approaching the normal physiological state. Two years later, Kety and Schmidt[4] published details of a procedure for the determination of cerebral blood flow in humans. The method, which made use of the Fick principle, consisted of the administration by inhalation of nitrous oxide for 10 min. During that time, samples of blood going to the brain (femoral artery) and leaving the brain (internal jugular bulb) were drawn simultaneously at regular intervals. They were analyzed for nitrous oxide content, arterial and venous curves were drawn, and the integrated arterio–venous nitrous oxide difference was obtained. This value was divided into the amount of nitrous oxide taken up by the brain. The latter figure was obtained by multiplying the nitrous oxide content of the jugular venous blood at 10 min by the blood–brain partition coefficient (unity) for nitrous oxide. This method, which estimates cerebral blood flow in terms of ml whole blood/100 g brain/min, has been described in detail by Kety.[5]

In the procedure summarized above, some blood samples were analyzed for O_2, CO_2, and glucose content. The product of the cerebral blood flow (CBF) multiplied by the arterio–venous (A–V) or venous–arterial (V–A) difference of these substances gave an estimation of the consumption of oxygen and glucose and of the production of carbon dioxide. For normal, healthy young men, average values have been reported[6] as 3.3 ml O_2/100 g brain/min and 5.4 mg glucose/100 g brain/min. The average CBF was 54 ml/-100 g brain/min. Using an average weight of whole brain as 1400 g, the O_2 and glucose consumption for total brain were estimated as 46 ml O_2 and 76 mg glucose/min, respectively.

B. Determination of Cerebral Metabolism in Human Subjects

In 1956, a report[7] from the author's laboratory gave details of a method for determining cerebral metabolism in humans *in vivo*. It made use of the arterio–venous technique of Kety and Schmidt and, in addition, employed, the intravenous injection of a labeled substrate (such as ^{14}C-glucose).[8] It may be described briefly as follows: After the intravenous injection of a ^{14}C- or ^3H-labeled substrate, blood samples were drawn simultaneously over a period of 90–120 min. They were analyzed for radioactivity of the injected substrate, of the end product of metabolism of the substrate (i.e., $^{14}CO_2$ or ^3HOH), and of certain intermediates (i.e., lactate and pyruvate). From the data could be calculated the cerebral uptake of labeled substrate, the production of $^{14}CO_2$ (or ^3HOH) by brain, and the per cent of labeled substrate taken up which was oxidized to $^{14}CO_2$ (or ^3HOH) by brain. In addition, by comparing the specific activity of venous blood (which was considered to represent brain) with the specific activity of brain $^{14}CO_2$* (or ^3HOH), the per cent of brain carbon dioxide (or water) derived by oxidation of the substrate could be determined.

C. Other Arterio–Venous Methods

A technique similar to that just described was employed by Coxon and Robinson[9,10] to determine the venous–arterial specific activity ratios of carbon dioxide for different organs of dogs and monkeys after injection of ^{14}C-glucose. They examined the ability of various organs to produce $^{14}CO_2$

---

* Brain CO_2 specific activity as used in this chapter in reference to the author's studies refers to the "added increment" CO_2 specific activity,[8] i.e., the specific activity of the CO_2 produced by the brain and added to the venous blood leaving the brain. Specific activity equals the per cent of injected activity per milligram of either CO_2 carbon or glucose carbon.

from blood ^{14}C-glucose and found that brain, liver, and kidney could raise the specific activity in CO_2 in blood passing through them, but that resting muscle exerted a lowering effect. On the basis of their experimental evidence and model systems,[11,12] they suggested the use of continuous infusion (rather than a single injection) of the isotope to reduce the complexities of the problem of determining the rate of oxidation of a labeled substrate in an intact animal by keeping the specific activity of the injected isotope in the arterial blood relatively constant.

II. BRAIN PERFUSION METHODS

The methods presently in use for brain perfusion studies attempt to isolate the cerebral circulation while allowing the brain to retain as many of its physiological functions as possible.

A. Isolated Cat Brain Perfusion

The perfusion of cat brain *in situ*, developed by Geiger and Magnes,[13] depends upon the selective isolation of the cerebral venous outflow with simultaneous blocking of venous outlets of extracerebral tissues of the head. It also includes a partial isolation of the arterial side of brain circulation. A "simplified blood" consisting essentially of washed bovine erythrocytes suspended in Krebs–Ringer solution containing bovine serum albumin was used. Usually, it contained glucose at a concentration of 100 mg %. Cat blood was ruled out because of the large quantities of blood needed, frequent occurrence of immunological reactions, and the quick sedimentation of cat erythrocytes. The apparatus included a flow meter, an oximeter, a pH meter, a heater, and equipment for automatically regulating CBF. CBF was usually about 100 ml/100 g brain/min (man is about 54 ml/100 g brain/min). Geiger found it necessary to add cytidine and uridine[14] to "simplified blood" to maintain physiological activity of brain for suspended periods (4–5 hr). Allweis and Magnes,[15] on the other hand, apparently found the addition of these substances unnecessary. In Geiger's laboratory, a successful preparation exhibited the following physiological functions: normal corneal reflex, pupillary reactions to light, natural respiration, maintenance of systemic blood pressure and of vasomotor responses, normal electrocorticogram, spontaneous blinking, movements of facial structures, and easily elicited electric reactions in the brain to stimulation of one of the extremities.

The original apparatus[13] incorporated a heater and presumably the perfusion blood was maintained at 37°C. In the report by Allweis and Magnes,[15] precautions were taken to minimize the metabolism of glucose by erythrocytes. The blood reservoir was immersed in ice and blood, prior to its entrance into the carotids, was heated rapidly to 37°C with a silver heating coil.

Some indication that the perfused cat brain may have been altered from its physiological state can be seen in the high consumption of glucose and oxygen during the experiments. Oxygen consumption of brain was very high,[13] especially at the start of perfusion, and sometimes reached 6 ml O_2/100 g brain/min. It declined within 5–10 min and was then usually maintained at a level of over 4 ml as long as cerebral activity remained excellent. Thereafter, oxygen consumption declined slowly with progressive deterioration of cerebral activity. Although one could calculate an average value for glucose consumption of 5.9 mg/100 g brain/min from their data, these investigators found that glucose uptake fluctuated considerably during an experiment and from one experiment to another i.e., 4.1–8.7 mg/100 g brain/min). Allweis and Magnes[15] gave somewhat higher values for these parameters; i.e., 9.2 mg glucose/100 g brain/min and 6.9 ml O_2/100 g brain/min. It was seen above that average values for normal human brain were 5.4 mg glucose/100 g brain/min and 3.3 ml O_2/100 g brain/min. In addition to the elevated uptake of O_2 and glucose, there was an excessive production of lactic acid by the perfused isolated cat brain. Allweis and Magnes[15] found that lactic acid was liberated into blood at a mean rate of 4.9 mg/100 g brain/min and that the mean lactic acid content of brain cortex at the end of their experiments was 177 mg/100 g. As will be seen below, the normal human brain *in vivo* adds little if any lactic acid to cerebral venous blood.[16–23]

Assuming cat brain *in vivo* does not differ widely in its metabolic requirements from human brain, the abnormal consumption of O_2 and glucose and production of lactic acid suggest that the perfusion procedure may have changed the metabolic activity of the cat brain.

B. Other Perfusion Techniques

A comparatively simple isolated, perfused rat brain preparation has recently been developed by Sloviter's group.[24] With a simpler surgical procedure than that of Geiger and Magnes, their preparation consisted of the skull and its contents, having the upper cervical vertebrae and small remnants of muscle attached. The rats were anesthetized by deep hypothermia, thus eliminating the problem of influence of chemical agents on cerebral metabolism and function. As in the perfused cat brain method, it was found necessary to employ an artificial blood. It consisted of canine erythrocytes in a solution of bovine albumin in Krebs–Ringer bicarbonate buffer. Spontaneous electroencephalographic (EEG) activity, which persisted for up to 5 hr with open-circuit perfusion and about 2 hr with recirculation, was used to indicate brain viability. The isolated rat brain and perfusion blood were allowed to reach room temperature before starting the perfusion and remained there throughout the experiment. From the curves in Fig. 6 of Ref. 24, one can approximate glucose consumption and lactate production of this preparation. It can be calculated that the isolated, perfused rat brain consumed about 2.5 mg glucose/100 g brain/min. when the initial concentration of perfusing blood was 200 mg %, and about 7.3 mg/100 g brain/min with an initial blood level of 300–400 mg % glucose. Values for lactate production were about 0.31 mg/100 g brain/min with the 200 mg % blood glucose, and about 0.36 mg/100 g brain/min with the 300–400 mg % blood glucose. These approximations indicate that in this perfusion preparation, done at lower than the animal's body temperature, the consumption of glucose varied considerably and did not differ widely from the isolated, perfused cat brain; however, production of lactic acid was less than in the Geiger–Magnes preparation.

The extracorporeal perfusion of the isolated dog head has been described by Gilboe *et al.*[25–27] With the brain left within the skull, perfusion was accomplished by heparinized, compatible donor blood, which entered through the common carotids and exited through a connector inserted into a small hole in the occipital bone to tap the confluent sinus. Electrocortical activity was used as an index of brain viability. A heat exchanger was incorporated into the system to maintain the brain at 37°C. Glucose consumption varied from 3.4 to 5.7 mg/100 g brain/min and oxygen consumption averaged 3.6 ml/100 g brain/min.

Another method in current use is that of White, Albin, and Verdura,[28] in which the brain of a monkey was surgically freed from all contiguous tissue except a small basal plate of bone and a thin bony strip over the sagittal sinus for supporting the brain and the electrodes. All extracranial vasculature was removed, vertebral arteries were ligated, and carotid arteries were cannulated, permitting autogenous cerebral perfusion until the start of extracorporeal circulation. The perfusion blood was freshly drawn, heparinized, compatible monkey blood diluted one-third with dextran. Intracerebral temperatures were maintained between 28 and 32°C to reduce cerebral metabolism. Therefore, (A–V) O_2 and (V–A) CO_2 differences were only 2.8 and 2.4 vol. %, respectively. Glucose uptake after 2 hr of closed circuit perfusion was about 2.3 mg/100 gm brain/min. At that time, the perfusion blood had high levels of lactate (75.5 mg %) and of pyruvate (8.4 mg %). As perfusion progressed, there were elevations in free hemoglobin and a gradually developing acidosis.

Recently the same laboratory[29] reported the successful transplantation, with survivals lasting from 6 hr to 2 days, of six isolated canine brains into the cervical vasculature of dogs. The brain was surgically isolated within the skull and internalization of the brain graft was accomplished by connecting the carotid arteries of isolated brain to the proximal carotid artery of recipient and by fixing the torcula cannula of isolated brain within the jugular vein of recipient. In addition to instrumentation for continuous monitoring of electrocortical activity, CBF, and temperature, catheters were implanted for sampling arterial and venous blood from the transplanted brain. Average (A–V) O_2 and (V–A) CO_2 differences were 11.1 and 10.2 vol %, respectively, and the average (A–V) glucose difference was 24.2 mg %. These values, though high, are consistent with the low average CBF of 24.2 ml/100 g brain/min. Oxygen and glucose consumption averaged 2.7 ml O_2/100 g brain/min and 5.6 mg glucose/100 g brain/min, respectively. Blood lactate determinations revealed no significant differences between arterial and venous samples.

A unique method was developed by Moss[30,31] for the benign, arterially isolated cerebral perfusion of the bull calf. This was accomplished by perfusion of a single carotid artery at 20 mm Hg pressure above systemic arterial pressure, which resulted in a pressure in the circle of Willis greater than aortic pressure causing the collaterals to flow in a retrograde direction. The perfusate consisted of fresh donor–compatible blood. The procedure was apparently innocuous, since, after 6 hr of perfusion, no clinical, EEG, or histopathologic evidence of cerebral dysfunction or damage could be detected. Unfortunately, no biochemical data were given in these reports, so that the preparation could not be evaluated with the usual metabolic parameters.

C. Use of Filters in Perfusion

An inherent difficulty in perfusion techniques has been a steady increase in perfusion pressure and a build-up of lactic acid. Swank and Hissen[32] have shown recently that this may be due to a blockage of the microvasculature with platelet aggregates, since normal perfusion pressure could be maintained by the use of Dacron wool or glass filters. Gilboe, Glover, and Cotanch[33] tested the hypothesis that occlusion of small blood vessels of brain limits the access of brain tissue to blood and oxygen and thereby causes increased dependence on anaerobic metabolism. Two groups of isolated dog brains were used: (1) perfused with blood filtered through Dacron wool to remove platelet aggregates and (2) perfused with blood filtered through a stainless steel screen to remove fibrin clots. No significant differences could be found in either glucose consumption (4.9 and 3.7 mg/100 g brain/min for Groups 1 and 2, respectively) or production of lactic acid (0.30 and 0.43 mg/100 g brain/min for Groups 1 and 2, respectively). Perfusion pressure increased more rapidly in the second group. The authors concluded that the excessive production of lactic acid observed by others with the isolated cat brain was probably due to the use of "simplified blood."

The effects of filtering "simplified blood" through glass wool in cat brain perfusion have been reported recently by Allweis, Abeles, and Magnes.[34] Use of the filter reduced to about 10% the incidence of progressive rise in perfusion pressure which occurred in 75% of experiments done without the glass wool filter. The pressure necessary to perfuse blood through the brain at a rate of 120 ml/100 g brain/min was about 30 mm Hg lower in experiments using the filter. It was found that CO_2 production by brain was significantly increased and that samples of cortex taken at the end of experiments showed greater glucose content and lower lactate content than in experiments not employing the glass wool filter.

III. CEREBRAL METABOLISM OF CARBOHYDRATES *IN VIVO*

A. Glucose as the Primary Substrate

The brain is unique among the organs of the body in that it has the highest rate of utilization of oxygen. In normal man, the cerebral oxygen consumption has been estimated at about 3.5 ml/100 gm brain/min or approximately 50 ml/min for total brain. This value represents about 20% of the body's total oxygen requirements under basal conditions. Kety[35] calculated the rate of energy utilization of human brain as about 20 W from the oxygen consumption and the respiratory quotient (which is usually very close to unity). The evidence suggests the principal substrate of brain is carbohydrate. In the human, glucose is taken up by brain from blood at about 5 mg/100 g brain/min. Or, stated another way, for each 100 ml of whole blood traveling through it, the brain will "extract" about 10 mg of glucose. That glucose is the primary foodstuff of the brain in humans is supported by studies showing that it alone will arouse patients in insulin hypoglycemic coma. Other substrates (i.e., glycerol and glutamic acid) reported to be capable of restoring contact during insulin hypoglycemia have been shown to cause an increase in blood glucose, so that it is doubtful that they alone can serve to support brain metabolism.

B. Cerebral $^{14}CO_2$ Derived from ^{14}C-Glucose

Making use of the arterio–venous technique and intravenous injections of variously labeled ^{14}C-glucose into normal human subjects, the author, in 1957, presented evidence suggesting that only part of cerebral CO_2 was derived from glucose.[8] Following a single injection of [U-^{14}C] glucose (uniformly labeled), it was found that brain $^{14}CO_2$ specific activity was relatively constant and that venous blood glucose specific activity (representing brain glucose specific activity) was considerably higher than brain $^{14}CO_2$ for most of the experiment (Fig. 1). Brain $^{14}CO_2$ specific activity curves were found following injections of [1-^{14}C] glucose, [2-^{14}C] glucose, and [6-^{14}C] glucose and these values were used to derive the theoretical brain $^{14}CO_2$ specific activity curve for carbons 3 and 4 (C_3 and C_4) of [U-^{14}C] glucose (Fig. 2). Since the derived curve (Curve D) was parallel to the venous (brain) glucose specific activity curve (Curve G), but was about one-half as high, it was thought that only about 50% of brain $^{14}CO_2$ was derived from ^{14}C-glucose. Also, it was shown that only about 40% of ^{14}C-glucose taken up by brain in 90 min came out as $^{14}CO_2$, and it was suggested that the ^{14}C remaining in brain became incorporated into the brain metabolic carbon pool consisting of glucose, glycogen, glutamine, glutamate, and N-acetyl-aspartate. Furthermore, comparison of cerebral $^{14}CO_2$ produced following injection of [1-^{14}C] glucose with that found after administration of [6-^{14}C] glucose led to the conclusion that practically all of the glucose-derived CO_2 arose via glycolysis.

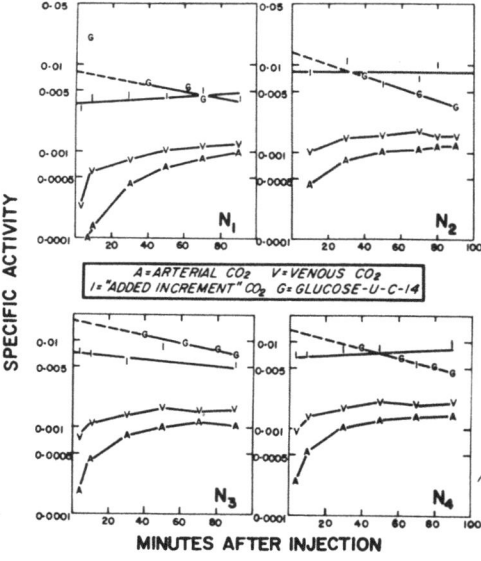

Fig. 1. Specific activity–time curves (plotted semilogarithmically) for four normal subjects given [U-^{14}C]glucose intravenously. Curve G is venous glucose decay curve. Curves V and A represent specific activity of $^{14}CO_2$ in venous (internal jugular) and arterial (femoral) blood, respectively. Curve I represents specific activity of "added increment" (brain) $^{14}CO_2$. Sacks.[8]

The question of the extent of cerebral CO_2 derived from glucose has been examined by others. Allweis and Magnes[15] employed [U-^{14}C] glucose in their perfused cat brain studies and found the radioactivity of CO_2 produced by brain reached an approximately constant value after about 30 min

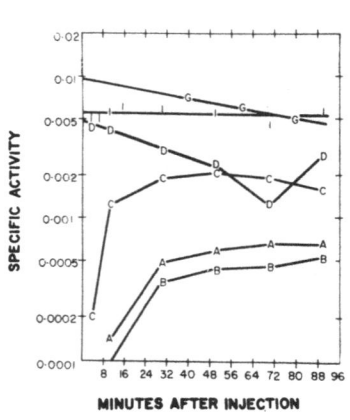

Fig. 2. Composite curves (plotted semilogarithmically) of [U-^{14}C] glucose specific activity (Curve G) and specific activity of "added increments" of (brain) $^{14}CO_2$ with variously labeled glucose substrates. Curves A and B represent specific activity of "added increment" $^{14}CO_2$ following injection of [1-^{14}C]glucose and [6-^{14}C]glucose, respectively. One-sixth of the actual specific activity was plotted. Curve C represents specific activity of "added increment" $^{14}CO_2$ for [2-^{14}C]glucose plus [5-^{14}C]glucose. Since the curve for [5-^{14}C]glucose was taken to be the same as that for [2-^{14}C]glucose, one-third of actual specific activity of [2-^{14}C]glucose was used in plotting Curve C. Curve I represents specific activity of "added increment" $^{14}CO_2$ with [U-^{14}C]glucose as substrate. Curve D resulted when points on Curves A, B, and C were added together and subtracted from corresponding experimental points of Curve I. Curve D theoretically defines rate of decline of specific activity of "added increments" of $^{14}CO_2$ derived from carbons 3 and 4 of [U-^{14}C]glucose. Taken from Sacks.[72]

of perfusion. Their data suggested that only 22% of glucose taken up from blood was oxidized to CO_2. Most of the balance, they believed, could be accounted for as lactate accumulated in brain and liberated into blood. Similar results were found by Geiger, Kawakita, and Barkulis.[36] After the addition of [U-^{14}C] glucose to "simplified blood", the concentration of ^{14}C in brain CO_2 increased steadily during the first 25–30 min to a concentration of 29–32% and remained at that level for the rest of the experiment. The results were interpreted to indicate that both glucose carbon and "cold" carbon from endogenous noncarbohydrate sources contributed to form respiratory CO_2. That brain glucose equilibrated quickly with glucose in the perfusing blood was established by showing that glucosazones prepared from blood and from extracts of brain cortex pieces had similar specific activities.

Because of an investigation[37] which indicated that 30–40% of CO_2 produced by rat whole cerebrum slices was derived from endogenous sources, Gainer, Allweis, and Chaikoff[38] studied the intact anesthetized dog. The animals were given a continuous intravenous infusion of [U-^{14}C] glucose to maintain a constant plasma glucose specific activity. At intervals, arterial and venous blood from the superior sagittal sinus were taken simultaneously for determination of specific activities of plasma glucose and brain CO_2. The specific activity of brain CO_2 rose during the first hour to a value about the same as that of blood glucose and then stayed level for the remaining 4.5 hr of the experiment. The authors concluded that almost all of the metabolic CO_2 was derived from blood glucose.

In view of these results, Barkai and Allweis[39] thought that differences between experiments with the intact dog and the perfused cat brain were due to some peculiarities in the latter preparation (i.e., use of "simplified blood"). Therefore, they employed the continuous infusion technique with the intact narcotized cat. In these studies, at apparent isotopic equilibrium, it was observed that brain CO_2 specific activity was only 45% that of blood glucose specific activity. This value, which did not differ widely from that found with the perfused cat brain (29–32%), was much lower than that obtained using the intact narcotized dog. The authors suggested that the latter dissimilarity may have been due to either a species variation or to different origins of the cerebral venous blood sampled in the two experiments (i.e., in the dog, blood was obtained from the anterior superior sagittal sinus, which drains blood mainly from anterior parts of cortex). In similar experiments performed upon intact narcotized cats, Gombos and co-workers[40] reported that brain CO_2 specific activity attained a value about 60% that of blood glucose after 4–5 hr of perfusion.

Thus, there appears to be some controversy as to the extent of brain CO_2 derived from glucose. It has been the opinion of this author for some time that differences between results found in humans and those obtained using the cat brain perfusion preparation were due in part to use of "simplified blood." The author's recent findings of mutarotase (believed to be involved in glucose transport) in erythrocytes and of considerable species difference in activity of this enzyme[41] and of evidence that red blood cells may play some role in the transfer of glucose from blood into brain[42] suggest that it may not be desirable to use bovine cells in the cat preparation. In addition, cat's normal plasma probably would be more likely to favor glucose transfer than the Krebs–Ringer plus bovine albumin used in "simplified blood."

Most likely the main factor in determining the relationship of brain CO_2 specific activity to that of glucose is the dilution of ^{14}C (derived from ^{14}C-glucose) in traveling the Krebs cycle. Following a series of experiments with $[3-^{14}C]$ glucose as injected substrate (in humans), production of $^{14}CO_2$ almost kept pace with the net uptake of ^{14}C by brain.[43] In addition, brain CO_2 specific activities practically coincided with venous blood (brain) glucose specific activities, suggesting little or no dilution of ^{14}C in traversing the Embden–Meyerhof–Parnas pathway. Since intermediates of the tricarboxylic acid cycle occur at concentrations in brain less than those of the glycolytic intermediates, brain CO_2 specific activity curves following injections of $[2-^{14}C]$ glucose, $[1-^{14}C]$ glucose, or $[6-^{14}C]$ glucose indicated a significant dilution of ^{14}C in traveling the Krebs cycle. From the extent of the estimated dilution (about 40-fold) and experimental data obtained in the author's laboratory and by others, a scheme for human cerebral metabolism of glucose *in vivo* was proposed. It included a small, metabolically active pool of glutamate, γ-aminobutyrate, and succinic semialdehyde directly in the Krebs cycle, which was thought to be open pool, i.e., with a constant outflow and a balancing inflow. The outflow was postulated to be due to a slow equilibration of the small pool of glutamate with a large inactive pool of glutamate and to formation of glutamine from the small glutamate pool. N-acetylaspartate was considered a source of inflow into the Krebs cycle in this hypothetical pathway. It was further thought to supply substrate under emergency conditions, i.e., hypoglycemia.

As a result of this dilution of ^{14}C in traveling the Krebs cycle,[43] significant amounts of ^{14}C taken up by brain in the form of ^{14}C-glucose (when labeled on C_1, C_2, C_5 or C_6) remained in brain for long periods of time. Indeed, following a single intravenous injection of $[2-^{14}C]$ glucose into a human subject, after 5 hr about 21 % of the ^{14}C-glucose uptake still remained in brain. Therefore, in cat brain perfusion studies, where $[U-^{14}C]$ glucose was always used, one could expect about two-thirds of the radioactivity to remain in brain for significant lengths of time. With incorporation of this ^{14}C into amino acids, proteins, fatty acids, and lipids (see below), one would not be surprised to find brain CO_2 specific activities only a fraction of blood glucose specific activities. In addition, it might well be that the extent of dilution varies in different structures of brain and in various animal species.

C. Incorporation of ^{14}C from Labeled Glucose into Brain Carbon Pool

Geiger *et al.*[44] excised brain cortex slices at various time intervals following addition of $[U-^{14}C]$ glucose to "simplified blood" in cat brain perfusion studies to determine specific activities of free amino acids. In the first samples (about 20 min), the combined amino acids had about 5–10 % of the specific activity of the perfused glucose. Although the radioactivity of amino acids in the next samples (about 40 min) remained about the same, that of the third samples usually showed a pronounced increase (as much as fourfold) in radioactivity compared to the first samples and reached a specific activity 15–18 % that of glucose, thereafter remaining constant. Glutamic and aspartic acids, which were further separated, showed the same trend in activities, but leveled off at 23–24 % of the glucose specific activity. In a subsequent paper,[45] it was found that ^{14}C derived from glucose was incorporated into proteins of brain cortex considerably more slowly than into free amino acids. It was further demonstrated[46] that ^{14}C (from glucose) was incorporated into N-acetyl-L-aspartic acid at a rate only about 1–5 % that of its incorporation into aspartic acid in brain. The findings on incorporation of ^{14}C derived from glucose into various intermediary metabolites have been summarized by Gombos, Geiger, and Otsuki[47] and are given in Table I.

D. Pyruvic and Lactic Acids

In normal human brain, arterio–venous (A–V) differences of oxygen and glucose are about 6.0–6.7 ml O_2/100 ml whole blood and about 10 mg glucose/100 ml whole blood, respectively. An uptake of 10 mg of glucose by brain would require about 7.5 ml O_2; or, stated another way, an oxygen consumption of 6.0–6.7 ml would be satisfied by 8.0–8.9 mg glucose. This disparity between theoretical and actual values has usually been explained by the production and release by brain of 1.6 mg lactic acid/100 ml whole blood[48] and of 0.22 mg pyruvic acid/100 ml whole blood.[49]

The author has studied the human cerebral metabolism of [1-^{14}C] pyruvate[16] and various ^{14}C-labeled DL-lactic acids, D-[1-^{14}C] lactic acid, and L-[1-^{14}C] lactic acid.[16,23] With [1-^{14}C] pyruvate as injected substrate, brain CO_2 specific activity curves resembled closely those found using [3-^{14}C] glucose (as would be expected, theoretically). With variously ^{14}C-labeled DL-lactic acid, percentages of injected ^{14}C oxidized to $^{14}CO_2$ by

TABLE I

Amounts of Radioactivity Derived from ^{14}C-Glucose in the Various Metabolites in Perfused Cat Brain, Expressed as per cent of their Total Carbon[a]

Free lactic acid	47–50
Respiratory CO_2	28–30
Free glutamic acid	31–35
Free aspartic acid	24–26
Free γ-aminobutyric acid	11–16

[a] Taken from Gombos, Geiger, and Otsuki.[47]

brain varied in magnitude, depending upon the position of the isotope in the lactic acid injected, the order being as follows: [2-^{14}C] lactic > [1-^{14}C] lactic > [3-^{14}C] lactic > [U-^{14}C] lactic acid. With individual D- and L-isomers, L-[1-^{14}C] lactic acid had a value less than that for DL-[2-^{14}C] lactic, but greater than the value for DL-[1-^{14}C] lactic acid. D-[1-^{14}C] lactic acid gave percentages lower than any, using DL-^{14}C-lactate. Blood lactate and pyruvate determinations indicated that both these substances were added to blood as it flowed through brain. For lactic acid, the average (V–A) difference for normal subjects was 0.55 mg %; for pyruvate, the average (V–A) difference was 0.04 mg %.

Other investigators have found that small quantities of lactic and pyruvic acids were added to cerebral venous blood,[17,19–22] although there was considerable variation in reported values. The study by Gottstein and co-workers[17] was particularly noteworthy, since substrate-specific enzymic methods were used. These investigators reported that the cerebral production in normal humans was 0.24 mg lactate/100 g brain/min (i.e., about 0.46 mg %) and 0.055 mg pyruvate/100 g brain/min (i.e., about 0.106 mg %). A contradictory report was that by Scheinberg, Bourne, and Reinmuth,[18] in which no consistent differences were found in arterial and cerebral venous values for pyruvate and lactate in 25 normal resting subjects. Also, White and co-workers[29] observed no significant (A–V) or (V–A) blood lactate differences in transplanted canine brains (see above).

The apparently paradoxical situation in which the brain catabolizes lactate and pyruvate[16,23] while excreting both into the venous return blood suggests that compartmentalization of brain includes metabolism as well as function, and that total production of lactic and pyruvic acids through glycolysis exceeds the amounts that are oxidized by way of the tricarboxylic acid cycle. With the more specific methods available today, it would seem that earlier measurements of brain lactate and pyruvate production were probably too high and that the "extraction" of glucose does indeed exceed the production of CO_2, lactate, and pyruvate. Much data exist pointing to a significant incorporation of glucose carbon into lipids and proteins of brain (see above).

E. The Pentose Cycle

Most of the evidence obtained with *in vivo* methods supports the concept that the pentose cycle (or hexose monophosphate shunt) plays little or no role in the metabolism of glucose by brain.[8,43,50,51] However, Moss,[31] using the arterially isolated brain perfusion technique described above, concluded that the hexose monophosphate shunt was probably the major pathway *in vivo* for bovine cerebral glucose oxidation. The procedure consisted of a constant infusion of [6-^{14}C] glucose for the first 26 min, an interruption for the next 2 min (no isotope given), and then a constant infusion of [1-^{14}C] glucose for the remaining 23 min of the experiment. No priming dose was used. The ratio of $^{14}CO_2$ produced by brain at about 50 min to that at about 25 min was used to indicate the ratio of $^{14}CO_2$ formed from [1-^{14}C]

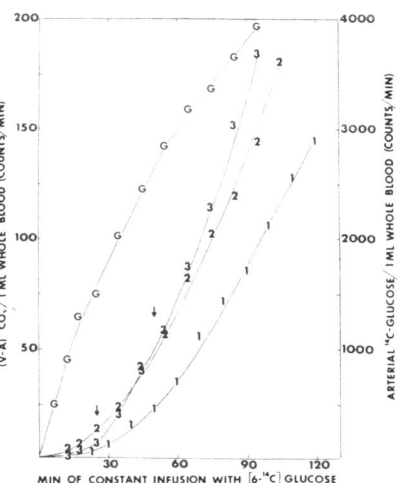

Fig. 3. V-A (brain) $^{14}CO_2$ activity–time curves with constant intravenous infusion of [6-^{14}C]glucose in three human subjects. No priming dose was used. Curve G represents typical levels for arterial blood ^{14}C-glucose in these experiments. The average value for the ratio of brain $^{14}CO_2$ at 50 min and 25 min (arrows) was 5.8:1. Similar curves were found with constant infusion of [1-^{14}C]glucose. These data illustrate the main reason for erroneous estimation of extent of pentose cycle activity by Moss[31] (see text).

MIN OF CONSTANT INFUSION WITH [6-^{14}C] GLUCOSE

glucose to that from [6-^{14}C] glucose, respectively. In one study, [6-^{3}H] glucose was given throughout the experiment and brain ^{3}HOH was used to indicate brain $^{14}CO_2$ produced from [6-^{14}C] glucose. Since an average ratio of 6.5:1 (ranging from 5.0–7.2:1) was found, the author decided that the pentose cycle most likely represented the primary mechanism in cerebral glucose metabolism.

This investigation by Moss is in apparent contradiction with results and conclusions from the author's laboratory,[8,43] which indicate a minor role (estimated at about 1%) for the pentose cycle. However, there were some basic flaws in the experimental procedures employed by Moss. One of these can be illustrated by an examination of some of the author's unpublished experiments, which used a constant intravenous ^{14}C-glucose infusion with no previous priming dose. Thus, conditions were similar to those of Moss, except that only one isotope of glucose was used in each experiment. From Fig. 3 it is apparent that when [6-^{14}C] glucose was given by constant infusion, the brain $^{14}CO_2$ curve showed very low radioactivity for the first 20–30 min and then rose sharply. If we take the ratios of brain $^{14}CO_2$ at 50 min to that at 25 min (arrows), we get values of 3.6:1, 7.1:1, and 6.6:1, giving an average of 5.8:1, which does not differ widely from that of 6.5:1 reported by Moss. It is obvious then that the ratios found by Moss would have been about the same even if he had not changed to [1-^{14}C] glucose, but kept giving [6-^{14}C] glucose. Brain $^{14}CO_2$ curves found following constant infusion of [1-^{14}C]glucose (author's unpublished results) were similar to those described for [6-^{14}C] glucose.

Another fallacy in Moss' experimental technique can be shown by a previous study on human cerebral metabolism from the author's laboratory,[43] where it was reported that the brain ^{3}HOH specific activity curve found after a single injection of $[1\text{-}^3H]$ glucose did not resemble the brain $^{14}CO_2$ specific activity curve from similar studies with $[1\text{-}^{14}C]$ glucose, but closely approximated that which resulted after $[3\text{-}^{14}C]$ glucose injection. Similarly, with $[6\text{-}^3H]$ glucose, brain ^3HOH curves (unpublished) resembled brain $^{14}CO_2$ specific activity curves found with $[3\text{-}^{14}C]$ glucose. Thus, it is evident that $[6\text{-}^3H]$ glucose cannot be used, as Moss did, to indicate production of $^{14}CO_2$ by brain from $[6\text{-}^{14}C]$ glucose.

A recent study by Hostetler et al.[51] has given a quantitative estimation of the extent of functioning of the pentose cycle in specific areas of perfused isolated monkey brain. With the technique of White (described above), $[2\text{-}^{14}C]$ glucose was added to the perfusion blood and, after perfusion, samples of brain tissue were taken for analyses. Glycogen was isolated from these brain samples, the glycogen was hydrolyzed, and the resulting glucose degraded to indicate the extent of labeling on each carbon atom. Calculations made from the data showed a 5–10% functioning of the pentose cycle in certain specific areas (i.e., hypothalamus, brain stem) of monkey brain.

F. The Tricarboxylic Acid Cycle

In the author's laboratory, experiments have been performed in which the following were employed as the injected substrates: $[2\text{-}^{14}C]$ fumarate, $[1,4\text{-}^{14}C]$ fumarate, $[1,4\text{-}^{14}C]$ maleate, and $[5\text{-}^{14}C]$ α-oxoglutarate. Maleic acid was included because of evidence of the existence of maleic hydratase (which converts maleic acid to D-malic acid) in humans.[52–55] When $[2\text{-}^{14}C]$ fumarate was given intravenously to normal human subjects, it was readily oxidized to $^{14}CO_2$ by brain.[7] Studies with $[1,4\text{-}^{14}C]$ fumarate (unpublished) showed a delay of 6–30 min in production of $^{14}CO_2$ by brain and more fluctuations in $(V\text{–}A)^{14}CO_2$ differences than in $[2\text{-}^{14}C]$ fumarate experiments. With $[1,4\text{-}^{14}C]$ maleate, there was little or no production of $^{14}CO_2$ by brain, although $^{14}CO_2$ appeared in arterial blood very quickly after injection, indicating the body readily oxidized this compound.[53] Following injection of $[5\text{-}^{14}C]$ α-oxoglutarate, the data (unpublished) showed a significant and consistent production of $^{14}CO_2$ by brain after an initial lag of 10–20 min. In these experiments with Krebs cycle intermediates, no significant ^{14}C labeling was found in glucose, amino acids, lactate, or pyruvate.

G. Conversion of ^{14}C-Glucose-Phosphate to ^{14}C-Glucose by Brain

Evidence is appearing in the literature contradicting the general belief that hexose phosphates do not penetrate cells.[56–58] In the author's studies of cerebral glucose metabolism,[8,16,43] a contradiction was found between actual glucose uptake and uptake in terms of ^{14}C-glucose. The data suggested that a labeled glucose intermediate was formed which had significant radioactivity. Therefore, experiments were undertaken using ^{14}C-glucose phosphate as injected substrate.[42] Following a single injection of $[1\text{-}^{14}C]$ glucose-6-phosphate, the cerebral production of $^{14}CO_2$ exceeded that

TABLE II

Comparison of Cerebral-C^{14} Uptake and $C^{14}O_2$ Production Following a Single Injection of Glucose-6-Phosphate-1-C^{14} with Similar Glucose-1-C^{14} Experiments (120 min cumulative results)[a]

Glucose-C^{14} uptake (% of Inj C^{14})	$C^{14}O_2$ produced by brain (% of Inj C^{14})	Glucose-C^{14} uptake oxidized to $C^{14}O_2$ by brain (%)	glucose-phos-phate-C^{14} uptake (% of Inj C^{14})	Total C^{14} uptake oxidized to $C^{14}O_2$ by brain (%)
Glucose-6-phosphate-1-C^{14} experiments				
20.7	18.1	87.6	19.9	44.7
29.5	15.8	53.6	14.7	35.8
Glucose-1-C^{14} experiments				
37.3	18.3	49.2	0	49.2
35.0	15.0	42.8	0	42.8

[a] Taken from Sacks.[42] For calculation purposes average cerebral blood flow taken as 52 ml/100 g brain/min; total brain weight assumed to be 1400 g.

expected from the uptake of ^{14}C-glucose. However, when the cerebral uptake of ^{14}C-glucose phosphate was added to that of ^{14}C-glucose, the net ^{14}C uptake by brain approximated that which occurred in similar [1-^{14}C] glucose experiments with about the same cerebral production of ^{14}CO$_2$ (Table II). In studies in which arterial blood levels of ^{14}C-glucose phosphate were elevated rapidly by constant infusion for 6–10 min, it was seen that there was a consistent cerebral uptake of ^{14}C-glucose phosphate (i.e., (A–V) ^{14}C-glucose phosphate differences) with a corresponding constant production of ^{14}C-glucose (as evidenced by (V–A) ^{14}C-glucose differences), while ^{14}C-glucose phosphate blood values were rising. In the latter studies (Fig. 4), addition of ^{3}H-glucose to the infusion demonstrated that isotopic glucose was taken up by brain in the usual manner while it was converting ^{14}C-glucose phosphate to ^{14}C-glucose. Thus, the evidence suggested that human brain *in vivo* exhibited glucose phosphatase activity. An experiment done with [U-^{14}C] glucose-1-phosphate showed similar results except that, as would be expected from [U-^{14}C] glucose studies, there was a much greater cerebral production of ^{14}CO$_2$ than in [1-^{14}C] glucose-6-phosphate experiments. From these studies it could not be determined if the brain must convert glucose phosphate to glucose prior to oxidizing it to CO$_2$.

H. Influence of Arterial Blood Glucose Level on Cerebral Uptake of Glucose

Using the perfused cat head, Geiger and co-workers[59] saw that when the blood glucose concentration was increased to 0.8 %, the brain contained only 0.3 % and that this concentration was increased only slightly by doubling

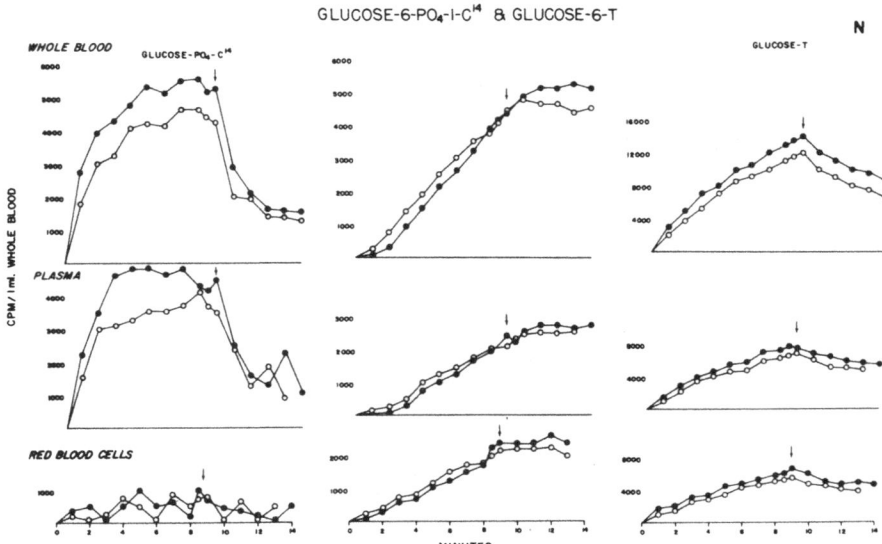

Fig. 4. Activity–time curves of arterial and venous whole blood, plasma, and red blood cell (RBC) ^{14}C-glucose-phosphate, ^{14}C-glucose, and ^{3}H-glucose (glucose-T) during and shortly after constant intravenous infusion (8.5 min) of [1-^{14}C]glucose-6-phosphate plus [6-^{3}H]glucose (glucose-6-T). Results for RBC were derived by subtracting values for plasma from those for whole blood making use of hematocrit. The arrows indicate the end of the constant infusion.

blood glucose. Sato et al.[60] in studying cerebral glucose consumption by the Kety–Schmidt cerebral blood flow method, stated that they could not demonstrate a parallelism between arterial glucose content and the (A–V) glucose difference observed in healthy subjects. Gottstein et al.[61] concluded that the cerebral uptake of glucose could not be improved in cerebral arteriosclerosis by elevating levels of arterial glucose. Intravenous infusions of 60 ml of 50% glucose influenced neither CBF nor cerebral metabolism significantly, in spite of an elevated blood sugar level from 96 to 265 mg%. In the author's laboratory, studies (unpublished) which combined a glucose tolerance test with our usual arterio–venous technique indicated no correlation between cerebral uptake of glucose and arterial glucose concentration. In these experiments, there seemed to be no unusual results in either brain ^{14}C-glucose uptake or ^{14}CO$_2$ production, although, in some cases, arterial glucose values were above 200 mg%. The reports cited fail to corroborate the findings of Rowe and co-workers[22] that larger quantities of glucose were taken up by brain when arterial glucose levels were rising. In the latter study, patients were fed a standard breakfast, which raised arterial blood glucose levels, on the average, from 105 to about 150 mg%.

With the exception of the report by Rowe et al., these investigations lead one to conclude that cerebral uptake of glucose was unaltered in hyperglycemia. What about the effect of hypoglycemia? In a study in which cerebral blood was sampled during therapeutic insulin treatment of schizophrenic patients, Himwich and co-workers[62] observed that oxygen uptake was reduced to an average value of 3.1 vol. % (from a control value of 6.7 vol. %) and glucose uptake fell to an average of 4.2 mg % (from a control value of 10 mg %) during the ensuing hypoglycemia. Gottstein and Held[63] recently expressed the belief that even a minor hypoglycemia with blood sugar values of 50 mg % could result in a marked reduction of cerebral glucose uptake. It would seem, therefore, that although hyperglycemia probably had no effect, the cerebral glucose uptake could be significantly lowered by hypoglycemia. In regard to this matter, Butterfield *et al.*[64] proposed the concept of a brain glucose threshold similar to although slower acting than the peripheral glucose threshold they established previously.

I. Influence of Insulin on Uptake and Utilization of Glucose

It has been unequivocally stated for many years that the brain is insensitive to insulin; however, recent studies *in vivo* have presented convincing evidence that insulin has some control on uptake and metabolism of glucose by brain. For details on these studies, the reader is referred to Chapter 21, "Insulin Action," by Rafaelsen and Mellerup in Vol. IV of this series.

J. Cerebral Uptake of Sugars Other Than Glucose

After continuous administration of fructose for about 1 hr preceding perfusion and then perfusing with 100 mg % fructose in "simplified blood," Geiger *et al.*[59] found an uptake of the sugar by cat brain; however, fructose apparently did not disappear from brain during the experiment. Brain glucose content and oxygen consumption declined rapidly as in other glucose-free perfusion blood experiments. Allweis and Magnes[15] perfused cat brain with blood containing fructose (100 mg %) and [U-^{14}C] fructose. Cerebral functional activity and oxygen consumption decreased rapidly and only about 5% of brain CO_2 was derived from fructose. It was concluded that the isolated perfused cat brain was able to oxidize fructose (in the absence of glucose) only to a very limited extent.

Using a simplified version of the Geiger–Magnes technique, Eidelberg, Fishman, and Hams[65] found an uptake of arabinose as determined by measuring brain content. The data indicated a higher preference towards the (−)-arabinose (2:1) than the (+)-arabinose. The uptake could be inhibited by ouabain and by glucose. The authors concluded that the mechanism was of the carrier-mediated or of the active transport type. These data were of special interest in view of the author's recent demonstration[41] of stereospecificity of erythrocyte mutarotase, in that the conversion of α-glucose to β-glucose was inhibited to a greater extent by L-arabinose than by D-arabinose.

In the author's laboratory, [1-^{14}C] galactose, [2-^{14}C] galactose and [1-^{14}C] ribose were used in studies (unpublished) in humans. Although there was some production of $^{14}CO_2$ by brain in all cases, the data indicated little or no cerebral uptake of either galactose or of ribose. It was felt that brain $^{14}CO_2$ arose from cerebral oxidation of ^{14}C-glucose which was synthesized by the body from the injected radioisotopes.

IV. CEREBRAL METABOLISM OF LIPIDS *IN VIVO*

A. Fatty Acids

An early report by the author showed cerebral oxidation to $^{14}CO_2$ of $[1-^{14}C]$ butyrate injected intravenously into normal human subjects.[66] A comparison of brain CO_2 specific activity with venous blood butyrate specific activity gave an average value of 9.8 % of cerebral CO_2 derived from oxidation of butyrate. Under similar conditions, there was no significant production of $^{14}CO_2$ by brain with $[1-^{14}C]$ octanoic acid[66] or with $[1-^{14}C]$ acetate (unpublished results) as substrates.

Allweis et al.[67] demonstrated that $[U-^{14}C]$ palmitic acid, which had been attached to purified bovine serum albumin and incorporated into the "simplified blood" used to perfuse the cat brain, was oxidized to $^{14}CO_2$. The relative specific activity of the $^{14}CO_2$ (compared to specific activity of palmitic acid in perfused blood) rose to an average value of 2.8 % and was increased slightly when glucose was omitted from the "simplified blood."

B. Glycerol

Glycerol is an integral part of all neutral fats and of phosphoglycerides. Reportedly its administration will alleviate the cerebral disorders produced by insulin hypoglycemia.[68] However, to date, no studies exist showing that this effect is due to glycerol *per se* and not to the rapid conversion of this compound to glucose by the body. The author has studied the cerebral oxidation of $[U-^{14}C]$ glycerol and of $[1,3-^{14}C]$ glycerol.[66] In estimating that 3.5 % of cerebral $^{14}CO_2$ was derived from injected ^{14}C-glycerol, it was necessary to subtract from the brain CO_2 specific activity those values attributed to ^{14}C-glucose synthesized from $[U-^{14}C]$ glycerol (which were calculated from venous blood glucose specific activities).

V. CEREBRAL METABOLISM OF AMINO ACIDS

It has been seen that, after perfusion with "simplified blood" containing $[U-^{14}C]$ glucose, ^{14}C was rapidly incorporated into free amino acids and, much more slowly, into proteins of brain (see above). However, there is a paucity of *in vivo* studies showing the oxidation of amino acids. A number of ^{14}C-labeled amino acids and derivatives have been used by the author as injected substrates with the usual arterio–venous procedure. With L-$[U-^{14}C]$ aspartate, (V–A) $^{14}CO_2$ differences showed a small but significant metabolism by brain; however, little or no oxidation of DL-[4-^{14}C] aspartate, L-[U-^{14}C] lysine, DL-[1-^{14}C] glutamate, or L-[U-^{14}C] glutamate by brain was observed.[66] [1-^{14}C] GABA, ^{14}C-serine, and ^{14}C-alanine were readily oxidized to ^{14}CO$_2$ by brain (unpublished results). With L-[U-^{14}C] serine, (V–A) differences were not nearly as large as with DL-[1-^{14}C] serine, indicating that C$_1$ contributed most of the activity of brain ^{14}CO$_2$. With DL-[1-^{14}C] alanine, the (V–A) ^{14}CO$_2$ differences were about the same as with L-[U-^{14}C] alanine; however, arterial curves were much higher with DL-[1-^{14}C] alanine, suggesting rapid body oxidation of C$_1$ of alanine. L-[U-^{14}C] glutamine and [carboxy-^{14}C] DOPA showed only isolated instances of ^{14}CO$_2$ production by brain (unpublished results), as did also ^{14}C-phenylalanine and ^{14}C-tyrosine.[69] The (A–V) and (V–A) ^{14}CO$_2$ differences suggested sporadic oxidation (or decarboxylation) or, more likely, compartmentalization of brain with varying rates of oxidation (or decarboxylation) in compartments.

Bianchi Porro and co-workers,[70] using the arterio–venous technique of Sacks, studied the cerebral metabolism of L-[U-^{14}C] glutamine and L-[U^{14}C] glutamic acid in chronic mental patients. Under basal conditions, they found a lag (4–10 min) in production of cerebral ^{14}CO$_2$ from both of these substrates. In one case, labeled GABA was found in cerebral venous blood after administration of L-[U-^{14}C] glutamine and in two cases after L-[U-^{14}C] glutamic acid injection. In this preliminary work, the actual data (personal communication with Dr. Bianchi Porro and Dr. Maiolo) corroborated an early study with ^{14}C-glutamic acid from the author's laboratory[66] and unpublished results with L-[U-^{14}C] glutamine (described above) indicating only sporadic production of ^{14}CO$_2$ by brain.

In an investigation to determine the relative rate of pyruvic acid formation in brain by decarboxylation of oxaloacetate or malate, Gombos, Geiger, and Otsuki[47] added ^{14}C-labeled aspartate to artificial blood in their cat brain perfusion experiments. It was seen that aspartic acid entered brain very slowly, since the specific activity of free aspartate in brain, after about 2 hr of perfusion, had only about 1% of the specific activity of aspartate in blood. By recalculating some of their data, one finds that after about 2 hr of perfusion with L-[U-^{14}C] aspartate, brain ^{14}CO$_2$ specific activity was about 31% that of brain free aspartate. Furthermore, the specific activity of free glutamate of brain at that time was about 29% that of brain free aspartate. Thus, it becomes apparent that although aspartate was taken up by brain from blood only very slowly, an unusually large per centage (31) of brain CO$_2$ was derived from the free aspartate of brain. That glutamate had nearly the same specific activity as brain respiratory CO$_2$ offers supporting evidence for the author's proposed scheme[43] for cerebral metabolism of glucose *in vivo* (see above). From the specific activity values found in cerebral lactate, it was concluded[47] that about 10% of pyruvate formed in brain was derived from decarboxylation of oxaloacetate or malate.

An important contribution to our knowledge of cerebral amino acid metabolism in man *in vivo* was made by Knauff, Gottstein, and Miller,[71] who obtained arterial (femoral) and venous (internal jugular bulb) blood

TABLE III
Summary of Cerebral Metabolism in Vivo

Animal species	Metabolite	Rate or evidence of utilization
Man	Oxygen	3.3 ml/100 g brain/min or 46 ml/min for total brain[6]a
	Glucose	5.4 mg/100 g brain/min or 76 mg/min for total brain[6]a
	DL-, D-, and L-^{14}C lactate	Oxidized to $^{14}CO_2$ [16]
	[1-^{14}C]pyruvate	Oxidized to $^{14}CO_2$ [16]
	[2-^{14}C]- and [1,4-^{14}C]fumarate	Oxidized to $^{14}CO_2$ [7]b
	[5-^{14}C]α-oxoglutarate	Oxidized to $^{14}CO_2$ c
	[1-^{14}C]glucose-6-phosphate and [U-^{14}C]glucose-1-phosphate	Oxidized to $^{14}CO_2$ [42]d
	[1-^{14}C]butyrate	Oxidized to $^{14}CO_2$ [66]
	[1,3-^{14}C] and [U-^{14}C]glycerol	Oxidized to $^{14}CO_2$ [66]
	L-[U-^{14}C]aspartate	Oxidized to $^{14}CO_2$ [66]
	[1-^{14}C]GABA	Oxidized to $^{14}CO_2$ c
	L-[U-^{14}C] and DL-[1-^{14}C]serine	Oxidized to $^{14}CO_2$ c
	L-[U-^{14}C] and DL-[1-^{14}C]alanine	Oxidized to $^{14}CO_2$ c
	Free amino acids	5 % (A-V) difference in plasma free total α-amino acid nitrogen[71]; 0.14 mg % (A-V) difference in whole blood α-amino acid nitrogen c
Cat (perfused brain)	Oxygen	Initially up to 6 ml/100 g brain/min, declining to 4 ml/100 g brain/min[13]; 6.9 ml/100 g brain/min[15]
	Glucose	4.1–8.7 mg/100 g brain/min[13]e; 9.2 mg/100 g brain/min[15]
	[U-^{14}C]fructose	Oxidized to $^{14}CO_2$ [15]
	[U-^{14}C]palmitic acid f	Oxidized to $^{14}CO_2$ [67]
Rat (perfused isolated brain)	Glucose	About 2.5 mg/100 g brain/min with 200 mg % blood glucose, about 7.3 mg/100 g brain/min with 300–400 mg % blood glucose[24]g
Dog (perfused isolated brain)	Oxygen	3.6 ml/100 g brain/min[25–27]
	Glucose	3.4–5.7 mg/100 g brain/min[25–27]
Dog (transplanted brain)	Oxygen	2.7 ml/100 g brain/min[29]h
	Glucose	5.6 mg/100 g brain/min[29]h
Monkey (perfused isolated brain)	Oxygen	2.8 ml/100 ml blood[28]i
	Glucose	2.3 mg/100 g brain/min[28]i

a Total weight of brain assumed to be 1400 g.
b Reference for [2-^{14}C]fumarate: [1.4-^{14}C]fumarate studies were unpublished (see above text).
c Unpublished results (see above text).
d Prior conversion to ^{14}C-glucose might be necessary.
e Fluctuated considerably during experiment and from one experiment to another.
f Attached to bovine serine albumin.
g With brain and perfusion blood at room temperature.
h With average CBF of 24.2 ml/100 g brain/min.
i At 28–32°C.

samples simultaneously for determination of plasma levels of 19 free amino acids, taurine, and urea in 25 patients. Cerebral oxygen consumption and CBF were measured, and utilization of amino acids/100 g brain/min was calculated. Significant (A–V) as well as (V–A) differences were found for all of the substances examined, although in no case were all 19 amino acids simultaneously consumed or released. The authors discovered that invariably uptake of certain amino acids was combined with release of others and concluded that there apparently existed in brain an exchange mechanism going on in both directions whereby significant amounts of free amino acids could be metabolized. An average value for (A–V) free total α-amino acid nitrogen of 5 % was found.

In amino acid studies done in the author's laboratory, α-amino acid nitrogen determinations were performed routinely on whole blood samples. In accordance with the results observed in Ref. (71) with plasma samples, our data (unpublished) also showed considerable variations, in that both (A–V) and (V–A) differences were found. With 222 sets of determinations done on 67 individuals, average values were 6.16 mg % for arterial blood, 6.02 mg % for cerebral venous blood, and 0.14 mg % for (A–V) difference. This gave an average value of about 2 % for (A–V) free total α-amino acid nitrogen, a figure slightly lower than the 5 % reported in Ref. (71).

A brief summary of the available data on cerebral metabolite utilization is given in Table III.

VI. REFERENCES

1. H. Waelsch, *in Neurochemistry* (K. A. C. Elliott, I. H. Page, and J. H. Quastel, eds.), p. 290, Charles C. Thomas, Springfield, Ill. (1962).
2. A. Geiger, Correlation of brain metabolism and function by the use of a brain perfusion method *in situ. Physiol. Rev.* **38**:1–20 (1958).
3. P. R. Dumke and C. F. Schmidt, Quantitative measurements of cerebral blood flow in the macaque monkey. *Am. J. Physiol.* **138**:421–431 (1943).
4. S. S. Kety and C. F. Schmidt, The determination of cerebral blood flow in man by the use of nitrous oxide in low concentrations. *Am. J. Physiol.* **143**:53–66 (1945).
5. S. S. Kety, *in Methods in Medical Research*, Vol. 1, pp. 204–217, Year Book Publishers, Chicago (1948).
6. S. S. Kety, in *Neurochemistry* (K. A. C. Elliott, I. H. Page, and J. H. Quastel, eds.), p. 113, Charles C. Thomas, Springfield, Ill. (1962).
7. W. Sacks, Cerebral oxidation of fumarate-2-C^{14} in normal human subjects. *J. Appl. Physiol.* **9**:43–48 (1956).
8. W. Sacks, Cerebral metabolism of isotopic glucose in normal human subjects. *J. Appl. Physiol.* **10**:37–44 (1957).
9. R. V. Coxon and R. J. Robinson, Specific activity of carbon dioxide dioxide in arterial and venous blood following injection of ^{14}C-labelled glucose. *J. Physiol.* **132**:48–49P (1956).
10. R. J. Robinson and R. V. Coxon, Radioactivity of blood carbon dioxide in animals oxidizing glucose labelled with carbon-14 and other labelled substances. *Nature* **180**:1279–1281 (1957).
11. R. V. Coxon and R. J. Robinson, The transport of radioactive carbon dioxide in the blood stream of the dog after administration of radioactive bicarbonate. *J. Physiol.* **147**:469–486 (1959).

12. R. V. Coxon and R. J. Robinson, Movements of radioactive carbon dioxide within the animal body during oxidation of ^{14}C-labelled substances. *J. Physiol.* **147**:487–510 (1959).

13. A. Geiger and J. Magnes, The isolation of the cerebral circulation and the perfusion of the brain in the living cat. *Am. J. Physiol.* **149**:517–537 (1947).

14. A. Geiger and S. Yamasaki, Cytidine and uridine requirement of the brain. *J. Neurochem.* **1**:93–100 (1956).

15. C. Allweis and J. Magnes, The uptake and oxidation of glucose by the perfused cat brain. *J. Neurochem.* **2**:326–336 (1958).

16. W. Sacks, Cerebral metabolism of glucose-3-C^{14}, pyruvate-1-C^{14} and lactate-1-C^{14} in mental disease. *J. Appl. Physiol.* **16**:175–180 (1961).

17. U. Gottstein, A. Bernsmeirer, and J. Sedlmeyer, Der Kohlenhydratstoffwechsel des menschlichen Gehirns. I. Untersuchungen mit substratspezifischen enzymatischen Methoden bei normaler Hirndurchblutung. *Klin Wschr.* **41**:943–948 (1963).

18. P. Scheinberg, B. Bourne, and O. M. Reinmuth, Human cerebral lactate and pyruvate extraction. *Arch. Neurol.* **12**:246–250 (1965).

19. S. Sato, M. Tateyama, C. Sasamori, S. Kobayashi, Y. Chiba, and Y. Takeda, On the intermediate metabolism of carbohydrates in the brain of healthy persons. *Tohoku J. Exptl. Med.* **81**:215–221 (1963).

20. J. Kneinerman, S. M. Sancetta, and D. B. Hackel, Effect of high spinal anesthesia on cerebral circulation and metabolism in man. *J. Clin. Invest.* **37**:285–293 (1958).

21. H. I. Otani, Pathophysiological study on cerebral carbohydrate metabolism in essential hypertension and cerebral arteriosclerosis: I. Study on cerebral carbohydrate metabolism during rest. *Jap. Circ. J.* **27**:534–546 (1963).

22. G. G. Rowe, G. M. Maxwell, C. A. Castillo, D. J. Freeman, and C. W. Crumpton. A study in man of cerebral blood flow and cerebral glucose, lactate, and pyruvate metabolism before and after eating. *J. Clin. Invest.* **38**:2154–2158 (1959).

23. W. Sacks, The cerebral metabolism of L- and D-lactate-C^{14} in humans *in vivo. Ann. N. Y. Acad. Sci.* **119**:1091–1108 (1965).

24. R. K. Andjus, K. Suhara, and H. A. Sloviter, An isolated, perfused rat brain preparation, its spontaneous and stimulated activity. *J. Appl. Physiol.* **22**:1033–1039 (1967).

25. D. D. Gilboe, W. W. Cotanch, and M. B. Glover, Extracorporeal perfusion of the isolated head of a dog. *Nature* **202**:399–400 (1964).

26. D. D. Gilboe, W. W. Cotanch, and M. B. Glover, Isolation and mechanical maintenance of the dog brain. *Nature* **206**:94–96 (1965).

27. D. D. Gilboe, W. W. Cotanch, M. B. Glover, and V. A. Levin, Changes in electrolytes, pH, and pressure of blood perfusing isolated dog brain. *Am. J. Physiol.* **212**:589–594 (1967).

28. R. J. White, M. S. Albin, and J. Verdura, Preservation of viability in the isolated monkey brain utilizing a mechanical extracorporeal circulation. *Nature* **202**:1082–1083 (1964).

29. R. J. White, M. S. Albin, G. E. Locke, and E. Davidson, Brain transplantation: Prolonged survival of brain after carotid-jugular interposition. *Science* **150**:779–781 (1965).

30. G. Moss, Cerebral arterial isolation: The effects of differential pressure perfusion. *J. Surg. Res.* **4**:170–177 (1964).

31. G. Moss, The contribution of the hexose monophosphate shunt to cerebral glucose metabolism. *Diabetes* **13**:585–591 (1964).

32. R. L. Swank and W. Hissen, Isolated cat head perfusion by donor dog. *Arch. Neurol.* **13**:93–100 (1965).

33. D. D. Gilboe, M. B. Glover, and W. W. Cotanch, Blood filtration and its effect on glucose metabolism by the isolated dog brain. *Am. J. Physiol.* **213**:11–15 (1967).

34. C. Allweis, M. Abeles, and J. Magnes, Perfusion of cat brain with simplified blood after filtration through glass wool. *Am. J. Physiol.* **213**:83–86 (1967).

35. S. S. Kety, *in Neurochemistry* (K. A. C. Elliott, I. H. Page, and J. H. Quastel, eds.), p. 119, Charles C. Thomas, Springfield, Ill. (1962).
36. A. Geiger, Y. Kawakita, and S. S. Barkulis, Major pathways of glucose utilization in the brain in brain perfusion experiments *in vivo* and *in situ*. *J. Neurochem.* 5:323–338 (1960).
37. C. L. Allweis, H. Gainer, and I. L. Chaikoff, Method for kinetic study of in vitro conversion of a C^{14}-labeled substrate to CO_2. *J. Appl. Physiol.* 15:949–952 (1960).
38. H. Gainer, C. L. Allweis, and I. L. Chaikoff, Precursors of metabolic CO_2 produced by the brain of the anaesthetized, intact dog: The effect of electrical stimulation. *J. Neurochem.* 10:903–908 (1963).
39. A. Barkai and C. Allweis, The contribution of blood glucose to the carbon dioxide produced by the narcotized brain of the intact cat. *J. Neurochem.* 13:23–33 (1966).
40. G. Gombos, S. Otsuki, W. Scruggs, G. Whitney, A. Schmolinske, and A. Geiger, Brain metabolites of normal intact narcotized cats. *Fed. Proc.* 22:633 (1963).
41. W. Sacks, Isolation and properties of mutarotase in erythrocytes. *Arch. Biochem. Biophys.* 123:507–513 (1968).
42. W. Sacks, Conversion of glucose phosphate-^{14}C to glucose-^{14}C in passage through human brain *in vivo*. *J. Appl. Physiol.* 24:817–827 (1968).
43. W. Sacks, Cerebral metabolism of doubly labeled glucose in humans *in vivo*. *J. Appl. Physiol.* 20:117–130 (1965).
44. S. S. Barkulis, A. Geiger, Y. Kawakita, and V. Aguilar, A study of the incorporation of ^{14}C derived from glucose into the free amino acids of the brain cortex. *J. Neurochem.* 5:339–348 (1960).
45. A. Geiger, N. Horvath, and Y. Kawakita, The incorporation of ^{14}C derived from glucose into the proteins of the brain cortex, at rest and during activity. *J. Neurochem.* 5:311–322 (1960).
46. R. U. Margolis, S. S. Barkulis, and A. Geiger, A comparison between the incorporation of ^{14}C from glucose into N-acetyl-L-aspartic acid and aspartic acid in brain perfusion experiments. *J. Neurochem.* 5:379–382 (1960).
47. G. Gombos, A. Geiger, and S. Otsuki, The metabolic pattern of the brain in brain perfusion experiments *in vivo*—II. Pyruvate and lactate formation from ^{14}C-labelled aspartate. *J. Neurochem.* 10:405–413 (1963).
48. E. L. Gibbs, W. G. Lennox, L. F. Nims, and F. A. Gibbs, Arterial and cerebral venous blood, arterial–venous differences in man. *J. Biol. Chem.* 144:325–332 (1942).
49. W. A. Himwich and H. E. Himwich, Pyruvic acid exchange of the brain. *J. Neurophysiol.* 9:133–136 (1946).
50. R. V. Coxon, *in Metabolism of the Nervous System* (D. Richter, ed.), pp. 303–322, Pergamon Press, London (1957).
51. K. Y. Hostetler, B. R. Landau, R. J. White, M. S. Albin, and D. Yoshon, Pentose cycle contribution to glucose metabolism in isolated, perfused monkey brain (in preparation).
52. W. Sacks and C. O. Jensen, Malease, a hydrase from corn kernels. *J. Biol. Chem.* 192:231–236 (1951).
53. W. Sacks, Evidence for the metabolism of maleic acid in dogs and human beings. *Science* 127:594 (1958).
54. S. Angielski, The effect and metabolism of maleic acid in the kidney. *Acta Biol. Med. Soc. Sc. Gedan.* 7:61–97 (1963).
55. S. Englard, J. S. Britten, and I. Listowsky, Stereochemical course of the maleate hydratase reaction. *J. Biol. Chem.* 242:2255–2259 (1967).
56. E. Figueroa and A. Pfeifer, Incorporation of ^{14}C-glucose and ^{14}C-glucose-6-phosphate into glycogen and CO_2 by rat liver slices. *Nature* 204:576–577 (1964).
57. B. R. Landau and E. A. H. Sims, On the existence of two separate pools of glucose 6-phosphate in rat diaphragm. *J. Biol. Chem.* 242:163–172 (1967).

58. D. G. Fraenkel, F. Falcoz-Kelly, and B. L. Horecker, The utilization of glucose 6-phosphate by glucokinaseless and wild-type strains of *Escherichia coli*. *Proc. Natl. Acad. Sci.* **52**:1207–1213 (1964).

59. A. Geiger, J. Magnes, R. M. Taylor, and M. Veralli, Effect of blood constituents on uptake of glucose and on metabolic rate of the brain in perfusion experiments. *Am. J. Physiol.* **177**:138–149 (1954).

60. S. Sato, M. Tateyama, C. Sasamori, S. Kobayshi, Y. Chiba, and Y. Takeda, On the intermediate metabolism of carbohydrates in the brain of hypertensive and postapoplectic patients. *Tohoku J. Exptl. Med.* **81**:207–214 (1963).

61. U. Gottstein, K. Held, H. Sebening, and G. Walpurger, Der Glucoseverbrauch des menschlichen Gehirns unter dem Einfluss intravenöser Infusionen von Glucose, Glucogon und Glucose-Insulin. *Klin. Wschr.* **43**:965–975 (1965).

62. H. E. Himwich, K. M. Bowman, J. F. Fazekas, and W. Goldfarb, Biochemical changes occurring in the cerebral blood during the insulin treatment of schizophrenia. *J. Nerv. Ment. Dis.* **89**:273–293 (1939).

63. U. Gottstein and K. Held, Insulinwirkung auf den menschlichen Hirnmetabolismus von Stoffwechselgesunden und Diabetikern. *Klin. Wschr.* **45**:18–23 (1967).

64. W. J. H. Butterfield, R. A. Sells, M. E. Abrams, G. Sterky, and M. J. Whichelow, Insulin sensitivity of the human brain. *Lancet* **1**:557–560 (1966).

65. E. Eidelberg, J. Fishman, and M. L. Hams, Penetration of sugars across the blood–brain barrier. *J. Physiol.* **191**:47–57 (1967).

66. W. Sacks, Cerebral metabolism of isotopic lipid and protein derivatives in normal human subjects. *J. Appl. Physiol.* **12**:311–318 (1958).

67. C. Allweis, T. Landau, M. Abeles, and J. Magnes, The oxidation of uniformly labelled albumin-bound palmitic acid to CO_2 by the perfused cat brain. *J. Neurochem.* **13**:795–804 (1966).

68. C. Voegtlin, E. R. Dunn, and J. W. Thompson, The antagonistic action of certain sugars, amino acids, and alcohol on insulin intoxication. *Am. J. Physiol.* **71**:574–582 (1925).

69. W. Sacks, Phenylalanine metabolism in control subjects, mental patients and phenylketonurics. *J. Appl. Physiol.* **17**:985–992 (1962).

70. G. Bianchi Porro, A. T. Maiolo, P. Della Porta, E. Rossella, and E. Polli, Cerebral metabolism of L-[U-^{14}C] glutamine and of L-[U-^{14}C] glutamic acid in chronic mental disease and in therapeutic insulin coma, *in Abstracts of 1st International Society for Neurochemistry*, p. 25 (1967).

71. H. G. Knauff, U. Gottstein, and B. Miller, Untersuchungen über den Austausch von freien Aminosäuren und Harnstoff zwichen Blut und Zentralnervensystem. *Klin. Wschr.* **42**:27–39 (1964).

72. W. Sacks, Cerebral metabolism of isotopic glucose in chronic mental disease. *J. Appl. Physiol.* **14**:849–854 (1959).

Chapter 16
ENZYMES

N. Seiler

Max Planck Institut für Hirnforschung
Arbeitsgruppe Neurochemie
Frankfurt am Main, Germany

I. INTRODUCTION

Cerebral enzymes have been studied for quite some time now. Oppen-heimer,[1] in his review of the enzymes found in the brain, mentioned the presence of lipases, amylases, inulinase, maltase, lactase, saccharase, catalase, oxidases, reductases, proteases, guanase, and urease. The autolyses of proteins, nucleic acids, and phosphatides have also been described. This review[1] does not cover enzymes with special functional importance in the CNS, such as enzymes of the ACh-metabolism, Glu-metabolism, or phospholipid metabolism. Until the present time most enzyme studies isolated and characterized enzymes from organs with high activity, with only a few on cerebral enzymes. Although Warburg[2] postulated in 1948 that enzymes characterizing the same metabolic step should vary according to various organs, the first direct evidence for this appeared 10 years later in the occurrence of different types of lactate dehydrogenase (LDH) (1.1.1.27), in the different organs of one animal.[3] Up to the present time few enzymes found in the brain could be called typical cerebral enzymes. This is not surprising, since the principal metabolic reactions, such as energy production, are similar in all cells and only a few enzymes are known to occur almost exclusively in the brain. However, enzymes, even if not completely restricted to the CNS, may be responsible for a specific function in the nervous system, i.e., in generation and propagation of electric impulses. An enzyme that illustrates these points may be ATP:pyridoxal 5-phosphotransferase (2.7.1.35), which is present in brain and in yeast,[4] with the brain enzyme, unlike the yeast enzyme, being activated more by Zn^{2+} than by Mg^{2+} [5]. This enzyme produces pyridoxal-5-phosphate, the coenzyme for glutamic acid decarboxylase (4.1.1.15) (GAD). GAD is present in bacteria, plants, and, as shown by Roberts and Frankel,[6,7] in brain, mostly in cortex. GAD is responsible for the formation of GABA, and is a participant in the "GABA-shunt,"[8,9] a process of succinate formation that parallels oxidative decarboxylation of α-ketoglutaric acid.

GABA α-ketoglutarate transaminase (2.6.1.19) is found not only in the brain but also in other organs.[10] This enzyme could result in the formation of GABA from glutamate and succinate semialdehyde. Recently there has been found some evidence for this pathway of GABA formation.[11,12]

Among enzymes important to brain function and studied in detail are acetylcholinesterase (3.1.1.7) (AchE), choline acetylase (2.3.1.6) (ChA), enzymes of amine metabolism such as tyrosine hydroxylase, tryptophan-5-hydroxylase, dopa-decarboxylase (4.1.1.26), dopamine hydroxylase (1.14.2.1), and catechol methyltransferase (2.1.1.6), and mitochondrial monoamine oxidase (1.4.3.4) (MAO). In contrast to extracellular amine oxidases, mitochondrial MAO needs FAD as a prosthetic group.[13] The special role of some enzymes is shown by their unusually high activity or by their unusual localization in the CNS, even though they may be present in other organs. An example for this is the Mg^{2+}-dependent, Na^+ and K^+ activated ATPase (3.6.1.3).[14] Highest activities of this enzyme are in the cortex,[15] and their localization in membranes suggests their importance in ion transport.[16] Hexokinase (2.7.1.1) is another example; in rat, according to Long,[17] activities are as follows: brain, 1.00; heart, 0.53; spleen, 0.31; kidney, 0.29; muscle, 0.27; pancreas, 0.21; lung, 0.16; and liver, 0.05. In contrast to other organs, cerebral hexokinase is bound to a large extent to particles; hexokinases in the other organs are also different enzymes.[18] A selective compilation of enzymes found in the CNS is given in the summary table that comprises Section V; it indicates occurrence, purification, isolation, characterization, distribution, and developmental changes. A number of facts are omitted because some of the old results are no longer of value, and comparing others is difficult because different standards are used, e.g., expression of units in terms of wet weight or protein content. Since few highly purified enzymes were obtained from brain, pH optima mentioned are useful only as a guide. Among missing data about cerebral enzymes is mechanism of action, since this has been mostly studied on enzymes from outside the brain. The presence of a number of other enzymes in the brain is only indirectly shown, especially those participating in lipid metabolism.

II. THE SUBCELLULAR LOCALIZATION OF ENZYMES

A. Particulate Fractions

The separation of tissue homogenates by differential centrifugation in isotonic or hypertonic solution, to separate nuclei, mitochondria, microsomes, and the supernatant fluid, does not give uniform fractions for the brain nor differentiate between neural and glial elements.* Recent methods may make it possible to separate pure nerve cell perikarya and glial cells on a preparative scale.[63,71,1433] Earlier papers are of limited value because

---

* The micromethods of Lowry,[1430] Hyden,[1431] and Roots[1432] furnish pure nerve cell perikarya and glial cells, but in so little numbers that only special techniques can be employed for their biochemical investigation.

morphological controls of subfractions were seldom used. The following paragraph deals with more recent results on enzymes of the CNS, and the table gives references to other investigations of the subcellular location of enzymes.

1. Nuclei

For the isolation of nuclei from brain homogenate, special methods have been reported.[1434–1436] Autoradiographic experiments show that 3H-thymidine is incorporated specifically into the nuclei of glial cells,[1437] giving evidence for the connection between mitosis and synthesis of DNA that occurs before cell division. Some enzymes of the Embden–Meyerhof pathway and the citric acid cycle have been observed in extracts of isolated nuclei from rat brain: hexokinase (2.7.1.1), aldolase (4.1.2.13), enolase (4.2.1.11), pyruvate kinase (2.7.1.40), lactate dehydrogenase (LDH) (1.1.1.27), triose-phosphate dehydrogenase (1.2.1.12), NADP, isocitrate dehydrogenase (1.1.1.42), NAD malate dehydrogenase (1.1.1.37) (MDH), adenylate kinase (2.7.4.3), and ATPase (3.6.1.3).[72] The role of $NADP^+$ and MDH in the nuclei is not known. The other enzymes are used for synthesis and regulation of the concentration of the adenosinemono- and polyphosphates. Mandel, Dravid, and Pete[676] have shown the ability of nuclei to synthesize poly-nucleotides. They found a CTP polymerizing enzyme (poly C synthetase) in nuclear fractions of the rat brain. Richter and Hullin observed a high activity of alkaline phosphatase (2.1.3.1) in the nuclei of human brain.[822] Another characteristic enzyme of the nuclei, the ATP:NMN-adenylyl-transferase (2.7.7.1), is used in the synthesis of NAD. This enzyme is found in higher concentration in neuronal nuclei than in glial cells.[669]

2. The Mitochondrial Fractions

The mitochondrial fractions obtained by the Brody and Bains method[1438] could be further separated by Gray and Whittaker[179,194, 1439,1440] using density gradient centrifugation. The fraction in the density of 0.8–1.2 M sucrose solution contained detached presynaptic nerve endings (synaptosomes); their projections had been torn off during homogenization. de Robertis and coworkers[177,1441] independently reported the identity of the nerve-ending particles with the particles of the submitochondrial fraction. The nerve-ending particles usually contain one or more mitochondria. They are filled with synaptic vesicles.[1442] Through osmotic shock the synaptic vesicles can be liberated from the synaptosomes, and by further density-gradient centrifugation they can be separated from the mitochondria, the synaptic membranes, and cytoplasmic particles of the nerve-ending particles.[632,1443]

3. Myelin

The regular concentric layers observed with electron microscope are characteristic for the myelin disc.[1444] They are made up of layers of lipids and proteolipid protein.[1445] In recent years it was possible to isolate pure myelin fractions of the brain by repeated density-gradient centrifugation.[1210,1446,1447] The lipid components in the different myelin fractions are almost in a stoichiometric relationship: cholesterin:phospholipid:galactolipid = 4:4:2. The single fractions differ only in the protein:lipid relationship.[1446]

Myelin seems to have a poor enzyme complement. Besides proteinase and alkaline phosphatase activity[1448] only the presence of Mg^{2+} dependent and Na^+ and K^+ activated ATPase could be shown in purified myelin fractions.[1209,1210]

B. The Embden–Meyerhof Pathway and Oxidative Metabolism

The enzymes involved in the Embden–Meyerhof pathway are generally soluble, and they are found in the supernatant fluid after centrifugation; 7–28% of these enzymes are bound to particles.[28] Earlier investigations seemed to indicate that brain mitochondria showed glycolytic activity, in contrast to mitochondria of other organs. By separation of mitochondrial fraction in a density gradient it was shown that mitochondria of either mature or of embryonic rat brain have no glycolytic activity.[102,581,1039]

LDH was investigated more closely in respect to its subcellular distribution. The bound fraction was localized in the cytoplasm of the synaptosomes.[74] Pfleiderer and Wachsmuth[41] found 5 isoenzymes of the LDH in the human brain. The soluble and the particle-bound LDH contains the same isoenzymes.[74] LDH is used during cell fractionation as a marker enzyme because of its occurrence exclusively in cytoplasm. An exception to the rule that the glycolytic enzymes are cytoplasmic is hexokinase, which is bound 85–90% to particles.[28,578] Hexokinase content of mitochondria varies from 30 to 75%. Bachelard[181] found 25% in the microsomal fraction, mainly bound to membrane fragments, and 40% in the primary mitochondrial fraction. In the density gradient, hexokinase moved together with the succinic dehydrogenase (1.3.99.1) (SDH); it is likely that the hexokinase is a mitochondrial enzyme. The soluble hexokinase is not an artifact; a soluble form different in its kinetic and electrophoretic properties from the mitochondrial enzyme has been shown.[181] The mitochondrion is a complete functional unit, containing all necessary enzymes, coenzymes, and cofactors. It needs only to be supplied with oxygen, inorganic phosphate, and Mg^{2+}. The enzymes of the oxidative phosphorylation, like the cytochrome oxidase system, are almost all localized in the mitochondria[183,298] even in the brain of newborn rats.[99] The enzymes of the citric acid cycle are mainly found in these organelles.[102] The SDH,[99,175] NAD$^+$ isocitric dehydrogenase,[99,104] and L-malate hydrolyase (4.2.1.2)[76,102] were found almost exclusively in mitochondria. Other enzymes like MDH,[96,99] aconitase (4.2.1.3),[99,1342] and NADP isocitric dehydrogenase[99,1342] were found to a large extent in the cytoplasm. The cytoplasmic enzymes often show different electrophoretic and kinetic properties than the ones bound to particles. For example, the mitochondrial aconitase has a pH optimum of 5.8, while the soluble aconitase works optimally at pH 7.4.[99] The enzyme pattern of the brain mitochondria does not differ greatly from the pattern of mitochondria from other tissue.[1142] Contrary to widespread opinion, cerebral mitochondria are able to oxidize a number of fatty acids at rates similar to that of liver mitochondria.[203,204] Brain homogenates oxidise amino acids as well.[226]

All mitochondria analyzed gave a constant molar relationship of cytochrome a per cytochrome c.[100] In relation to the cytochrome c level, MDH, SDH, pyruvate oxidase (1.2.3.3), and glutamate-oxaloacetate transaminase (2.6.1.1) activities are almost constant; also, there are large differences in the absolute activities.[100] However, there are large differences in the proportion of other enzymes, like NAD^+ isocitrate dehydrogenase, glutamate dehydrogenase (1.4.1.4) GDH, and glycerol-3-phosphate oxidase (1.1.1.8).[100] The enzymes of the citric acid cycle are so geared that the pyruvate oxidation (*in vitro*) leads to an accummulation of α-oxoglutarate. In the presence of both substrates, brain mitochondria oxidize pyruvate preferentially and liver mitochondria oxidize α-oxoglutarate. *In vivo*, α-ketoglutaric acid in the brain will be quickly metabolized to glutamic acid. The high glutamate level is caused by the characteristic enzyme pattern of the citric acid cycle in brain mitochondria.[1449]

Mitochondria are biochemically and morphologically inhomogeneous in the brain. Cytochrome-oxidase activity per milligram protein is less in mitochondria of the thalamus than in mitochondria of the cerebral cortex and the cerebellar cortex and is even less in mitochondria of the corpus callosum.[312] Weiner[197] showed that the relationship of monoamine oxidase activity to SDH activity is not constant in the different regions of the brain. MAO is more active in the thalamus-hypothalamus region and SDH in the more cellular areas of the brain. The exclusively mitochondrial localization of MAO has been confirmed by Rodriguez de Lores Arnaiz and de Robertis using an improved method.[172] With fetal chicken brain, MAO is found in the supernatant fraction as well as in mitochondria.[259]

The enzymes of the pentosephosphate-shunt have been found in liver and kidney only in the supernatant fraction. Yamade and Shimazono[112] have shown that 10% of the activity of glucose-6-phosphate dehydrogenase (1.1.1.44) (G-6-DH) and 6-phosphogluconate dehydrogenase (1.1.1.43) was in particles of the microsome and mitochondria fractions in homogenates of brains from guinea pigs, rats, and rabbits. With improved methods of fractionation these enzymes in the subcellular structure can be more accurately located.

C. Glutamic Acid Metabolism

The reversible step of the α-oxoglutarate to glutamate is catalyzed by GDH as well as by the aminotransferases.[101,209,530] In brain, liver, and kidney GDH is mainly associated with the $NADPH^-$ system.[101] It is located only in the mitochondria and follows almost the same distribution pattern as the SDH.[147] The alanine aminotransferase (2.6.1.2) controls the equilibrium between citric acid cycle and glycolytic activity and supplies alanine from pyruvate. It shows a bimodal distribution; one part is soluble and one is bound to mitochondria. A small part is enclosed in the synaptic complexes. Aspartate aminotransferase (2.6.1.1) (GOT) also shows this bimodal distribution.[147,211,489] It is likely that soluble enzyme keeps the equilibrium between glutamate and aspartate, while the mitochondrial one is responsible for the formation of aspartate and acetyl-CoA.[147]

The so-called GABA-shunt plays an important role in the metabolism of cerebral gray matter. About 10% (0.03 M GABA/g) of the turnover rate of the citric acid cycle of the energy production in the brain is contributed by it.[1450] McKhann et al.[1451] assumed that this pathway is confined to the neurons. There is a reciprocal connection in the activity between citric acid cycle and GABA-shunt, at least in vitro.[1452]

The location of single enzymes of the GABA-shunt is contradictory. Glutamate decarboxylase (4.1.1.15) was found by Løvtrup[1267] in nuclei and by Albers[553] in the mitochondrial fraction. Salganicoff and de Robertis[549] showed, as did Weinstein, Roberts, and Kakefuda,[1269] that the main part of the GAD was located in nerve endings, mainly in fractions of the nerve ending with little AChE. By osmotic shock the GAD could be liberated from synaptosomes. It seems to be located in the axoplasm of the nerve ending. Fixation experiments with Ca^{2+} indicated that the GAD is possibly loosely bound to the membranes of the synaptic vesicles.[147] The experiments by van Kempen et al.[315] suggest a specific location of the GAD in comparison with other enzymes. They found that the GAD peak moves continuously in a density gradient when the time of centrifugation is prolonged. These authors suggest the possibility of a special "GAD particle." Baláfz, Dahl, and Harwood,[211] on the other hand, showed a bimodal distribution of the GAD after separation of the primary mitochondrial fraction in a continuous density gradient; partly the enzyme was in the supernatant fraction, partly in the nerve-ending particles. GABA aminotransferase (2.6.1.19)[211,315, 549,553] and succinate semialdehyde dehydrogenase[147] (1.2.1.16), the two other enzymes of the GABA-shunt, are localized in mitochondria. Their activity in the mitochondria of the cell soma, in comparison to the SDH activity, is much higher than in the mitochondria of nerve ending particles. Numerous observations indicate the inhibitor role of GABA.[1453–1455] The production of this amino acid in the nerve ending particles, the transport into the adjacent neuron, and the reduction by enzymes in the somatic mitochondria fit into the picture which we have of the mechanism of a synaptic transmitter. It will be shown that the localization of enzymes of the metabolism of some amines is similar to that of the enzymes of the GABA-shunt. According to Waelsch[1419] the Glu-synthesis occurs in the microsome fraction of the brain homogenate. Salganicoff and de Robertis[147] confirmed these results. Because of the complex nature of the microsome fraction it has not been possible to give an exact location of the glutamine synthetase [L-glutamate:ammonia ligase, (6.3.1.2) (GS)]. A small GS activity could also be located in noncholinergic nerve endings.[258] The changes in the nerve endings originated by methioninsulphoximine, a convulsant that inhibits the GS, and the location of this enzyme in the synaptosome fraction suggests a role of the GS in the synaptic apparatus in noncholinergic nerves.[1456]

Glutaminase I (3.5.1.2) has been shown in mitochondria fractions by different authors, but only Salganicoff and de Robertis[147] have shown that this enzyme is localized exclusively in mitochondria and follows the distribution of the SDH in the subcellular fractions closely. All results indicate that glutaminase regulates the Glu content in the nerve ending.

D. Choline Acetylase, Acetylcholine Esterase, and Biogenic Amine Metabolism

The best criterion for a cholinergic neuron is the presence of high concentrations of ACh in its ending. Already Feldberg and Vogt[1457] and other authors[414] showed a fairly good correlation among the enzymes of the ACh metabolism, choline acetylase (2.3.1.6) (ChA), and acetylcholine esterase (3.1.1.7) (AChE) and the ACh content in the different regions of the brain. There are some exceptions to this rule, and an earlier reference states that the ChA and the AChE are spatially separated.[388] Histochemical work showed AChE in the cells, axons, and endings of all choline-containing neurons, while the unspecific choline esterases have been found mainly in glia cells, some neurons, and also in nonnervous structure.[900,1458]

The experiments on the intracellular localization of the enzymes of ACh metabolism gave contradictory results. Hebb and Whittacker[421] were able to separate the ChA-containing particles from mitochondria. A small amount of the enzyme was in microsomes, 20–28 % in soluble state in the supernatant. The AChE activity was shown to be up to 70 % in the large cytoplasmic particles and in the mitochondria of the caudate nucleus.[820] Toschi[816] as well as Aldridge and Johnson[171] showed that the highest specific activity was in the microsomes and mitochondria. In general de Robertis et al.[174,175] found parallelism in the activity of the three components of the ACh system in mitochondrial subfractions of the brain. In the primary fractions there is a difference in the distribution: in contrast to ACh and ChA the microsomes contain AChE. All three components were found in the fraction containing the nerve endings and also in an additional subfraction with membranes and fragmented nerve endings. The particles of the second fraction containing nerve endings, which did not have any of the enzymes of the ACh system, were recognized as the endings of the system without choline. These particles contain the GAD. After treatment of the primary mitochondrial fraction with distilled water and density-gradient centrifugation, the AChE was shown to be bound tightly to membranes, which are considered to be cholinergic synaptic membranes.[258] The incorporation of C^{14}-tubocurarine by the subfraction parallels the AChE activity. This observation strengthens the opinion that distribution of the cholinergic receptor is similar to that of AChE. McCaman, Rodriguez de Lores Arnaiz, and de Robertis[256] found a bimodal distribution for ChA within the nerve endings, in contrast to AChE and MAO, which were concentrated in only one fraction. The amount of enzyme dissolved in the cytoplasm of the synaptosomes and the amount bound to the synaptic vesicles varied greatly, depending on the animal species. In rat brain the soluble part of ChA was the smallest and the binding to vesicles were very strong. Relationships in pigeon brain were reversed while rabbit and guinea pig brains were in between in respect to the soluble and the bound amount of ChA. Similar observations have been made by Tuček.[69]

Acetyl-CoA is necessary for the formation of ACh from choline. The acetyl-CoA synthetase is responsible for the formation of acetyl-CoA from acetate. The location of this enzyme is complicated by the presence of a thermolabile and fluoride-sensitive inhibitor, which is localized in the particulate fractions. Its distribution in the single fractions is comparable with that of the acid phosphatase. This enzyme can be extracted, along with the LDH, from acetone powder and can be determined in the extract.[179] Schuberth,[1395] using C^{14}-acetate as substrate, showed that the highest acetyl-CoA synthetase activity was in mitochondrial fractions of brain homogenates (using the weight of the acetone-dried fractions as a reference). Tuček[76] could show the bimodal distribution of the acetyl-CoA synthetase with an improved method: 17–25 % of the enzyme was in the supernatant fraction and the rest was mainly bound to mitochondria.

ATP :citrate lyase (4.1.3.8) is an enzyme that catalyzes the synthesis of acetyl-CoA from citrate. In all species investigated, more ACh was synthesized from citrate than from acetate in synaptosome fractions.[1321] Two factors in this fraction seem to be responsible for the higher utilization of the citrate in the ACh synthesis: the extramitochondrial location of the ATP: citrate lyase in comparison to the intramitochondrial binding of the acetyl-CoA synthetase and its higher activity in nerve endings.

As early as 1957 Brody and Shore[1459] made a hypothesis that noradrenaline (NA) and serotonin (5-HT) serve in the synaptic transmission of two antagonistic kinds of neurons. Today it is quite certain that the function is one of neurotransmission. It is likely that dopamine acts as a transmitter, especially in the neurons of the neostriatum, the olfactory tuberculum, and the nucleus acumbens.[1460,1461] While we are well informed about the metabolism of these amines and their location in the synaptosomes,[355, 1461,1463] there has been relatively few investigations into the location of those enzymes that take part in the metabolism of the biogenic amines. Tyrosine hydroxylation to DOPA is the first step, and it determines the turnover rate of the NA biosynthesis. Tyrosine hydroxylase, catalyzing this reaction, is loosely bound to particles of the primary mitochondrial fraction and goes into solution very easily with prolonged homogenization.[331] The hydroxylase that oxidizes L-Try to 5-HTP shows its highest activity in the brain of the dog in the hypothalamus, thalamus, and midbrain.[344] It is present in brain mitochondria of different animal species.[339,345] The enzyme needs NADPH and a pteridin-like cofactor. The decarboxylation of dopa to dopamine, like the decarboxylation of 5-HTP to 5-HT, is probably accomplished by the same decarboxylase (4.1.1.26).[1286,1462] A large part of this decarboxylase is in solution. After very careful homogenization of the tissue, 50 % of the enzyme of the rat brain was found to be bound to particles by Rodriguez and de Robertis.[1298] The main part of the bound enzyme was found in nerve ending particles, the highest concentration being in the fraction of the cholinergic nerve ending particles. It is assumed that the dopa-decarboxylase is dissolved in the axoplasm of the nerve ending particles and is not bound to any special intrasynaptic structure.[355] The enzyme that converts dopamine to NA, dopamine-β-hydroxylase (1.14.2.1), is in the guinea pig partly dissolved in the cytoplasm, with more bound to particles.[325] A more accurate location in the synaptic apparatus is necessary. While MAO is bound exclusively to mitochondria,[172,240] an accurate location in the cell of the other catabolic enzymes of the catecholamine metabolism, the catechol-O-methyl transferase (2.1.1.6), has been difficult because of its poor binding strength. Catechol-O-methyl transferase is partly localized in the synaptosomes, from which it can be released by hypertonic solutions.[257,355] The increase of the concentration of the biogenic amines in the brain after inhibition of MAO[1464] and the localization in the mitochondria of the synaptosomes indicates a function of this enzyme in the intraneuronal metabolism of the biogenic amines, comparable with the model suggested by Costa and Brody for the mechanism of NA transmission.[1465] Axelrod[1466] postulated a function of the catechol-O-methyl transferase in the synaptic mechanism. Its importance in the inactivation of the NA by 3-O-methylation in the peripheral sympathetic nerve system[1467] and in the CNS[1468,1469] is certain.

E. Na$^+$ and K$^+$ Activated ATPase

Na$^+$-K$^+$-ATPase was found in a number of tissues;[1176] however, the highest activity was found in the gray matter of the brain.[15] The principal localization at the outer or inner cell walls indicates the biological function of this enzyme. It is considered to be active in the active transport of Na$^+$ and K$^+$ through the cell walls. The conventional fractionation of brain homogenates by differential centrifugation showed a large part of the Na$^+$-K$^+$-ATPase (as well as Mg^{2+}-ATPase) in the microsomal fraction.[180] A considerable amount of the enzyme was located in the primary mitochondrial fraction.[1197] After further separation of the mitochondrial fractions a high specific activity of Na$^+$-K$^+$-ATPase was found in the fraction with the nerve ending particles.[75,1143,1210] Earlier observations on isolated nerve cells strengthened the assumption that the Na$^+$-K$^+$-ATPase is bound to neuronal and glial membranes.[1198] The DNP-stimulated ATPase is mainly collected in the pure mitochondrial fraction.[1143] After hyposmotic disruption of the synaptosomes an extremely interesting result was obtained. The Na$^+$-K$^+$-ATPase was fixed to the outer synaptic membrane, while the synaptic vesicles contained only the Mg^{2+}-ATPase and no Na^+-K^+-ATPase.[1143] The Mg^{2+}-ATPase, as well as the Na^+-K^+-ATPase, was present in all subfractions, also with different activity.[1210] The importance of the Mg^{2+}-ATPase of the synaptic vesicles is not clear yet. Reserpine and Prenylamine inhibit this ATPase, isolated from NA-accumulating granula (from the sympathic stellate ganglion of cattle) at a concentration of 140 mM/ml, 22% and 54%, respectively. At the same time the ATP-Mg^{2+} activated NA entry into the granule as the spontaneous liberation of NA is inhibited. The ATPase inhibition could be the mechanism of action of reserpine or Prenylamine.[1470]

The properties of the Na^+-K^+-ATPase bound to the synaptosome membrane are influenced by ions, ouabaine, or other inhibitors similar to those of the microsomal ATPase. The distribution in the subfractions of the synaptosome follows closely the distribution of AChE, but AChE occurs preferentially in the smaller membrane fragments.[258,923] This has not been explained.

Another enzyme that is bound firmly to the synaptic membrane is the K^+-stimulated p-nitrophenyl phosphatase.[258,923] This unspecific enzyme has been found in higher concentrations on the membranes of noncholinergic nerve endings in contrast to the Na^+-K^+-ATPase. The different membranes of the nerve ending particles differ not only in respect to their enzyme patterns, but also in their structural components characteristic of other lipoprotein-membranes, like the membranes of mitochondria, the microsomes, and the synaptic vesicles. The results obtained on the chemical analysis of components and the observations of the ultrastructure and of the enzymatic arrangement of the nerve ending particles show a fairly clear picture of the structural and biochemical organization of the synaptic apparatus and indicate the way that neurotransmission functions.[632]

F. Ribosomes

The microsomes constitute a heterogeneous fraction of the whole vacuole system of the neurons and glia cells. They contain the membrane fragments of the endoplasmic reticulum, with the ribosomes, the membranes of the Golgi complex, and some neuronal and glial membranes. The conventional fractionation of a brain homogenate shows a number of enzymes in the microsome fraction.

It is interesting that the microsomes of the nerve cells of various species contain, besides cytochrome b 5, a number of flavine enzymes (NADH (1.6.99.3) and NADPH-cytochrome c reductase), where b 5 acts as acceptor, a dichlornaphthoquinone-dependent NADPH oxidase.

These enzymes are insensitive to respiratory poisons (cyanide, antimycin A, and amylobarbitone). They take part in an electron transfer pathway, which serves the oxidation of the reduced pyridine nucleotides.[281,282] The first evidence for this mechanism in the brain was shown by Giuditta and

Strecker.[292] The Nissl granules in the perikarya of the nerve cells are nucleo-protein particles with a diameter of 150–250 Å, which are partly attached in a specific way to the endoplasmic reticulum. They correspond to the ribosomes in other organs. Deoxycholate will release them from the membrane, and centrifugation in sucrose density gradient will separate the membrane fragments of the microsome fractions.[697] The ribosomes consist of about 30 % RNA, about 70 % protein, and practically no DNA. The ribosomes from nerve cells, as well as other organs, contain an amino acid-incorporating enzyme system and are the sites of protein synthesis in the cell.[697] The enzyme content of the ribosome fractions has not been studied intensively. Datta and Ghosh[695,696] found a high activity of acid and alkaline ribonuclease (2.7.7.16) (RNase) in ribosomes of goat brain. The acid RNase showed a pH optimum of 5.4, while the alkaline RNase worked optimally at pH 7.9. In the same preparation an acid and alkaline phosphomonoesterase (3.1.3.2 ; 3.1.3.1) could be shown. The enzyme with a pH optimum at 8.1 had the higher activity. It worked with nucleoside mono-, di-, and triphosphates and the same way with 3'- and 5'-adenylic acid. A phosphodiesterase (3.1.4.1) with a wide pH optimum in the alkali range was found, and an amylase.[964] Freshly prepared ribosomes from rat brain showed no RNase activity, using the procedure of Murthy and Rappoport.[691] The function of the ribosomal enzymes is speculative. The phosphomono- and diesterases, together with the RNase in the ribosomes, could be responsible for regulation and concentration of the nucleotides, which are liberated from nucleic acids. The role of the nucleotides in the synthesis of nucleic acids, and as cofactors, is well known. The esterases referred to might also have a role in the process of the chromatolytic changes that happen in the Nissl granules under stress.[697]

G. Lysosomes

A number of other cell organelles are collected in the mitochondrial portion of brain homogenates using the conventional separation by gravitational field. These "lysosomes" are responsible for numerous catabolic pathways in the cell. Until now brain lysosomes could not be obtained without mitochondria and other cell parts.

In homogenates of rat brain, acid phosphatase (3.1.3.2), RNase, deoxyribonuclease (3.1.4.5) (DNase), cathepsin (3.4.4.9), and β-glucuronidase (3.2.1.31) were found. These enzymes were bound to particles, as in the liver. They showed full activity to added substrates only after treatment with triton-100. So far it has not been possible to separate the hydrolase-containing particles from the mitochondria by differential centrifugation.[702] The β-glucuronidase and the acid phosphatase of the mitochondrial fraction of the guinea pig brain did not move together in a density gradient.[194] Numerous electronmicroscopic and histochemical observations on the acid phosphatase,[882,1471–1473] β-glucuronidase, β-galactosidase (3.2.1.23),[1474] acid DNase,[1475] and a diisopropylfluorphosphate-resistant esterase reminiscent of cathepsin C(3.4.4.9)[1476] indicated that these esterases are contained in certain spherical inclusions present in the neurons.[701] The enzymes have been recognized as the enzymes of the liver lysosomes. A finer fractionation of brain homogenate showed distribution patterns for acid phosphatase and β-D-glucosaminidase (3.2.1.30) (another lysosomal enzyme in liver) that were completely alike.[499] It was clearly different from the one for the aspartate transaminase (2.6.1.1), which was found in a soluble state and in the mitochondrial fraction.

Koenig *et al.*[178] investigated the distribution of hydroxylases in a discontinuous density gradient: acid glycerophosphatase (3.1.3.2), acid *p*-nitrophenyl phosphatase, acid DNase, acid RNase, cathepsin, and β-glucuronidase in the primary mitochondrial and microsomal fractions from rat brain homogenates. These enzymes did not show homogeneous behavior in the gravitational field; they had one thing in common: The highest activity was found in the heaviest fraction (spec. grav., 1.18) which was practically free of mitochondrial SDH. In this fraction the number of the native yellow fluorescing particles with a diameter of 0.15–1 μ was the highest. These cytoplasmic particles are considered to be lysosomes.[701] The acid proteinases, too, are collected in a sucrose density gradient in the heaviest fraction.[1071] Observations made by Sellinger and Rucker[912] with a discontinuous density gradient (1, 2; 1, 0; 0, 8 and 0, 6 M sucrose) gave similar results. They found a slightly different distribution of acid phosphatase, β-acetylaminodeoxy glucosidase (3.2.1.30), and β-galactosidase after separation of the so-called light mitochondria fraction of rat brain. The highest enzyme activity was found at 1.0 M sucrose, the second highest, however, was in the pellet on the bottom of the centrifuge tube. These observations, together with the fact that the activity of different enzymes is not the same in untreated lysosome fractions, although these differences disappear after disruptive treatment (prolonged homogenization, osmotic shock, detergent treatment, etc.), indicate the heterogeneity of the lysosome population in respect to the quantitative composition of their enzyme pattern. One can conclude that, in the brain as well as in other organs, polyanionic lipoprotein granules exist; their isolation and purification from brain was not yet successful, but they can be classified as lysosomes because of histochemical and morphological data and their characteristic enzyme content. Cerebral lysosomes could be responsible for binding of biogenic amines.[1477] Lysosomal enzymes are capable of decomposing proteins, nucleoproteids, carbohydrates, and numerous other cell components as soon as lysosome membrane disruption causes a liberation of enzymes. The optimal condition is in the acidic range. In fact, a number of physiological and pathological processes are closely linked to its function, which are needed in the decomposition of normal intracellular components, as well as foreign substances in the cell (bacteria, toxins, viruses, etc.).[924,1478] Lysomal enzymes seem to have an influence in the origin of convulsions.[1479]

III. ENZYME CHANGES DURING DEVELOPMENT

A. Introduction

With growth and maturation, in the brain as in other organs, morphological changes are closely associated with biochemical changes. Important information can be obtained about the function of single enzymes or enzyme systems by studying the changes in the enzyme pattern during morphogenesis. In addition, information can be obtained on the state of the biological maturation of an organ at a certain time by a comparison with another organ or with the same organ of another species having a different speed of development, because maturation involves the successive release of enzymes. Generally enzymes, which are involved in the process of growth or the synthesis of structural components, like proteins, lipids, and nucleic acids, should be found in the early stages of life. Enzymes concerned in functional activity tend to develop later, as physiological function matures.[1532] It is necessary to correlate the results of morphological and physiological observations with enzymatic data. Most of the enzyme tests have been made on the whole brain or on relatively large areas that are heterogeneous in their function. In general, however, important conclusions could be drawn from these observations.

It is to be expected, and it is clear from the present data, that a direct correlation between enzyme activity and functional development is better when the morphological unit being investigated is smaller. While most of the time the maximal turnover rate has been measured, little is known about the activity under natural conditions, making it harder to obtain conclusions about connections between function and enzyme activity.

B. Periods of Brain Maturation

It is necessary to have a basis to be able to compare the observations on the brain of different species. The morphological development of the brain has been used as such a basis. Koch and Koch and others[1445] distinguished four different periods of brain maturation that go smoothly from one into the other; however, it is important to know that these periods differ in various parts of the brain and that there is a considerable overlap. This classification is still quite useful for the comparison of the functional state of the brain of different species that vary in the speed of development and the state of maturation of the brain at the time of birth. The first period covers the time of the beginning of cell division until almost the complete number of cells is reached. The neuroblasts and spongioblasts developing by mitosis (out of which the neurons and glial will be developed later) move through an intermediary zone into a peripheral layer to build up the cortical layer. In this period the weight of the brain is only a small percentage of its final weight. There is no indication yet of spontaneous electric activity. In the rat this period covers the time until birth; in the guinea pig and in the human being it takes about three fourths of that time. The second period is distinguished by the growth of the cells and by the development of axons and dendrites. In the rat, this state of development is reached 10 days after birth; in the human being, by the time of birth. Guinea pigs, which have a gestation period of 66 days, form dendrites from the forty-first day on. The nucleus stops growing in the critical time from the forty-first to forty-fifth day. The nucleus, which in the beginning is homogeneous using the Feulgen staining method, forms in this time interval the vacuole.[1480] Only in this interval of the maturation of the nucleus do the fundamental changes start in the cytoplasm, which is indicated by the fast increase of the Nissl-substance and the growth of nerve endings.[1481] In the third period, the growth rate of the brain decreases. Characteristic of this interval, which in the rat is between the tenth and twentieth day, being in the human the first months after birth, is the formation of the myelin layer around the axon. Now it is possible to detect an EEG, and the neuromuscular control starts to function.

The longest interval is covered by the fourth period, in which the myelin formation is still active. The increase in the size of the brain is slower even than in the preceding period. In the last part of this period, size and composition of the brain are fairly constant.

C. Respiration

The long-known fact that newborn animals show a higher resistance to anoxia than older and adult animals of the same species, independent of the cause of the anoxia,[1482] coupled with the fact that the brain is the organ most sensitive to the lack of oxygen, suggested interesting connections between development and enzyme activity in the energy turnover. With increasing age the oxygen consumption of the brain tissues goes up. In rat brain slices the oxygen consumption in newborn is about 60% of the value for adult animals. In the first days after birth it decreases slightly; from the tenth day on it increases quickly. It reaches a maximum in animals 30–40 days old (110%) and decreases slowly to the value of adult animals (100%).[1483] The curve of the survival time of rats in pure nitrogen is almost reciprocal to the respiration curve of brain tissue.[1482] There were considerable differences in the respiration rate of different parts of the brain, depending on the age of the animals, and differences in respect to the level at which the O_2 consumption starts in relation to the time of the steep increase of the O_2 consumption curve. Medullar tissue of the newborn shows the highest O_2 consumption in the rat as well as in the dog.[1483,1484] The decrease of the respiration rate is considerable during myelination. Later on respiration is increased in midbrain and thalamus and last in the caudate nucleus and in the cortex. In these parts it stays almost constant after reaching the maximal value. The cerebral respiration rate increases in the investigated species always in the state of development in which the growth of the nucleus has been completed. This is indicated by the beginning electric activity of the brain. Parallel to the increased respiration rate of the brain tissue, the number of mitochondria increases.[1485]

D. Oxidative Metabolism

During fetal development, the brain is usually supplied with enough glucose and oxygen. The assumption is that the respiration rate is limited by the enzyme activity of the brain. The observations mentioned suggest that the enzymes linked to oxidative metabolism increase at the same rate in activity as the oxygen consumption, as the brain develops. Most authors relate the enzyme activities to the wet weight of the tissue. These and all other standards have to be criticized. In some cases the increase of enzyme activity could be attributed to the loss of water of the brain tissue. The increase of activity of different enzymes during development is generally quite large, so that loss of water from the tissue might distort the curves but not cause a principal adulteration of results.

The determination of the cytochrome oxidase activity (CYO)[184] and the SDH[1509] activity in the rat brain gave curves parallel to a large extent to the respiration curves of the brain tissue. Although the enzyme concentration is lower at the time of birth (30 % of the concentration of the brain of adult animals) than the value for the oxygen consumption (60 % of the value of adult animals), Flexner et al.[1481] found at the critical time (forty to forty-fifth day of pregnancy) in the guinea pig brain a steep increase in the SDH and ATPase activity. In the fetal development the activity of these enzymes is at a constant low level up to the critical period.

The CYO system is, during the whole development, at a higher level than the succinate-oxidase-complex (SOX) or the SDH. This is understandable, since succinic acid is only one of the substrates supplied by the CYO system with hydrogen.

A constant low enzyme level during the first period of development is common for a number of enzymes. The level is low up to the start of the second period, when a steep increase in the enzyme activity occurs, up to the level of the adult animals. Some examples are ATPase,[1509] NAD glycohydrolase,[1038] (3.2.2.5) and MDH[82] in the rat brain, the CYO-system and the SDH in the brain of the pig,[190,317] the SOX and the SDH in rabbit brain,[189] and the GS in chicken brain and in the retina.[1422] The increase of the carboanhydrase activity (4.2.2.1) in the rat brain occurs much more slowly and reaches the value for adult animals only in the fourth period, about 80 days after birth. The enzyme activity increases sixfold between the tenth and twenty-fifth day after birth in this animal.[1336,1337]

Corresponding to its morphological maturity the guinea pig brain shows at birth almost normal carboanhydrase-activity (in contrast to the rat). In the human brain, as in the calf brain, this enzyme is missing in the earlier stages of fetal development, except in the spinal cord extension and in the medulla oblongata. In the human brain it is formed right after birth in the brainstem, but not in the hemispheres; in the newborn calf a normal enzyme pattern is observed at this time, but at a lower activity level than in the adult animal. The biochemical maturation takes place in these brains in a caudo-cranial direction.[900]

The increase of enzyme activity does not happen in all species in the same way. For example, the CYO does not increase in the guinea pig brain in the same dramatic way as that observed in the rat during the critical period of development; it happens shortly before birth.[1481] Kuhlman and Lowry[82] found with microchemical determination of MDH, LDH, and G-6-DH in the cortex of the rat that the overall change of enzyme activities in the cortex is the sum of changes in the enzyme activity that goes on in individual layers side by side, in a rather complicated way. The histochemical data of the change of the SDH in the rat brain correspond to the microchemical observations.[1486] In the single laminal of the cerebellar cortex of rat, Robins and Lowe[83] determined changes of activities of MDH, LDH, and isocitrate NADP oxidoreductase (ICDH), (1.1.1.42) correlated with the morphogenesis. In this part of the brain, too, the chronological pattern of the MDH activity corresponded to the histochemically determined changes of the SDH activity.[1487] On the second day after birth a high SDH and CYO activity occurred in the bodies of the Purkinje cells. On the ninth day, short, straight dendrites with SDH activity grew into the molecular layer. The neuropil surrounding the dendrite, contained a similar SDH activity.

ICDH activity shows a completely different chronological pattern. It starts growing at the time of birth, continues until the fourteenth day, and then decreases rapidly.[83] In the brain of the chicken embryo the enzyme bound to particles increases slowly until hatching, while soluble ICDH shows its maximal activity at the fifteenth to eighteenth day of embryonic development. At the time of hatching the activity has decreased to about one-half the value. The hypothesis of Baker and Newburgh[109] suggests that the soluble enzyme provides NADPH for reductive synthesis, while the activity of the particle-bound enzyme is an index of the activity of the oxidative metabolism. With increasing age the activity of the enzyme bound to brain mitochondria decreases.[106] Supposedly it is of special importance to the development of the cerebellum.

The general trend of the caudo-cranial maturation of the brain, as indicated by the respiration rate of brain tissue, is the sum of the metabolic changes in numerous nuclei with their own critical period for the beginning of enzymatic changes, as shown in observations by Friede[1486,1488–1491] and Knolle[1492] on SDH. Knolle investigated six bird species and seven mammals. Birds that leave the nest after hatching and those that stay in the nest and need feeding differ in the time of enzyme formation. Numerous histochemical observations on different species showed that in the time after the proliferation of the glial cell, but before the start of the myelination, dramatic changes happen in the activity of many enzymes of oxidative metabolism, especially in the glial cells of the white matter.[900] An increase of indophenol oxidase, SDH, and CYO activity was shown in these glial cells. NAD diaphorase (1.6.4.13) activity increases in the same way in the human brain.[288]

While glyceraldehyde-3-phosphate dehydrogenase (1.2.1.12), like most of the enzymes already mentioned, reaches a maximal enzyme activity during the second period of development of the rat brain, the activity of NAD glycerol-3-phosphate dehydrogenase (1.1.1.8) increases slowly until it reaches its highest activity. This happens in a later state of myelination, about 40 to 50 days after birth.[29] This enzyme is necessary to supply glycerin-3-phosphate for the lipid synthesis. The importance of this enzyme for the myelination process is obvious. Glyceraldehyde-3-phosphate dehydrogenase and glycerol-3-phosphate dehydrogenase compete for the triosephosphates. Their chronologically successive appearance suggests that in earlier states of development glycolytic processes play an important role, while in later development periods, after the respiratory apparatus is developed, lipid synthesis can start, which is necessary for the myelin formation.

E. The Pentosephosphate Cycle

G-6-DH behaves differently during the course of brain maturation from the enzymes already discussed. In the different cortical layers the activity of the enzyme is constant from birth on in the rat brain.[82] In the mitochondria of the brain the activities of G-6-DH and of 6-phospho gluconate dehydrogenase decrease during the course of development to about one third of the value at birth.[115] These enzymes are important for the alternate pathway of glucose oxidation, not using the citric acid cycle. Enzymes of the pentose phosphate cycle have been shown in all brains investigated.[111,118] The amount of glucose oxidized in the cortical tissue in this way is small in comparison with the amount metabolized by glycolysis and the citric acid cycle.[1495,1496] The pentose phosphate pathway supplies the pentoses necessary for the RNA synthesis, and it could be important for the formation of NADPH.

In the chicken embryo G-6-DH shows peak activity on the third day and a second peak between the sixth and ninth day of incubation. Later on, the activity decreases slowly. At the time of hatching it reaches the level of the adult animal. These changes go parallel to the mitotic activity and occur before changes in the rate of synthesis of RNA.[121,122] Experiments by Liuzzi and Angeletti[1497] showed that the pentose phosphate pathway is active *in vivo* in the embryonic chicken brain. After 2 weeks of incubation, its activity decreases slowly and in the adult brain it has practically ceased to function.

F. Glycolysis

Geiger and Magnes[1498] have observed that glycolysis decreases in the brain during the postnatal period. Newborn animals are much more sensitive to inhibitors of glycolysis, like iodacetate, than adult animals.[1499,1500] In the brain of the rat the amount of glucose oxidation increases from 36% for 3-day-old animals to 56% for adult animals, while at the same time the rate of glycolysis decreases from 45 to 33%.[1501] The rate of glycolysis is maximal in the medulla after birth (dog and cat) and decreases with increasing age. In other parts of the brain the highest rates of glycolysis are reached at a later time and are followed by a decrease in later periods of development. The increase of the glycolytic activity in the caudate nucleus and in the cortex is higher than in the medulla, so that in these parts of the brain the glycolytic activity is higher in the adult animal than in the medulla. The trend of the glycolytic activity during development in different parts of the brain is similar to the observed chronological changes of the enzyme activities of oxidative metabolism. While the relative amount of glycolysis to energy production decreases in the brain with increasing age, the absolute amount of glycolytic activity increases until a certain stage of development is reached, and it decreases slowly at a later age. This trend is noticeable from lactic acid formation in rat brain sections (*in vivo* experiments have not been made). In 3-week-old rat brain twice as much lactic acid is produced per unit weight as in newborn rat brain.[1502]

G. Glycogen Metabolism and Glycolysis

The cortex of young dogs and cats contains about 20 mg/100 g glycogen. This increases in adult animals about threefold. In the medulla and other phylogenetically older parts glycogen concentration decreases from 120–130 mg/100 g to 20–40 mg/100 g with increasing age.[1503] Glycogen synthesizing activity also increases 70 % in the first 3 weeks of development of the rat. Directly before the appearance of glycogen in chicken and rat embryos UDPglucose:glycogen glycosyltransferase (2.4.1.11) can be observed.[465] Its activity increases in the brain of the chicken embryo until the fourteenth day of incubation and then decreases sharply. The coincidence of the appearance of the UDP glucose:glycogen glucosyltransferase and glycogen indicates that glycogen is synthesized through UDP glucose.[456,460] The phosphorylase activity in the brain of newborn rats is very small. Only in the meninges and some other structures that do not belong morphologically to the CNS is phosphorylase activity noticeable.[1504] Between the tenth and twentieth day after birth, which is the critical period of development of the rat brain, the phosphorylase activity increases very sharply. It is the same for the phosphoglucomutase (2.7.5.1).[454] The appearance of the phosphorylase seems to occur directly before formation of myelin layers. Its development is similar to that of many other enzymes in that it appears first in the pons and the medulla and later in the mesencephalon and the diencephalon and last in the telencephalon and the cerebellum.[459] The phosphoglucomutase concentration is supposedly high in embryonic brain. Its activity seems to be the limiting factor of the glycogen pathway.[454]

Just as many enzymes of oxidative metabolism and the glycogen pathway increase the enzymes of the Embden–Meyerhof pathway, hexokinase[556,582] and aldolase[185,1313] are maximal during the second period of brain development. The activity of the aldolase is higher in the gray matter of the cerebrum and cerebellum of the rat than in the white matter and in the medulla oblongata.[82,1313] Three pure isoenzymes of the aldolase are known: aldolase A (ketose-1-phosphate aldehyde lyase) (4.1.2.7), aldolase B (fructose-1.6-diphosphate-D-glyceraldehyde-3-phosphate lyase) (4.1.2.13), and aldolase C, which is between A and B in respect to its immunological properties and the substrate specificity.[1505] In the rat brain the differentiation occurs from an aldolase similar to A in the direction of aldolase B or C. Aldolase C is observed in the rat brain 1–2 days before birth. Hybrid aldolases are found earlier in the brain (they are built up of A and C subunits). In all other organs the fetal relationship of the activity with fructose 1-phosphate and fructose 1.6 diphosphate is very distinctly different from the adult relationship, and the adult relationship is reached soon after birth. This relationship is almost constant in the brain during the whole period of development.[1315]

The changes of the activity of the LDH in the brain are somewhat similar to the changes of the MDH. In the cortex of the mouse the LDH activity is constant until the twelfth day after birth and then increases sharply to three-fold the activity.[84] In the rat brain the LDH activity doubles as the brain matures.[82] During development of the chicken embryo a fairly constant LDH activity has been observed. The MDH activity, however, increases from the eighth day until the time of hatching.[97] Another exception is the outer granular section in the fetal cerebellum, which compared with the SDH and MDH shows a higher LDH activity.[83]

Bonavita[80] showed five isoenzymes of LDH in the rat brain. Three of these enzymes underwent a change in their concentration in the course of development. The enzyme pattern of the brain of adult animals was reached only after the fiftieth day. Determination of LDH-isoenzymes in the brain of a number of animal species showed that there is parallelism between phylogenetic and ontogenetic development in respect to these enzymes; even so, not all results fall into this general order.

H. The GABA-Shunt

Experiments of van den Berg and coworkers[66] show that the enzymes participating in the GABA-shunt behave very similarly during the development of the brain of rat, chicken, and rabbit. The enzymes referred to are GAD, GABA amino transferase (GABAT), and succinate semialdehyde dehydrogenase. The enzyme activities increase in the rat on the fifth day, in the rabbit, from birth on; and in the chicken, 14 days after the start of incubation. In chicken cerebellum the rise in GAD activity coincides with a gross acceleration in the formation of dendritic arborizations.[1506] While the relationship of GABAT to GAD stays rather constant in the course of brain maturation of the rat, the relationship decreases in the chicken brain with increasing age of the embryo. Also, in the rabbit brain a decreasing tendency of the GABAT/GAD ratio is observed; only in the spinal cord is a small increase noticeable. The supposition has not been confirmed that these enzymes as well as aspartate amino transferase, LDH, and alkaline phosphatase would show a different distribution in the subcellular elements in different stages of maturation. Only the GAD from immature brains showed a slightly slower sedimentation than from adults.

I. Choline Acetylase, Carnitine Acetyltransferase, and Choline Esterases

The extreme importance of ACh in nerve excitation and nerve conductivity suggests a special significance for the connection between functional development of the CNS and enzyme activities that have a role in ACh metabolism. The location of these enzymes and ACh in the synaptic vesicles of the nerve ending particles and the synaptic membranes increases this expectation. It is not easy to interpret the experimental results, and many more details are necessary to see the connection between enzyme activity and functional development.

ChA can be observed in the rabbit brain at the eighteenth to twentieth day of gestation and in the brain of the guinea pig at the thirty-second day. At the time of birth the activity in the guinea pig brain amounts to 30–50% and in the brain of the rabbit only 30% of the adult activity. In the rabbit only on the twentieth to twenty-first day is 50% of the adult level reached.[425] In all investigated parts ChA activity increases steadily from the time of birth. If the activity is related to the DNA-P value and not to the wet weight of the brain, a good correlation is obtained to the cell maturation. The highest activity increase is observed in the caudate nucleus (620%) and the smallest increase in the medulla oblongata (160%). In the cortex ChA activity shows, in all stages of development, the smallest value.[419] The biochemical brain maturation also occurs for this enzyme caudorostrally. The activity of the carnitine acetyltransferase (2.3.1.7) is quite similar to that of ChA during development in the rabbit brain until the twentieth postnatal day. This enzyme occurs mainly bound to mitochondria. Newer theories suggest that it has a function in the transport of the acetyl group through the barrier of the mitochondrial membrane.[1507,1508] In the brain of 4-day-old rabbit 10% of the carnitine acetyltransferase activity of adult animals has been found. Between the twelfth and sixteenth day the enzyme activity remains constant. While the ChA reaches 90% of the adult activity in the cortex at the thirty-second day[419] only 50% could be observed for the carnitine acetyltransferase.[422] An important difference between these enzymes is that the ChA is mainly found in structures rich in nerve ending particles, while the carnitine acetyltransferase is equally distributed in the brain. These observations argue against the assumption that the carnitine acetyltransferase acts directly as a neurotransmitter or that it participates indirectly in the synthesis of ACh.

20–25 ACh-splitting fractions could be observed in the human brain by electrophoresis.[852] Three isoenzymes of the brain AChE are known.[840] During development of the rat brain, the total esterase activity is constant in the first 10 days after birth and then increases rapidly until the twentieth day. The further increase to the level of the adult animals is slow. Altogether the total esterase activity in the rat brain doubles from birth on, while single esterase fractions change in a characteristic way. However, only the amount of the bound enzymes increases; the soluble part remains constant during all of development.[853] Elkes and Todrick[826] observed a similar course of activities in the unspecific acylcholine acylhydrolases in the rat brain during postnatal development. At the time of birth pseudocholinesterase and AChE show about the same activity; later on the AChE activity is preponderant. Between the third and twenty-second day after birth its activity doubles and reaches 75% of the value for adult animals. After that the activity increases slowly.[1510] This course is similar to that of many other enzymes. In the cortex of the rat this increase is observed 10 days after birth.[185] The maximal value is reached after 100 days. A slow decrease of the activity occurs afterward. In the subcortical structure the maximal AChE activity is reached earlier than in the cortex.[753] Ishii[1511] made very detailed histochemical observations on the AChE activity in the rat brain during development and on the distribution of this enzyme in the brain of adult animals. These observations on brain parts of the rat show that the maturation occurs in caudo-cranial directions. In principle, the same results were obtained on rabbit,[419, 832] guinea pig,[833] human,[841,843] and chicken brain,[845] namely, a steep increase of the esterase activity during the critical period of brain development and a caudo-cranial gradient of development. In the chicken embryo the AChE activity can be observed histochemically only 15 hr after incubation starts, even before the differentiation of neuroblasts is histochemically demonstrable.[1512,1514] Before hatching the esterase activity increases strongly. In the medulla oblongata of the chicken embryo the AChE activity increases fivefold in the short time that corresponds to the start of nerve function.[1515] The chronological sequence of functioning of the CNS of the chicken corresponds in general with the ontogenesis of the AChE.[1516] The early appearance of AChE in the chicken embryo (and in amphibian) 2–3 days before reflexes are observed cannot be correlated with neurotransmission. For the meaning of these observations, and for the extraembryonic esterase activity, explanations have to be found.

When ChA and AChE activity have reached about 50% of the adult value in the brain of various species, the EEG corresponds to that of adult animals. In the guinea pig this occurs at birth, while in rats and rabbits such electric activity of the brain is only reached 2 weeks after birth. One would expect that the enzyme level at this time should have reached the adult level. It is known from inhibition experiments on adult animals that 50% of the AChE activity in the brain is enough for physiological control[1517] and that 50% of the ChA can supply sufficient ACh. The meaning of the further increase of enzyme activity in the later development is not yet clear. It is questionable whether it is only a safety reserve.

J. Synthesis and Catabolism of Serotonin and Noradrenaline

5 HT is supposedly the neurotransmitter in the trophotrope system, an hypothesis postulated by Brodie and Shore.[1459] It includes the central parasympathetic functions, while NA controls the ergotrope system. That these amines are correlated with the mature behavior of mammals is indicated

by the fact that normal behavior is observed only after the adult amine levels are reached in the brain.[261] Newborn guinea pigs already have an adult level, the brains of newborn rats have only 20–30 % of the adult amine content.[260,1518]

An important step in the biogenesis of these amines is the decarboxylation of 5-HTP to 5-HT, or DOPA to dopamine, which occurs in the CNS. The relationship of the decarboxylase activities against DOPA and 5-HTP is constant in the different regions of the brain; presumably one decarboxylase acts on both substrates. In most areas of the brain of adult animals the 5-HT amount corresponds to 5-HTP decarboxylase activity (except in the pyriform cortex and in the amygdala).[1296] It is the same for the NA concentration, with some exceptions. The 5-HTP-decarboxylase activity, however, is almost as high in the brain of newborn rats and guinea pigs as in the brain of adult animals.[263,265] Obviously it is not a lack of decarboxylase activity in the fetal brain that causes the low amine level, but the lesser ability of the fetus to hydroxylate Try to 5-HTP.[263] In the rabbit brain 5-HTP-decarboxylase activity increases noticeably in all investigated regions of the brain during the postnatal period. The steepest increase within the cortex is observed between the twentieth and thirtieth day. The highest enzyme increase is in the caudate nucleus (760 %). The ontogenetic development of the 5-HTP decarboxylase is almost parallel to the AChE and the ChA activity; it reaches about 50 % of the adult value during the critical period of development in the rabbit brain.[419]

MAO behaves differently during development. An increase of this enzyme in the postnatal development has been observed in primary (rat)[261–263] and secondary (human being)[266] species whose young are fed in the nest after birth, whereas the young of the species that leave the nest after birth (guinea pig)[261,265] already show at birth nearly adult MAO activities. In the rat brain the increase of MAO activity is higher in the first 20 days of life than the increase of 5-HT concentration.[262] In the cortex of the rabbit brain the MAO activity reaches the adult value after 15 days. Until the twentieth day there is an even higher increase of the enzyme activity. During the critical period of development between the twentieth and thirtieth day, the MAO activity keeps the maximal value of 138 % of the value for adult animals. After that it slowly decreases in its activity to the adult level. This course has been observed when DNA-P is used as a reference of enzyme activity. Referred to the wet weight of the brain, an increase in the MAO activity has been observed, with a maximum at about the nineteenth postnatal day, after which there is a gentle decrease of activity to the adult value, which in the cortex and thalamus is below the value of 3-day-old animals. This has been observed in all investigated parts of the brain. An exception is the medulla, where until the fifteenth day a fairly constant MAO activity has been found, which then decreases to 50 % of the value for 3-day-old animals.[419] A caudal-rostral gradient in the development has been observed in the case of 5-HTP-decarboxylase and MAO in rabbit brain.

In addition to the changes in the concentration of the biogenic amines,

Bourne[259] investigated MAO activity during development of the chicken. Measured with 5-HT and NA as substrates, MAO activity increased steadily from the eighth to the twentieth day, but it showed tremendous changes with dopamine as substrate. NA was not at all oxidized by the fetal brain, but it was by the brain of adult chickens. The results suggest that the MAO is not used for the inactivation of NA during fetal development; instead, it may play an important role in the regulation and synthesis of this amine.

The different behavior of the MAO activity during development, in comparison with other enzymes of amine biogenesis, is explained by its localization in mitochondria. The regional distribution of MAO activity differs considerably from that of other mitochondrial enzymes, e.g. SDH. This is an important indication of the heterogeneity of the mitochondria.[172,197]

Other important enzymes of amine metabolism, such as DOPA-β-hydroxylase and catecholamine-O-methyltransferase, have not been investigated regarding their connection with the morphological and functional development of the brain during ontogenesis.

K. Phosphatases

Phosphoric acid esters participate in all important metabolic processes. Of special importance are the nucleotides ATP and GTP, as are ADP, GDP, and phosphocreatine, as a link between the energy-producing and energy-using processes. It is not surprising that a large number of phosphoric acid ester-splitting enzymes exist, with important biological functions.

1. Acid and Alkaline Phosphatases

We know little about the biological functions of a group of enzymes that occurs in all cells and split β- and α-glycerophosphate (alkaline and acid phosphomonoesterases). Their regional and subcellular distribution, like their activity, differs in the brain of adults from species to species. These changes could not be correlated with either the evolutionary development or the size of the animal. Presumably these enzymes have different functions at the same time, which makes experimental data unclear and difficult to interpret.[900]

In contrast to most enzymes whose activity increases during the critical period of development, alkaline phosphatase (pH optimum 9.3) (3.1.3.1) has been found in high concentrations in all vertebrate brains in early fetal development and shows a decrease of activity in adults.[894] A high activity of alkaline phosphatase was found, e.g., in the neural plate of mice and rat embryos. The activity stayed high for the longest period in the structures with long active differentiation, such as the hemispheres, the cerebellum, and the corpus striatum.[900] Between the fourteenth and the twentieth day after birth the alkaline phosphatase activity was twice as high in rat brain as in the brain of adult animal.[827] In the spinal cord and the medulla oblongata of dog and rabbit a postnatal increase of this enzyme was observed,[895] while no clear trend in the activity of the alkaline phosphatase was observed during the development of the guinea pig brain.[894] In the cerebral blood vessels of the mouse the alkaline phosphatase develops slowly, mainly in a caudocranial direction.[1519] It is already detectable in the capillaries of 1.5–4.6 mm-long rat embryos. In the chicken brain the maturation goes in a caudo-cranial direction.[898] In the nerve tissue of chicken embryo the activity of the alkaline phosphatase is high; from the sixth day to the eighth day it decreases, it increases again at the time of hatching, and then it decreases to the adult value. The enzyme activity is observed in fetal brain mainly in the cytoplasm and in small amounts in the nuclei.[845,892]

In general the above results are the same for human brain. In 1.8–4.5-mm-long embryos (18–26 days old) high activities of alkaline phosphatase have been observed; 4–5-month-old fetuses already have much less activity in the nerve tissue.[1520,1521] The enzyme distribution pattern of the adult has nearly been reached when the alkaline phosphatase appears in the blood vessels, in the choroid plexus, and in the meninges.

There are three different types of acid phosphatases: type I with a pH-optimum at 5.4, type II at 3.8. Both of these prefer β-glycerophosphate to α-glycerophosphate as substrate. Type III, however, splits at a pH optimum of 5.7, the α-ester being faster than the β-ester. While Type II is inhibited by Mg^{2+}, Type III is activated by this cation.[869] In papers concerned with enzyme formation and localization during development (mainly with histochemical methods), no attention is paid in general to the differentiation of the acid phosphatases. The concentration of acid phosphatases as well is higher in the embryonic brain than in the adult. It appears already in a 1-day-old chicken embryo, as does the alkaline phosphatase, but has lower activity.[1522] It is detectable very early in the fetal nerve tissue of mouse and rat. In the newborn rat enzyme activity was observed in the choroid plexus, in the ependyma cells, and also in some nerve cells, but not in the neuroblasts.[1523] The activity of acid phosphatase shows a maximum between the seventh and fifteenth postnatal day in the rat brain. On the fortieth day adult values are reached.[827] It is possible that these activity changes are caused by the decrease of water content of the tissue. In the guinea pig brain the activity is almost constant from the thirty-second fetal day until birth and is about twice as high as in adult brain. (In these brains the acid phosphatases were observed in the neuro-blasts).[920] In dog and rabbit brain, Chirovskaya[895] showed a very similar decrease of the enzyme activity in the late fetal and early postnatal development. In the medulla of the cat it might be possible to correlate the activity of the acid phosphatase with the myelination or with the development of the functional activity.[921]

In human fetus (16 mm long) the acid phosphatases were observed first in the ependyma and later on (26–145 mm long) in migratory neuroblasts. Only in embryos of 140–230 mm length were they found in nerve cells—first in the purkinje cells.[1524]

2. 5-Nucleotidase

5-Nucleotidase (3.1.3.5) splits mainly AMP and IMP. Histochemically this enzyme was observed mostly in the myelin layers of the nerve fibers in the brain of mouse and rat. It increases rapidly during the period of myelination.[935,1525] In the cerebellum of the rat it was observed in the myelin and also in the purkinje cells and in the trigeminal ganglion.[882,1526] The histochemical localization of this enzyme in the cerebellum however, is, doubtful.[900]

In the rat brain the 5-nucleotidase activity increases from the time of birth almost linearly until reaching the adult level. In the cerebellum its activity increases from birth on and reaches about twice the value on the twelfth day. A very rapid increase occurs until the nineteenth postnatal day. The adult value is about six times as high as the value at the time of birth.[934]

3. Adenosinetriphosphatases

It has been mentioned that the ATPase activities in the brain change during the course of development in a similar way to the enzymes of the citric acid cycle and oxidative phosphorylation. The Ca^{2+}-activated ATPase starts to increase the sixth day after birth in rat brains. The increase of the Mg^{2+} ATPase is parallel, however, on a higher level. The adult values do not differ from the values of 20-day-old rats.[934] Experiments, using cell fractionation, showed that the percentage of Ca^{2+}- and Mg^{2+}-activated ATPases is higher in the supernatant fraction of immature brains than in that of mature brains.[1216] Côté[1211] observed a low Na^+-K^+-ATPase level in the fetal brain of the rat until the sixteenth day of the intrauterine life and an increase in activity before birth. A much steeper increase in enzyme activity has been noticed in the rat brain between the fifth and twenty-first postnatal day (about sixfold).[1212] It is assumed that most of the increase of the energy turnover in this period of development is correlated with the increase of the ATPase system. There seems to be a connection between the known stimulating effect of the respiration of brain tissue *in vitro* by Na^+ and K^+ and the Na^+-K^+-ATPase activity, because only tissue cultures of brains of adult rats show stimulation of the rate of respiration by these ions. It is not possible to stimulate brain sections of newborn animals, which show very little activity of this enzyme.[1212] The location of the Na^+-K^+-ATPase in the nerve ending particles[923,1210] suggests that the increase of Na^+-K^+-ATPase is due to the increase in the number of nerve endings during brain maturation. Speculations about the function of the Na^+-K^+-ATPase in the synapses are not discussed.[1212]

L. Some Enzymes of Amino Acid and Protein Metabolism

1. Enzymes of Glutamic Acid Metabolism

Data on the changes of activity of the enzymes participating in amino acid metabolism during development are rather meager. Most is known about the enzymes of Glu-metabolism.

Glutamate decarboxylase has been mentioned in connection with the enzymes of the GABA-shunt. The activity of glutamate dehydrogenase (1.4.1.3) increases from the thirtieth to the one hundred and fourteenth day of fetal development of the pig; that means that until birth it increases in the pig embryo about threefold. In the same period of time the activity of the GOT increases fivefold and the GS twentyfold. The steepest increase in the activity of this enzyme has been observed in the pig in the last third of intrauterine life. Similar are the relationships during the embryonic development of the rhesus monkey. However, the highest increase of activity of the three enzymes occurs during the early postnatal stage of development. The increase up to the two hundred and twenty-fifth day after conception is almost linear. Further increase is minimal. (Data on the enzyme activities during postnatal development of the pig are not available.) The earlier increase of the enzyme activity in the brain of the pig and the higher enzyme level at the time of birth in comparison with the rhesus monkey is correlated with the advanced maturation of the brain of the newborn pig.[212]

The chicken brain has the first peak of GDH activity on the fourteenth day and the second increase in enzyme activity at the time of hatching.[601] The chicken brain contains little GS activity during morphogenesis. Only on the eleventh to twelfth day does the GS concentration start to increase. It increases almost equally in all investigated parts of the brain (cerebrum, diencephalon, optic lopes, cerebellum, medulla, and spinal cord) and two to three times more than the total amount of protein. Between the sixteenth and nineteenth day of embryonic development a sudden increase of the GS activity has been observed, which is three to five times larger than the increase in the amount of protein. The GS is the enzyme in which the highest activity increases have been observed during ontogenesis. From the ninth day until hatching it increases in the chicken brain thirtyfold; in the retina the increase is even higher. The time of the sudden increase in the GS activity in the chicken brain corresponds with the development of the blood-brain barrier, as shown by Lajtha.[1422] It is the time at which the brain is furnished with an independent nitrogen metabolism. The GS is a link in this system. In the experiments discussed the GS activity was determined with transfer of glutamyl residues to hydroxylamine. The enzyme responsible for this reaction was called glutamine transferase. Today most authors believe that the formation of glutamine from Glu, ATP, and NH_3, the transfer of glutamyl residues to hydroxylamine by releasing NH_3, and the arsenolysis of glutamine are all catalyzed by the same enzyme.[1403] Also, the rate of the transfer reaction in chicken[1422] and cat[1417] brain is completely parallel with the rate of glutamine synthesis during all stages of development. (There is a bacterial glutamyl transferase that does not show any synthetic ability.)[1403]

In the cortex of the cat the GS activity increases very little until the second week after birth, when a very steep increase occurs. In the fifth week adult values have been observed which are five to six times higher than the values right after birth. The enzyme increase in diencephalon and hippocampus is much smaller. In the cerebellum the activity increases from birth on. It reaches adult values after four weeks. In the pons and medulla, however, a flat maximum was observed in the first week (about double the natal activity); in the third to fourth week, a minimum; and then a second, slow increase to the adult enzyme level.

At birth the neurons in the hippocampus of the cat are more mature than the neurons of the neocortex; the purkinje cells of the cerebellum are 2–3 weeks less mature.[1527] In all three regions of the brain the activities of the GS are found to be very slight after birth. Results by Waelsch and Rudnick[1422] on the retina of the chicken suggest that there is a very close correlation between GS increase and functional maturation of certain neurons connected with vision. From the total activity of the GS in the brain only indications of connections between functional brain development and enzyme activity are recognized. Berl's assumption[1418] that there was a coincidence between the time of enzyme increase and the formation of special sections of the Glu-metabolism was true for the cortex but not for other brain areas.[1528]

2. The pH 5 Enzymes

The activation of the carboxyl group of the amino acids is the first step in protein biosynthesis the equation of the reaction is

$$\text{AE} + \text{ATP} + R-\underset{\underset{\text{NH}_2}{|}}{\text{CH}}-\text{COOH} \longrightarrow \text{AE}-\text{AMP}-\overset{\overset{\text{O}}{\|}}{\text{C}}-\underset{\underset{\text{NH}_2}{|}}{\text{CH}}-R + \text{PP}_i$$

(where AE is the activating enzyme). The second step obtains the linkage between amino acid and s-RNA:

$$\text{AE}-\text{AMP}-\overset{\overset{\text{O}}{\|}}{\text{C}}-\underset{\underset{\text{NH}_2}{|}}{\text{CH}}-R + s\text{-RNA} \longrightarrow \text{AE} + \text{AMP} + R-\overset{\overset{\text{O}}{\|}}{\underset{\underset{\text{NH}_2}{|}}{\text{CH}}}-s\text{-RNA}$$

The activation is done by cytoplasmic enzymes with an isoelectric point of about pH 5, the so-called pH 5 enzymes. (The reaction steps together are considered an effect of the amino acid: s-RNA-ligases (6.1.1), every amino acid having a specific ligase). This is not the only way for the incorporation of amino acids into proteins in the CNS, but very little is known about other ways (without first activating the carboxyl group).

Lipmann[1363] was the first to show the specific activation of Try, Phe, Gly, Val, Leu, and Lys by the enzymes in the supernatant fraction of the immature rat brain. Investigations of the amino acid activation by an acetone dry powder prepared from guinea pig brain gave the following results.[1369] In the fetal white and gray matter Lys, His, Glu, Gly, Tyr, and Try were activated to the same extent. The rate of the Glu-activation, in the fetal phase development, was much higher than for the other amino acids. It increased in the postnatal period, while the Cys-activation decreased slowly to zero.

The speed of activation of the amino acids could not be correlated with their content in the proteins of white and gray matter. Piha et al.[1367], with other methods, observed in mouse brain subfractions the quick activation of a number of amino acids in the supernatant fraction. The microsomal fraction was active in the presence of Leu, Lys, Thr, and Met, while the mitochondrial fraction was ineffective in amino acid activation. The activity of amino acid : s-RNA ligase was different with each amino acid. Leu had the highest rate of activation. Glu and Arg activation did not occur in these preparations, in contrast to results by Wender and Hierowski.[1367]

During the first 14 postnatal days the activation rate decreased in general about 30 % and increased thereafter in a week to the level observed at birth. This rate of activation remained constant during further development. It is difficult to decide what causes the differences reported in two different species.

Yamagami et al.[1388] investigated some connections between the development of the brain and ribosomal protein synthesis in the rat brain. Until the fortieth day after birth the L-C^{14}-Leu incorporation into the brain ribosomes of these animals was maximal. At about the fiftieth day a sharp decrease of protein synthesis was observed and also a decrease in the activity of the pH 5-soluble fractions from rat brain. The addition of nuclear RNA to reaction mixtures, especially RNA from nuclei of very young animals, activated the protein synthesis. There is a connection between the distribution pattern of components of the nuclear RNA in the density gradient, the activity of the fractions soluble at pH 5, the activity of the ribosomal amino acid-incorporating system, and the age of the animals. The concentration of the pH 5 enzymes remained almost constant in the brain throughout the development of rats.[1368]

3. RNA-Polymerase

Because of the importance of nucleic acids for the protein synthesis, a few results about the RNA-polymerase (2.7.7.6) will be added.

Bondy and Waelsch[1529] found in the nuclei of the cortex of adult rabbits an in vitro activity of this enzyme twice as high as in the nuclei of the cortex of newborn animals. The RNA-polymerase activity also was twice as high in the cortex as in the corpus callosum, in the thalamus, and in the liver. Barondes,[674] however, showed that enzyme preparation from the brain of 12-day-old rats was more active than preparations from brains of adult animals.

M. Cholesterol Esterases

It has been known for a long time that the brains of adult animals and the human being do not contain cholesterol esters. They are observed, however, in fetal and immature brains; that is the reason for connecting them with myelin formation.[1530,1531] The question was obvious: Does the brain contain cholesterol-ester-splitting enzymes (3.1.1.13) and at which stage of development can they be observed?

Although the solubility is very unfavorable, cholesterin oleate can be split by rat brain homogenates hydrolytically.[860,862] The cholesterol esterase activity is by far the highest in the hemispheres. All subcellular fractions contain this enzyme. It is mainly in the mitochondrial fraction, supposedly bound to membrane structures. Its activity in the rat brain increases rapidly from the seventh to the twentieth day, which means during the time of myelination, and remains constant until the third month. Only in the course of 9 months does the activity decrease to one half of its original value.[860] Supposedly this enzyme is used during the course of development for the structural synthesis to release essential fatty acids from cholesterol ester.

IV. SUMMARY

Our knowledge of the enzymes of the nervous system is rather like a mosaic finished in its main outlines but unfinished in further detail. The summary in the table of the presently available knowledge about enzymes makes it clear that many pieces of the mosaic are still needed to complete the picture.

Most of the enzymes shown to be present in the nervous system are in their main characteristics similar to such enzymes in other organs. However, some important differences begin to emerge. An example of this may be the enzymes responsible for the production of energy in the Embden–Meyerhof and the tricarboxylic acid cycles. The subcellular localization of these enzymes in brain and in other organs is similar in that the glycolytic enzymes, with the exception of hexokinase, are soluble while those of the tricarboxylic acid cycle and electron transport are localized in mitochondria. However, some of the activities are different. Brain shows the highest hexokinase activity for all organs. The enzymes of the tricarboxylic acid cycle are organized in the brain in such a way as to preferentially form α-ketoglutarate and thereby to form large amounts of glutamic acid. Specific for the central nervous system is the production of energy through the GABA-shunt. Glutamic acid decarboxylase, according to the needs of the organ, can form GABA, which may act as a neurotransmitter; and through the action of GABA-aminotransferase and succinate semialdehyde dehydrogenase, this compound can be further metabolized. Further typical enzymes for nervous tissue are the ones participating in the generation and transmission of electrical impulses. These are enzymes participating in the metabolism of acetylcholine and in the metabolism of biogenic amines. The distribution of these enzymes may have an important role in the function of the nervous system. Although a number of such enzymes have a bimodal distribution (some particulate bound, some soluble), the ratio may change from one species to another. The interaction of such enzymes as acetylcholine synthetase and ATP citrate lyase is then responsible for the formation of acetylcholine in the brain. Choline esterase is bound in different membrane structures, in the microsomes, and in the membranes of the nerve endings. Among enzymes localized in specific membranes is Na^+-K^+-ATPase, which was also shown in myelin. The fraction of nerve endings which is free of choline esterase and which contains glutamic acid decarboxylase seems to be the nerve endings of the noncholinergic systems.

Because of their weak binding the proper localization of the enzymes participating in the biosynthesis of noradrenalin and serotonin is not well established. The main portion of DOPA-decarboxylase is in the cholinergic nerve endings, where the catechol-O-methyltransferase also can be found. Monoamine oxidase, which is an enzyme participating in the metabolism of neurotransmitters, is strongly bound to mitochondria. These points illustrate that the biochemical and enzymatic composition of a specific structure, as well as its morphological description, are important in establishing the functional mechanisms of these structures in the nervous system.

Important clues to the functional roles of enzymes are gained by the measurement of changes of enzyme activities during morphogenesis. It is to be expected that in the early stages of development the enzymes responsible for the formation of structural components of the proteins, lipids, and nucleic acids, and enzyme systems that are specially connected with the mechanism of growth, are predominant first. These are followed by the enzymes responsible for cell differentiation, while enzymes responsible for the final function of the brain appear later. In general terms, the phylogenetically older parts of the brain are developed first, and the enzyme complement of the phylogenetically younger parts, which are more responsible for function, are developed at later stages. Corresponding to this general picture, species in which the brain at birth is functionally at a later developmental stage show more advanced development of their enzyme components as compared to other species, as was shown, e.g., for carboanhydrase.

In general, the profile of enzyme activity changes is that in the early stages of embryonal development the activities are low and in later stages there is a steep increase in enzyme activity nearly to adult levels, in connection with the development of specific structures or functions, for many enzymes. This change in activity is found for many enzymes at the stage of vacuolization of the cell nuclei. Some enzymes seem to be constant from birth on, like glucose-6-phosphate dehydrogenase in rat cortex. Others, like acid and alkaline phosphatases, are more active in the embryonic brain than in adults. This may be an indication that these enzymes play a part in the differentiation of neurons and glia. Cholesterol esterase, which participates in the myelination, can be found only in the early developmental stages. The isoenzymes also change during development as shown by lactate dehydrogenase and aldolase, where three of the five isoenzymes of lactate dehydrogenase show characteristic developmental alterations. Although these give us important clues, we are still far from clearly understanding the function from changes in enzyme levels. This partially may be due to the fact that we have a few methods available to measure the actual *in vivo* activities of the enzymes. Most of the present measurements tell us about enzyme levels *in vitro*, which still do not give good estimates of living rates. New techniques, however, are becoming available, e.g., those of Lowry and Hyden with micro techniques, which bring us new understanding and new correlation among enzyme composition, enzyme structure, and the complex functioning of the nervous system.

V. SUMMARY TABLE OF ENZYMES OF THE NERVOUS SYSTEM

E.C. number	Systematic name	Trivial name	Reaction
	5-Hydroxytryptophol:NADP oxidoreductase	Hydroxytryptophol dehydrogenase	5-Hydroxytryptophol + NADP = 5-hydroxy-indoleacetaldehyde + NADPH$_2$
	Occurrence, properties: rat, human[20]		
1.1.1.8	L-Glycerol-3-phosphate: NAD oxidoreductase	Glycerol-3-phosphate dehydrogenase	L-Glycerol 3-phosphate + NAD = dihydroxyacetone phosphate + NADH$_2$
	Occurrence: rat,[22–25] mouse,[23,25] calf,[26] Japanese quail[27]		
	Subcellular localization: (supernatant and mitochondrial fraction) rat,[28] calf,[26] Japanese quail[28]		
1.1.1.27	L-Lactate:NAD oxidoreductase	Lactate dehydrogenase	L-Lactate + NAD = pyruvate + NADH$_2$
	Occurrence: in all cells with glycolytic activity[30]		
	Activity (total): rat,[25,31] mouse[25,31–33]		
	Isoenzyme pattern: rat (5 isoenzymes),[34–36] rabbit (5 isoenzymes),[34,35,37] sheep (5 isoenzymes),[38] beef (3 isoenzymes),[39] dog (3 isoenzymes),[40] primate and human (5 isoenzymes),[34,37,38,41–43] other vertebrates,[42,44] turtle, fish.[42]		
	Purification: human (pure crystalline isoenzymes),[19] beef (100-fold)[45,46]		
	Properties: rabbit,[47] beef[39,45] other vertebrates,[44] human,[48] (*p*H optimum for the isoenzymes)[19]		
	Regional distribution: rat,[49,50] rabbit (in cerebellar layers and hippocampus),[51] beef (isoenzymes),[52,53] human and primate (cortex and subadjacent white matter),[54] human (whole brain),[55–58] isoenzyme pattern in different regions,[58,60,61] review[62]		
	Cytological differences: cerebellar Purkinje cells (beef)[63]		
	Subcellular localization: (highest activity in the cytoplasma) rat,[28,58,64–68] in nerve ending cytoplasma,[69,70] in neuron and glia cells,[71] in nuclei,[72] rabbit,[66,69,70,73] guinea pig (in nerve ending cytoplasma),[70,74–77] cat,[69] dog,[69,78] calf,[26] sheep (in nerve ending cytoplasma),[69,76] pigeon,[69,70] chicken,[66] Japanese quail,[27] other vertebrates[79]		
	Developmental changes: rat (whole brain),[36,49,66,80,81] cortex,[82] cerebellum,[83] mouse,[80,84] rabbit,[66] guinea pig,[84] chicken,[66,80,97,475] other animals,[80] *Rana temporaria*,[85] human (isoenzyme pattern)[61,86]		
1.1.1.30	D-3-Hydroxybutyrate:NAD oxidoreductase	3-Hydroxybutyrate dehydrogenase	D-3-Hydroxybutyrate + NAD = acetoacetate + NADH$_2$

Occurrence: rat,[87] human[58]
Properties: rat[87]
Regional distribution: human[58]
Subcellular localization: rat (mitochondria)[87]
Developmental changes: rat (mitochondria)[88]

| 1.1.1.35 | L-3-Hydroxyacyl-CoA:NAD oxidoreductase | 3-Hydroxyacyl-CoA dehydrogenase | L-3-Hydroxyacyl-CoA + NAD = 3-oxo-acyl-CoA + $NADH_2$ |

Occurrence: rat[89,90] sheep, pig, calf, rabbit, human[89]
Subcellular localization: soluble fraction[89,90]

| 1.1.1.37 | L-Malate:NAD oxidoreductase | Malate dehydrogenase | L-Malate + NAD = oxaloacetate + $NADH_2$ |

Occurrence: In all cells with citrate cycle[91]
Activity: rat,[92-95] mouse,[92,93] rabbit,[96] sheep,[38] beef,[93] dolphin[94]
Isoenzyme pattern: rat (2 isoenzymes),[24,96] human (6 isoenzymes)[38]
Purification and properties: rat,[96] rabbit,[47] beef[45]
Regional distribution: rat,[50] rabbit,[98] monkey,[54,57] human,[55,59] review (cerebellum)[62]
Subcellular localization: rat (45% in the supernatant, 40% in mitochondria, low in the microsomal fraction);[96,99] localization in nuclei and nuclear membranes,[72] activity in mitochondria,[100,101] localization in mitochondrial subfractions,[102] Japanese quail[27]
Developmental changes: rat,[82,83] chicken[97] (isoenzyme pattern),[103] frog[85]

| 1.1.1.41 | threo-D_s-Isocitrate:NAD oxidoreductase (decarboxylating) | Isocitrate dehydrogenase (decarboxylating) | threo-D_s-Isocitrate + NAD = 2-oxoglutarate + CO_2 + $NADH_2$ |

Occurrence: rat,[104] guinea pig[105]
Subcellular localization: rat (exclusively in mitochondria)[99,104]
Developmental changes: rat[106]

| 1.1.1.42 | threo-D_s-Isocitrate:NADP oxidoreductase (decarboxylating) | Isocitrate dehydrogenase (NADP) | threo-D_s-Isocitrate + NADP = 2-oxoglutarate + CO_2 + $NADPH_2$ |

Occurrence: in all cells with citrate cycle[107]

Summary Table of Enzymes of the Nervous System — (continued)

E.C. number	Systematic name	Trivial name	Reaction
	Activity: rat,[24,95,104] mouse,[25] human[95] *Regional distribution:* rat,[97] rabbit[51] *Subcellular localization:* (in mitochondria and supernatant) rat,[72,99–102,104] guinea pig,[108] chicken,[109] Japanese quail[27] (in nuclei) rat[72] *Developmental changes:* rat,[83,106] chicken[109,110]		
1.1.1.44	6-Phospho-D-gluconate:NADP oxidoreductase (decarboxylating) *Occurrence, activity:* rat[95,111] *Properties:* guinea pig[112] *Regional distribution:* rat,[50] monkey,[57] human[58,95] *Subcellular localization:* rat (10% in mitochondria and microsomes),[112] guinea pig[112,114] *Developmental changes:* rat (mitochondria),[115] chicken[110]	Phosphogluconate dehydrogenase (decarboxylating)	6-Phospho-D-gluconate + NADP = D-ribulose 5-phosphate + CO_2 + $NADPH_2$
1.1.1.47	β-D-Glucose:NAD(P) oxidoreductase *Occurrence:* rabbit[116]	Glucose dehydrogenase	β-D-Glucose + NAD(P) = D-glucono-δ-lactone + $NAD(P)H_2$
1.1.1.49	D-Glucose-6-phosphate:NADP oxidoreductase *Occurrence, activity:* rat,[25,31,95,117] mouse,[25] rabbit,[117,118] opossum, skunk, chicken, sparrow, snake, alligator, turtle, collared lizard, salamander, bullfrog, goldfish, lamprey[117] *Regional distribution:* rat,[50] rabbit,[119,120] monkey,[54,57] human[58,95] *Subcellular localization:* (cytoplasma, 10% in mitochondrial and microsomal fractions) rat (mitochondrial subfractions),[102] guinea pig,[112,114] calf,[26] Japanese quail[27] *Developmental changes:* rat,[82,123] mitochondria,[115] chicken[121,122]	Glucose-6-phosphate dehydrogenase	D-Glucose-6-phosphate + NADP = D-glucono-δ-lactone 6-phosphate + $NADPH_2$

1.1.1.55	1,2-Propanediol:NADP oxidoreductase	Lactaldehyde reductase	1,2-Propanediol + NADP = L-Lactaldehyde + NADPH$_2$
	Occurrence, properties: beef[21]		
1.1.1.61	4-Hydroxybutyrate:NAD oxidoreductase	4-Hydroxybutyrate dehydrogenase	4-Hydroxybutyrate + NAD = succinate semialdehyde + NADH$_2$
	Occurrence: rat[11]		
1.1.99.5	L-Glycerol-3-phosphate:(acceptor) oxidoreductase	Glycerolphosphate dehydrogenase	L-Glycerol-3-phosphate + acceptor = dihydroxyacetone phosphate + reduced acceptor

Activity: rat (mitochondria),[100] (mitochondria and cytoplasma),[124] mouse,[33] pig (mitochondria)[125]
Purification: (from pig mitochondria)[126-128]
Properties: pig (mitochondrial enzyme),[126-128] activation by pyridoxalphosphate[124]
Subcellular localization: pig (soluble and mitochondrial fraction)[124]

1.2.1.3	Aldehyde:NAD oxidoreductase	Aldehyde dehydrogenase	Aldehyde + NAD + H$_2$O = acid + NADH$_2$

Occurrence: rat,[129,130] beef, monkey[131]
Purification: beef (100-fold)[132]
Properties: beef (wide substrate specificity)[131,132]
Regional distribution: beef[132]
Subcellular localization: rat (microsomes),[129] (mitochondria and supernatant),[130] beef, monkey (mitochondria and supernatant)[131,132]

1.2.1.8	Betaine-aldehyde:NAD oxidoreductase	Betaine aldehyde dehydrogenase	Betaine aldehyde + NAD + H$_2$O = betaine + NADH$_2$
	Occurrence: rat[133]		

Summary Table of Enzymes of the Nervous System—(continued)

E.C. number	Systematic name	Trivial name	Reaction
1.2.1.10	Aldehyde:NAD oxidoreductase (acylating CoA)	Aldehyde dehydrogenase (acylating)	Aldehyde + CoA + NAD = acyl-CoA + $NADH_2$
	Occurrence: rat (formation of palmitaldehyde)[134,135]		
	Properties: (requirement for Mn^{2+} and pyridoxalphosphate)[134]		
	Subcellular localization: rat (supernatant and microsomes)[134,135]		
1.2.1.12	D-Glyceraldehyde-3-phosphate:NAD oxidoreductase (phosphorylating)	Glyceraldehyde phosphate dehydrogenase, triose-phosphate dehydrogenase	D-Glyceraldehyde-3-phosphate + phosphate + NAD = 1.3-diphospho-D-glyceric acid + $NADH_2$
	Occurrence: in all cells[136]		
	Activity: rat[25,31,95] (extramitochondrial),[124,137] mouse,[25,33] beef,[138] human[95]		
	Regional distribution: rat,[50] rabbit,[139] human[95]		
	Subcellular localization: (highest activity in the cytoplasma) rat,[28,64,65] calf,[26] in mitochondrial subfractions (rat),[102] in cell nuclei (rat),[72] (beef)[138]		
	Developmental changes: rat[29]		
1.2.1.16	Succinate-semialdehyde:NAD(P) oxidoreductase	Succinate semialdehyde dehydrogenase	Succinate semialdehyde + NAD(P) = succinate + NAD(P)H$_2$
	Occurrence, activity: rat,[140,141] mouse,[141,142] rabbit,[141,142] guinea pig,[141,142] monkey,[142] human[144]		
	Purification: rat,[140] monkey (150-fold),[143] (3000-fold),[141] human,[144] lobster[145]		
	Properties: rat,[140] monkey,[141] human,[141] lobster[145]		
	Regional distribution: rat,[140] monkey,[141] human[146]		
	Subcellular localization: (highest activity in the soluble fraction, mitochondria) rat[147]		
	Developmental changes: rat[148]		
1.2.3.2	Xanthine:oxygen oxidoreductase	Xanthine oxidase	Xanthine + H_2O + O_2 = urate + H_2O_2
	Occurrence: rat, rabbit,[149] dog, sheep[150]		

1.2.3.3	Pyruvate:oxygen oxidoreductase (phosphorylating)	Pyruvate oxidase	Pyruvate + phosphate + O_2 = acetyl-phosphate + CO_2 + H_2O

Occurrence: in all cells with the ability to oxidize pyruvate[151]
Activity, properties: pigeon,[152] rat (mitochondria)[100]

1.2.4.1	Pyruvate:lipoate oxidoreductase (acceptor-acetylating)	Pyruvate dehydrogenase	Pyruvate + oxidized lipoate = 6-S-acetylhydrolipoate + CO_2

Occurrence: pigeon[153]
Properties: (a multienzyme complex requiring thiamine pyrophosphate, NAD, CoA, and α-lipoic acid as cofactors)[153]

1.2.4.2	2-Oxoglutarate:lipoate oxidoreductase (acceptor-acylating)	Oxoglutarate dehydrogenase	2-Oxoglutarate + oxidized lipoate = 6-S-succinylhydrolipoate + CO_2

Properties: (a multienzyme complex requiring thiamine pyrophosphate, NAD, CoA, and α-lipoic acid as cofactors)[147]
Subcellular localization: rat (strictly mitochondrial)[147]

1.3.99.1	Succinate:(acceptor) oxidoreductase	Succinate dehydrogenase	Succinate + acceptor = fumarate + reduced acceptor

Occurrence: in all aerobic cells[154]
Activity: rat[94] (mitochondria),[100] mouse,[155] dolphin[94]
Properties: (a flavoprotein) pig[156]
Regional distribution: rat[157] (cerebellum)[158] (Formatio reticularis)[159] (Nucleus supraopticus)[160] (Plexus chorioidei)[161] (synapse),[162] mouse,[157] rabbit,[157,163–165,1278] (synapse),[162] guinea pig,[157,162,163] cat,[1278] (Formatio reticularis)[159] (optical and motor analyzers),[165] dog,[164,166] bee,[167] hedgehog,[164] reptiles,[168] monkey[57,169] (Formatio reticularis)[159] (optical and motor analyzers),[165] human[170]
Subcellular localization: rat (mitochondria)[28,113,147,171–173] (synaptosomes)[147,174–178] (nuclei)[171] (neurons and glia cells),[71] guinea pig[194] (mitochondria)[74,180] (synaptosomes),[75,179,181] rabbit (mitochondria),[182] cat (mitochondria),[182,183] beef (neurons)[63]
Developmental changes: rat[81,123,184–186] (mitochondria)[193] (mesencephalon),[187] guinea pig,[163,188] rabbit[163,189] (mesencephalon),[187] pig,[190] chicken,[191] rook[192]

Summary Table of Enzymes of the Nervous System—(continued)

E.C. number	Systematic name	Trivial name	Reaction
	Succinate oxidase system		Succinate + $\frac{1}{2}O_2$ = fumarate + H_2O
	Activity: rat,[93,195] mouse, beef,[93] rabbit (single neurons and glia cells),[196] *Regional distribution:* dog,[197] beef[167,197] *Subcellular localization:* (highest activity in mitochondria) rat[198] (neonatal),[99] beef[199] *Developmental changes:* rat,[186,200] guinea pig,[188] pig,[190] chicken[201]		
1.3.99.3	Acyl-CoA:(acceptor) oxidoreductase	Acyl-CoA dehydrogenase	Acyl-CoA + acceptor = 2.3-dehydroacyl-CoA + reduced acceptor
	Occurrence: (oxidation of fatty acids by brain tissue): golden hamster[202] (oxidation of long-chain fatty acids by mitochondria): rat,[203] beef[204,205]		
	Dihydrosphingosine:(acceptor) Δ^4-oxidoreductase		Dihydrosphingosine + FAD = sphingosine + $FADH_2$
	Occurrence: rat[129,134]		
1.4.1.3	L-Glutamate:NAD(P) oxidoreductase (deaminating)	Glutamate dehydrogenase [NAD(P)]	L-Glutamate + H_2O + NAD(P) = 2-oxoglutarate + NH_3 + NAD(P)H_2
	Occurrence: In all animal cells[206] *Activity:* rat[95] (mitochondria),[100,101,207–209] rabbit[47] (mitochondria),[182] cat (mitochondria),[182] hamster[210] *Properties:* rat (mitochondria),[208] rabbit (contains Zn^{2+})[47] *Regional distribution:* rat,[50] rabbit,[47] beef,[167] human[95] (4 isoenzymes, no regional differences in the isoenzyme pattern)[59] Japanese quail[27] *Subcellular localization:* rat (mitochondria),[102,147,211] Japanese quail[27] *Developmental changes:* rat,[82] pig,[212] monkey,[212] chicken[601]		

L-Thyroxine:NAD oxidoreductase

L-Thyroxine + H_2O + NAD = $HO-C_6H_2J_2-O-C_6H_2J_2-CH_2-CO-COOH$ + NH_3 + $NADH_2$

Occurrence: rat[213,214]
Subcellular localization: rat (mitochondria)[213,214]

| 1.4.3.2 | L-Amino-acid:oxygen oxidoreductase (deaminating) | L-Amino-acid oxidase | An L-amino acid + H_2O + O_2 = a 2-oxo-acid + NH_3 + H_2O_2 |

Occurrence: rabbit (L-Ala, -Arg, -Leu, -Ile, -Glu, -Tyr, -Val),[215] rat (L-Leu, -Val),[217] Scyllium canicula, Mugil cephalus, Gobius niger (L-Ala, -His)[216]

L-Amino-acid oxidase system (dissimilation of L-amino acids)

Occurrence: rat (Gly),[218] (D,L-Ala),[219] mouse (L-Leu, -Ile, -Val, -Phe, -Thr),[220] guinea pig (L-Glu)[221-223]
Properties: rat (activation and inhibition)[224] (a flavin enzyme)[218]
Developmental changes: rat (L-Glu),[225] mouse (L-Leu, -Ile, -Val, -Phe, -Thr)[226]

| 1.4.3.3 | D-amino-acid:oxygen oxidoreductase (deaminating) | D-Amino-acid oxidase | A D-amino acid + H_2O + O_2 = a 2-oxo-acid + NH_3 + H_2O_2 |

Occurrence: rat,[256,227-229] dog,[228] beef,[228,229] horse, rabbit, guinea pig,[228] mouse, sheep, duck, dogfish, frog,[229] human[230]
Purification, properties: dog[231] (a flavoprotein); (inhibitors)[229]
Regional distribution: beef, human[232]

| 1.4.3.4 | Monoamine:oxygen oxidoreductase (deaminating) | Monoamine oxidase | A monoamine + H_2O + O_2 = an aldehyde + NH_3 + H_2O_2 |

Occurrence, activity: rat,[25,233-236] (activity during the estrous cycle),[237,238] mouse (activity in different strains),[239] guinea pig,[233,235] (mitochondria),[240] rabbit,[234] dog,[241] sheep,[242] beef,[233] human,[20] (physiologic significance),[268] other animals[243,244]

Summary Table of Enzymes of the Nervous System—(continued)

E.C. number	Systematic name	Trivial name	Reaction
	Purification: rat (100-fold),[245] pig,[13] beef[250] *Properties: Substrate specificity:* rat,[233,247,248] rabbit,[248,249] mouse, cat, dog, human,[248] guinea pig,[233] sheep,[242] beef[233,250] *Other properties:* pig (FAD-content),[13] rabbit,[250] beef[250] (inhibitors)[244] *Regional distribution:* rat,[237] rabbit,[251] cat,[252] dog,[197,252] beef,[167,197] human,[197,253] pigeon,[254] carp,[255] reptiles[168] *Subcellular localization:* (highest activity in mitochondria and mitochondrial subfractions (nerve endings, synaptic membranes, myelin),[172,177,256–258] rabbit,[252,256] guinea pig,[256] cat, dog, human,[197] chicken (embryo; activity in mitochondria and cytoplasma),[259] pigeon[256] *Developmental changes:* rat,[260–264] guinea pig,[261,265] chicken,[259] human,[266] *Xenopus laevis*[267]		
1.4.3.5	Pyridoxaminephosphate:oxygen oxidoreductase (deaminating) *Occurrence, properties:* rabbit[269,270]	Pyridoxamine phosphate oxidase	Pyridoxamine phosphate + H_2O + O_2 = pyridoxal phosphate + NH_3 + H_2O_2
1.4.3.6	Diamine: oxygen oxidoreductase (deaminating) *Occurrence, activity:* rat,[271] mouse,[271] guinea pig,[272,273] human,[274,275] carp and other fish,[255] other animals[276] (no activity was found with putrescine and cadaverine as substrates in rat, guinea pig, and rabbit)[255,277] *Properties:* (substrate specificity) rabbit[249] *Regional distribution:* human,[274] carp[277] *Subcellular localization:* rabbit (microsomes and mitochondria)[278]	Diamine oxidase, histaminase	A diamine + H_2O + O_2 = an aminoaldehyde + NH_3 + H_2O_2
1.5.1.1	L-Proline:NAD(P) 2-oxidoreductase *Occurrence:* rat[279]	Pyrroline-2-carboxylate reductase	L-Proline + NAD(P) = Δ^1-pyrroline-2-carboxylate + NAD(P)H_2
1.5.1.2	L-Proline:NAD(P) 5-oxidoreductase *Occurrence:* rat[279]	Pyrroline-5-carboxylate reductase	L-Proline + NAD(P) = Δ^1-pyrroline-5-carboxylate + NAD(P)H_2

1.5.1.4	7,8-Dihydrofolate:NADP oxidoreductase	Dihydrofolate dehydrogenase	7,8-Dihydrofolate + NADP = folate + NADPH$_2$
	Occurrence: rat[280]		
1.6.1.1	NADPH$_2$:NAD oxidoreductase	NAD(P) transhydrogenase	NADPH$_2$ + NAD = NADP + NADH$_2$
	Occurrence, activity: rat[104,124]		
	Subcellular localization: rat (mitochondria)[104]		
1.6.2.2	NADH$_2$:ferricytochrome b_5 oxidoreductase	Cytochrome b_5 reductase	NADH$_2$ + 2 ferricytochrome b_5 = NAD + 2 ferrocytochrome b_5
	Occurrence: rabbit,[281] fowl, toad, carp[282]		
	Subcellular localization: rabbit (microsomes)[281]		
	NADPH$_2$:ferricytochrome b_5 oxidoreductase		NADPH$_2$ + 2 ferricytochrome b_5 = NADP + 2 ferrocytochrome b_5
	Occurrence: rabbit, guinea pig, beef,[333] fowl, toad, carp[334]		
1.6.4.2	NAD(P)H$_2$:oxidized glutathione oxidoreductase	Glutathione reductase	NAD(P)H$_2$ + oxidized glutathione = NAD(P) + 2 glutathione
	Occurrence: rat, guinea pig[283,284]		
	Purification: beef (43-fold)[285]		
	Properties: rat, guinea pig,[284] beef (no tetrazolium reductase activity)[285]		
	Distribution: rat[50]		
	Subcellular localization: rat (supernatant, mitochondria)[171]		
1.6.4.3	NADH$_2$:lipoamide oxidoreductase	Lipoamide dehydrogenase	NADH$_2$ + lipoamide = NAD + dihydro-lipoamide
	Occurrence: Widely distributed in nature		
	Purification: beef (800-fold)[285,286]		
	Properties: beef[286]		

Summary Table of Enzymes of the Nervous System—(continued)

E.C. number	Systematic name	Trivial name	Reaction
1.6.99.1	NADPH$_2$:(acceptor) oxidoreductase	NADPH$_2$ dehydrogenase	NADPH$_2$ + acceptor = NADP + reduced acceptor

Regional distribution: rat,[159] rabbit,[287] cat,[159] monkey,[57,159] human[170,288]
Subcellular localization: (mitochondria, supernatant) rat,[285,287] guinea pig, rabbit, pigeon[287]
Developmental changes: (highest activity in the adult cerebellum) rat, rabbit, guinea pig, pigeon[287]

a. Amylobarbitone and antimycin A sensitive enzyme

Occurrence: rat,[289] *Octopus vulgaris, Sargus sargus, Rana esculenta, Thalassochelis caretta*[290]
Properties: (mitochondria) rat[291]
Subcellular localization: rat (mitochondria)[65]
Developmental changes: rat (mitochondria)[106,291]

b. Amylobarbitone and antimycin A insensitive enzyme

Occurrence: rat,[292] rabbit, beef, guinea pig,[281] fowl, toad, carp,[282] *Thisanozoon thisanozoon, Octopus vulgaris, Loligo vulgaris, Sargus sargus, Rana esculenta, Thalassochelis caretta*[290]
Properties: rat,[292] beef, rabbit, guinea pig,[281] fowl, toad carp[282]
Subcellular localization: microsomal fraction, (rat)[292] (beef, guinea pig, rabbit, carp, toad, fowl)[281,282]

| 1.6.99.2 | NAD(P)H$_2$:(acceptor) oxidoreductase | NAD(P)H$_2$ dehydrogenase, Menadione reductase | NAD(P)H$_2$ + acceptor = NAD(P) + reduced acceptor |

Occurrence: "Diaphorase I and II", rat, beef,[285,293,294] guinea pig,[293] rabbit, human, pigeon[294]
Purification: "Diaphorase I", beef (200-fold),[294,295] rat[285]
"Diaphorase II", beef (300-fold)[296]

Properties: "Diaphorase I," beef (a flavine enzyme, extractable with H$_2$O, dicoumarol sensitive),[294,295] rat[285]
"Diaphorase II," beef (a flavine containing particulate bound enzyme),[296] rat ("Diaphorase I and II")[297]

| 1.6.99.3 | NADH$_2$:(acceptor) oxidoreductase | NADH$_2$ dehydrogenase | NADH$_2$ + acceptor = NAD + reduced acceptor |

a. Amylobarbitone and antimycin A sensitive enzyme

Occurrence: rat,[218,289,292] guinea pig,[281,284] rabbit, beef,[281] *Thalassochelis caretta, Rana esculenta, Sargus sargus, Gobius paganellus, Coris iulis, Scyllium canicola, Torpedo ocellata, Sepia officinalis, Loligo vulgaris, Octopus vulgaris, Thisanozoon thisanozoon*[290]

Properties: rat (mitochondria),[291,292,299] guinea pig[284]

Subcellular localization: rat (mitochondria)[65,298,299]

Developmental changes: rat (mitochondria)[106,291]

b. Amylobarbitone and antimycin A insensitive enzyme

Occurrence: rat,[292] rabbit, guinea pig, beef,[281] *Gallus gallus, Thalassochelis caretta, Rana esculenta, Sargus sargus, Gobius paganellus, Coris iulis, Scyllium canicola, Torpedo ocellata, Spyrographis spallanzani, Sepia officinalis, Loligo vulgaris, Octopus vulgaris, Thisanozoon thisanozoon*[290]

Properties: rat,[292] rabbit, guinea pig, beef[281]

Subcellular localization: (microsomal fraction)[292,281]

NADH$_2$-oxidase-system (antimycin A sensitive)

Occurrence: rat,[290,300] *Gallus gallus, Thalassochelis caretta, Rana esculenta, Sargus sargus, Gobius paganellus, Coris iulis, Scyllium canicola, Torpedo ocellata, Peneus caramote, Spyrographis spallanzani, Sepia officinalis, Loligo vulgaris, Octopus vulgaris, Thisanozoon thisanozoon*[290]

Properties: rat[290,300]

Subcellular localization: (mitochondria) rat[300]

NADH$_2$-oxidase-system (dichloronaphthoquinone dependent)

Occurrence: rabbit, guinea pig, beef,[281] fowl, toad, carp[282]

Properties, subcellular localization: (microsomes) beef, rabbit, guinea pig, fowl, toad, carp[281,282]

Summary Table of Enzymes of the Nervous System—(continued)

E.C. number	Systematic name	Trivial name	Reaction
	NAD(P)H$_2$-ferricyanide oxidoreductase	NAD(P)H$_2$-ferricyanide reductase	NAD(P)H$_2$ + ferricyanide = NAD(P) + ferrocyanide

5 different enzymes:
1. Dicoumarol sensitive-NADH$_2$-ferricyanide reductase (O$_2$ an acceptor)
2. Dicoumarol sensitive-NADH$_2$- and NADPH$_2$-ferricyanide-reductase, (O$_2$ not an acceptor)
3. Dicoumarol insensitive-NADH$_2$-ferricyanide reductase
4. Dicoumarol insensitive-NAD(P)H$_2$-ferricyanide reductase (other electron acceptors also acting, O$_2$ not an acceptor)
5. Dicoumarol insensitive-NAD(P)H$_2$-ferricyanide reductase (other electron acceptors and O$_2$ acting)

Occurrence: rat,[289] beef,[285] *Gallus gallus, Thalassochelis caretta, Rana esculenta, Sargus sargus, Gobius pagallenus, Coris iulis, Scyllium canicola, Torpedo ocellata, Peneus caramote, Spirographis spallanzani, Sepia officinalis, Loligo vulgaris, Octopus vulgaris, Thisanozoon thisanozoon*[290]

Preparation: (dicoumarol sensitive enzymes) *Sepia officinalis*[301] (dicoumarol insensitive enzymes) rat[289] *Sepia officinalis*[302]

Properties: (dicoumarol sensitive enzymes) *Sepia officinalis*[301] (dicoumarol insensitive enzymes) rat,[289] beef,[285] *Sepia officinalis*[301,302]

Subcellular localization: Soluble fraction (rat),[289] *Sepia officinalis*[301,302]

E.C. number	Systematic name	Trivial name	Reaction
1.7.3.3	Urate:oxygen oxidoreductase	Urate oxidase (uricase)	Urate + O$_2$ = unidentified products

Occurrence, regional distribution: rat, rabbit[303]

E.C. number	Systematic name	Trivial name	Reaction
1.9.3.1	Ferrocytochrome c: oxygen oxidoreductase	Cytochrome oxidase	4 Ferrocytochrome c + O$_2$ = 4 ferricytochrome c + 2 H$_2$O

Activity: rat,[94] sheep,[304,305] dolphin,[94] birds, amphibia, fish, reptiles, mammals;[306] mitochondria : (rat),[299,307] (beef),[308] (fish);[307] neuron and glia cells : rat,[71] rabbit[196]

Regional distribution: rat (cortex)[309] (cerebellum),[158] rabbit,[165,310,311,1278] cat,[165,1278] dog,[311] monkey,[57,165,312] human (isocortex)[313]

Subcellular localization: (mitochondria) rat,[67,99,292,298,299,314,315] cat,[315] monkey[312]

Developmental changes: rat,[88,184,185,187,316] guinea pig,[163] rabbit,[163,187,189,311] pig,[317] dog,[311] chicken,[191] rook[192]

1.10.3.1	o-Diphenol:oxygen oxidoreductase	o-Diphenol oxidase (tyrosinase)	$2\ o\text{-Diphenol} + O_2 = 2\ o\text{-quinone} + 2\ H_2O$

Occurrence: rat,[318] amphibian larvae[319]
Properties: (contains Cu^{2+}) rat[318]
Developmental changes: cat (*Substantia nigra*)[320]

1.11.1.6	Hydrogen-peroxide:hydrogen peroxide oxidoreductase	Catalase	$H_2O_2 + H_2O_2 = O_2 + 2\ H_2O$

Occurrence: little activity, if any, in sheep brain,[304] (but see [996])

1.11.1.7	Donor:hydrogen peroxide oxidoreductase	Peroxidase	Donor $+ H_2O_2 =$ oxidized donor $+ 2\ H_2O$

Occurrence: trace amounts of activity (rabbit, hog, sheep, calf, horse)[321,996]

1.14.2.1	3,4-Dihydroxyphenylethylamine, ascorbate:oxygen oxidoreductase (hydroxylating)	Dopamine hydroxylase	3,4-Dihydroxyphenylethylamine + ascorbate $+ O_2 =$ noradrenaline + dehydro-ascorbate $+ H_2O$

Regional distribution: rat,[322–323] dog, beef,[323] monkey [324] (highest in N. caudatus and hypothalamus)
Subcellular localization: guinea pig (higher in the particulate than in the soluble fraction)[325]

1.14.3.1	L-Phenylalanine, tetrahydropteridine: oxygen oxidoreductase	Phenylalanine 4-hydroxylase, tyrosine hydroxylase	L-Phenylalanine + tetrahydropteridine + O_2 = L-tyrosine + dihydropteridine + H_2O L-tyrosine + tetrahydropteridine + O_2 = 3,4-dihydroxyphenylalanine + dihydropteridine + H_2O

Occurrence: rat,[326,327] rabbit, guinea pig, beef,[327] human[328]
Purification: rat,[329] guinea pig,[327] dog[330]
Properties: rat (cofactor: 2-amino-4-hydroxy-6, 7-dimethyl-tetrahydropteridin),[329] guinea pig[327] dog[394,399]
Regional distribution: cat[332,333] monkey[324]
Subcellular localization: (mitochondrial fraction, supernatant) guinea pig,[326] pig,[331] beef[334]

Summary Table of Enzymes of the Nervous System—(continued)

E.C. number	Systematic name	Trivial name	Reaction
	L-Tryptophan, tetrahydropteridine: oxygen oxidoreductase (5-hydroxylating)	Tryptophan 5-hydroxylase	L-Tryptophan + tetrahydropteridine + O_2 = 5-hydroxytryptophan + dihydropteridine + H_2O
	Occurrence: rat,[335–338] rabbit,[339,340] pigeon,[335,341] pig, beef[339]		
	Purification: rat,[342] rabbit[91,343,344]		
	Properties: rat,[339,342,345] rabbit (cofactors)[343]		
	Regional distribution: dog,[91,346] rabbit[346]		
	Subcellular localization: (cytoplasma, mitochondria, nerve endings) rat,[339,345] rabbit[91,344]		
		Fatty acid α-hydroxylase	
	Occurrence, properties: rat[347]		
		γ-Aminobutyrate α-hydroxylase	
	Occurrence, properties: guinea pig[348]		
		Thyroxine deiodinase	
	Occurrence: rat[1352,1353]		
	Properties: rat (FMN-dependence)[1248,1354]		
	Subcellular localization: rat (mitochondria)[213,214]		
2.1.1.4	S-Adenosylmethionine: N-acetylserotonin O-methyltransferase	Acetylserotonin methyltransferase	S-Adenosylmethionine + N-acetylserotonin = S-adenosylhomocysteine + N-acetyl-5-methoxytryptamine
	Occurrence: Rana pipiens, Xenopus laevis,[249] pineal glands of cat, cow, monkey[350]		
	Developmental changes: Xenopus laevis[351]		

2.1.1.6	S-Adenosylmethionine:catechol O-methyltransferase *Occurrence*: rat[236,238,352,353] (activity during the sexual cycle)[238] *Regional distribution*: monkey (*Macaca mulatta*)[350,354] (green monkey)[324] *Subcellular localization*: rat (synaptic vesicles)[257,355] *Developmental changes*: rat (constant activity)[264]	Catechol methyltransferase	S-Adenosylmethionine + catechol = S-adenosylhomocysteine + guiacol
2.1.1.8	S-Adenosylmethionine:histamine N-methyltransferase *Occurrence*: rat, rabbit, guinea pig,[276,356,357] mouse,[271,276,356] cat,[276,356,358-360] dog, pig,[359] chicken, *Rana temporaria*, *Perca fluviatilis*[361] *Purification*: guinea pig (30-fold)[356] *Properties*: guinea pig[356] *Regional distribution*: rabbit, cat,[356] *Macaca mulatta*[350]	Histamine methyltransferase	S-Adenosylmethionine + histamine = S-adenosylhomocysteine + 1-methyl-histamine
	S-Adenosylmethionine:phenylethanolamine N-methyltransferase *Occurrence*: rabbit,[362,363] cat, monkey[363] *Purification*: rabbit[362] *Regional distribution*: rabbit[362,363]	Phenylethanolamine methyltransferase	S-Adenosylmethionine + norepinephrine = S-adenosylhomocysteine + epinephrine
	S-Adenosylmethionine:phosphatidyl ethanolamine N-methyltransferase *Occurrence*: rat[364] *Subcellular localization*: (rat) microsomes[364]	Phosphatidyl ethanolamine methyltransferase	3 S-Adenosylmethionine + phosphatidyl ethanolamine = 3 S-adenosylhomocysteine + phosphatidyl choline
	Occurrence: rat, kitten[365,366]	RNA-methylase	

Summary Table of Enzymes of the Nervous System—(continued)

E.C. number	Systematic name	Trivial name	Reaction
	Purification: rat[366] *Developmental changes*: rat, cat[366]		
2.1.3.3	Carbamoylphosphate:L-ornithine carbamoyltransferase *Occurrence*: beef, chicken,[367] human[368]	Ornithine carbamoyltransferase	Carbamoylphosphate + L-ornithine = orthophosphate + L-citrulline
2.1.4.1	L-Arginine:glycine amidinotransferase *Purification, properties*: rabbit[369]	Glycine amidinotransferase	L-Arginine + glycine = L-ornithine + guanidinoacetate
	L-Arginine:γ-amino acid amidinotransferase *Occurrence*: rat[370]	γ-aminobutyrate amidinotransferase	L-Arginine + γ-aminobutyrate = L-ornithine + γ-guanidinobutyrate
2.2.1.1	Sedoheptulose-7-phosphate: D-glyceraldehyde-3-phosphate glycolaldehyde transferase *Occurrence*: rat[371] *Regional distribution*: rat[372,373]	Transketolase, glycolaldehyde transferase	Sedoheptulose 7-phosphate + D-glyceraldehyde-3-phosphate = D-ribose 5-phosphate + D-xylulose-5-phosphate
2.3.1.1	Acetyl-CoA:L-glutamate N-acetyltransferase *Occurrence*: rat[374]	Amino acid acetyltransferase	Acetyl-CoA + L-glutamate = CoA + N-acetyl-L-glutamate

2.3.1.4	Acetyl-CoA: 2-amino-2-deoxy-D-glucose-6-phosphate N-acetyltransferase	Glucosamine-phosphate acetyltransferase	Acetyl-CoA + 2-amino-2-deoxy-D-glucose-6-phosphate = CoA + 2-acetamido-2-deoxy-D-glucose 6-phosphate

Occurrence, purification, properties: sheep (75-fold)[375]

2.3.1.6	Acetyl-CoA: choline O-acetyltransferase	Choline acetyltransferase	Acetyl-CoA + choline = CoA + O-acetylcholine

Occurrence: rat,[376-379] mouse,[380,381] guinea pig,[376,377,382-384] rabbit,[385,386] cat,[377,382] dog,[385] beef,[376,382] pigeon,[377,] frog,[377,387] fish, reptiles,[387] *Electrophorus electricus*,[377] lobster,[386] review[388]

Purification: rat,[389] mouse,[390] guinea pig,[377,384,389,391-392] rabbit[394] (42-fold),[395] squid (head ganglia) (58-fold)[396,397]

Properties: Reaction mechanism (rat, guinea pig, beef)[377,393,398-401] (squid),[402] rat: (substrate specificity)[400,403] (inhibitors, activators and other properties),[389,404-407] guinea pig,[391] rabbit[395] (mol wt.),[408] squid (head ganglia) (substrate specificity, pH-optimum)[409] (other properties);[393,396,397,410] Temperature optimum: (rabbit, *Labrus viridis*)[411] (*Scyllium canicola, Rana temporaria, Lacerta viridis, Gallus domesticus*)[412] Review.[413-415]

Regional distribution: rat,[416,417] guinea pig,[70,412,417,418] rabbit,[70,412,417,418,419] cat,[417,418] dog,[166,418,420] pig,[418] sheep,[418] human,[412,418,420] chicken,[412] pigeon[254,412,417] *Scyllium canicola, Rana temporaria, Lacerta viridis, Abramis blicca, Cavia porcellus*[412]

Subcellular localization: (mitochondrial subfractions: nerve endings, nerve-ending cytoplasma and synaptic vesicles) rat,[70, 175-177,256,257,407,421-423] guinea pig,[70,76,256] rabbit,[70] sheep,[76] pigeon[69,70,256,423]

Developmental changes: guinea pig,[425] rabbit,[419,425] chicken[426]

2.3.1.7	Acetyl-CoA: carnitine O-acetyltransferase	Carnitine acetyltransferase	Acetyl-CoA + carnitine = CoA + O-acetylcarnitine

Occurrence: rat,[422,427] rabbit[427]
Properties: rabbit[422]
Regional distribution: rat,[428] rabbit[422]
Subcellular localization: rat (mitochondria)[422,427]
Developmental changes: rabbit[422]

Summary Table of Enzymes of the Nervous System—(continued)

E.C. number	Systematic name	Trivial name	Reaction
2.3.1.9	Acetyl-CoA:acetyl-CoA C-acetyltransferase *Occurrence*: rat, rabbit, calf, pig, sheep, human[89]	Acetyl-CoA acetyltransferase, acetyl-CoA thiolase	Acetyl-CoA + acetyl-CoA = CoA + acetoacetyl-CoA
2.3.1.15	Acyl-CoA:L-glycerol-3-phosphate O-acyltransferase *Occurrence*: rat[429]	Glycerolphosphate acyltransferase	Acyl-CoA + L-glycerol 3-phosphate = CoA + monoglyceride phosphate
	Acyl-CoA:lysophosphatidic acid O-acyltransferase *Occurrence, properties*: rat[430] *Subcellular localization*: guinea pig (microsomes)[431]	Lysophosphatidic acid acyltransferase	Acyl-CoA + lysophosphatidic acid = CoA + phosphatidic acid
	Acyl-CoA:lysolecithin O-acyltransferase *Occurrence*: rat, human[430] *Properties*: rat, human[430,432] *Subcellular localization*: rat (mitochondria)[432]	Lysolecithin acyltransferase	Acyl-CoA + lysolecithin = CoA + lecithin
2.3.1.17	Acetyl-CoA:L-aspartate N-acetyltransferase *Occurrence*: mouse, rabbit[433] *Purification, properties*: rat (soluble enzyme),[434] cat (particulate bound enzyme)[98]	Aspartate acetyltransferase	Acetyl-CoA + L-aspartate = CoA + N-acetyl-L-aspartate
	Acetyl-CoA:histidine N-acetyltransferase	Histidine acetyltransferase	Acetyl-CoA + L-histidine = CoA + α-N-acetyl-L-histidine

Occurrence: fish[435,436]
Purification: killfish (*Fundulus heteroclitus*)[437]
Properties: killfish,[437] fish[436]

Choline succinyltransferase
Succinyl-CoA:choline
O-succinyltransferase
Occurrence: beef[438]

Succinyl-CoA + choline = CoA + O-succinylcholine

Sphingosine acyltransferase
Acyl-CoA:sphingosine
N-acyltransferase
Occurrence: rat[439]

Acyl-CoA + sphingosine = CoA + N-acylsphingosine (Ceramide)

Sphingosylphosphorylcholine acyltransferase
Acyl-CoA:sphingosylphosphorylcholine
N-acyltransferase
Occurrence: rat[440]

Acyl-CoA + sphingosylphosphorylcholine = CoA + sphingomyelin

Acetylcarnitine:choline
O-acetyltransferase
Occurrence: rat, guinea pig[441]

Acetylcarnitine + choline = carnitine + O-acetylcholine

Acetylcarnityl-CoA:choline
O-acetylcarnityltransferase
Occurrence: rat[406]

Acetylcarnityl-CoA + choline = CoA + O-acetylcarnitylcholine

Palmityladenylate:CoA
S-palmityltransferase
Occurrence: rat[442]

Palmityladenylate + CoA = palmityl-CoA AMP

Glutathione acetyltransferase
Acetyl-CoA:glutathione
acetyltransferase
Occurrence: brain[443]

Acetyl-CoA + glutathione = CoA + acetylglutathione

Summary Table of Enzymes of the Nervous System—(continued)

E.C. number	Systematic name	Trivial name	Reaction
	Palmityl-CoA:glutathione palmityltransferase *Occurrence*: rat[442]	Glutathione palmityltransferase	Palmityl-CoA + glutathione = CoA + palmitylglutathione
2.4.1.1	α-1,4-Glucan:orthophosphate glucosyltransferase	Phosphorylase a	$(\alpha\text{-}1,4\text{-Glucosyl})_n$ + orthophosphate = $(\alpha\text{-}1,4\text{-glucosyl})_{n-1}$ + α-D-glucose 1-phosphate
	Occurrence: rat,[444,445,457,584] mouse,[445] guinea pig, horse, human,[457] rabbit,[444,446,447,457] cat,[444,457] beef,[444] hamster[448] *Purification*: rat,[445] mouse,[449] rabbit (100-fold)[449] *Properties*: rat,[445] mouse[449] (conversion of phosphorylase *b* to phosphorylase *a*),[450] rabbit[447,449,451,452] *Regional distribution*: rat, mouse, guinea pig,[459] rabbit,[120,454–456] human,[58] monkey,[54] rat, rabbit, cat (cerebellum),[453] (review)[62] *Developmental changes*: rat[454,459]		
2.4.1.11	UDP glucose:glycogen α-4-glucosyltransferase	glycogen-UDP glucosyltransferase	UDP glucose + $(\text{glycogen})_n$ = UDP + $(\text{glycogen})_{n+1}$
	Occurrence: rat,[460] rabbit,[461,462] human[463] *Purification*: sheep[464] (50-fold) *Properties*: rabbit,[461] sheep[464] *Regional distribution*: rabbit[456] *Subcellular localization*: rabbit[461] *Developmental changes*: rat[465]		
2.4.1.18	α-1,4-Glucan:α-1,4-glucan 6-glycosyltransferase *Occurrence*: rabbit[466] *Regional distribution*: rabbit[466]	Q-enzyme, branching factor	Transfers part of a 1,4-glucan chain from a 4- to a 6-position

2.4.1.23	UDP galactose:sphingosine O-galactosyltransferase	UDP galactose-sphingosine galactosyltransferase, psychosine-UDP galactosyl-transferase	UDP galactose + sphingosine = UDP + psychosine
	Occurrence: guinea pig, rat[467]		
	Properties: guinea pig[467]		
	Subcellular localization: microsomes[467]		
	UDP galactose: glycolipid galactosyltransferase		Incorporates galactose into glycolipids
	Occurrence: rat[468]		
	Subcellular localization: rat (microsomes)[468]		
	Developmental changes: rat[468]		
	UDP galactose: Tay–Sachs ganglioside galactosyltransferase		UDP galactose + Tay–Sachs ganglioside = UDP + monosialoganglioside
	Occurrence: rat, pig, chicken (embryo and young animals)[469]		
	Properties: chicken embryo (requirement for Mn^{2+})[469]		
	Subcellular localization: chicken embryo (particulate fraction)[469]		
	Developmental changes: chicken, pig[469]		
	UDP-N-acetyl galactosamine: hematoside N-acetylgalactosaminyl-transferase		UDP-N-acetylgalactosamine + hematoside = UDP + Tay-Sachs-ganglioside
	Occurrence: chicken[469]		
	Properties: chicken embryo (requirement for Mn^{2+})[469]		
	CMP-N-acetylneuraminic acid: ceramide disaccharide N-acetyl-neuraminyl transferase	CMP-N-acetylneuraminic acid ceramide disaccharide sialyltransferase	CMP-N-acetylneuraminic acid + ceramide disaccharide = CMP + monosialo-ceramide disaccharide
	Occurrence: rat, guinea pig, sheep, pig, calf, chicken[469]		

Summary Table of Enzymes of the Nervous System—(continued)

E.C. number	Systematic name	Trivial name	Reaction
	Properties: chicken embryo[469] *Developmental changes*: rat, pig, chicken[469]		
	CMP-*N*-acetylneuraminic acid: ceramide tetrasaccharide *N*-acetyl-neuraminyltransferase *Occurrence*: rat, guinea pig, sheep, pig, calf, chicken[469] *Preparation, properties*: chicken embryo[469] *Subcellular localization*: chicken embryo (particulate fraction)[469] *Developmental changes*: rat, pig, chicken[469]	CMP-*N*-acetylneuraminic acid-ceramide tetrasaccharide sialyltransferase	CMP-*N*-acetylneuraminic acid + ceramide tetrasaccharide = CMP + mono- and disialoganglioside
	CMP-*N*-acetylneuraminic acid: ganglioside *N*-acetylneuraminyl-transferase *Occurrence*: rat[470] *Properties*: rat[470] *Subcellular localization*: rat (mitochondria, microsomes, supernatant)[470]	CMP-*N*-acetylneuraminic acid-ganglioside sialyltransferase	CMP-*N*-acetylneuraminic acid + an asialoganglioside = CMP + a monosialo-ganglioside (lactose, ceramide-lactose and sphingosine-lactose are also acceptors)
2.4.2.1	Purine-nucleoside: orthophosphate ribosyltransferase *Occurrence*: rat, rabbit, guinea pig[471] *Properties*: monkey[472] *Regional distribution*: monkey (cerebellum)[472]	Purine nucleoside phosphorylase	Purine nucleoside + orthophosphate = α-D-ribose 1-phosphate + purine
2.4.2.3	Uridine: orthophosphate ribosyltransferase *Occurrence*: mouse[473]	Uridine phosphorylase	Uridine + orthophosphate = uracil + D-ribose 1-phosphate

2.5.1.6	ATP:L-methionine S-adenosyltransferase	Methionine adenosyltransferase	ATP + L-methionine + H_2O = orthophosphate + pyrophosphate + S-adenosyl-methionine
	Occurrence: rat, monkey, human[474]		
		Sulfate activating enzyme	ATP + K_2SO_4 = adenosine-3'-phosphate-5'-phosphosulfate + orthophosphate
		Occurrence, properties: rat (pH-optimum Mg^{2+}, Mn2⁺, or Zn^{2+} requirement)[476] Developmental changes: rat[476]	
2.6.1.1	L-Aspartate:2-oxoglutarate aminotransferase	Aspartate aminotransferase	L-Aspartate + 2-oxoglutarate = oxaloacetate + L-glutamate
	Occurrence, activity: rat[25,95,209,477-479] (isoenzymes)[480,501,503] (mitochondria),[100] mouse,[25,481,482] horse (medulla),[513] catfish (Parasilurus asotus),[483] turtle (Clemmys japonica),[484] frog (Rana nigromaculata)[485]		
	Purification: beef[486] (100-fold),[487] human[486,488]		
	Properties: rat,[489-491] beef[486] (mol. wt.),[492] human[486,488]		
	Physiological significance.[493]		
	Regional distribution: rat,[494] rabbit,[495,496] human[95,497,498]		
	Subcellular localization: (cytoplasma, mitochondria) rat,[66,67,79,102,147,207,211,315,491,494,499,500] rabbit,[66] beef,[502] chicken,[66] Japanese quail[27]		
	Developmental changes: rat[66,504,505] (isoenzymes),[501,503] rabbit,[66,506] dog,[507] pig,[212] monkey,[212] human,[508] chicken[66,475,509-512]		
2.6.1.2	L-Alanine:2-oxoglutarate aminotransferase	Alanine aminotransferase	L-Alanine + 2-oxoglutarate = pyruvate + L-glutamate
	Occurrence: rat,[95,478,480,514] mouse,[482] horse (medulla),[513] catfish (Parasilurus asotus),[483] turtle (Clemmys japonica),[484] frog[485]		
	Regional distribution: dog,[507] human,[95] frog (Rana tigrina)[515]		
	Subcellular localization: (cytoplasm, mitochondria) rat,[102,147,207,315,491] Japanese quail[27]		
	Developmental changes: rat,[504] rabbit,[506] human,[508] chicken[509-512]		

Summary Table of Enzymes of the Nervous System—(continued)

E.C. number	Systematic name	Trivial name	Reaction
2.6.1.4	Glycine:2-oxoglutarate aminotransferase *Occurrence*: rat, rabbit, pigeon,[516] catfish (*Parasilurus asotus*),[483] frog (*Rana nigromaculata*)[485]	Glycine aminotransferase	Glycine + 2-oxoglutarate = glyoxalate + L-glutamate
2.6.1.5	L-Tyrosine:2-oxoglutarate aminotransferase *Occurrence*: rat,[517] dog,[518] catfish (*Parasilurus asotus*),[483] turtle (*Clemmys japonica*)[484] *Purification*: rat[519] *Properties*: rat[519] *Developmental changes*: chicken[509-511]	Tyrosine aminotransferase	L-Tyrosine + 2-oxoglutarate = p-hydroxyphenylpyruvate + L-glutamate
2.6.1.6	L-Leucine:2-oxoglutarate aminotransferase *Occurrence*: rat,[478] catfish (*Parasilurus asotus*),[483] turtle (*Clemmys japonica*),[484] frog (*Rana nigromaculata*)[485] *Developmental changes*: chicken[509-511]	Leucine aminotransferase	L-Leucine + 2-oxoglutarate = 2-oxoiso-caproate + L-glutamate
2.6.1.8	2,5-Diaminovalerate:2-oxoglutarate aminotransferase *Occurrence*: mouse[481]	Diamino-acid aminotransferase	2,5-Diaminovalerate + 2-oxoglutarate = 5-amino-2-oxovalerate + L-glutamate
2.6.1.13	L-Ornithine:2-oxoacid aminotransferase *Occurrence*: rat,[520] mouse,[481] catfish (*Parasilurus asotus*),[483] turtle (*Clemmys japonica*),[484] frog (*Rana nigromaculata*)[485] *Developmental changes*: chick[509-511]	Ornithine-ketoacid aminotransferase	L-Ornithine + a 2-oxoacid = L-glutamate γ-semialdehyde + an L-amino acid

2.6.1.15	L-Glutamine:2-oxoacid aminotransferase	Glutamine-ketoacid amino-transferase "Glutaminase II"	L-Glutamine + a 2-oxoacid = 2-oxoglutarate + an amino acid
	Occurrence: rat,[521] rabbit,[521–523] guinea pig,[521,523] cat, dog, sheep, goat,[521] pig,[523] pigeon[524] *Properties:* rat[521] *Subcellular localization:* rat (mitochondria, nuclei)[521,525]		
		Glutamine-glyoxylic acid aminotransferase	L-Glutamine + glyoxylic acid = 2-oxoglutarate + glycine
	Occurrence: rat, rabbit, pigeon[516]		
2.6.1.16	L-Glutamine:D-fructose-6-phosphate aminotransferase	Glutamine-fructose-6-phosphate aminotransferase	L-Glutamine + D-fructose-6-phosphate = 2-amino-2-deoxy-D-glucose 6-phosphate + L-glutamate
	Occurrence: rat,[526,527] cat[526] *Properties:* rat[527] *Regional distribution:* rabbit[528] *Developmental changes:* rat,[526] rabbit[528]		
2.6.1.18	L-Alanine:malonate-semialdehyde aminotransferase	β-Alanine aminotransferase	L-Alanine + malonate-semialdehyde = pyruvate + β-alanine
	Occurrence: mouse,[481] pig,[529] catfish (*Parasilurus asotus*),[483] frog (*Rana nigromaculata*)[484] *Developmental changes:* chick[509–511]		
	L-Glutamate:2-oxoglutarate aminotransferase	Glutamate-oxoglutarate aminotransferase	L-Glutamate + 2-oxoglutarate = 2-oxoglutarate + L-glutamate
	Occurrence: rat[517,530] *Properties:* rat[530]		

Summary Table of Enzymes of the Nervous System—(continued)

E.C. number	Systematic name	Trivial name	Reaction
	β-aminoisobutyrate:2-oxoglutarate aminotranferase *Occurrence*: pig[529]	β-aminoisobutyrate-oxoglutarate aminotranferase	β-aminoisobutyrate + 2-oxoglutarate = methylmalonate-semialdehyde + L-glutamate
	δ-aminolevulinic acid:2-oxoacid aminotranferase *Occurrence*: rat[531]	δ-aminolevulinic acid aminotransferase	δ-aminolevulinic acid + an 2-oxoacid = 2-oxoglutarate-semialdehyde + an amino acid
	L-Phenylalanine:2-oxoacid aminotransferase *Occurrence*: rat,[517,532] pig[533] *Purification, properties*: rat[519]	Phenylalanine-ketoacid aminotransferase	L-Phenylalanine + an 2-oxoacid = phenylpyruvate + an amino acid
	L-Thyroxine:2-oxoglutarate aminotransferase *Occurrence*: rat[214]	L-Thyroxine-ketoglutarate aminotransferase	L-Thyroxine + 2-oxoglutarate = $HO-C_6H_2J_2-O-C_6H_2J_2-CH_2-CO-COOH$ + L-glutamate
	L-3,4-Dihydroxy-phenylalanine:2-oxoacid aminotransferase *Occurrence*: rat[517,532] *Purification*: rat,[519] guinea pig (38-fold)[534] *Properties*: rat,[519] guinea pig[534]	DOPA-ketoacid aminotransferase	L-3,4-Dihydroxy-phenylalanine + an 2-oxoacid = 3,4-dihydroxy-phenylpyruvate + an amino acid

	Systematic name	Common name	Reaction / details
	L-Tryptophan:2-oxoacid aminotransferase	Tryptophan-ketoacid aminotransferase	L-Tryptophan + an 2-oxoacid = indolylpyruvate + an amino acid
	Occurrence: rat[517]		
	Purification, properties: rat[519]		
	L-5-Hydroxytryptophan:2-oxoacid aminotransferase	5-Hydroxytryptophan-ketoacid aminotransferase	L-5-Hydroxytryptophan + an 2-oxoacid = 5-hydroxy-indolylpyruvate + an amino acid
	Occurrence: rat[517,532]		
	Purification, properties: rat[519]		
	L-Serine-O-phosphate:2-oxoglutarate aminotransferase	Serine phosphate-oxoglutarate aminotransferase	L-Serine-O-phosphate + 2-oxoglutarate = hydroxypyruvatephosphate + L-glutamate
	Occurrence: mouse,[535] beef, dog[536]		
	Purification, properties: sheep (500-fold)[537]		
2.6.1.19	4-Aminobutyrate:2-oxoglutarate aminotransferase	Aminobutyrate aminotransferase	4-Aminobutyrate + 2-oxoglutarate = succinate semialdehyde + L-glutamate
	Occurrence, activity: rat,[538-540] mouse,[481,541] rabbit, calf,[538] cat, monkey,[540] catfish (*Parasiluris asotus*),[483] turtle (*Clemmys japonica*),[484] frog (*Rana nigromaculata*)[485]		
	Purification: mouse,[542] lobster[543]		
	Properties: rat,[140,538,544] mouse,[542] beef, (coenzyme: pyridoxalphosphate)[545] monkey,[544] lobster[543]		
	Regional distribution: rat[546] (cerebellar layers),[547] mouse, guinea pig, rabbit,[547] monkey,[544] human[548]		
	Subcellular localization: (mitochondria and mitochondrial subfractions) rat,[66,140,147,211,315,549] rabbit,[66] chicken[66,475]		
	Developmental changes: rat,[550] chicken[551,552] (subcellular localization rat, rabbit, chicken);[66,475] reviews (distribution, properties)[8,553]		
2.7.1.1	ATP:D-hexose 6-phosphotransferase	Hexokinase	ATP + D-hexose = ADP + D-hexose 6-phosphate
	Occurrence, activity: rat[25,95,554,555] (isoenzymes),[556,557] mouse,[25,33] guinea pig,[585] rabbit, dog,[558] sheep,[559,585] chicken, pigeon, mollusca, teleosts, amphibia,[555] fish[307]		

Summary Table of Enzymes of the Nervous System—(continued)

E.C. number	Systematic name	Trivial name	Reaction
	Purification: rat[560–564] sheep,[561] beef[565,578] (1129-fold)[566]		
	Properties: rat,[560–562,564,567–570,574,583] guinea pig,[571,572,584] sheep[561] beef,[565,566,570,573,578] human[574]		
	Regional distribution: rat,[575] rabbit[51,120] (cortical layers)[455] (Ammons horn),[576] dog,[577] beef[578] monkey[54,579] (cerebellum),[580] human[95]		
	Subcellular localization: (cytoplasma and mitochondria) rat[28,64,68,569,578,581] (nuclei),[72] beef,[26,578] guinea pig[181]		
	Developmental changes: rat (isoenzymes)[556] review[582]		
2.7.1.2	ATP:D-glucose 6-phosphotransferase	Glucokinase	ATP + D-glucose = ADP + D-glucose 6-phosphate
	Occurrence: rat, beef,[586] human[1279]		
	Properties: rat,[587] human[1279]		
	Subcellular localization: rat (cytoplasm, 45% particle bound),[587] human (cytoplasm)[1279]		
2.7.1.4	ATP:D-fructose 6-phosphotransferase	Fructokinase	ATP + D-fructose = ADP + D-fructose 6-phosphate
	Activity: mouse[33]		
2.7.1.6	ATP:D-galactose 1-phosphotransferase	Galactokinase	ATP + D-galactose = ADP + α-D-galactose-1-phosphate
	Occurrence, properties: rat (D-galactosamine can also act as acceptor)[588]		
2.7.1.7	ATP:D-mannose 6-phosphotransferase	Mannokinase	ATP + D-mannose = ADP + D-mannose 6-phosphate
	Occurrence: rat[587]		
2.7.1.8	ATP:2-amino-2-deoxy-D-glucose phosphotransferase	Glucosamine kinase	ATP + 2-amino-2-deoxy-D-glucose = ADP + 2-amino-2-deoxy-D-glucose phosphate

			Occurrence: rat[589] *Properties*: beef[590]
	ATP:2-acetamido-2-deoxy-D-glucose phosphotransferase	N-Acetylglucosamine kinase	ATP + 2-acetamido-2-deoxy-D-glucose = ADP + 2-acetamido-2-deoxy-D-glucose-6-phosphate *Occurrence*: human[463] *Purification, properties*: sheep (25-fold)[591] *Regional distribution*: sheep[592] *Developmental changes*: rat[593]
2.7.1.11	ATP:D-fructose-6-phosphate 1-phosphotransferase	Phosphofructokinase	ATP + D-fructose 6-phosphate = ADP + D-fructose 1,6-diphosphate *Occurrence*: rat,[25,568] mouse[25] *Activity*: sheep[595] *Purification, properties*: rat,[569] guinea pig,[571] dog, beef[594] *Regional distribution*: rabbit,[51,120,455] monkey[54] *Subcellular localization*: rat[28] (in mitochondria),[65] beef[26]
2.7.1.21	ATP:thymidine 5'-phosphotransferase	Thymidine kinase	ATP + thymidine = ADP + thymidine 5'-phosphate *Occurrence*: rabbit[596]
2.7.1.25	ATP:adenylylsulphate 3'-phosphotransferase	Adenylylsulphate kinase	ATP + adenylylsulphate = ADP + 3'-phospho-adenylylsulphate *Occurrence*: sheep[672]
2.7.1.31	ATP:D-glycerate 3-phosphotransferase	Glycerate kinase	ATP + D-glycerate = ADP + 3-phospho-D-glycerate *Occurrence*: mouse[33]

Summary Table of Enzymes of the Nervous System—(continued)

E.C. number	Systematic name	Trivial name	Reaction
2.7.1.32	ATP:choline phosphotransferase *Occurrence*: rat,[597] rabbit, beef[598] *Purification, properties*: rabbit[599] *Regional distribution*: rat, rabbit[600] *Subcellular localization, developmental changes*: rat, rabbit[600]	Choline kinase	ATP + choline = ADP + phosphocholine
2.7.1.35	ATP:pyridoxal 5-phosphotransferase *Occurrence*: mouse[7] *Purification*: rat, beef,[604] human (200-fold)[5] *Properties*: rat, beef (Zn^{2+}-activated),[603,604] mouse,[7,602] human[5] *Regional distribution*: beef[604] *Subcellular localization*: rat (supernatant, microsomes, nuclei),[604] guinea pig (cytoplasm of cell bodies and synaptosomes)[605]	Pyridoxal kinase	ATP + pyridoxal = ADP + pyridoxal 5-phosphate
2.7.1.36	ATP:mevalonate 5-phosphotransferase *Occurrence*: rat[606]	Mevalonate kinase	ATP + mevalonate = ADP + 5-phospho-mevalonate
2.7.1.37	ATP:protein phosphotransferase *Preparation*: beef[607] *Properties*: guinea pig[608–610] (microsomal enzyme),[611,612] beef[609] (microsomal enzyme)[613,614] *Subcellular localization*: guinea pig (cytoplasm, microsomes, mitochondria)[609]	Protein kinase	ATP + a protein = ADP + a phosphoprotein
2.7.1.38	ATP:phosphorylase phosphotransferase *Occurrence*: mouse[450]	Phosphorylase kinase	4 ATP + 2 phosphorylase *b* = 4 ADP + phosphorylase *a*

2.7.1.40	ATP:pyruvate phosphotransferase *Activity*: rat,[25,33] mouse[25,33] *Regional distribution*: human[95] *Subcellular localization*: (mitochondria, supernatant) rat[28,615] (mitochondrial subfractions)[102] (nuclei),[72] calf[126]	Pyruvate kinase	ATP + pyruvate = ADP + phosphoenol-pyruvate
2.7.1.48	ATP:uridine 5'-phosphotransferase *Occurrence*: mouse[473]	Uridine kinase	ATP + uridine = ADP + uridine 5'-phosphate
	ATP:monoglyceride phosphotransferase *Occurrence*: guinea pig, beef[616] *Subcellular localization*: (microsomal fraction) guinea pig, beef[616]	monoglyceride kinase	ATP + a monoglyceride = ADP + a monoglyceride phosphate
	ATP:2,3-diglyceride phosphotransferase *Occurrence*: rat,[617,618] guinea pig[619,620] *Properties*: rat,[618] guinea pig[610] *Subcellular localization*: (microsomes) rat,[617] guinea pig[620]	diglyceride kinase	ATP + a diglyceride = ADP + a diglyceride phosphate
	ATP:phosphatidylinositol phosphotransferase *Occurrence*: rabbit, beef[621] *Properties*: rat[622,623] *Subcellular localization*: rat (microsomes)[622] (nerve endings, nuclei)[113]	Phosphatidylinositol kinase	2 ATP + phosphatidylinositol = 2 ADP + diphosphoinositide
	ATP:diphosphoinositide phosphotransferase *Occurrence*: rat,[623] rabbit,[621,624] guinea pig,[625] beef[624]	Diphosphoinositide kinase	ATP + diphosphoinositide = ADP + triphosphoinositide

Summary Table of Enzymes of the Nervous System—(continued)

E.C. number	Systematic name	Trivial name	Reaction
	Properties: rat[623,626] *Subcellular localization*: rat (supernatant)[623,626]	Cyclic 3', 5'-adenosine monophosphate synthesizing enzyme, adenylcyclase	ATP = 3', 5'-AMP + pyrophosphate
	Occurrence: dog,[627,628,629] sheep, calf, beef, pig[629,630] *Purification*: dog, sheep, calf, beef, pig[629] *Properties*: dog,[627] calf[629-631] *Regional distribution*: cat[630] *Subcellular localization*: rat (microsomes, synaptic membranes, myelin)[632,633]		
	Glycerol-2-phosphate:riboflavine 5'-phosphotransferase *Occurrence*: chicken[634] *Developmental changes*: chicken embryo[634]		Glycerol-2-phosphate + riboflavine = glycerol + riboflavine-5'-phosphate
	L-Serine-O-phosphate:L-serine phosphotransferase *Purification, properties*: mouse[635]	Phosphoserine phosphotransferase	L-Serine-O-phosphate + L-serine = L-serine-O-phosphate + L-serine
	Occurrence properties: rat[246]	Nucleoside phosphotransferase	Phenylphosphate + a nucleoside = phenol + a 5'-nucleotide
2.7.2.1	ATP:acetate phosphotransferase *Occurrence*: rat[636] *Subcellular localization*: rat (mitochondria)[636]	Acetate kinase	ATP + acetate = ADP + acetylphosphate

2.7.2.3	ATP:3-phospho-D-glycerate 1-phosphotransferase	ATP + 3-phospho-D-glycerate = ADP + 1,3-diphospho-D-glyceric acid
	Phosphoglycerate kinase	
	Occurrence: In all cells with glycolytic activity	
	Activity: rat,[137] rabbit[637]	
	Subcellular localization: rat,[28,102] calf[26]	
2.7.3.2	ATP:creatine phosphotransferase	ATP + creatine = ADP + phosphocreatine
	Creatine kinase	
	Occurrence: rat[638,639,649] (2 isoenzymes),[640-642] mouse (2 isoenzymes),[640] guinea pig,[639] rabbit (2 isoenzymes),[640,641, 643,646] beef (2 isoenzymes),[642,644] calf, lamb, dog,[643] horse,[513] chicken (2 isoenzymes),[646-648] human[643,645] (isoenzymes) [642,643,647]	
	Purification: beef (50-fold),[644] chicken[648]	
	Properties: rat,[639,650] guinea pig,[639] (mitochondrial enzyme, soluble enzyme)[77] (microsomal enzyme),[651] rabbit,[646] beef[652]	
	Regional distribution: rabbit[640] (cortex)[51]	
	Subcellular localization: (cytoplasm, mitochondria, microsomes) rat, rabbit,[641] guinea pig,[77,651] beef,[653] human[654]	
	Developmental changes: rat (isoenzymes)[655]	
2.7.4.3	ATP:AMP phosphotransferase	ATP + AMP = ADP + ADP
	Adenylate kinase, myokinase	
	Occurrence: rat[568,638,656]	
	Regional distribution: rabbit[51]	
	Subcellular localization: rat (nuclei)[72]	
2.7.4.6	ATP:nucleosidediphosphate phosphotransferase	ATP + a nucleoside diphosphate = ADP + a nucleoside triphosphate
	Nucleosidediphosphate kinase	
	Occurrence: rabbit[596]	
2.7.4.8	ATP:GMP phosphotransferase	ATP + GMP = ADP + GDP
	Guanylate kinase	
	Purification, properties: hog[658]	

Summary Table of Enzymes of the Nervous System—(continued)

E.C. number	Systematic name	Trivial name	Reaction
2.7.4.9	ATP: thymidinemonophosphate phosphotransferase *Occurrence*: rabbit[596]	Thymidinemonophosphate kinase	ATP + thymidine monophosphate = ADP + thymidine diphosphate
2.7.4.10	GTP: AMP phosphotransferase *Occurrence*: sheep[657]	GTP-adenylate kinase	GTP + AMP = GDP + ADP or ITP + AMP = IDP + ADP
2.7.5.1	α-D-Glucose-1,6-diphosphate:α-D-glucose-1-phosphate phosphotransferase *Activity*: rat,[458] rabbit,[447] beef[659] *Purification*: beef,[662] monkey[661] *Properties*: rabbit,[662] beef[660,663] *Regional distribution*: rabbit,[121,455,456,664] monkey[54] *Subcellular localization*: rabbit,[664,665] Japanese quail[27] *Developmental changes*: rat,[454] mouse[666]	Phosphoglucomutase	α-D-glucose 1,6-diphosphate + α-D-glucose 1-phosphate = α-D-glucose 6-phosphate + α-D-glucose 1,6-diphosphate
2.7.5.2	2-Acetamido-2-deoxy-D-glucose 1,6-diphosphate:2-acetamido-2-deoxy-D-glucose-1-phosphate phosphotransferase *Occurrence*: rat[667]	Acetylglucosamine phosphomutase	2-Acetamido-2-deoxy-D-glucose 1,6-diphosphate + 2-acetamido-2-deoxy-D-glucose 1-phosphate = 2-acetamido-2-deoxy-D-glucose 6-phosphate + 2-acetamido-2-deoxy-D-glucose 1,6-diphosphate
2.7.5.3	2,3-Diphospho-D-glycerate:2-phospho-D-glycerate phosphotransferase	Phosphoglyceromutase, glycerate phosphomutase	2,3-Diphospho-D-glycerate + 2-phospho-D-glycerate = 3-phospho-D-glycerate + 2,3-diphospho-D-glycerate

Occurrence, activity: in all cells with glycolytic activity,[671] rat,[25,137] mouse,[25] sheep, beef[668]
Subcellular localization: (cytoplasma) calf[26]

EC	Systematic name	Trivial name	Reaction
2.7.5.4	1,3-Diphospho-D-glyceric acid:3-phospho-D-glycerate phosphotransferase *Activity:* mouse[33]	Diphosphoglyceromutase	1,3-Diphospho-D-glyceric acid + 3-phospho-D-glycerate = 3-phospho-D-glycerate + 2,3-diphospho-D-glycerate
	Ribose-1,5-diphosphate:ribose-1-phosphate phosphotransferase *Occurrence:* monkey[471,661]	Phosphoribomutase	Ribose-1,5-diphosphate + ribose-1-phosphate = ribose-5-phosphate + ribose-1,5-diphosphate
2.7.7.1	ATP:NMN adenylyltransferase *Occurrence:* guinea pig (glia and neuronal nuclei)[669]	NAD pyrophosphorylase	ATP + nicotinamide ribonucleotide = pyrophosphate + NAD
2.7.7.2	ATP:FMN adenylyltransferase *Occurrence:* mouse, rabbit[670]	FAD pyrophosphorylase	ATP + FMN = pyrophosphate + FAD
2.7.7.4	ATP:sulphate adenylyltransferase *Occurrence, regional distribution:* sheep[672]	Sulphate adenylyltransferase, sulfurylase	ATP + sulphate = pyrophosphate + adenylylsulphate
2.7.7.6	Nucleosidetriphosphate:RNA nucleotidyltransferase *Occurrence:* rat[673] *Properties:* rat[674] *Regional distribution:* rat,[674,675] rabbit, monkey[675] *Subcellular localization:* rat, rabbit, monkey (nuclei)[675] *Developmental changes:* rat[674,675]	RNA nucleotidyltransferase	m-Nucleoside triphosphate + RNA_n = m-pyrophosphate + RNA_{n+m}

Summary Table of Enzymes of the Nervous System—(continued)

E.C. number	Systematic name	Trivial name	Reaction
		CTP-poly-C-synthetase	
	Occurrence, properties: rat (Mg^{2+} and Mn^{2+} activated)[676] *Subcellular localization*: (nuclei) rat[676]		
		ATP-poly-A-synthetase	
	Occurrence: rat[677] *Subcellular localization*: (soluble fraction) rat[677]		
2.7.7.9	UTP: α-D-glucose-1-phosphate uridylyltransferase	Glucose-1-phosphate uridylyltransferase, UDPG pyrophosphorylase	UTP + α-D-glucose 1-phosphate = pyrophosphate + UDP glucose
	Occurrence: rat,[458,678,679] rabbit[680] *Purification*: human[681] *Properties*: rabbit,[680] human[681] *Regional distribution*: rabbit[456]		
2.7.7.12	UDP glucose: α-D-galactose-1-phosphate uridylyltransferase	Hexose-1-phosphate uridylyltransferase	UDP glucose + α-D-galactose 1-phosphate = α-D-glucose 1-phosphate + UDP galactose
	Occurrence: rat[678,679,682] *Developmental changes*: rat[682]		
2.7.7.14	CTP: ethanolaminephosphate cytidylyltransferase	Ethanolaminephosphate cytidylyltransferase	CTP + ethanolamine phosphate = pyrophosphate + CDP ethanolamine
	Occurrence: in all animal tissues[683,684] *Properties*:[684]		

2.7.7.15	CTP: cholinephosphate cytidylyltransferase	Cholinephosphate cytidylyltransferase	CTP + choline phosphate = pyrophosphate + CDP choline
	Occurrence: rat, calf, and other animals[683–685]		
2.7.7.16	Ribonucleate pyrimidinenucleotide-2′-transferase (cyclizing)	Ribonuclease	Transfers the 3′-phosphate of a pyrimidine nucleotide residue of a polynucleotide from the 5′-position of the adjoining nucleotide to the 2′-position of the pyrimidine nucleotide itself, forming a cyclic nucleotide
	a. acid RNAse *Occurrence:* in all cells[689] *Activity:* rat[690] (ribosomes),[691] fish, amphibia[692] *Properties:* rat (mitochondrial enzyme),[693] cat,[694] goat (ribosomal enzyme)[695–697] *Regional distribution:* rat,[698] rabbit,[699] cat,[694,700] sheep,[305] human[701] *Subcellular localization:* (microsomes, mitochondria, nuclei, supernatant) rat,[693] (lysosome-like particles),[178,701 702] mouse,[703] goat (ribosomes)[696] b. alkalic RNAse *Occurrence:* in all cells[689] *Properties:* mouse, cat,[694] goat (ribosomal enzyme)[695,696] *Regional distribution:* cat[694] *Subcellular localization:* mouse (microsomes, mitochondria, nuclei, supernatant),[703] goat (ribosomes)[695,696]		
2.7.7.23	UTP: 2-acetamido-2-deoxy-α-D-glucose-1-phosphate uridylyltransferase	UTP acetylglucosamine pyrophosphorylase	UTP + 2-acetamido-2-deoxy-D-glucose 1-phosphate = pyrophosphate + UDP-2-acetamido-2-deoxy-D-glucose
	Occurrence: human[463] *Purification:* sheep (46-fold)[686] *Properties:* sheep[686] *Regional distribution:* sheep[592] *Developmental changes:* rat[593]		

Summary Table of Enzymes of the Nervous System—(continued)

E.C. number	Systematic name	Trivial name	Reaction
	CTP:N-acetylneuraminic acid cytidylyltransferase *Purification*: sheep (39-fold)[687] *Properties*: sheep[687] *Regional distribution*: sheep[688]	CMP-N-acetylneuraminic acid synthesizing enzyme	CTP + N-acetylneuraminic acid = pyrophosphate + CMP-N-acetylneuraminic acid
2.7.8.1	CDP ethanolamine:1,2-diglyceride cholinephosphotransferase *Occurrence*: rat[683,705]	Ethanolamine phosphotransferase	CDP ethanolamine + 1,2-diglyceride = CMP + phosphatidylethanolamine
2.7.8.2	CDP choline:1,2-diglyceride cholinephosphotransferase *Occurrence*: rat[622,683,705–708] *Properties*: rat[600,683,707] *Regional distribution*: rabbit[600] *Subcellular localization*: (mitochondria, microsomes) rat[600,706] *Developmental changes*: rat[600]	Cholinephosphotransferase	CDP choline + 1,2-diglyceride = CMP + a phosphatidylcholine
2.7.8.3	CDP choline:ceramide cholinephosphotransferase *Occurrence*: rat[709]	Ceramide cholinephosphotransferase	CDP choline + ceramide = CMP + sphingomyelin
	CDP diglyceride:inositol diglyceridephosphotransferase *Occurrence*: rat,[710,711] guinea pig,[711] cat[712]		CDP diglyceride + inositol = CMP + phosphatidylinositol

| 2.8.1.1 | Thiosulphate : cyanide sulphotransferase | Thiosulphate sulphotransferase, rhodanese | Thiosulphate + cyanide = sulphite + thiocyanate |

Occurrence: rat,[713–716] rabbit, monkey,[713] guinea pig,[715] dog,[713,718] pig, sheep, horse,[716] beef,[716,717] human,[714,716] pigeon[714,715]
Regional distribution: dog[718]
Developmental changes: rat,[719] chicken[720]

| 2.8.2.5 | 3′-Phosphoadenylylsulphate chondroitin sulphotransferase | Chondroitin sulphotransferase | 3′-Phosphoadenylylsulphate + chondroitin = adenosine 3′,5′-diphosphate + chondroitin 4-sulphate |

Occurrence: rat[721]
Properties: rat[721]
Developmental changes: rat[721]

| 2.8.2.5 | 3′-Phosphoadenylylsulphate galactocerebroside sulphotransferase | Galactocerebroside sulphotransferase | 3′-Phosphoadenylylsulphate + a galacto-cerebroside = adenosine 3′,5′-diphosphate + a galactocerebroside sulphate |

Occurrence: rat,[722,723] sheep[723]
Regional distribution: sheep[723]
Subcellular localization: rat (microsomes)[722]
Developmental changes: rat[723,724]

| 3.1.1.1 | Carboxylic-ester hydrolase | Carboxylesterase | A carboxylic ester + H_2O = an alcohol + a carboxylate |

Occurrence: ubiquituous[725]
Properties: rat,[726] human[727,728]
Regional distribution: human[727]
Subcellular localization: (soluble and particulate bound isoenzymes) rat,[729] human[730]
Developmental changes: mouse,[731] chicken[733]

Summary Table of Enzymes of the Nervous System—(continued)

E.C. number	Systematic name	Trivial name	Reaction
3.1.1.2	Aryl-ester hydrolase *Occurrence*: rat,[734] guinea pig,[735] human[863] *Properties*: rat,[734] human[727,728] *Regional distribution*: human[727] *Subcellular localization*: rat (lysosomes)[729,734] *Developmental changes*: rat, guinea pig, rabbit, human[737]	Arylesterase	A phenyl acetate + H_2O = a phenol + acetate
3.1.1.3	Glycerol-ester hydrolase *Occurrence*: rat,[738] dog,[739] hog, sheep, beef, horse, human,[1] chicken,[736] and other mammals[740–742] *Properties*: dog,[739] chicken[736] *Regional distribution*: guinea pig,[743] dog,[744,745] cat, monkey,[745] human[745,746] *Subcellular localization*: rat (cytoplasm)[729] *Developmental changes*: chicken,[747] silkworm[748] *Review*.[749]	Lipase	A triglyceride + H_2O = a diglyceride + a fatty acid ion
3.1.1.4	Phosphatide acyl-hydrolase *Occurrence*: rat,[751] human[750]	Phospholipase A Lecithinase A	A lecithin + H_2O = a lysolecithin + an unsaturated fatty acid ion
3.1.1.5	Lysolecithin acyl-hydrolase *Regional distribution*: human[752]	Lysophospholipase, phospholipase B	A lysolecithin + H_2O = glycerol-phosphocholine + a fatty acid ion
3.1.1.7	Acetylcholine hydrolase *Occurrence*: in all nervous structures	Acetylcholinesterase	Acetylcholine + H_2O = choline + acetate

Activity: rat,[753–757] mouse,[239] guinea pig,[757] rabbit,[758] beef,[757] dog,[757,840] elephant,[760] hamster,[448,849] mammals,[769] killifish,[762] torpedo (electric organ),[763] lobster (abdominal nerve chain),[764] selachian and teleost fish,[764] honey bee[765]

Activity: in single neurons and ganglion cells[766–768]

Purification: rat,[770] rabbit,[771] sheep,[773,776] calf[772] (35-fold),[773] beef,[774,775] human[773]

Properties: rat,[777–779] mouse[777,780] (2 isoenzymes),[781] rabbit,[771,782] sheep, calf,[773] beef, cat, lobster, squid,[777] dog,[783] human[816,838] (2 isoenzymes),[773,839,840] mammalian[769] torpedo (electric organ),[763] fish[784]

Physiological significance:[268,388,785,786]

Regional distribution: rat[787–793] (cortical layers)[794] (cerebellum),[417,795] mouse[800] (cerebellum),[795] guinea pig[792] (cortex)[819] (cerebellum)[417,795] (amygdaloid complex),[796] rabbit[787,792,797,798] (Formatio reticularis),[799] (hippocampus)[576] (cerebellum)[417,795] hamster,[792] cat[792] (*Formatio reticularis*),[799] (Corpus geniculatum),[842] (cerebellum),[417,795,801] (Sup. cervical ganglia),[802,803] dog[166,416,744,787] (cerebellum),[795] beef[167,274,792] horse (cerebellum),[804] squirrel,[792] *Myocastor coypus,*[814] monkey[792,805–807] (cerebellum),[795] human[266,808–811] (cortex),[274,313,812] (extrapyramidal system),[813] pigeon[254] (cerebellum),[417] duck,[274,795] *Strix aluco, falco timnunculus, Pica pica, Sturmus vulgaris, Chrysolophus pictus,*[274] frog[815]

Subcellular localization: rat[171,173,177,256,314,779,816,817] (nerve endings, nerve ending membranes),[174–176,181,258,632,633,848] guinea pig,[179,180,256,818] rabbit,[256,820,821] cat,[795] bccf (microsomes, nerve endings),[818] human,[822] pigeon,[256] fish, amphibia[387,823]

Developmental changes: rat,[49,185,753,824–831] mouse,[800,836] guinea pig,[187,833] rabbit,[187,311,797,832] cat,[834,835] dog[311] sheep,[837] human,[841,843] chicken,[201,426,824,828,844–846] cecropis silkmoth,[847] review[62]

3.1.1.8	Acylcholine acyl-hydrolase	Cholinesterase	An acylcholine + H_2O = choline + an anion

Activity: rat[829,850] (sex differences),[851] mouse[836]

Properties: rat,[62,850] dog,[783] human[728,838,852]

Regional distribution: rat,[790,792,795] mouse,[239,795] guinea pig,[792,795] rabbit,[795,799] cat,[792,795,799,842] dog,[166,744,795] sheep,[795] horse (cerebellum),[804] hamster, squirrel,[792] monkey,[792,795,805] human[746,792,810,843]

Subcellular localization: (soluble and particulate) rat[28,171,173,729]

Developmental changes: rat,[826,829,853] mouse,[836] cat,[834] human,[843] chicken[846]

Procain esterase	Procain + H_2O = *p*-aminobenzoate + diethylamino-ethanol

Occurrence: human[856]

Properties: human[856]

Summary Table of Enzymes of the Nervous System—(continued)

E.C. number	Systematic name	Trivial name	Reaction
		Nortropacocain esterase	Nortropacocain + H_2O = benzoate + norpseudotropine
	Properties: human[857] *Regional distribution*: human[857]		
3.1.1.13	Sterol-ester hydrolase	Cholesterol esterase	A cholesterol ester + H_2O = cholesterol + an anion
	Occurrence: rat,[858] guinea pig[859] *Subcellular localization*: (soluble and particulate bound) rat[860] *Developmental changes*: rat[860–862]		
3.1.2.1	Acetyl-CoA hydrolase *Occurrence*: rabbit[395]	Acetyl-CoA deacylase	Acetyl-CoA + H_2O = CoA + acetate
3.1.2.2	Palmitoyl-CoA hydrolase *Occurrence*: rat[442,863] *Purification, properties*: pig[863]	Palmitoyl-CoA hydrolase	Palmitoyl-CoA + H_2O = CoA + palmitate
3.1.2.5	3-Hydroxy-3-methylglutaryl-CoA hydrolase *Occurrence*: pigeon[864]	Hydroxymethylglutaryl-CoA hydrolase	3-Hydroxy-3-methylglutaryl-CoA + H_2O = CoA + 3-hydroxy-3-methylglutarate
3.1.2.6	S-2-Hydroxyacylglutathione hydrolase *Occurrence*: nearly in all cells[865]	Hydroxyacylglutathione hydrolase, Glyoxalase II	S-2-Hydroxyacylglutathione + H_2O = glutathione + a 2-hydroxyacid anion

3.1.2.7	S-Acylglutathione hydrolase	S-Acylglutathione + H_2O = glutathione + an anion

Occurrence: rat,[442] beef[866,867]
Properties: beef[867,868]

S-acylmercaptane hydrolase

Occurrence.[854,855]

3.1.3.1	Orthophosphoric monoester phosphohydrolase — Alkaline phosphatase	An orthophosphoric monoester + H_2O = an alcohol + orthophosphate

Occurrence: in all cells of animal origin[869]
Activity: rat,[870] guinea pig, human[871]
Purification: rabbit[771]
Properties: rat,[872,873] guinea pig (2 isoenzymes),[874] rabbit,[771,873] sheep[873,875] (2 isoenzymes),[876] goat (ribosomal enzyme),[877] dog, hedgehog, pigeon, sparrow, grass snake,[872] human[873,878] (2 isoenzymes),[874] chicken[879]
Regional distribution: rat[880,881] (cerebellum),[882] guinea pig[883] (choroid plexus),[884] rabbit[880,881] (cortex)[885] (choroid plexus)[890] (hippocampus),[455,576] sheep,[886] beef,[887] cat,[880,881] dog,[744,880,881] monkey[888] (hypothalamus)[889] human,[886,890] chicken[845,881]
Subcellular localization: rat,[66.314,315] goat (ribosomes),[697,877] beef,[891] human (nuclei),[822] chicken (cytoplasm, nuclei)[201,892]
Developmental changes: rat,[66,827,870,880,893] guinea pig,[894] rabbit,[66,880,895] mouse,[896] cat, dog,[880] chicken,[66,201,475,720, 845,879,892,897,898] frog,[899] review[62,900]

3.1.3.2	Orthophosphoric monoester phosphohydrolase — Acid phosphatase	An orthophosphoric monoester + H_2O = an alcohol + orthophosphate

Occurrence: in all cells of animal origin[869]
Activity: rat, cat, dog, birds,[881] guinea pig,[901] rabbit,[881,901] human[902]
Purification: rabbit[771]
Properties: rat[872,873] (isoenzymes),[903,904] guinea pig,[884,901,905] rabbit,[771,872,873,884,901] sheep,[873,875] beef,[891] dog, hedgehog, frog, grass snake, pigeon, sparrow,[872] goat (ribosomes),[697,877] human,[873,878] chicken[879]

Summary Table of Enzymes of the Nervous System—(continued)

E.C. number	Systematic name	Trivial name	Reaction
			Regional distribution: rat (cerebellum)[882] (neurosecretory cells)[906] (intracellular distribution),[907] guinea pig[883] (choroid plexus),[884] rabbit[908] (cortex)[576,885] (choroid plexus),[884] sheep,[886] beef,[887] dog,[744] monkey[909] (hypothalamus),[889] human,[886,890] chicken[879,897]
			Subcellular localization: rat[66,67,314,315,632,910] (lysosome-like particles),[499,702,911,912] guinea pig (lysosome-like particles), [194,913] rabbit[66] (lysosome-like particles)[914] (nuclei),[915] goat (ribosomes),[877] beef[891] cat (lysosome-like particles) [701,916–918] (nuclei),[915] human (cytoplasma, nuclei),[822] chicken,[66] fish, amphibia (lysosome-like particles)[692]
			Developmental changes: rat,[66,827,880,893] mouse,[919] guinea pig,[187,920] rabbit[66,187,880,895] (hippocampus),[455] cat,[880,921] dog,[880] chicken[66,110,879,896,897]
		K^+-activated *p*-nitro-phenylphosphatase	*p*-Nitrophenylphosphate + H_2O = *p*-nitrophenol + orthophosphate
			Properties: rat[922]
			Subcellular localization: rat (nerve endings, nerve ending membranes),[258,923] mouse[924]
3.1.3.3	Phosphoserine phosphohydrolase	Phosphoserine phosphatase	L-(or D-) Phosphoserine + H_2O = L- (or D-) serine + orthophosphate
			Purification, properties: mouse[535,635]
3.1.3.4	L-α-Phosphatidate phosphohydrolase	Phosphatidate phosphatase	An L-α-phosphatidate + H_2O = a D-2,3 (or L-1,2) diglyceride + orthophosphate
			Occurrence: rat,[706,925] guinea pig, beef, pig[925]
			Purification: pig[926]
			Properties: rabbit,[927] pig[926]
			Regional distribution: rabbit[927]
			Subcellular localization: (microsomes) rat,[928] chicken[929]
			Developmental changes: chicken[929,930]

3.1.3.5	5'-Ribonucleotide phosphohydrolase	5'-Nucleotidase	A 5'-ribonucleotide + H_2O = a ribonucleoside + orthophosphate

Occurrence: rat[656]
Purification: sheep[931]
Properties: mouse,[932] sheep,[931] human[933]
Regional distribution: rat,[698,882,934-936] mouse,[932,937] guinea pig,[883] rabbit (hippocampus),[455] human[890,932]
Subcellular localization: rat (microsomes, myelin, nerve endings),[113,292] rabbit,[938] goat (ribosomes)[695]
Developmental changes: rat[934,935]

3.1.3.6	3'-Ribonucleotide phosphohydrolase	3'-Nucleotidase	A 3'-ribonucleotide + H_2O = A ribonucleoside + orthophosphate

Occurrence: rat, mouse[940]

3.1.3.9	D-Glucose-6-phosphate phosphohydrolase	Glucose-6-phosphatase	D-Glucose 6-phosphate + H_2O = D-glucose + orthophosphate

Occurrence: rat,[941,942] mouse, guinea pig, rabbit, goat, pigeon, chicken, fish, tortoise, toad[942]
Regional distribution: rat[943] (localization in neurons)[944]

3.1.3.10	D-Glucose-1-phosphate phosphohydrolase	Glucose-1-phosphatase	D-Glucose 1-phosphate + H_2O = D-glucose + orthophosphate

Occurrence: rabbit[447]
Properties: rabbit[447]

3.1.3.11	D-Fructose-1,6-diphosphate 1-phosphohydrolase	Hexosediphosphatase	D-Fructose 1,6-diphosphate + H_2O = D-fructose 6-phosphate + orthophosphate

Occurrence: rat,[941] beef,[945] guinea pig, human[871]

3.1.3.16	Phosphoprotein phosphohydrolase	Phosphoprotein phosphatase	A phosphoprotein + n H_2O = a protein + n orthophosphate

Occurrence: rat, rabbit, cat,[946] guinea pig[947]

Summary Table of Enzymes of the Nervous System—(continued)

E.C. number	Systematic name	Trivial name	Reaction
	Purification: beef (100-fold)[948] *Properties*: rabbit,[949] beef[608,948] *Regional distribution*: beef[950] *Subcellular localization*: (supernatant, mitochondria) rabbit,[949] guinea pig[950] *Developmental changes*: chicken;[951] review[62]		
3.1.3.19	2-Phosphoglycerol phosphohydrolase	Glycerol-2-phosphatase	2-Phosphoglycerol + H_2O = glycerol + orthophosphate
	Occurrence: mouse,[952] beef[953] *Properties*: mouse,[952] beef[953]		
	Ribose-5-phosphate phosphohydrolase	Ribose-5-phosphatase	Ribose-5-phosphate + H_2O = ribose + orthophosphate
	Occurrence: goat (ribosomes)[954]		
	Pyridoxal 5′-phosphate phosphohydrolase	Pyridoxalphosphate phosphatase	Pyridoxal 5′-phosphate + H_2O = pyridoxal + orthophosphate
	Purification: human (70-fold)[954] *Properties*: human[954]		
	Thiaminemonophosphate phosphohydrolase *Occurrence, properties*: rat[955]	Thiamine-monophosphatase	Thiaminemonophosphate + H_2O = thiamine + orthophosphate
	Diphosphoinositide phosphohydrolase	Diphosphoinositide hydrolase	A diphosphoinositide + H_2O = a monophosphoinositide + orthophosphate
	Occurrence: rat,[928] guinea pig,[957,958] beef[956] sheep, rabbit, monkey[958]		

	Enzyme name	Details	Reaction
		Regional distribution: rabbit[958] *Subcellular localization*: rat (cytoplasma, nuclei, mitochondria, microsomes)[928]	
	Triphosphoinositide phosphohydrolase Triphosphoinositide hydrolase	*Occurrence*: beef[956] *Purification*: beef[959] *Properties*: beef[959,960] *Subcellular localization*: rat (cytoplasm, nuclei, mitochondria, microsomes)[928]	A triphosphoinositide + H_2O = a diphosphoinositide + orthophosphate
	Neuraminic acid 9-phosphate phosphohydrolase	*Occurrence*: sheep[961]	Neuraminic acid 9-phosphate + H_2O = neuraminic acid + orthophosphate
3.1.4.1	Orthophosphoric diester phosphohydrolase Phosphodiesterase	*Purification*: lamb[962] (70-fold)[963] *Properties*: lamb[962] goat (ribosomal enzyme)[697,964] *Subcellular localization*: goat (ribosomes)[964]	A phosphoric diester + H_2O = a phosphoric monoester + an alcohol
3.1.4.2	L-3-Glycerylphosphorylcholine glycerophosphohydrolase Glycerophosphorylcholine diesterase	*Occurrence*: rat, guinea pig, rabbit, dog, human, chicken[965] *Properties*: rat[965] *Regional distribution*: human[965]	L-3-Glycerylphosphorylcholine + H_2O = choline + glycerol 1-phosphate
3.1.4.3	Phosphatidylcholine cholinephosphohydrolase Phospholipase C, lecithinase	*Occurrence*: rabbit, dog, beef[966] rat[967] *Purification, properties*: rat[967]	A phosphatidylcholine + H_2O = a 1,2-diglyceride + choline phosphate

Summary Table of Enzymes of the Nervous System—(continued)

E.C. number	Systematic name	Trivial name	Reaction
	Purification: rat (18-fold)[968] *Properties*: rat (does not hydrolyze lecithin)[968]	Sphingomyelinase	A phosphosphingoside + H_2O = ceramide + choline phosphate
	Occurrence, properties: rat[969]	Monophosphoinositide hydrolase	A monophosphoinositide + H_2O = inositol monophosphate + a diglyceride
	Occurrence: guinea pig,[957,958] rabbit, cat, sheep, monkey, human[958] *Properties*: rabbit (Ca^{2+} activated)[958] *Regional distribution*: rabbit[958]	Diphosphoinositide phosphodiesterase	A diphosphoinositide + H_2O = inositol diphosphate + a diglyceride
	Occurrence: beef[956] *Purification, properties*: beef[970]	Triphosphoinositide phosphodiesterase	A triphosphoinositide + H_2O = inositol triphosphate + a diglyceride
3.1.4.5	Deoxyribonucleate oligonucleotide-hydrolase *Properties*: rabbit,[971] calf[972] *Developmental changes*: rat[973]	Deoxyribonuclease I	DNA + $(n - 1)$ H_2O = n oligodeoxyribonucleotides

3.1.4.6	Deoxyribonucleate 3'-nucleotidohydrolase	Deoxyribonuclease II	Forms 3'-nucleotides from DNA
	Occurrence: rat,[690,940,974] mouse, guinea pig[974] *Properties*: rat,[940,974] mouse, guinea pig,[974] calf[972] *Subcellular localization*: rat (lysosome-like particles),[701,702] mouse[924] *Developmental changes*: rat,[973,975] rabbit[699]		
		Cyclic 2',3'-ribonucleotide phosphodiesterase	A cyclic 2',3'-ribonucleotide + H_2O = a ribonucleoside-2'-phosphate
	Occurrence: dog[976] *Properties, regional distribution*: dog, beef[977]		
		Cyclic 3',5'-nucleotide phosphodiesterase	Adenosine-3',5'-cyclic phosphate + H_2O = adenosine-5'-phosphate
	Occurrence: rat,[976] dog,[978] beef[978] *Purification*: rat,[979] dog[978] *Properties*: rat,[976,979] rabbit[976] *Regional distribution*: rabbit[976,980] *Subcellular localization*: rat (mitochondria, supernatant),[633,981] beef[629]		
		Tabunase	$(CH_3)_2N-PO-CN + H_2O = (CH_3)_2N-PO-CN$ $\quad\quad\quad\mid\quad\quad\quad\quad\quad\quad\quad\quad\quad\quad\quad\mid$ $\quad\quad\quad O-C_2H_5 \quad\quad\quad\quad\quad\quad\quad OH$ $+ C_2H_5OH$
	Occurrence: rabbit[982]		
3.1.6.1	Aryl-sulphate sulphohydrolase	Arylsulphatase	A phenol sulphate + H_2O = a phenol + sulphate
	Occurrence: rat,[983] rabbit,[984] human[985,986]		

Summary Table of Enzymes of the Nervous System—(continued)

E.C. number	Systematic name	Trivial name	Reaction
	Purification: beef (4 isoenzymes),[987] human[988] *Properties*: beef (4 isoenzymes),[987,1225] human[988] *Regional distribution*: guinea pig,[989] sheep,[672] calf (2 enzymes)[990] *Subcellular localization*: chicken,[991] human[985] *Developmental changes*: chicken,[991,992] human; review[62]		
3.2.1.1	α-1,4-Glucan 4-glucanohydrolase	α-Amylase	Hydrolyzes α-1,4-glucan links in polysaccharides containing three or more α-1,4-linked D-glucose units
	Occurrence: brain,[996] rat,[444,994] guinea pig,[994,995] rabbit,[444,994] hog, sheep, horse,[1] cat,[444] dog,[995] goat[695] beef,[1,444,995] human,[1,997] chicken, pigeon, bullfrog[994] *Purification*: brain[998] *Regional distribution*: dog[744]		
		γ-Amylase	
	Occurrence: rabbit[999]		
3.2.1.7	β-1,2-Fructan fructanohydrolase *Occurrence*: brain[996]	Inulase	Hydrolyzes β-1,2-fructan links in inulin
3.2.1.11	α-1,6-Glucan 6-glucanohydrolase *Occurrence*: brain[1000]	Dextranase	Hydrolyzes α-1,6-glucan links
3.2.1.17	Mucopeptide N-acetylmuramyl-hydrolase *Regional distribution*: dog[1001]	Mucopeptide glucohydrolase, lysozyme	Probably hydrolyzes β-1,4-links between N-acetylmuramic acid and 2-acetamido-2-deoxy-D-glucose residues in a mucopolysaccharide or mucopeptide

3.2.1.18	Mucopolysaccharide N-acetylneuraminylhydrolase	Neuraminidase	Probably hydrolyzes terminal α-2,6-links between N-acetylneuraminic acid and 2-acetamido-2-deoxy-D-galactose residues in various mucopolysaccharides[1005]
	Occurrence: rat,[1002,1003] guinea pig,[1003,1004] rabbit,[1003] beef, human,[1004] chicken[1005]		
	Purification, properties: guinea pig, beef, human,[1004] chicken[1005]		
3.2.1.20	α-D-Glucoside glucohydrolase	α-Glucosidase	An α-D-glucoside $+ H_2O =$ an alcohol $+$ D-glucose
	Occurrence: rabbit, beef, dog[996]		
3.2.1.21	β-D-Glucoside glucohydrolase	β-Glucosidase	A β-D-glucoside $+ H_2O =$ an alcohol $+$ D-glucose
	Occurrence: rat, mouse, guinea pig, dog, human[1006]		
	Purification: rat,[1009] beef (10-fold)[1007,1008]		
	Properties: rat (hydrolyzes ceramide glucoside),[1009] beef[1007,1008]		
3.2.1.22	α-D-Galactoside galactohydrolase	α-Galactosidase	An α-D-Galactoside $+ H_2O =$ an alcohol $+$ D-galactose
	Occurrence: rabbit[1010]		
3.2.1.23	β-D-Galactoside galactohydrolase	β-Galactosidase	A β-D-galactoside $+ H_2O =$ an alcohol $+$ D-galactose
	Occurrence: rat,[996,1010,1011] mouse,[1012] pig[1013,1016]		
	Purification: rat (20-fold),[1007,1014] beef (40-fold),[1007] pig[1013]		
	Properties: rat[1007] (ceramide lactoside hydrolyzing)[1008] (ganglioside degrading),[1014,1015,1017] beef,[1007] pig[1013,1016]		
	Subcellular localization: rat (lysosome-like particles)[912] (ganglioside degrading, nerve cell nuclei),[1017] mouse[924]		
	Developmental changes: rat[1018]		

Summary Table of Enzymes of the Nervous System—(continued)

E.C. number	Systematic name	Trivial name	Reaction
3.2.1.24	α-D-Mannoside mannohydrolase *Occurrence*: rat, mouse[1012]	α-Mannosidase	An α-D-mannoside + H_2O = an alcohol + D-mannose
3.2.1.26	β-D-Fructofuranoside fructohydrolase *Occurrence*: brain[996]	β-Fructofuranosidase	A β-D-fructofuranoside + H_2O = an alcohol + D-fructose
3.2.1.30	β-2-Acetamido-2-deoxy-D-glucoside acetamidodeoxyglucohydrolase *Occurrence*: rat,[1012] beef[1019] *Purification*: calf (soluble enzyme)[1020] *Properties*: mouse (particulate bound enzyme), calf (soluble enzyme)[1019,1020] *Subcellular localization*: rat (lysosome-like particles),[499,912] mouse[924,1021]	β-Acetylglucosaminidase	β-Phenyl-2-acetamido-2-deoxy-D-glucoside + H_2O = phenol + 2-acetamido-2-deoxy-D-glucose
	α-2-Acetamido-2-deoxy-D-glucoside acetamidodeoxyglucohydrolase *Occurrence, properties*: rat[1249]	α-Acetylglucosaminidase	α-Phenyl-2-acetamido-2-deoxy-D-glucoside + H_2O = phenol + 2-acetamido-2-deoxy-D-glucose
	β-2-Acetamido-2-deoxy-D-galactoside acetamidodeoxygalactohydrolase *Purification, properties*: calf (soluble enzyme)[1019,1020]	β-Acetylgalactosaminidase	β-Phenyl-2-acetamido-2-deoxy-D-galactoside + H_2O = phenol + 2-acetamido-2-deoxy-D-galactose

	β-N-acetyl-hexosaminidase	Hydrolyzes ceramide glycosides
	Purification, properties: calf[1020,1022]	
3.2.1.31	β-D-Glucuronide glucuronohydrolase	β-Glucuronidase	A β-D-glucuronide + H_2O = an alcohol + a D-glucuronate

Activity: mouse[1012]
Regional distribution: rat,[1023] dog[744,1024]
Subcellular localization: (lysosome-like particles) rat,[178,702,910] mouse,[924,1025] guinea pig[194,913]
Developmental changes: rat[1018]

| 3.2.2.5 (3.2.2.6) | NAD(P) glycohydrolase | NAD(P) nucleosidase | $NAD(P) + H_2O$ = nicotinamide + $R(P)$ |

Occurrence: rat,[1026-1028] guinea pig,[1026,1028,1029] mouse,[1027-1029] rabbit,[1027-1029] beef,[1030,1031] pig,[1028,1032] sheep,[1029,1031] pigeon, hamster,[1027] human[1026]
Purification: pig (8000-fold),[1034] sheep, beef[1030]
Properties: rat,[1035-1037] mouse, guinea pig, rabbit, sheep,[1029,1037] beef, dog, horse,[1037] pig[1034,1037]
Subcellular localization: (microsomes) rat[1027,1028,1038,1039]
Developmental changes: rat,[1038] guinea pig[1040]

β-Aspartylglucosylamine amidohydrolase — 1-L-β-aspartamido-(2-acetamido)-1,2-dideoxy-β-D-glucose + 2 H_2O = 2-acetamido-2-deoxy-D-glucose + NH_3 + aspartic acid

Occurrence, properties: rat[1041]

Ethanolamine-plasmalogen hydrolase — An ethanolamine-plasmalogen + H_2O = a fatty aldehyde + lysophosphatidyl-ethanolamine

Occurrence, properties: rat[1042]

Summary Table of Enzymes of the Nervous System—(continued)

E.C. number	Systematic name	Trivial name	Reaction
3.4.1.1	L-Leucyl-peptide hydrolase *Regional distribution*: rat[1044] *Subcellular localization*: calf[1043]	Leucine aminopeptidase	An L-leucyl-peptide + H_2O = L-leucine + a peptide
3.4.1.2	Amino-acyl-oligopeptide hydrolase *Occurrence*: rat,[1045] human[1046] *Preparation, properties*: rat[1045,1047]	Aminopeptidase	An amino-acyl-oligopeptide + H_2O = an amino acid + an oligopeptide
3.4.1.3	Amino-acyl-dipeptide hydrolase *Occurrence*: mouse,[1049] calf,[1048] human[1049,1050] *Properties*: mouse,[1047,1049] calf[1048] *Subcellular localization*: calf[1043]	Aminopeptidase	An amino-acyl-dipeptide + H_2O = an amino acid + a dipeptide
	Occurrence, properties: beef[1081]	Thiol activated aminopeptidase	
3.4.2.1	Peptidyl-L-amino-acid hydrolase *Preparation, properties*: brain[1047] *Subcellular localization*: calf[1043]	Carboxypeptidase A	A peptidyl-L-amino acid + H_2O = a peptide + an L-amino acid
	Occurrence: mouse, rabbit, beef[1051]	Insulinase	

3.4.3.1	Glycyl-glycine hydrolase *Occurrence:* mouse,[1052] human[1050] *Properties:* mouse,[1052,1053] human[1050]	Glycyl-glycine dipeptidase	Glycyl-glycine + H_2O = 2 glycine
3.4.3.2	Glycine-L-leucine hydrolase *Properties:* mouse[1052]	Glycyl-leucine dipeptidase	Glycyl-L-leucine + H_2O = glycine + L-leucine
3.4.3.3	Amino-acyl-L-histidine hydrolase *Properties:* rabbit (microsomal enzyme, mitochondrial enzyme)[1054]	Amino-acyl-histidine dipeptidase, carnosinase	Amino-acyl-L-histidine + H_2O = an amino acid + L-histidine
3.4.3.6	L-Prolyl-amino acid hydrolase *Occurrence:* brain (low activity)[1055]	Iminodipeptidase	An L-prolyl-amino acid + H_2O = L-proline + an amino acid
	Occurrences: rat,[1056,1057] mouse,[1058] rabbit,[1059] sheep,[1060] calf,[1048] cat,[1059,1060] human,[1050] chicken,[1061] frog, bullock,[1060] *Tinca tinca, Testudo graeca,*[1062] *Natrix taxispilota, Ptyas mucosus, Dispholidus typus, Agama hispidadistanti, Cordylus giganteus, Varanus bengalesis*[1063] *Properties:* mouse,[1049,1058] human[1049] *Regional distribution:* rat,[794,1064] human (cortical layers)[313,1065] *Subcellular localization:* calf[1043] *Developmental changes:* rat,[83] review[62]	Dipeptidases	A dipeptide + H_2O = 2 amino acids
3.4.4.9		Cathepsins	Hydrolyze peptides, especially at bonds involving an aromatic amino acid adjacent to a free α-amino group

Summary Table of Enzymes of the Nervous System—(continued)

E.C. number	Systematic name	Trivial name	Reaction
	Occurrence: rat (low activity)[1066]	Cathepsin B	
	Occurrence: rat,[1066,1067] calf,[1048] beef, human,[1057] rabbit,[1057,1067] pigeon[1067] *Purification:* rat,[1068] calf,[1048] beef[663,1069,1070] *Properties:* rat,[1068,1071] calf,[1048] beef[663,1047,1057,1070] *Regional distribution:* beef,[663] sheep, monkey,[1072] human[1050] *Subcellular localization:* rat (mitochondrial fraction)[1071,1072] (in lysosome-like particles),[178,701,702] rabbit[665] *Developmental changes:* rat,[1072] rabbit[1073]	Cathepsin C and acid proteinase	
	Occurrence: rat[1074,1075] *Purification:* rat[1047,1068,1076] *Properties:* rat[1047,1068,1071,1074,1075,1076] *Regional distribution:* rat,[1074] sheep, monkey[1072] *Subcellular localization:* rat (mitochondrial fraction)[1071,1072,1075,1077]	Neutral proteinase	
3.4.4.19	*Occurrence:* rat[1078]	Clostridiopeptidase A collagenase	Hydrolyzes peptides containing proline, including collagen and gelatin
3.4.4.23	*Occurrence, purification, properties:* beef[1079,1080]	Cathepsin D	Hydrolyzes peptides; its specificity is similar to pepsin

	Oxytocin peptidase	
	Occurrence: dog[1084]	
	γ-Glutamyl transpeptidase	γ-D,L-Glutamyl-α-aminopropionitril + $H_2O = \alpha$-aminopropionitril + glutamate
	Occurrence: rat, mouse, rabbit, guinea pig, human[1082]	
	Regional distribution: rabbit, human[1083]	
	Nervesidase	Inactivates nerveside
	Occurrence, properties: dog[1085]	
3.5.1.1	L-Asparagine amidohyrolase	L-Asparagine + H_2O = L-aspartate + NH_3
	Occurrence: guinea pig,[1086] sheep,[1087] beef, pig[1088]	
	Properties: beef[1088]	
	Regional distribution: beef, pig,[1088] guinea pig[1086]	
3.5.1.2	L-Glutamine amidohydrolase	L-Glutamine + H_2O = L-glutamate + NH_3
	Occurrence: rat,[1089–1091] mouse, guinea pig, rabbit[1090]	
	Properties: rat[208,1089,1090,1092]	
	Regional distribution: guinea pig,[1086] cat,[1094] beef[1093]	
	Subcellular localization: (mitochondria) rat,[147,207,208,1095] guinea pig,[1086] rabbit[665,1093]	
	Developmental changes: rat[1096]	
3.5.1.13	Aryl acylamidase	An N-acyl-anilide + H_2O = a fatty acid ion + aniline
	Aryl-acylamide amidohydrolase	
	Occurrence, properties: brain[1047]	
	Nicotinamidase	Nicotinamide + H_2O = nicotinic acid + NH_3
	Occurrence: rat, rabbit[1097]	

Summary Table of Enzymes of the Nervous System—(continued)

E.C. number	Systematic name	Trivial name	Reaction
3.5.1.14	N-Acylamino-acid amidohydrolase *Occurrence:* rat,[1056,1098] human[1099] *Properties:* rat[1098]	Aminoacylase (dehydropeptidase II)	An N-acyl-amino acid + H_2O = a fatty acid ion + an amino acid
3.5.1.15	N-Acylaspartate amidohydrolase *Occurrence:* mouse, rabbit, hamster, cat, human[1100] *Properties:* mouse, rabbit[1100]	Aspartoacylase	N-Acyl-aspartate + H_2O = a fatty acid ion + aspartate
	α-N-Acetyl-L-histidine amidohydrolase *Occurrence, properties:* Skipjack tuna (not found in rat and mouse)[1101]	N-Acetylhistidine amidohydrolase	α-N-Acetyl-L-histidine + H_2O = acetate + L-histidine
	 Occurrence, purification: rat (100-fold)[1102] *Properties:* rat[1102,1103]	Ceramidase	Ceramide + H_2O = sphingosine + fatty acid ion
3.5.2.6	Penicillin amido-β-lactam hydrolase *Occurrence:* rat,[1104,1105] guinea pig[1105] (not found in rat and guinea pig)[1106]	Penicillinase	Penicillin + H_2O = penicilloate
3.5.3.1	L-Arginine amidinohydrolase *Occurrence:* in all tissues of vertebrates having a uriotelic metabolism,[1107] rat,[1108–1110] guinea pig, beef, monkey, human,[1109] chicken, frog[1110] *Regional distribution:* pig[1111]	Arginase	L-Arginine + H_2O = L-ornithine + urea

3.5.4.3	Guanine aminohydrolase	Guanine deaminase	Guanine + H_2O = xanthine + NH_3

Occurrence: rat,[940,1112] mouse,[940,1113,1114] rabbit,[1115] beef,[663] human[1114]
Purification: rat (80–100-fold)[1116] (particulate enzyme),[1117,1118] sheep[1116]
Properties: rat[1116] (particulate enzyme),[1117,1118] sheep, monkey[1116]
Regional distribution: monkey[1116]
Subcellular localization: (cytoplasm, mitochondria) rat,[198,1118,1119] rabbit[665]
Developmental changes: rat[1120]

	8-Azaguanine deaminase	8-Azaguanine + H_2O = 8-azaxanthine + NH_3

Occurrence: human[1121]

3.5.4.4	Adenosine aminohydrolase	Adenosine deaminase	Adenosine + H_2O = inosine + NH_3

Occurrence: rat,[656,1122] mouse,[1113] rabbit, guinea pig, cat, dog[1122]
Purification: beef[1123]
Properties: rabbit,[1122] beef[663,1123]
Regional distribution: rabbit[938,1124]
Subcellular localization: (cytoplasma, mitochondria) rat,[1120] rabbit[665,938,1120,1124]

3.5.4.5	Cytidine aminohydrolase	Cytidine deaminase	Cytidine + H_2O = uridine + NH_3

Occurrence: rat[1125]

	Guanosine aminohydrolase	Guanosine deaminase	Guanosine + H_2O = xanthosine + NH_3

Occurrence: rat,[939,940] mouse,[940] rabbit, cat, pig[1115]
Subcellular localization: (cytoplasm) rat,[1119,1120] rabbit[665,1119]
Developmental changes: rat[1120]

3.5.4.6	AMP aminohydrolase	AMP deaminase	AMP + H_2O = IMP + NH_3

Occurrence: rat,[568] dog[1126]
Purification: (particulate enzyme) beef (10–20-fold)[1127,1128] (180-fold)[1129] (soluble enzyme) sheep (5-fold),[1130] dog[1131]

Summary Table of Enzymes of the Nervous System—(continued)

E.C. number	Systematic name	Trivial name	Reaction
	Properties: rabbit,[1133] beef,[1127–1129,1132] sheep,[1130] dog[1126,1131] *Regional distribution*: rabbit[938,1124] *Subcellular localization*: rat (mitochondria)[656,1119] (microsomes),[1134] rabbit,[1119,1133] beef[1127]		
	GMP aminohydrolase	GMP deaminase	$GMP + H_2O$ = xanthosinemonophosphate $+ NH_3$
	Occurrence: rat, mouse[940]		
		Deaminases for RNA and DNA	
	Occurrence: rat, mouse[939,940]		
	NAD aminohydrolase	NAD deaminase	$NAD + H_2O$ = nicotinic acid-adenine dinucleotide $+ NH_3$
	Occurrence, properties: rat[1135]		
3.6.1.1	Pyrophosphate phosphohydrolase *Occurrence*: in all cells[1136] *Activity*: rat, rabbit, sheep, human,[873] guinea pig[1137] *Purification*: pig (165-fold)[1138] *Properties*: rat,[873,886,1139] guinea pig,[1137] rabbit,[1140] pig[1138] *Regional distribution*: rabbit[1140] (cortical layers, hippocampus),[51,455] sheep, human[886] *Subcellular localization*: beef[629] *Developmental changes*: rabbit[1141]	Inorganic pyrophosphatase	Pyrophosphate $+ H_2O$ = 2 orthophosphate
3.6.1.3	ATP phosphohydrolase a. DNP activated enzyme *Properties*: calf (mitochondria)[1142] *Subcellular localization*: rat (mitochondria),[1144] guinea pig (mitochondria)[1143]	ATPase	$ATP + H_2O$ = ADP + orthophosphate

b. Mg^{2+} activated enzymes

Occurrence: rat,[1145,1146] guinea pig,[1147] rabbit,[1148-1150] mouse, cat, beef, human,[1145] dog[1149]

Purification: rabbit[771]

Properties: rat (mitochondrial enzyme),[886,1151,1152,1153] (microsomal enzyme),[1250] (enzyme in synaptic vesicles),[1143] guinea pig (mitochondrial enzyme),[1154] rabbit (mitochondrial enzyme),[1148,1155,1156,1157] (microsomal enzyme),[1157,1158] (nuclear enzyme)[1157] beef (mitochondrial enzyme),[1159] pig, [1221,1222] *Rana temporaria*[1223]

Regional distribution: rat (cortex),[1160] rabbit[1161] (hippocampus),[455,576] cat,[1161] hamster,[1162] human (cortex),[890,1163] beef[1174]

Subcellular localization: rat (mitochondria)[65,1152,1153,1164-1166] (synaptic vesicles),[1167,1168] guinea pig (mitochondria)[75,1154,1169] (synaptic vesicles),[1143] mouse (mitochondria, microsomes),[1170] rabbit (mitochondria, microsomes, nuclei),[665,1157] beef (mitochondria, microsomes)[1159]

Developmental changes: rat,[934,1039,1171,1172] chicken,[1173] rabbit[1174]

Review:[62,1175]

c. Na^+ and K^+-activated enzymes

Occurrence: on cell surfaces[1176]

Purification: rat,[922,1177-1180] guinea pig,[1178] mouse,[1181] rabbit,[1178] beef (microsomal enzyme)[1182,1183]

Properties: rat (microsomal enzyme),[922,1151,1177-1180,1184-1190,1219,1250] guinea pig (microsomal enzyme),[77,180,1138,1154,1178,1191-1195,1220] (mitochondrial enzyme),[1169] (enzyme in nerve endings),[75] mouse,[1181] rabbit (microsomal enzyme)[1196,1197] (neuronal enzyme, glial enzyme),[1198] pig (microsomal enzyme),[1199,1221,1222] beef (microsomal enzyme),[1159,1182,1200,1224] human (microsomal enzyme),[1201] *Rana temporaria*,[1223] crab nerve[1203]

Physiological significance: (ion transport)[1184,1191,1203,1204]

Regional distribution: rat (cortex),[936,1205] cat[15]

Subcellular localization: rat (microsomes)[1165,1166,1206,1207] (nerve endings, synaptic membranes)[113,258,632,633,923,1164] (myelin),[632,633,1209] (nuclei),[77,1214] mouse (microsomes, mitochondria),[1170] guinea pig (microsomes)[88,180,651,1154] (nerve endings, synaptic membranes)[75,818,1138,1143,1210] (myelin),[1210] rabbit (microsomes),[1197] (neurons, neuroglia),[1198] beef (microsomes, cell membranes)[1159]

Developmental changes: rat,[1039,1211,1212] chicken[732,1213]

d. Ca^{2+}-activated enzymes

Occurrence, activity: rat,[568,1145,1146] mouse,[1145] rabbit,[1156] cat, beef, human[1145]

Properties: rat,[1208,1215] rabbit (microsomal enzyme[1245,1246] (nuclear enzyme)[1245]

Regional distribution: rat (cortex),[1160] human[890,1163]

Subcellular localization: rat (microsomes)[1206,1208,1216] (synaptic vesicles)[1167,1168] rabbit (microsomes, nuclei)[665,1161]

Developmental changes: rat,[934,1216,1217] guinea pig,[920] chicken[1218]

Summary Table of Enzymes of the Nervous System—(continued)

E.C. number	Systematic name	Trivial name	Reaction
3.6.1.6	Nucleosidediphosphate phosphohydrolase	Nucleosidediphosphatase	A nucleoside diphosphate + H_2O = a nucleoside + orthophosphate
	Regional distribution: rabbit[908]		
3.6.1.7	Acylphosphate phosphohydrolase	Acylphosphatase	An acylphosphate + H_2O = an anion + orthophosphate
	Occurrence: in all animal tissues[1136]		
	Purification: calf (1750-fold)[1226]		
	Properties: beef[1224] calf, guinea pig, rabbit[1227]		
	Regional distribution: hamster[1162]		
	Subcellular localization: rat (cytoplasma),[1228] beef[1183]		
3.6.1.9	Dinucleotide nucleotidohydrolase	Nucleotide pyrophosphatase	A dinucleotide + H_2O = 2 mononucleotides
	Occurrence: rat, mouse, hamster, pigeon, rabbit[1027]		
	Properties, subcellular localization: rat, rabbit, mouse, hamster, pigeon (microsomes)[1027]		
3.6.1.11	Polyphosphate phosphohydrolase	Exopolyphosphatase	$(Polyphosphate)_n + H_2O =$ $(polyphosphate)_{n-1}$ + orthophosphate
	Regional distribution: chicken[1229]		
	Developmental changes: chicken[1229,1230]		
		UDP-acetylglucosamine pyrophosphatase	UDP-2-acetamido-2-deoxy-D-glucose + H_2O = 2-acetamido-2-deoxy-D-glucose 1-phosphate + UMP
	Occurrence: rat, sheep[1231]		
	Purification, properties: sheep (3-fold)[1231]		
	Regional distribution: sheep[592]		

	Enzyme	Details	Reaction
	Thiamine pyrophosphatase	*Properties:* rat[955,1232] *Regional distribution:* rat,[1232,1233] rabbit,[908] chicken[1233] *Subcellular localization:* rabbit (Golgi apparatus)[1234]	Thiamine pyrophosphate + H_2O = thiamine monophosphate + orthophosphate
3.9.1.1	Phosphoamidase Phosphoamide hydrolase	*Properties:* rat (2 enzymes)[1235] *Regional distribution:* rat,[1236] pigeon[1237]	Phosphocreatine + H_2O = creatine + orthophosphate
4.1.1.1	Pyruvate decarboxylase 2-Oxoacid carboxylyase	*Occurrence, activity:* rat,[1238,1239] guinea pig, rabbit, cat, beef[1239] *Properties:* rabbit[1240,1241] *Regional distribution:* rat[1238] *Subcellular localization:* beef[1239]	A 2-oxoacid = an aldehyde + CO_2
	Hydroxypyruvate decarboxylase 2-Oxo-fatty acid decarboxylase	*Occurrence, properties:* rat, rabbit, beef[1242] *Occurrence, properties:* pig[1243]	Hydroxypyruvate = glycolaldehyde + CO_2
	α-Hydroxy acid decarboxylases	*Occurrence:* rat,[1244-1247] beef[1247] *Properties:* rat (requiring ATP and NAD)[1245,1246] *Subcellular localization:* rat (microsomes)[1244-1247]	

Summary Table of Enzymes of the Nervous System—(continued)

E.C. number	Systematic name	Trivial name	Reaction
4.1.1.15	L-Glutamate 1-carboxylyase *Occurrence, activity*: rat,[479] mouse,[6,239,1251,1252] hamster,[849] beef,[1253] human,[1254] honey bee[1255] *Purification*: mouse,[1256] cat[1257] *Properties*: rat,[1258] mouse[7,1252,1259] *Regional distribution*: rat,[546,1260] guinea pig,[1260] rabbit,[1252,1260,1261] cat,[1260,1261] dog,[1261] monkey,[553,1252,1260,1262,1263] human,[1264,1265] chicken[1260,1266] *Subcellular localization*: rat (supernatant, mitochondrial fraction)[207,500,1267,1268] (noncholinergic nerve endings),[147,211,315,549] mouse (nerve endings),[1269] beef[1268] lobster (*Homarus americanus*)[1270] *Developmental changes*: rat,[66,550,1271] mouse,[10,1261,1271] rabbit,[66,1271,1272] dog,[1271] chicken,[10,66,475,551,1266] bullfrog[10]	Glutamate decarboxylase	L-Glutamate = 4-aminobutyrate + CO_2
4.1.1.16	3-Hydroxy-L-glutamate 1-carboxy-lyase *Occurrence*: brain[1273]	Hydroxyglutamate decarboxylase	3-Hydroxy-L-glutamate = 4-amino-3-hydroxybutyrate + CO_2
	4-Hydroxy-L-glutamate 1-carboxy-lyase *Occurrence*: rat[1274]	γ-Hydroxyglutamate decarboxylase	4-Hydroxy-L-glutamate = 4 amino-2-hydroxybutyrate + CO_2
4.1.1.22	L-Histidine carboxy-lyase *Occurrence*: rat,[1275] mouse,[1347] cat, dog, pig,[359,360] human (fetal brain)[1348] *Purification, properties*: dog[1293] *Regional distribution*: beef,[1276,1277] cat, pig, dog[358] *Subcellular localization*: (mitochondria, microsomes) rabbit,[278] beef[1277]	Histidine decarboxylase	L-Histidine = histamine + CO_2
4.1.1.26	3,4-Dihydroxy-L-phenylalanine carboxy-lyase *Occurrence, activity*: rat,[1254,1280,1281] mouse,[1282] guinea pig,[1254,1280,1281] rabbit,[1281] human[1254]	DOPA decarboxylase	3,4-Dihydroxy-L-phenylalanine = 3,4-dihydroxyphenylethylamine + CO_2

Regional distribution: rat,[1283] rabbit,[1280,1284,1285] cat,[1280,1285,1286] dog, sheep,[1280] beef,[1276,1287] pig,[1280,1284] human,[1254] chicken[1288]

Subcellular localization: guinea pig (supernatant)[325]

Developmental changes: rat, guinea pig[261,1289]

4.1.1.28	5-Hydroxy-L-tryptophan carboxy-lyase	Hydroxytryptophan decarboxylase	5-Hydroxy-L-tryptophan = 5-hydroxy-tryptamine + CO_2

Occurrence, activity: rat,[1254,1281,1290,1291] mouse,[1290,1292] guinea pig,[1254,1281,1290,1293] rabbit,[1281,1290] cat, dog, hamster,[1290] human[1254,1294]

Properties: mouse,[1290] rabbit[1251]

Regional distribution: rabbit,[251,419,1284] pig,[1284] cat,[252,1286] dog,[73,252,1296,1297] human,[1254] chicken,[1254] pigeon,[254] carp[255]

Subcellular localization: rat (cytoplasma, nerve endings, mitochondria, microsomes, nuclei),[177,256,355,1298] rabbit,[252,256] guinea pig, pigeon[256]

Developmental changes: rat,[261,263,1289,1299] guinea pig,[261,265,1289,1299,1300] goat[1301]

4.1.1.29	L-Cysteinsulphinate carboxy-lyase	Cysteinesulphinate decarboxylase	L-Cysteine sulphinate = hypotaurine + CO_2

Occurrence: rat,[1302,1303] dog, human, cat, chicken, toad[1303]

Properties: rat[1303]

Regional distribution: beef[1304]

	L-Serine carboxy-lyase	Serine decarboxylase	L-Serine = ethanolamine + CO_2

Occurrence: rat[1305]

	Phosphatidylserine carboxy-lyase	Phosphatidylserine decarboxylase	Phosphatidylserine = phosphatidylethanolamine + CO_2

Occurrence: rat (mitochondria)[1305,1306]

	Aminobutyric acid carboxy-lyase	Aminobutyrate decarboxylase	Aminobutyrate = n-propylamine + CO_2

Occurrence: rat[1307]

Summary Table of Enzymes of the Nervous System—(continued)

E.C. number	Systematic name	Trivial name	Reaction
	L-Leucine carboxy-lyase *Occurrence:* guinea pig[1308]	Leucine decarboxylase	L-Leucine = 2-methylbutylamine + CO_2
	L-Phenylalanine carboxy-lyase *Occurrence:* mouse,[239] guinea pig[1293]	Phenylalanine decarboxylase	L-Phenylalanine = β-phenylethylamine + CO_2
	L-Pipecolic acid carboxy-lyase *Occurrence:* guinea pig[1309]	Pipecolic acid decarboxylase	L-Pipecolic acid = piperidine + CO_2
4.1.2.7	Ketose-1-phosphate aldehyde lyase	Ketose-1-phosphate aldolase	A ketose 1-phosphate = dihydroxyacetone phosphate + an aldehyde
4.1.2.13	Fructose-1,6-diphosphate D-glyceraldehyde-3-phosphate lyase	Fructosediphosphate aldolase	Fructose-1,6-diphosphate = dihydroxy-acetone phosphate + D-glyceraldehyde 3-phosphate
	Occurrence: in all cells with glycolytic activity *Activity:* mouse,[25,33] rat,[25,1310] rabbit[576] *Properties:* rat (isoenzymes),[1311] rabbit[771] (5 isoenzymes),[1312] beef[663] *Regional distribution:* rat (cortex),[82] rabbit[1313] (cortex),[455] dog,[1313] monkey,[54] human,[58,95] review (cerebellum)[62] *Subcellular localization:* (cytoplasma, mitochondrial fraction) rat,[28,64,65,102,1314] (nuclei),[72] rabbit,[665] calf,[26] Japanese quail[27] *Developmental changes:* rat[123,185] (isoenzymes),[1315] guinea pig,[894] mouse,[33] rabbit[1313]		
4.1.3.3	N-Acetylneuraminate pyruvate-lyase *Occurrence:* rat,[1316] guinea pig, dog[1317]	N-Acetylneuraminic acid aldolase	N-Acetylneuraminate = 2-acetamido-2-deoxy-D-mannose + pyruvate

	Neuraminate pyruvate-lyase	Neuraminic acid aldolase	Neuraminate = 2-amino-2-deoxy-D-mannose + pyruvate
	Occurrence: rat, rabbit[1003]		
		N-Acetylneuraminic acid synthesizing enzyme	2-Acetamido-2-deoxy-D-mannose-6-phosphate + phosphoenolpyruvate = N-acetyl-neuraminate-9-phosphate + orthophosphate
	Purification: sheep (20-fold), human (9-fold)[961]		
	Properties: sheep (Mg^{2+}-dependent)[961]		
	Regional distribution: sheep[961]		
		Sphingosine synthetase	Palmitylaldehyde + serine = dihydrosphingosine + CO_2
	Occurrence, properties: rat[129,134]		
4.1.3.7	Citrate oxaloacetate-lyase (CoA-acetylating)	Citrate synthase	Citrate + CoA = acetyl-CoA + H_2O + oxaloacetate
	Occurrence: beef[1318]		
	Subcellular localization: guinea pig, sheep, cat, dog (supernatant, mitochondria, nerve endings)[76,393] rabbit[395]		
4.1.3.8	ATP:citrate oxaloacetate-lyase (CoA-acetylating and ATP-dephosphorylating)	ATP citrate lyase	ATP + citrate + CoA = ADP + orthophosphate + acetyl-CoA + oxaloacetate
	Occurrence: rat, pig,[1319] guinea pig, rabbit,[398,1319] beef[1319,1320]		
	Purification: pig, beef[1319]		
	Subcellular localization: (cytoplasmic fraction) sheep,[76,1318] guinea pig,[76] (nerve endings) rabbit, cat, sheep, dog, pig[1321,1322]		
4.2.1.1	Carbonate hydro-lyase	Carbonate dehydratase, carbonic anhydrase	H_2CO_3 (or H^+ + HCO_3^-) = CO_2 + H_2O

Summary Table of Enzymes of the Nervous System—(continued)

E.C. number	Systematic name	Trivial name	Reaction
	Occurrence, activity: rat,[1323,1324] guinea pig,[1324] cat, beef, dog,[1324,1325] sheep, pig, horse, monkey, human,[1325] rabbit, chicken,[1324,1325,1326] fish, reptiles, amphibia[1326] *Properties:* cat[1327] *Regional distribution:* rabbit,[1328] cat, dog, pig,[1329] human[1329–1332] *Subcellular localization:* rat (mitochondria, cytoplasm)[1333] (neurons, glia cells),[71,1334] mouse,[1335] human[822] *Developmental changes:* rat, guinea pig,[1336,1337] cat, beef,[1336] rabbit, dog,[311,1336] pig,[1338] chicken,[720,1339,1340] rook,[1339] monkey[1341]		
4.2.1.2	L-Malate hydro-lyase *Purification, properties:* rabbit[771] *Regional distribution:* monkey[54] *Subcellular localization:* (mitochondria) rat,[102,500] rabbit,[1342] guinea pig, sheep[76]	Fumarate hydratase	L-Malate = fumarate + H_2O
4.2.1.3	Citrate (isocitrate) hydro-lyase *Occurrence:* pigeon[1343] *Subcellular localization:* (cytoplasm, mitochondria) rat,[99] rabbit[1342] *Properties:* rat (mitochondrial and cytoplasmic enzyme)[99]	Aconitate hydratase	Citrate = cis-aconitate = H_2O
4.2.1.11	2-Phospho-D-glycerate hydro-lyase *Occurrence, activity:* rat[25,95] (extramitochondrial compartment),[137] mouse,[25,33] beef[771] *Purification, properties:* beef (3,5-fold)[1344] *Regional distribution:* human[95] *Subcellular localization:* rat (cytoplasma, mitochondria)[28,65,102] (nuclei),[72] calf[26]	Phosphopyruvate hydratase	2-Phospho-D-glycerate = phosphoenol-pyruvate + H_2O

4.2.1.13	L-Serine hydro-lyase (deaminating) L-Serine dehydratase, cystathionine synthetase *Occurrence*: rat,[1345] monkey,[474] human[474,1346]	L-Serine = pyruvate + NH_3 L-Serine + L-homocysteine = L-cystahionine + H_2O
4.2.1.15	L-Homoserine hydro-lyase (deaminating) Homoserine dehydratase, cystathionase *Occurrence*: rat, monkey,[474] human[474,1346]	L-Homoserine = 2-oxobutyrate + NH_3 L-cystathionine + H_2O = 2-oxobutyrate + L-cysteine + NH_3
4.2.1.17	L-3-Hydroxyacyl-CoA hydro-lyase Enoyl-CoA hydratase, crotonase *Occurrence*: rat[1349]	An L-3-hydroxyacyl-CoA = a 2,3- (or 3,4-) trans-enoyl-CoA + H_2O
4.2.1.22	L-Serine hydro-lyase (adding hydrogen sulphide) cysteine synthase *Occurrence*: rat, chicken[1350]	L-Serine + H_2S = L-cysteine + H_2O
4.2.99.1	Hyaluronate lyase Hyaluronidase *Occurrence*: rat, mouse, guinea pig, rabbit, sheep, beef, horse, cat, pig, cock, frog[1351]	Hyaluronate = n 3(β-D-gluco-4,5-en-urono)-2-acetamido-2-deoxy-D-glucose
4.3.2.1	L-Argininosuccinate arginine-lyase Argininosuccinate lyase *Occurrence*: rat,[1109,1110] guinea pig[1109] *Regional distribution*: cat, beef, monkey, human,[1109] pig[1111]	L-Argininosuccinate = fumarate + L-arginine
4.4.1.1	L-Cysteine hydrogensulphide-lyase (deaminating) Cysteine desulphhydrase *Occurrence*: rat, guinea pig, pigeon[715]	L-Cysteine + H_2O = pyruvate + NH_3 + H_2S

Summary Table of Enzymes of the Nervous System—(continued)

E.C. number	Systematic name	Trivial name	Reaction
4.4.1.5	S-Lactoyl-glutathione methylglyoxal-lyase (isomerizing) *Occurrence*: in nearly all animal cells[1355]	Lactoyl-glutathione-lyase, glyoxalase I	S-Lactoyl-glutathione = glutathione + methylglyoxal
5.1.3.3	Aldose 1-epimerase *Occurrence, properties*: rat[1356,1357]	Aldose mutarotase	α-D-Glucose = β-D-glucose
5.3.1.1	D-Glyceraldehyde-3-phosphate ketol-isomerase *Occurrence, activity*: rat[137,1358] *Properties*: rat[1358] *Subcellular localization*: rat,[28] calf[26]	Triosephosphate isomerase	D-Glyceraldehyde 3-phosphate = dihydroxyacetone phosphate
5.3.1.9	D-Glucose-6-phosphate ketol-isomerase *Occurrence*: in all cells with glycolytic activity *Activity*: rat,[25] mouse,[25,33] rabbit[495] *Properties*: rabbit[662] *Regional distribution*: rabbit[120,496] (cortex),[455] monkey,[54] human[58] *Subcellular localization*: (cytoplasm, mitochondrial fraction) rat[28,65]	Glucosephosphate isomerase	D-Glucose 6-phosphate = D-fructose 6-phosphate
5.3.1.10	2-Amino-3-deoxy-D-glucose-6-phosphate ketol-isomerase (deaminating) *Occurrence*: rat,[1359] human[463] *Purification*: human (66-fold)[1360] *Properties*: rat,[589,1361] human[1360] *Regional distribution*: sheep[592] *Developmental changes*: rat[593]	Glucosaminephosphate isomerase	2-Amino-2-deoxy-D-glucose 6-phosphate + H_2O = D-fructose 6-phosphate + NH_3

	Enzyme	Reaction	Occurrence / data
	Δ⁵-3-Keto-isomerases	Androst-5-ene-3,17-dione = androst-4-ene-3,17-dione	*Occurrence, properties:* rat[1362]
6.1.1	Ligases forming amino-acyl-RNA (pH 5 enzymes)	ATP + an L-amino acid + sRNA = AMP + pyrophosphate + amino-acyl-sRNA	*Occurrence:* rat[1363] *Properties:* guinea pig,[1364] rabbit[1365] *Regional distribution:* rabbit[1366] *Developmental changes:* mouse,[1367] rat,[1368] rabbit[1369] Ribosomal amino acid incorporating systems *Occurrence:* rat,[1370,1371] guinea pig[1372,1373] *Properties:* rat,[697,1374–1380] mouse,[1381] guinea pig,[1364,1380,1382] rabbit,[1365,1380,1383,1384] human[1385] *Regional distribution:* monkey[1386] *Developmental changes:* rat,[1365,1368,1380,1387–1390] guinea pig, rabbit[1380]
6.2.1.1	Acetyl-CoA synthetase Acetate:CoA ligase (AMP)	ATP + acetate + CoA = AMP + acetyl-CoA + pyrophosphate	*Occurrence:* rat, guinea pig, cat, pigeon, frog, *Electrophorus electricus,*[377] beef[1391] *Purification:* beef,[1392,1393] pigeon[1394] *Properties:* beef,[1392,1393] sheep, rabbit (nerve endings),[1322] pigeon[1394] *Subcellular localization:* (cytoplasm, mitochondria, nerve endings) rat,[1395] rabbit,[393,395] guinea pig,[76] sheep,[76,393] cat, dog, pig, goat[393] *Developmental changes:* chicken[426]
6.2.1.3	Acyl-CoA synthetase Acid:CoA ligase (AMP)	ATP + an acid + CoA = AMP + an acyl-CoA + pyrophosphate	*Occurrence, properties:* beef (mitochondria)[1396–1398]

Summary Table of Enzymes of the Nervous System—(continued)

E.C. number	Systematic name	Trivial name	Reaction
6.2.1.5	Succinate:CoA ligase (ADP)	Succinyl-CoA synthetase	ATP + succinate + CoA = ADP + succinyl-CoA + orthophosphate
	Occurrence, properties: calf[438,1399]		
		Fatty acid synthesizing system	
	Occurrence: rat[1400]		
	Purification, properties: rat[90,1401]		
	Subcellular localization: rat (mitochondria)[636]		
	Developmental changes: rat[1402]		
6.3.1.2	L-Glutamate:ammonia ligase (ADP)	Glutamine synthetase, glutamotransferase	a. ATP + L-glutamate + NH_3 = ADP + orthophosphate + L-glutamine
			b. L-glutamine + NH_2OH = γ-glutamyl-hydroxamate + NH_3
			c. L-glutamine + H_2O $\xrightarrow{\text{arsenate}}$ L-glutamate + NH_3

Identity of the enzymes catalyzing reactions a–c[1403]

Occurrence, activity: in 17 vertebrate species,[1404] rat,[479,1089] guinea pig, pig, pigeon,[1089] sheep,[1405] chicken[1406]

Purification: sheep[1407,1408] (1000-fold),[1409] pig (200-fold)[1410]

Properties: rat,[1411] sheep[1412,1413] (Mn^{2+} activation)[1407,1414] (substrate and stereospecificity)[1415] (pH optimum),[1408,1409] pig,[1410,1413] beef[1416]

Regional distribution: rabbit,[1417] cat[1094,1417,1418]

Subcellular localization: rat (microsomes, supernatant)[67,147,207,1207,1417,1419] rabbit, cat[1417] (mitochondrial subfractions, nerve endings),[258] rabbit, cat[1417]

Developmental changes: rat,[1420] rabbit,[1421] cat,[1418] pig, monkey,[212] chicken[1422]

6.3.4.5	L-Citrulline:L-aspartate ligase (AMP)	Argininosuccinate synthetase	ATP + L-citrulline + L-aspartate = AMP + pyrophosphate + L-argininosuccinate
	Occurrence, regional distribution: rat,[1109,1110] beef, monkey, human[1109]		
6.4.1.1	Pyruvate: carbon-dioxide ligase (ADP)	Pyruvate carboxylase	ATP + pyruvate + CO$_2$ + H$_2$O = ADP + orthophosphate + oxaloacetate
	Occurrence: rat, guinea pig,[1423] cat[1423-1426] *Purification*: beef (mitochondria)[1423] *Properties*: beef[1423,1427]		
		N-demethylases (microsomal)	(CH$_3$)$_2$N—NO + O$_2$ $\longrightarrow$ 2 CH$_2$O + H$_2$O + N$_2$
	Occurrence: rat (microsomes)[1428,1429]		

Acknowledgments. I would like to thank Mrs. J. Wiechmann, Mr. M. Wiechmann, and my wife. Without their help the compilation of the Summary Table of Enzymes of the Nervous System would have been impossible for me. I am particularly indebted to Dr. A. Lajtha who subjected himself to the laborious task of translating my contribution. In addition, I would like to thank him for his expert advice. Furthermore, I am indebted to Dr. Werner for the interest he has taken in this work.

VI. REFERENCES

1. C. Oppenheimer, *Die Fermente und ihre Wirkungen*, 5th ed. p. 401, Georg Thieme, Stuttgart (1925).
2. O. Warburg, *Wasserstoffübertragende Fermente*, p. 54, Verlag Dr. Werner Sänger, Berlin (1948).
3. Th. Wieland and G. Pfleiderer, *Angew. Chem.* **74**:261–270 (1962).
4. J. Hurwitz, *J. Biol. Chem.* **205**:935–947 (1953).
5. D. B. McCormick and E. E. Snell, *Proc. Natl. Acad. Sci. U.S.* **45**:1371–1379 (1959).
6. E. Roberts and S. Frankel, *J. Biol. Chem.* **187**:55–63 (1950).
7. E. Roberts and S. Frankel, *J. Biol. Chem.* **190**:505–512 (1951).
8. E. Roberts, Free amino acids of nervous tissue: Some aspects of metabolism of gamma-aminobutyric acid, *in Inhibition in the Nervous System and γ-Aminobutyric Acid* (E. Roberts, ed.), p. 144, Pergamon Press, New York (1960).
9. C. F. Baxter and E. Roberts, *in The Neurochemistry of Nucleotides and Amino Acids*, p. 127 (R.O. Brady and D. B. Tower, eds.), John Wiley, New York (1960).
10. E. Roberts, I. P. Lowe, L. Guth, and B. Jelinek, *J. Exptl. Zool.* **138**:313–328 (1958).
11. M. Wollemann and T. Dévényi, *Rev. Agressol.* **IV**:593–595 (1963).
12. G. Della Pietra, G. Illiano, V. Capano, and R. Rava, *Nature* **210**:733–734 (1966).
13. K. F. Tipton, *Biochem. J.* **104**:36P–37P (1967).
14. J. C. Skou, *Biochim. Biophys. Acta* **23**:394–401 (1957).
15. S. L. Bonting, K. A. Simon, and N. M. Hawkins, *Arch. Biochem. Biophys.* **95**:416–423 (1961).
16. H. McIlwain, *Exploration of the brain*, Elsevier, London, (1963).
17. C. Long, *Biochem. J.* **50**:407–415 (1952).
18. C. F. Lange, Jr., and P. Kohn, *J. Biol. Chem.* **236**:1–5 (1961).
19. E. D. Wachsmuth und G. Pfleiderer, *Biochem. Z.* **336**:545–556 (1962/63).
20. D. Eccleston, A. T. B. Moir, H. W. Reading, and J. M. Ritchie, *Brit. J. Pharmacol. Chemotherap.* **28**:367–377 (1966).
21. N. K. Gupta and W. G. Robinson, *J. Biol. Chem.* **235**:1609–1612 (1960).
22. H. L. Young and N. Pace, *Arch. Biochem.* **76**:112–121 (1958).
23. G. E. Boxer and C. E. Shonk, *Cancer Res.* **20**:85–91 (1960).
24. M. U. Tsao, *Arch. Biochem. Biophys.* **90**:234–238 (1960).
25. R. v. Fellenberg, H. Eppenberger, R. Richterich, and H. Aebi, *Biochem. Z.* **336**:334–350 (1962).
26. D. S. Beattie, H. R. Sloan, and R. E. Basford, *J. Cell Biol.* **19**:309–316 (1963).
27. K. D. Martin, L. Z. McFarland, and R. A. Freedland, *Poultry Sci.* **45**:588–594 (1966).
28. M. K. Johnson, *Biochem. J.* **77**:610–618 (1960).
29. R. H. Laatsch, *J. Neurochem.* **9**:487–492 (1962).
30. G. Pfleiderer, *in Hoppe-Seyler/Thierfelder, Handbuch der physiologisch- und pathologisch-chemischen Analyse* (K. Lang and E. Lehnartz, eds.), Vol. 6A, pp. 356–363, Springer-Verlag, Berlin, Göttingham and Heidelberg, New York (1966).
31. A. Delbrück, H. Schimassek, K. Bartsch, und T. Bücher, *Biochem. Z.* **331**:297–311 (1959).
32. A. Meister, *J. Natl. Cancer Inst.* **10**:1263–1271 (1950).
33. O. H. Lowry and J. V. Passonneau, *J. Biol. Chem.* **239**:31–42 (1964).
34. Th. Wieland, G. Pfleiderer, J. Haupt, und W. Wörner, *Biochem. Z.* **332**:1–10 (1959).
35. W. N. Fishbein and S. P. Bessman, *J. Biol. Chem.* **239**:357–361 (1964).
36. M. S. Kanungo and S. N. Singh, *Biochem. Biophys. Res. Commun.* **21**:454–459 (1965).
37. P. G. W. Plagemann, K. F. Gregory, and F. Wroblewski, *J. Biol. Chem.* **235**:2282–2287 (1960).
38. A. Lowenthal, M. van Sande, and D. Karcher, *Ann. N.Y. Acad. Sci.* **94**:988–995 (1961).

39. V. Bonavita and R. Guarneri, *Biochim. Biophys. Acta* **59**:634–642 (1962).
40. M. Battistacci and M. C. Compagnoni, *Atti Soc. Ital. Sci. Vet.* **20**:249–254 (1966).
41. G. Pfleiderer under E. D. Wachsmuth, *Biochem. Z.* **334**:185–198 (1961).
42. M. van Sande, *Ingr. Chimiste* **45**:61–73 (1963).
43. F. N. Syner and M. Goodman, *Nature* **209**:426–428 (1966).
44. V. Bonavita and R. Guarneri, *J. Neurochem.* **10**:743–753 (1963).
45. A. D. Winer, *Biochem. J.* **76**:5P (1960).
46. L. B. Flexner, G. de la Haba, and J. B. Flexner, *J. Neurochem.* **9**:313–320 (1962).
47. J. L. Strominger and O. H. Lowry, *J. Biol. Chem.* **213**:635–646 (1955).
48. N. Robinson, *Clin. Chim. Acta* **11**: 293–297 (1965).
49. E. L. Bennett, D. Krech, N. R. Rosenzweig, H. Karlsson, N. Dye, and A. Ohlander, *J. Neurochem.* **3**:153–160 (1958).
50. A. Brunnemann und H. Coper, *Naunyn-Schmiedebergs Arch. Expth. Pathol. Pharmakol.* **246**:493–503 (1964).
51. O. H. Lowry, *Morphol. Biochem. Correlates Neural Activity.* **1964**:178–191.
52. V. Bonavita and R. Guarneri, *J. Neurochem.* **10**:755–764 (1963).
53. V. Bonavita and R. Guarneri, *Naturwiss.* **50**:597–598 (1963).
54. E. Robins, D. E. Smith, and M. K. Jen, *Progr. Neurobiol* **2**:205–214 (1957).
55. H. Tyler, *Proc. Soc. Exptl. Biol. Med.* **104**:79–83 (1960).
56. R. L. Friede and L. M. Fleming, *Am. J. Anat.* **113**:215–234 (1963).
57. R. L. Friede, L. M. Fleming, and M. Knoller, *J. Neurochem.* **10**:263–277 (1963).
58. N. Robinson and B. M. Phillips, *Biochem. J.* **92**:254–259 (1964).
59. H. J. van der Helm, *J. Neurochem.* **9**:325–327 (1962).
60. W. Gerhardt and C. Petri, *Acta Neurol. Scand.* Suppl. 13, **41**:609 (1965).
61. W. Gerhardt, B. Ovlisen, J. Clausen, and H. Andersen, *Protides Biol. Fluids Proc. Colloq.* **12**:203–206 (1965).
62. L. Arvy, *Intern. Rev. Cytol.* **20**:277–359 (1966).
63. V. Bocci, *Nature* **212**:826–827 (1966).
64. R. Balázs and J. R. Lagnado, *J. Neurochem.* **5**:1–17 (1959).
65. L. G. Abood, E. Brunngraber, and M. Taylor, *J. Biol. Chem.* **234**:1307–1311 (1959).
66. C. J. van den Berg, G. M. J. van Kempen, J. P. Schadé, and H. Veldstra, *J. Neurochem.* **12**:863–869 (1965).
67. F. de Balbian Verster, O. Z. Sellinger, and J. C. Harkin, *J. Cell. Biol.* Suppl., **25**:69–80 (1965).
68. D. Biesold and E. Canzler, *Wiss. Z. Karl-Marx-Univ. Leipzig, Math.-Naturwiss. Reihe* **14**:691–693 (1965) (CA **65**, 2536 e; 1966).
69. S. Tuček, *J. Neurochem.* **13**:1317–1327 (1966).
70. F. Fonnum, *Biochem. J.* **103**:262–270 (1967).
71. S. P. R. Rose, *Biochem. J.* **102**:33–43 (1967).
72. D. A. Rappoport, R. R. Fritz and A. Moraczewski, *Biochim. Biophys. Acta* **74**:42–50 (1963).
73. D. G. Grahame-Smith, *Biochem. J.* **105**:351–360 (1967).
74. M. K. Johnson and V. P. Whittaker, *Biochem. J.* **88**:404–409 (1963).
75. M. Kurokawa, T. Sakamoto, and M. Kato, *Biochem. J.* **97**:833–844 (1965).
76. S. Tuček, *J. Neurochem.* **14**:531–545 (1967).
77. P. D. Swanson, *J. Neurochem.* **14**:343–356 (1967).
78. R. Laverty, J. A. Michaelson, D. F. Sharman, and V. P. Whittaker, *Brit. J. Pharmacol.* **21**: 482–490 (1963).
79. D. R. Dahl, *Dissertation Abstr.* **26**:667–668 (1965) (CA **63**, 16672 e; 1965).
80. V. Bonavita, *in Progress in Brain Research*, (D. P. Purpura and J. P. Schadé, eds.) Vol. 4, pp. 254–272, Elsevier, Amsterdam, London, New York (1964).

81. J. Fischer and J. Jilek, *Acta Univ. Carlinae, Med.* Suppl., 21:195–199 (1965) (CA 65, 14193 a; 1966).
82. R. E. Kuhlmann and O. H. Lowry, *J. Neurochem.* 1:173–180 (1956).
83. E. Robins and I. P. Lowe, *J. Neurochem.* 8:81–95 (1961).
84. L. B. Flexner, J. B. Flexner, R. B. Roberts, and G. de la Haba, *Develop. Biol.* 2:313–328 (1960).
85. J. N. R. Grainger and Y. W. Kunz, *Helgolaender Wiss. Meeresunters.* 14:335–342 (1966) (CA 67, 51567F; 1967).
86. M. Mino, Y. Ueda, K. Fujitani, T. Tsatsuta, and T. Takai, *Sabco J. (Osaka)* 1:28–36 (1965) (CA 65, 17455E; 1966).
87. A. L. Lehninger, H. C. Sudduth and J. B. Wise, *J. Biol. Chem.* 235:2450–2455 (1960).
88. C. B. Klee and L. Sokoloff, *J. Biol. Chem.* 242:3880–3883 (1967).
89. O. Wieland, D. Reinwein, and F. Lynen in *Biochemical Problems of Lipids*, Proc. 2nd Inter. Conf. Gent, 1955, (G. Popjak and E. Le Breton eds.), p. 155, London (1956).
90. R. O. Brady, *J. Biol. Chem.* 235:3099–3103 (1960).
91. F. B. Straub, Äpfelsäuredehydrogenase, in *Hoppe-Seyler/Thierfelder, Handbuch der physiologisch- und pathologisch chemischen Analyse* (K. Lang and E. Lehnartz, eds.), Vol. 6A, pp. 367–377, Springer-Verlag, Berlin, Göttingen, Heidelberg, and New York (1964).
92. V. R. Potter, *J. Biol. Chem.* 165:311–324 (1946).
93. G. H. Fried and S. R. Tipton, *Proc. Soc. Exptl. Biol. Med.* 82:531–532 (1953).
94. K. P. Du Bois, E. M. K. Geiling, A. F. McBride, and J. F. Thomson, *J. Gen. Physiol.* 31:347–359 (1948).
95. E. Schmidt und F. W. Schmidt, *Klin. Wschr.* 1960:957–962.
96. M. K. Johnson, *Biochem. J.* 84:25P–26P (1962).
97. J. B. Solomon, *Biochem. J.* 70:529–535 (1958).
98. H. Knizley Jr., *J. Biol. Chem.* 242:4619–4622 (1967).
99. M. R. V. Murthy and D. A. Rappoport, *Biochim. Biophys. Acta* 74:51–59 (1963).
100. D. Pette, M. Klingenberg and T. Bücher, *Biochem. Biophys. Res. Commun.* 7:425–429 (1962).
101. M. Klingenberg and D. Pette, *Biochem. Biophys. Res. Commun.* 7:430–432 (1962).
102. E. G. Brunngraber, V. Aguilar, and W. G. Occomy, *J. Neurochem.* 10:433–438 (1963).
103. J. L. Conklin and J. Nebel, *J. Histochem. Cytochem.* 13:510–514 (1965).
104. P. V. Vignais and P. M. Vignais, *Biochim. Biophys. Acta* 47: 515–528 (1961).
105. S. Banerjee, D. K. Biswas, and H. D. Singh, *J. Biol. Chem.* 234:405–411 (1959).
106. M. R. V. Murthy and D. A. Rappoport, *Biochim. Biophys. Acta* 74:328–339 (1963).
107. G. Siebert, in *Hoppe-Seyler/Thierfelder, Handbuch der physiologisch und pathologisch chemischen Analyse* (K. Lang and E. Lehnartz, eds.), Vol. 6A, pp. 387–414, Springer-Verlag, Berlin, Göttingen, Heidelberg, and New York (1964).
108. S. S. Hotta, R. H. Laatsch, and P. V. Myron Jr., *J. Neurochem.* 10:841–847 (1963).
109. W. W. Baker and R. W. Newburh, *Biochem. J.* 89:510–515 (1963).
110. R. H. Lee, P. U. Angeletti, and S. G. Caramia, *Growth* 25:393–400 (1961).
111. G. E. Glock and P. McLean, *Biochem. J.* 56:171–175 (1954).
112. K. Yamada and N. Shimazono, *Biochim. Biophys. Acta* 54:205–206 (1961).
113. M. Kai, G. L. White, and J. N. Hawthorne, *Biochem. J.* 101:328–337 (1966).
114. S. Uchida and H. Sugawara, *J. Fac. Sci. Univ. Tokyo*, Sect. IV, 9(3):349–355 (1962) (CA 59, 11992E; 1963).
115. G. Bagdasirian and D. Hulanicka, *Biochim. Biophys. Acta* 99:367–369 (1965).
116. S. Hayman, M. F. Lou, L. O. Merola, and J. H. Kinoshita, *Biochim. Biophys. Acta* 128:474–482 (1966).
117. E. Wenger and B. S. Wenger, *Taxonomic Biochem. Serol.* 1964:391–400 (CA 62, 5644B; 1965).

118. H. Z. Sable, *Biochim. Biophys. Acta* **8**:687–697 (1952).
119. O. H. Lowry, in *Biochemistry of the Developing Nervous System* (H. Waelsch, ed.), pp. 350–357, Academic Press, New York (1955).
120. V. Buell, O. H. Lowry, N. R. Roberts, M. W. Chang, and J. I. Kapphahn, *J. Biol. Chem.* **232**:979–993 (1958).
121. A. M. Burt and B. S. Wenger, *Develop. Biol.* **3**:84–95 (1961).
122. A. M. Burt, *Develop. Biol.* **12**:213–232 (1965).
123. T. Abe, *Arch. Histolog. Jap.* **25**:339–369 (1965).
124. M. Klingenberg, *Pyridinnucleotide und biologische Oxydation*, 11. Coll. of the Gesellschaft für Physiologische Chemie, Mosbach, 1960, pp. 82–106, Springer-Verlag, Berlin, Göttingen, and Heidelberg (1961).
125. R. L. Ringler and T. P. Singer, *Federation Proc.* **17**:297 (1958).
126. Kuo-Huang Ling, Shu-Mei Ting, and Ta-Cheng Tung, *Tai-wan i-hsüe-hui tsa-chih* **57**:232–237 (1958) (CA **53**, 8220e; 1959).
127. Kuo-Huang Ling, Chun-Cheng Tung. and Liang-Tien Chang, *Tai-wan i-hsüe-hui tsa-chih* **57**:342–348 (1958) (CA **53**, 13222 i; 1959).
128. R. L. Ringler, *J. Biol. Chem.* **236**:1192–1198 (1961).
129. R. O. Brady, I. V. Formica, and G. J. Koval, *J. Biol. Chem.* **233**:1072–1076 (1958).
130. R. A. Deitrich, *Biochem. Pharmacol.* **15**:1911–1922 (1966).
131. V. G. Erwin, *Dissertation Abstr.* **26**:6096 (1966).
132. V. G. Erwin and R. A. Deitrich, *J. Biol. Chem.* **241**:3533–3539 (1966).
133. H. A. Rothschild and E. S. G. Barron, *J. Biol. Chem.* **209**:511–523 (1954).
134. R. O. Brady and G. J. Koval, *J. Biol. Chem.* **233**:26 (1958).
135. P. V. Vignais and I. Zabin, *Biochem. Lipids Proc. 5th Intern. Conf.*, Vienna (1958), pp. 78–84 (1960) (CA **55**, 2854h; 1961).
136. G. Mohn, in *Hoppe-Seyler/Thierfelder, Handbuch der physiologisch- und pathologisch-chemischen Analyse* (K. Lang and E. Lehnartz, eds.), Vol. 6A, pp. 574–632, Springer-Verlag, Berlin, Göttingen, Heidelberg, and New York (1964).
137. D. Pette, W. Luh. and T. Bücher, *Biochem. Biophys. Res. Commun.* **7**:419–424 (1962).
138. G. Siebert, K.-H. Bässler, R. Hannover, E. Adloff, and R. Beyer, *Biochem. Z.* **334**:388–400 (1961).
139. R. T. Schimke, *Federation Proc.*, **16**:334 (1957).
140. K. Nakamura and F. Bernheim, *Jap. J. Pharmacol.* **11**:37–45 (1961).
141. F. N. Pitts, Jr., and E. Robins, *Federation Proc.* **24**:350 (1965).
142. F. N. Pitts Jr. and C. Quick, *J. Neurochem.* **12**:893–900 (1965).
143. R. W. Albers and G. J. Koval, *Biochim. Biophys. Acta* **52**:29–35 (1961).
144. L. J. Embree and R. W. Albers, *Biochem. Pharmacol.* **13**:1209–1217 (1964).
145. Z. W. Hall and E. A. Kravitz, *J. Neurochem.* **14**:55–61 (1967).
146. A. A. L. Miller and F. N. Pitts Jr., *J. Neurochem.* **14**:579–584 (1967).
147. L. Salganicoff and E. de Robertis, *J. Neurochem.* **12**:287–309 (1965).
148. F. N. Pitts, Jr., and C. Quick, *J. Neurochem.* **14**:561–570 (1967).
149. M. Mitolo and A. Loizzi, *Boll. Soc. Ital. Biol. Sper.* **32**:191 (1956).
150. U. A. S. Al-Khalidi and T. H. Chaglassian, *Biochem. J.* **97**:318–320 (1965).
151. F. H. Bruns, in *Hoppe-Seyler/Thierfelder, Handbuch der physiologisch- und pathologisch-chemischen Analyse* (K. Lang and E. Lehnartz, eds.) Vol. 6C, pp. 481–498, Springer-Verlag, Berlin, Göttingen, Heidelberg, and New York (1966).
152. R. A. Peters and R. W. Wakelin, *J. Physiol.* (London) **119**:421–427 (1953).
153. C. Long and R. A. Peters, *Biochem. J.* **33**:759–773 (1939).
154. E. Ch. Slater, in *Hoppe-Seyler/Thierfelder, Handbuch der physiologische- und pathologisch-chemischen Analyse* (K. Lang and E. Lehnartz, eds.), Vol. 6A, pp. 633–637, Springer-Verlag, Berlin, Göttingen, Heidelberg, and New York (1964).

155. T. Oda, S. Seki, T. Shibata, A. Sakai, and H. Okazaki, *Okayama Igakkai Zasshi* **70**:123–125 (1958) (CA **53**, 495a; 1959).
156. P. Cerletti, R. Strom, M. G. Giordano, F. Balestrero and M. A. Giovenco, *Biochem. Biophys. Res. Commun.* **14**:408–412 (1964).
157. N. Shimizu and N. Morikawa, *J. Histochem. Cytochem.* **5**:334–345 (1957).
158. H. B. Tewari and G. H. Bourne, *J. Histochem. Cytochem.* **10**:619–627 (1962).
159. E. I. Ilína-Kakueva, *Dokl. Akad. Nauk. SSSR* **152**:1267–1269 (1963) (CA **60**, 4549f; (1964).
160. J. D. Ifft, W. F. McNary, Jr., and L. Simoneit, *Proc. Soc. Exptl. Biol. Med.* **117**:170–171 (1964).
161. W. B. Quay, *Physiol. Zool.* **33**:206–212 (1960).
162. T. Kumamoto and G. H. Bourne, *Nature* **204**:295–296 (1964).
163. Z. D. Pigareva and D. A. Chetverikov, *Dokl. Akad. Nauk. SSSR* **78**:393–396 (1951) (CA **45**, 10272g; 1951).
164. M. M. Busnyuk, *Zh. Vysshei Nervnoi Deyatel'nosti im. I. P. Pavlova* **13**:731–739 (1963) (CA **60**, 2125e; 1964).
165. E. L. Dovedova and Z. D. Pigareva, *Vopr. Med. Khim.* **10**:370–376 (1964) (CA **61**, 15115g; 1964).
166. A. S. V. Burgen and L. M. Chipman, *J. Physiol.* (*London*) **114**:296–305 (1951).
167. P. A. Kometiani and E. E. Klein, *Ukrain. Biokhim. Zhur.* **22**:410–419 (1950) (CA **48**, 2791f; 1954).
168. H. Masai, T. Kusunoki, and H. Ishibashi, *Experientia* **22**:745–746 (1966).
169. W. Grimmer, *Z. Anat. Entwicklungsgesch.* **122**:141–140 (1961).
170. R. L. Friede and L. M. Fleming, *J. Neurochem.* **9**:179–198 (1962).
171. W. N. Aldridge and M. K. Johnson, *Biochem. J.* **73**:270–276 (1959).
172. G. Rodriguez de Lores Arnaiz, and E. de Robertis, *J. Neurochem.* **2**:503–509 (1962).
173. S. S. Parmar, M. C. Sutter, and M. Nickerson, *Can. J. Biochem. Physiol.* **39**:1335–1345 (1961).
174. E. de Robertis, A. Pellegrino de Iraldi, G. Rodriguez de Lores Arnaiz, and L. Salganicoff, *J. Neurochem.* **9**:23–35 (1962).
175. E. de Robertis, G. Rodriguez de Lores Arnaiz, L. Salganicoff, A. Pellegrino de Iraldi, and L. M. Zieher, *J. Neurochem.* **10**:225–235 (1963).
176. E. de Robertis, L. Salganicoff, L. M. Zieher, and G. Rodriguez de Lores Arnaiz, *Science* **140**:300–301 (1963).
177. E. de Robertis, *in Progress in Brain Research* (H. E. Himwich and W. A. Himwich, eds.), Vol. 8, pp. 118–136, Elsevier, Amsterdam, London, and New York (1964).
178. H. Koenig, D. Gaines, Th. McDonald, R. Gray, and J. Scott, *J. Neurochem.* **11**:729–743 (1964).
179. V. P. Whittaker, *in Progress in Brain Research* (H. W. Himwich and W. A. Himwich, eds.), Vol. 8, pp. 90–117, Elsevier, Amsterdam, London, and New York (1964).
180. A. Schwartz, H. S. Bachelard, and H. McIlwain, *Biochem. J.* **84**:626–637 (1962).
181. H. S. Bachelard, *Biochem. J.* **104**:286–292 (1967).
182. Z. D. Pigareva, N. N. Bogolepov, G. P. Gulidova, and E. L. Dovedova, *Mitokhondrii, Strukt. Funkts. Mater. Simp. Moscow*, 1965, pp. 25–27 (1966) (CA **66**, 16672; 1967).
183. R. L. Friede and R. A. Pax, *Histochemie* **2**:186–191 (1961).
184. Van R. Potter, W. C. Schneider, and G. J. Liebl, *Cancer Res.* **5**:21–24 (1945).
185. M. Hamburgh and L. B. Flexner, *J. Neurochem.* **1**: 279–288 (1957).
186. E. A. Sazonova, *Uch. Zap. Khar'kovsk. Gos. Univ.* 108, *Tr. Nauchn.-Issled. Inst. Biol.*; *Biol. Fak.* (29) 287–291 (1960) (CA **60**, 4553b; 1964).
187. M. Wawrczyniak, *Folia Histochem. Cytochem.* (*Kracow*) **1**:503–533 (1963) (CA **62**, 8187c; 1965).

188. L. B. Flexner, E. L. Belknap, Jr., and J. B. Flexner, *J. Cellular Comp. Physiol.* **42**:151–161 (1953).
189. S. Cassin and C. S. Herron, *Am. J. Physiol.* **201**:440–442 (1961).
190. L. B. Flexner and J. B. Flexner, *J. Cellular Comp. Physiol.* **27**:35–42 (1946).
191. Z. D. Pigareva and D. A. Chetverikov, *Dokl. Akad. Nauk. SSSR*, **78**:169–172 (1951) (CA **45**, 9150a; 1951).
192. Z. D. Pigareva and D. A. Chetverikov, *Biokhimiya* **15**:517–522 (1950) (CA **45**, 3440a; 1951).
193. R. J. Grabske, *Dissertation Abstr.* **25**:6032–6033 (1965).
194. V. P. Whittaker, *Biochem. J.* **72**:694–706 (1959).
195. K. A. C. Elliott and M. E. Greig, *Biochem. J.* **32**:1407–1423 (1938).
196. H. Hydén, S. Lovtrup and A. Pigon, *J. Neurochem.* **2**:304–311 (1958).
197. N. Weiner, *J. Neurochem.* **6**:79–86 (1960).
198. S. Kumar, K. K. Tewari, and P. S. Krishnan, *Biochem. J.* **95**: 797–802 (1965).
199. N. Chalazonitis and M. Otsuka, *Compt. Rend.* **243**:978–980 (1956).
200. V. V. Portugalov, Z. D. Pigareva, M. M. Busnyuk, E. L. Dovedova, and E. I. Ilina, *Tret'ya Vses. Konf. Biokhim Nervnoi Systemy, Akad. Nauk. Arm. SSR Konf. Biokhim., Sb. Dokl. Erevan, 1962,* 297–309 (1963) (CA **60**, 12466f; 1964).
201. A. Bonichon, *Bull. Soc. Lorraine Sci.* **2**:1–107 (1962) (CA **59**, 15649b; 1963).
202. B. Th. Cole, *Proc. Soc. Exptl. Biol. Med.* **93**:290–294 (1956).
203. P. M. Vignais, G. H. Gallagher, and I. Zabin, *J. Neurochem.* **2**:283–287 (1958).
204. D. S. Beattie and R. E. Basford, *J. Neurochem.* **12**:103–111 (1965).
205. D. S. Beattie and R. E. Basford, *J. Biol. Chem.* **241**:1412–1418 (1966).
206. K.-H. Bässler, *in Hoppe-Seyler/Thierfelder, Handbuch der physiologisch und pathologisch chemischen Analyse* (K. Lang and E. Lehnartz, eds.), Vol. 6A, pp. 638–653, Springer-Verlag, Berlin, Göttingen, Heidelberg, and New York (1964).
207. K. Uemura, *Seikagaku* **35**:880–892 (1963) (CA **61**, 3505f; 1964).
208. N. E. Lofrumento, G. de Gregorio, S. Papa, C. Serra, and E. Quagliariello, *Boll. Soc. Ital. Biol. Sper.* **40**:1452–1455 (1964).
209. R. Balázs, *Biochem. J.* **95**:497–508 (1965).
210. E. E. Klein, *Soobshcheniya Akad. Nauk. Gruzin. SSR* **13**:273–280 (1952) (CA **48**, 5234e; 1954).
211. R. Balázs, D. Dahl, and J. R. Harwood, *J. Neurochem.* **13**:897–905 (1966).
212. R. L. Jolley and D. H. Labby, *Arch. Biochem. Biophys.* **90**:122–124 (1960).
213. J. R. Tata, *Biochim. Biophys. Acta* **28**:95–99 (1958).
214. K. Yamamoto, S. Shimizu, and I. Ishikawa, *Japan J. Physiol.* **10**:594–601 (1960).
215. A. D. D. Rose, *Boll. Soc. Ital. Biol. Sper.* **22**:448–449 (1946).
216. F. Salvatore, V. Zappia, and C. Costa, *Comp. Biochem. Physiol.* **16**:303–309 (1965).
217. F. Salvatore and S. Papa, *Ital. J. Biochem.* **8**:231–247 (1959).
218. H. B. Burch, O. H. Lowry, A. M. Padilla, and A. M. Combs, *J. Biol. Chem.* **223**:29–45 (1956).
219. F. Friedberg and L. M. Marshall, *Biochim. Biophys. Acta* **10**:624–625 (1953).
220. B. Schepartz, *Biochim. Biophys. Acta* **53**:602–603 (1961).
221. J. H. Quastel and A. H. M. Wheatley, *Biochem. J.* **26**:725–744 (1932).
222. H. Weil-Malherbe, *Biochem. J.* **30**:665–676 (1936).
223. M. M. Cohen, G. R. Simon, J. F. Berry, and E. B. Chain, *Biochem. J.* **84**:43P–44P (1962).
224. F. Friedberg, *Biochim. Biophys. Acta* **11**:308–309 (1953).
225. Fan-shen Li, Sjun Sjui, *Seng Wu Hua Hsueh Yu Sheng Wu Wu Li Hsueh Pao* **4**:139–144 (1964) (CA **65**, 15873; 1966).
226. B. Schepartz and M. Turczyn, *J. Neurochem.* **10**:825–829 (1963).
227. S. Edlbacher und O. Wiss, *Helv. Chim. Acta* **27**:1060–1073 (1944).

228. S. Nakano, *Bitamin* (*Kyoto*) **22**:306–312 (1961).
229. D. B. Goldstein, *J. Neurochem.* **13**:1011–1016 (1966).
230. J. T. Dunn and G. T. Perkoff, *Biochim. Biophys. Acta* **73**:327–331 (1963).
231. M. Hori, *Seishin. Shinkeigaku Zasshi* **67**:548–553 (1965) (CA **65**, 4173b; 1966).
232. A. H. Neims, W. D. Zieverink, and J. D. Smilack, *J. Neurochem.* **13**:163–168 (1966).
233. C. E. M. Pugh and J. H. Quastel, *Biochem. J.* **31**:286–291 (1937).
234. A. Sjoerdsma, T. E. Smith, T. D. Stevenson, and S. Udenfriend, *Proc. Soc. Exp. Biol.* (*N.Y.*) **89**:36–38 (1955).
235. T. L. Sourkes and E. Townsend, *Congr. Intern. Biochim. Résumés Communs.* 3ᵉ Congr. *Brussels, 1956*, p. 56 (CA **51**, 8852c; 1957).
236. G. Milhaud and J. Glowinski, *Compt. Rend.* **256**:1033–1035 (1963).
237. A. Zolowick, R. Pearse, K. W. Boehlke, and B. E. Eleitheriou, *Science* **154**:649 (1966).
238. M. M. Salseduc, I. J. Jofre, and J. A. Izquierdo, *Med. Pharmacol. Exptl.* **14**:113–119 (1966).
239. G. T. Pryor, K. Schlesinger, und W. H. Calhoun, *Life Sci.* **5**:2105–2111 (1966).
240. T. Nukada, T. Sakurai, and R. Imaizumi, *Japanese J. Pharmacol.* **13**:124 (1963).
241. S. Udenfriend, E. D. Titus, H. Weissbach, and R. E. Peterson, *J. Biol. Chem.* **219**:335–344 (1956).
242. K. Baghvat, H. Blaschko, and D. Richter, *Biochem. J.* **33**:1338–1341 (1939).
243. H. Blaschko, *Pharmacol. Rev.* **4**:415:458 (1952).
244. A. Pletscher, K. F. Gey, and P. Zeller, in *Progress in Drug Research* (E. Jucker, ed.), Vol. 2, pp. 417–590, Birkhäuser-Verlag, Basel, Stuttgart, (1960).
245. L. S. Seiden and J. Westley, *Biochim. Biophys. Acta* **58**: 363–364 (1962).
246. G. Brawerman and E. Chargaff, *Biochim. Biophys. Acta* **16**:524–532 (1955).
247. C. E. M. Pugh and J. H. Quastel, *Biochem. J.* **31**:2306–2321 (1937).
248. N. Weiner, *Arch. Biochem. Biophys.* **91**:182–188 (1960).
249. T. Wakizaka, *Nichidai Igaku Zhassbi* **17**:669–679 (1958) (CA **55**, 21316b; 1961).
250. T. Nagatsu, *J. Biochem.* (*Tokyo*) **59**:606–612 (1966).
251. R. E. McCaman, M. W. McCaman, J. M. Hunt, and M. S. Smith, *J. Neurochem.* **12**:15–23 (1965).
252. D. F. Bogdanski, H. Weissbach, and S. Udenfriend, *J. Neurochem.* **1**:272–278 (1957).
253. H. Birkhäuser, *Schweiz. Med. Wschr.* **71**:750–752 (1941).
254. M. H. Aprison, R. Takahashi and T. L. Folkerth, *J. Neurochem.* **11**:341–350 (1964).
255. W. P. Burkhard, K. F. Gey and A. Pletscher, *J. Neurochem.* **10**:183–186 (1963).
256. R. E. McCaman, G. Rodriguez de Lores Arnaiz, and E. de Robertis, *J. Neurochem.* **12**:927–935 (1965).
257. M. Alberici, G. Rodriguez de Lores Arnaiz, and E. de Robertis, *Life Sci.* **4**:1951–1960 (1965).
258. G. Rodriguez de Lores Arnaiz, M. Alberici, and E. de Robertis, *J. Neurochem.* **14**:215–225 (1967).
259. B. B. Bourne, *Life Sci.* **4**:583–591 (1965).
260. V. T. Nachmias, *J. Neurochem.* **6**:99–104 (1960).
261. M. Karki, R. Kuntzman and B. B. Brodie, *J. Neurochem.* **9**:53–58 (1962).
262. St. Kurzepa and J. Bojanek, *Biol. Neonat.* **8**:216–221 (1965).
263. D. S. Bennet and N. J. Giarman, *J. Neurochem.* **12**:911–918 (1965).
264. A. J. Prange Jr., J. E. White, M. A. Lipton, and A. M. Kinkead, *Life Sci.* **6**:581 (1967).
265. A. Tissari, *Acta Physiol. Scand. Suppl.*, **265**:80 (1966).
266. H. Birkhäuser, *Helv. Chim. Acta* **23**:1071–1086 (1940).
267. P. C. Baker, *Develop. Biol.* **14**:267–277 (1966).
268. M. H. Aprison, in *Progress in Brain Research* (W. A. Himwich and J. P. Schadé, eds.), Vol. 16, pp. 48–80, Elsevier, Amsterdam, London, and New York (1965).

269. Y. Morino, H. Wada, T. Morisue, Y. Sakamoto, and K. Ichihara, *J. Biochem. (Tokyo)* **48**: 18–27 (1960).
270. T. Morisue, Y. Morino, Y. Sakamoto, and K. Ichihara, *J. Biochem. (Tokyo)* **48**:28–36 (1960).
271. E. C. Cotzias and V. P. Dole, *J. Biol. Chem.* **196**:235–242 (1952).
272. E. Maggio, *Boll. Soc. Ital. Biol. Sper.* **27**:789–790 (1951).
273. T. Koeda, *Nippon Yakurigaku Zasshi* **56**:1093–1102 (1960) (CA **55**, 24397g; 1961).
274. E. A. Zeller, H. Birkhäuser, H. Mislin, and M. Wenk, *Helv. Chim. Acta* **22**:1381–1395 (1939).
275. S. E. Lindell and H. Westling, in *Heffter-Heubner Handbuch der experimentellen Pharmakologie* (O. Eichler, O. Farah, H. Herken, and A. D. Welch, eds.), Vol. 18, pp. 734–788, Springer Verlag, Berlin, Göttingen, Heidelberg, and New York (1966).
276. F. Buffoni, *Pharmacol. Rev.* **18**:1163–1199 (1966).
277. W. P. Burkard, K. F. Gey, and A. Pletscher, *C.R. Suisse Physiol.* **58**:61–62 (1961).
278. T. Ito and M. Kawaai, *Nippon Univ. J. Med.* **1**:185–190 (1959) (CA **56**, 6505h; 1962).
279. A. Meister, A. N. Radhakrishnan, and S. D. Buckley, *J. Biol. Chem.* **229**:789–800 (1957).
280. H. Herken and R. Timmler, *Naunyn-Schmiedeberg's Arch. Exptl. Pathol. Pharmacol.* **250**: 293–303 (1965).
281. A. Inouye and Y. Shinagawa, *J. Neurochem.* **12**:803–813 (1965).
282. A. Inouye and Y. Shinagawa, *J. Neurochem.* **13**:385–396 (1966).
283. H. McIlwain, H. Martin and M. A. Tresize, *Biochem. J.* **65**:2P (1957).
284. H. McIlwain and M. A. Tresize, *Biochem. J.* **65**:288–296 (1957).
285. C. Vesco and A. Giuditta, *Biochim. Biophys. Acta* **113**:197–215 (1966).
286. A. Giuditta and H. J. Strecker, *Biochim. Biophys. Acta* **67**:316–318 (1963).
287. S. A. Millard, A. Kubose, and E. M. Gál, *J. Neurochem.* **14**:847–850 (1967).
288. R. L. Friede, *J. Neurochem.* **8**:17–30 (1961).
289. M. Banay-Schwartz and H. J. Strecker, *Biochem. Biophys. Res. Commun.* **8**:66–71 (1962).
290. A. Giuditta and E. Aloj, *J. Neurochem.* **12**:567–579 (1965).
291. M. R. V. Murthy and D. A. Rappoport, *Biochim. Biophys. Acta* **78**:71–76 (1963).
292. A. Giuditta and H. J. Strecker, *J. Neurochem.* **5**:50–61 (1959).
293. S. Englard and H. J. Strecker, *Federation Proc.* **15**:248 (1956).
294. W. Levine, A. Giuditta, S. Englard, and H. J. Strecker, *J. Neurochem.* **6**:28–36 (1960).
295. A. Giuditta and H. J. Strecker, *Biochim. Biophys. Acta* **48**:10–19 (1961).
296. E. Harper and H. J. Strecker, *J. Neurochem.* **9**:125–134 (1962).
297. R. Hess and A. G. E. Pearse, *Biochim. Biophys. Acta* **71**:285–294 (1963).
298. T. M. Brody, R. I. H. Wang, and J. A. Bain, *J. Biol. Chem.* **198**:821–826 (1952).
299. S. Lovtrup, *J. Neurochem.* **11**:377–386 (1964).
300. H. J. Strecker and G. di Prisco, *Acta Chem. Scand.* Suppl. 1, **17**:567–573 (1963).
301. E. Aloj and A. Giuditta, *J. Neurochem.* **14**:955–965 (1967).
302. A. Giuditta and E. Aloj, *J. Neurochem.* **14**:967–975 (1967).
303. D. Ruccia, *Boll. Soc. Ital. Biol. Sper.* **34**:68–72 (1958).
304. C. H. Gallagher and S. H. Buttery, *Biochem. J.* **72**:575–582 (1959).
305. C. F. Mills and R. B. Williams, *Biochem. J.* **85**:629–632 (1962).
306. N. A. Verzhbinskaya, *Fiziol. Zhur. SSSR* **39**:17–26 (1953) (CA **47**, 5032g; 1953).
307. Y. Takahashi, S. Shirayose, K. Nagano, and H. Sato, *Koso Kagaku Shinpojiumu* **13**:296–305 (1958) (CA **55**, 1734i; 1961).
308. P. Parsons and R. E. Basford, *J. Neurochem.* **14**:823–840 (1967).
309. A. Pope, H. H. Hess, J. R. Ware, and R. H. Thomson, *J. Neurophysiol.* **19**:259–270 (1956).
310. J. W. Ridge, *Biochem. J.* **102**:612–617 (1967).
311. Z. D. Pigareva, in *Problems of the Biochemistry of the Nervous System* (A. V. Palladin, ed.), pp. 186–195, Pergamon Press, New York (1964).
312. A. J. Tolani and G. P. Talwar, *Biochem. J.* **88**:357–362 (1963).

313. A. Pope, *Arch. Neurol.* **16**:351–356 (1967).
314. E. Makovskii, *Ukrain. Biokhim. Zhur.* **30**:18–26 (1958).
315. G. M. J. van Kempen, C. J. van der Berg, H. J. van der Helm, and H. Veldstra, *J. Neurochem.* **12**:581–588 (1965).
316. Y. Yamada, *Med. J. Osaka Univ.* **11**:383–400 (1961).
317. J. B. Flexner, L. B. Flexner, and W. L. Straus Jr., *J. Cellular Comp. Physiol.* **18**:355–368 (1941).
318. C. van der Wende, *Arch. Intern. Pharmacodyn.* **152**:433–444 (1964).
319. P. Weiss, *Proc. Soc. Exptl. Biol. Med.* **48**:343–346 (1941).
320. C. D. Marsden, *J. Anat.* **99**:175–179 (1965).
321. K. A. C. Elliott, *Biochem. J.* **28**:1911–1919 (1934).
322. B. Anagnoste and M. Goldstein, *Life Sci.* **6**:1535–1540 (1967).
323. S. Udenfriend and C. R. Creveling, *J. Neurochem.* **4**:350–352 (1959).
324. M. Goldstein, B. Anagnoste, W. S. Owen, and A. F. Battista, *Experientia* **23**:98–99 (1967).
325. T. Itoh, M. Matsuoka, K. Nakazima, K. Tagawa, and R. Imaizumi, *Jap. J. Pharmacol.* **12**:130–136 (1962).
326. N. T. Iyer, P. L. McGeer, and E. G. McGeer, *Can. J. Biochem. Physiol.* **41**:1565–1570 (1963).
327. T. Nagatsu, M. Levitt, and S. Udenfriend, *J. Biol. Chem.* **239**:2910–2917 (1964).
328. W. v. Studnitz, *Clin. Chim. Acta* **12**:597–599 (1965).
329. E. G. McGeer, S. Gibson, and P. L. McGeer, *Can. J. Biochem. Physiol.* **45**:1557–1563 (1967).
330. M. Ikeda, M. Levitt, and S. Udenfriend, *Biochem. Biophys. Res. Commun.* **18**:482–488 (1965).
331. S. Udenfriend, *Pharmacol. Rev.* **18**:43–51 (1966).
332. E. G. McGeer, G. M. Ling, and P. L. McGeer, *Biochem. Biophys. Res. Commun.* **13**:291–296 (1963).
333. D. T. Matsuory, H. F. Schott, and L. Petriello, *J. Pharmacol. Exptl. Therap.* **139**:73–76 (1963).
334. P. L. McGeer, S. P. Bagchi, and E. G. McGeer, *Life Sci.* **4**:1859–1867 (1965).
335. E. M. Gál, M. Morgan, S. K. Chatterjee, and F. D. Marshall Jr., *Biochem. Pharmacol.* **13**:1639–1653 (1964).
336. L. J. Weber, *Dissertation Abstr.* **25**:3026 (1964).
337. E. M. Gál, M. Morgan, and F. D. Marshall, Jr., *Life Sci.* **4**:1765–1772 (1965).
338. S. Consolo, S. Garattini, R. Ghielmetti, P. Morselli, and L. Valzelli, *Life Sci.* **4**:625–630 (1964).
339. E. M. Gál, J. C. Armstrong, and B. Ginsberg, *J. Neurochem.* **13**:643–654 (1966).
340. W. Lovenberg, E. Jequier, and A. Sjoerdsma, *Science* **155**:217–219 (1967).
341. E. M. Gál and F. D. Marshall, Jr., *in Progress in Brain Research* (H. E. Himwich and W. A. Himwich, eds.), Vol. 8, pp. 56–60, Elsevier, Amsterdam, London, and New York (1964).
342. H. Green and J. L. Sawyer, *Anal. Biochem.* **15**:53–64 (1966).
343. S. Nakamura, A. Ichiyama, and O. Hayaishi, *Federation Proc.* **24**:604 (1965).
344. D. G. Grahame-Smith and L. Moloney, *Biochem. J.* **96**:66P (1965).
345. E. M. Gál, *Federation Proc.* **24**:580 (1965).
346. D. G. Grahame-Smith, *Biochem. J.* **92**:52P (1964).
347. J. F. Mead and G. M. Levis, *Biochem. Biophys. Res. Commun.* **9**:231–234 (1962).
348. K. Inui, *Osaka Dalgaku Igaku Zhassi* **11**:681–685 (1959) (CA **53**, 15167h; 1959).
349. J. Axelrod, W. B. Quay, P. C. Baker, *Nature* **208**:386 (1965).
350. J. Axelrod, P. D. McLean, R. W. Albers, and H. Weissbach, *in Regional Neurochemistry*, Proc. 4th. Intern. Neurochem. Symp., Varenna, 1960 (S. S. Kety and J. Elkes, eds.), pp. 307–311, Pergamon Press, Oxford, London, New York, and Paris, (1961).

351. P. C. Baker, W. B. Quay, and J. Axelrod, *Life Sci.* 4:1981–1987 (1965).
352. J. Axelrod, *Science* 127:754–755 (1958).
353. R. E. McCaman, *Life Sci.* 4:2353–2359 (1965).
354. J. Axelrod, W. Albers, and C. D. Clemente, *J. Neurochem.* 5:68–72 (1959).
355. E. de Robertis, *Pharmacol. Rev.* 18:413–424 (1966).
356. D. D. Brown, R. Tomchick, and J. Axelrod, *J. Biol. Chem.* 234:2948–2950 (1959).
357. G. Valette, Y. Cohen, and W. Burkard, *Pharm. Acta Helv.* 31:282–390 (1956).
358. T. White, *J. Physiol.* (*London*) 149:34–42 (1959).
359. T. White, *J. Physiol.* (*London*) 152:299–308 (1960).
360. T. White, *Federation Proc.* 23:1103–1106 (1964).
361. A. Gustafsson and G. P. Forshell, *Acta Chem. Scand.* 18:2098–2102 (1964).
362. J. Axelrod, *J. Biol. Chem.* 237:1657–1660 (1962).
363. P. L. McGeer and E. G. McGeer, *Biochem. Biophys. Res. Commun.* 17:502–507 (1964).
364. J. Bremer and D. M. Greenberg, *Biochim. Biophys. Acta* 46:205–216 (1961).
365. P. R. Srinivasan and E. Borek, *Proc. Natl. Acad. Sci., U.S.* 49:529–533 (1963).
366. L. N. Simon, A. J. Glasky, and T. H. Rajal, *Biochim. Biophys. Acta* 142:99–104 (1967).
367. K. Sindelarova, V. Neumann, and Z. Svobodova, *Sb. Vysoke Skoly Zemedel. Brne, Rada B* 13:117–124 (1965) (CA 64, 14484C; 1966).
368. H. Reichard, *J. Lab. Clin. Med.* 56:218–221 (1960).
369. T. Kita and H. Kamiya, *Chem. Pharm. Bull.* (*Tokyo*) 10:1065–1070 (1962).
370. J. J. Pisano, D. Abraham, and S. Udenfriend, *Arch. Biochem. Biophys.* 100:323–329 (1963).
371. M. Bryn, *J. Nutrit.* 78:179–183 (1962)
372. F. v. Bruchhausen, *Naunyn-Schmiedebergs Arch. Exptl. Pathol. Pharmakol.* 246:330–337 (1964).
373. P. M. Dreyfus, *J. Neurpath. Exptl. Neurol.* 24:119–129 (1965).
374. M. Ciman and F. Olivo, *Giorn. Biochim.* 13:313–318 (1964) (CA 62, 8281E; 1965).
375. T. N. Pattabiraman and B. K. Bachhawat, *Biochim. Biophys. Acta* 59:681–689 (1962).
376. P. J. G. Mann, M. Tennenbaum, and J. H. Quastel, *Biochem. J.* 32:243–261 (1938).
377. D. Nachmansohn and A. L. Machado, *J. Neurophysiol.* 6:397–403 (1943).
378. R. Balbi, *Boll. Soc. Ital. Biol. Sper.* 18:285–287 (1943).
379. S. Okinaka, F. Nakata, I. Takeuchi, K. Nakao, M. Yoshikawa, K. Shizume, and H. Ibayashi, *Igaku To Seibutsugaku* (*Med. Biol.*) 22:97–100 (1952) (CA 46, 10362h; 1952).
380. P. J. G. Mann, M. Tennenbaum, and J. H. Quastel, *Biochem. J.* 33:822–835 (1939).
381. S. Yoshie and S. Kozawa, *Folia Pharmacol. Japan.* 48:10–11 (1952) (CA 48, 7150a; 1954).
382. E. Stedman and E. Stedman, *Biochem. J.* 31:817–827 (1937).
383. E. R. Trethewie, *Australian J. Exptl. Biol. Med. Sci.* 16:343–346 (1938).
384. M. Cohen, *Arch. Biochem. Biophys.* 60:284–296 (1956).
385. B. B. Dikshit, *Proc. Soc. Biol. Chemists, India* 3:63 (1938).
386. D. Nachmansohn and M. Berman, *J. Biol. Chem.* 165:551–563 (1946).
387. N. A. Verzhbinskaya and N. L. Leibson, *Izv. Akad. Nauk. SSSR, Ser. Biol.* 31:750 (1966).
388. C. O. Hebb, *Physiol. Rev.* 37:196–220 (1957).
389. D. Nachmansohn and H. M. John, *J. Biol. Chem.* 158:157–171 (1945).
390. B. K. Schriever and L. Shuster, *J. Neurochem.* 14:977–985 (1967).
391. H. Kumagai and S. Ebashi, *Nature* 173:871–872 (1954).
392. G. Rentsch, *Naturwissenschaften* 48:308 (1961).
393. S. Tuček, *Biochim. Biophys. Acta* 117:278–280 (1966).
394. R. W. Morris, *Biochim. Biophys. Acta* 73:511–513 (1963).
395. S. E. Severin and V. Artenie, *Biokhimiya* 32:125–132 (1967).
396. R. Berman, I. B. Wilson, and D. Nachmansohn, *Biochim. Biophys. Acta* 12:315–324 (1953).

397. R. Berman-Reisberg, *Biochim. Biophys. Acta* **14**:438–440 (1954).
398. M. A. Lipton and E. S. Guzman-Barron, *J. Biol. Chem.* **166**:367–380 (1946).
399. W. E. Balfour, *J. Physiol.* (*London*) **129**:81P–82P (1955).
400. M. Wollemann and G. Feuer, *Proc. Intern. Symp. Enzyme Chem. Tokyo Kyoto, 1957* **2**: 191–198 (1958) (CA **53**, 17208e; 1959).
401. E. T. Browning, *Dissertation Abstr.* **B27**:2271 (1967).
402. S. Korkes, A. del Campillo, S. Korey, J. R. Stern, D. Nachmansohn, and S. Ochoa, *J. Biol. Chem.* **198**:215–220 (1952).
403. A. S. V. Burgen, G. Burke, and M.-L. Desbarats-schonbaum, *Brit. J. Pharmacol.* **11**:308–312 (1956).
404. W. Feldberg and T. Mann, *J. Physiol.* (*London*) **104**:411–425 (1946).
405. H. McLennan and K. A. C. Elliott, *Am. J. Physiol.* **163**:605–613 (1950).
406. E. A. Hosein and J. M. Smoly, *Arch. Biochem. Biophys.* **114**:102–107 (1966).
407. S. Tuček, *J. Neurochem.* **13**:1329–1332 (1966).
408. G. Bull, A. Feinstein, and D. Morris, *Nature* **201**:1326 (1964).
409. S. R. Korey, B. de Braganza, and D. Nachmansohn, *J. Biol. Chem.* **189**:705–715 (1951).
410. R. B. Reisberg, *Biochim. Biophys. Acta* **14**:442–443 (1954).
411. A. S. Milton, *J. Physiol.* (*London*) **142**:25P (1958).
412. C. O. Hebb and D. Ratkovic, *in Comparative Neurochemistry* (D. Richter, ed.), Proc. 5th. Intern. Neurochem. Symp., St. Wolfgang, 1962, pp. 347–354, Pergamon Press, Oxford, London, New York, and Paris (1964).
413. D. Nachmansohn and I. B. Wilson, *Advan. Enzymol.* **12**:259–339 (1951).
414. C. O. Hebb, *in Heffter-Heubner, Handbook of Experimental Pharmacology* (G. B. Koelle, ed.), Suppl. XV, pp. 55–88, Springer-Verlag, Berlin, Göttingen, Heidelberg, and New York (1963).
415. F. C. McIntosh, *Can. J. Biochem. Physiol.* **41**:2555–2571 (1963).
416. F. Nakata, *Nisshin Igaku* **39**:493–497 (1952) (CA **48**, 8365e; 1954).
417. A. M. Goldberg and R. E. McCaman, *Life Sci.* **6**:1493–1500 (1967).
418. C. O. Hebb and J. Silver, *J. Physiol.* (*London*) **134**:718–728 (1956).
419. R. E. McCaman and M. H. Aprison, *in Progress in Brain Research* (W. A. Himwich and H. E. Himwich, eds.), Vol. 9, pp. 220–233, Elsevier, Amsterdam, London, and New York (1964).
420. G. Zetler and L. Schlosser, *Naunyn-Schmiedeberg's Arch. Exptl. Pathol. Pharmakol.* **224**: 159–175 (1955).
421. C. O. Hebb and V. P. Whittaker, *J. Physiol.* (*London*) **142**:187–196 (1958).
422. R. E. McCaman, M. W. McCaman, and M. L. Stafford, *J. Biol. Chem.* **241**:930–934 (1966).
423. D. Bellamy, *Biochem. J.* **72**:165–168 (1959).
424. V. P. Whittaker, *in Progress in Brain Research* (H. E. Himwich and W. A. Himwich, eds.), Vol. 8, pp. 90–117 (Elsevier, Amsterdam, London, and New York (1964).
425. C. O. Hebb, *J. Physiol* (*London*) **133**:566–570 (1956).
426. C. J. Burdick and C. F. Strittmatter, *Arch. Biochem. Biophys.* **109**, 293–301 (1965).
427. N. R. Marquis and I. B. Fritz, *J. Biol. Chem.* **240**:2193–2196 (1965).
428. N. R. Marquis and I. B. Fritz, *Recent Research on Carnitine, Its Relation to Lipid Metabolism, Symp. at M.I.T.*, 1964, (G. Wolf, ed.), p. 27, The M.I.T. Press, Cambridge, Massachusetts (1964).
429. E. Sanchez-Quintanar, *Dissertation Abstr.* **23**:824–825 (1962).
430. G. R. Webster, *Biochim. Biophys. Acta* **98**:512–519 (1965).
431. R. A. Pieringer and L. E. Hokin, *J. Biol. Chem.* **237**:659–663 (1962).
432. G. R. Webster and R. I. Alpern, *Biochem. J.* **90**:35–42 (1964).
433. K. B. Jacobson, *J. Gen. Physiol.* **43**:323–333 (1959).

434. F. B. Goldstein, *J. Biol. Chem.* **234**:2702–2706 (1959).
435. M. H. Baslow, *Am. Zool.* **5**:230 (1965).
436. A. Hanson, *Acta Chem. Scand.* **20**:159–164 (1966).
437. M. H. Baslow, *Brain Res.* **3**:210–213 (1966).
438. M. Wollemann and G. Feuer, *Acta Physiol. Acad. Sci. Hung.* **10**:445–447 (1956).
439. M. Sribney, *Federation Proc.* **21**:280 (1962).
440. R. O. Brady, R. M. Bradley, O. M. Young, and H. Kaller, *J. Biol. Chem.* **240**:3693–3694 (1965).
441. W. D. Tomitzek and E. Strack, *Acta Biol. Med. Ger.* **13**:110–125 (1964).
442. P. V. Vignais and I. Zabin, *Biochim. Biophys. Acta* **29**:263–269 (1958).
443. G. Feuer, *Acta Physiol. Acad. Sci. Hung.* **9**:393–398 (1956).
444. A. V. Palladin, *Fiziol. Zh. SSSR*, **35**:596–603 (1949) (CA **44**, 2099a; 1950).
445. B. M. Breckenridge and J. H. Norman, *J. Neurochem.* **9**:383–392 (1962).
446. C. F. Cori, S. P. Colowick, and G. T. Cori, *J. Biol. Chem.* **123**:375–380 (1938).
447. B. I. Khaikina, *Ukrain. Biokhim. Zh.* **20**:342–351 (1948) (CA **48**, 5896b; 1954).
448. L. C. Mokrasch and H. J. Grady, in *Comparative Neurochemistry* (D. Richter, ed.) Proc. 5th Intern. Neurochem. Symp., St. Wolfgang, 1962, pp. 213–223, Pergamon Press, Oxford, London, New York, and Paris (1964).
449. G. J. Drummond, J. Keith, and M. W. Gilgan, *Arch. Biochem. Biophys.* **105**:156–162 (1964).
450. B. M. Breckenridge and J. H. Norman, *J. Neurochem.* **12**:51–57 (1965).
451. G. T. Cori and C. F. Cori, *J. Biol. Chem.* **135**:733–756 (1940).
452. O. H. Lowry, D. W. Schulz, and J. V. Passoneau, *J. Biol. Chem.* **242**:271–280 (1967).
453. K. Iijima, *Bull. Tokyo Med. Dental Univ.* **11**:77–101 (1964) (CA **62**, 871e; 1965).
454. B. Shapiro and E. Wertheimer, *Biochem. J.* **37**:397–403 (1943).
455. O. H. Lowry, in *Biochemistry of the Developing Nervous System* (H. Waelsch, ed.), Proc. 1st Intern. Neurochem. Symp. Oxford, 1954, pp. 350–357, Academic Press, New York (1955).
456. B. M. Breckenridge and E. J. Crawford, *J. Neurochem.* **7**:234–240 (1961).
457. T. Takeuchi, K. Higashi, and S. Watanuki, *J. Histochem. Cytochem.* **3**:485–491 (1955).
458. C. Villar-Palasi and J. Larner, *Arch. Biochem. Biophys.* **86**:270–273 (1960).
459. N. Shimizu and M. Okada, *J. Histochem. Cytochem.* **5**:459–471 (1957).
460. L. F. Leloir, J. M. Olivarria, S. H. Goldemberg, and H. Carminatti, *Arch. Biochem. Biophys.* **81**:508–520 (1959).
461. B. M. Breckenridge and E. J. Crawford, *J. Biol. Chem.* **235**:3054–3057 (1960).
462. B. J. Khaikina and V. E. Yakushko, *Ukrain. Biochem. Zh.* **34**:876–882 (1962) (CA **58**, 8281g; 1963).
463. J. H. Austin, A. S. Balasubramanian, T. N. Pattabiraman, S. Saraswathi, D. K. Basu, and B. K. Bachhawat, *J. Neurochem.* **10**:805–816 (1963).
464. D. K. Basu and B. K. Bachhawat, *Biochim. Biophys. Acta* **50**:123–128 (1961).
465. T. A. I. Grillo, G. Okuno, S. Price, and P. P. Foa, *J. Histochem. Cytochem.* **12**:275–280 (1964).
466. T. Takeuchi, *J. Histochem. Cytochem.* **6**:208–216 (1958).
467. W. W. Cleland and E. P. Kennedy, *J. Biol. Chem.* **235**:45–51 (1960).
468. R. M. Burton, M. A. Sodd, and R. O. Brady, *J. Biol. Chem.* **233**:1053–1060 (1958).
469. B. Kaufman, S. Basu, and S. Roseman, in *Inborn Disorders of Sphingo-lipid Metabolism* (S. M. Aronson and B. W. Volk, eds.), pp. 193–213, Pergamon Press, Oxford, London, and New York (1967).
470. A. Arce, H. F. Maccioni, and R. Caputto, *Arch. Biochem. Biophys.* **116**:52–58 (1966).
471. D. A. Rappoport, R. R. Fritz, J. R. Allen, and A. Moraczewski, *J. Neurochem.* **6**:21–27 (1960).
472. E. Robins, D. E. Smith, and R. E. McCaman, *J. Biol. Chem.* **204**:927–937 (1953).

473. O. Sköld, *Biochim. Biophys. Acta* **44**:1–12 (1960).
474. S. H. Mudd, J. D. Finkelstein, F. Irreverre, and L. Laster, *J. Biol. Chem.* **240**:4382–4392 (1965).
475. J. Vos, J. P. Schadé, and H. J. Van der Helm, in *Progress in Brain Research* (C. G. Bernhard and J. P. Schadé, eds.), Vol. 26, pp. 193–213, Elsevier, Amsterdam, London, and New York (1967).
476. A. S. Balasubramanian and B. K. Bachhawat, *J. Sci. Ind. Res. (India)* **20C**:202–204 (1961) (CA **55**, 24885b; 1961).
477. P. P. Cohen and G. L. Heckhuis, *J. Biol. Chem.* **140**:711–724 (1941).
478. J. Awapara and B. Seale, *J. Biol. Chem.* **194**:497–502 (1952).
479. T. Oka, *Nippon Yakurigaku Zasshi* **56**:857–865 (1960) (CA **55**, 22523; 1961).
480. K. Yamada, S. Sawaki, A. Fukumura, and M. Hayashi, *J. Vitaminol. (Japan)* **8**:286:291 (1962).
481. E. Roberts, *Arch. Biochem. Biophys.* **48**:395–401 (1954).
482. S. Sumida, *Seikagaku* **28**:339–342 (1956).
483. A. Imai, *Okayama Igakkai Zasshi* **71**: 1641–1645 (1959) (CA **54**, 4939e; 1960).
484. A. Imai, *Okayama Igakkai Zasshi* **71**:1647–1650 (1959) (CA **54**, 4939g; 1960).
485. A. Imai, *Okayama Igakkai Zasshi* **71**: 1651–1654 (1959) (CA **54**, 4939g; 1960).
486. T. N. Pattabhiraman and B. K. Bachhawat, *Ann. Biochem. Exptl. Med. (Calcutta)* **19**:205–212 (1959) (CA **54**, 13217f; 1960).
487. G. Marino and V. Scardi, *Boll. Soc. Ital. Biol. Sper.* **40**:717–719 (1964).
488. V. Bonavita, *J. Neurochem.* **4**:275–281 (1959).
489. O. Z. Sellinger and D. L. Rucker, *Biochim. Biophys. Acta* **67**:504–507 (1963).
490. N. S. Nilova, *Vopr. Med. Khim.* **12**:514–517 (1966) (CA **65**, 20693B; 1966).
491. G. Ya. Kivman and R. P. Porfir'eva, *Dokl. Akad. Nauk. SSSR*, **170**:1209–1211 (1966).
492. G. Marino, R. Zito, and V. Scardi, *Boll. Soc. Ital. Biol. Sper.* **40**:720–722 (1964).
493. V. Bonavita, *Rass. Med. Sper.* **7**:241–260 (1960) (CA **55**, 12579g; 1961).
494. L. May, M. Miyazaki, and R. G. Grenell, *J. Neurochem.* **4**:269–274 (1959).
495. O. H. Lowry, N. R. Roberts, and M. W. Chang, *J. Biol. Chem.* **220**:879–892 (1956).
496. O. H. Lowry, N. R. Roberts, and M. W. Chang, *J. Biol. Chem.* **222**:97–107 (1956).
497. M. Miyazaki, *J. Nervous Mental Disease* **126**:169–175 (1958).
498. J. B. Green, H. A. Oldewurtel, and F. M. Forster, *Neurology* **9**:540–544 (1959).
499. O. Z. Sellinger, D. L. Rucker, and F. de Balbian Verster, *J. Neurochem.* **11**:271–280 (1964).
500. F. Fonnum, *Biochem. J.* **96**:66P–67P (1965).
501. A. Rendon, P. Mandel, and A. Waksman, *Hoppe-Seyler's Z. Physiol. Chem.* **348**:1242–1243 (1967).
502. E. Bombardelli and V. Zambotti, *Biochim. Biol. Sper.* **2**:287–291 (1963).
503. G. Amore and V. Bonavita, *Life Sci.* **4**:2417–2424 (1965).
504. L. Mallucci, *Boll. Soc. Ital. Biol. Sper.* **33**:1079–1081 (1957).
505. Fan-Shen Li and Sjun Sjui, *Sheng Wu Hua Hsueh Yu Sheng Wu Wu Li Hsueh Pao* **4**:139–144 (1964) (CA **65**, 15873B; 1966).
506. B. Majoie and J. Guidicelli, *Compt. Rend. Soc. Biol.* **160**:69–74 (1966).
507. D. S. Zaprudskaya and N. G. Dorosheva, *Byul. Eksperim. Biol. Med.* **56**:54–56 (1963) (CA **60**, 5971a; 1964).
508. T. F. Ponomareva, *Ukr. Biokhim. Zh.* **36**:513–520 (1964).
509. T. Yamada, *Okayama Igakkai Zasshi* **71**:7541–7545 (1959) (CA **55**, 739d; 1961).
510. T. Yamada, *Okayama Igakkai Zasshi* **71**:7547–7550 (1959) (CA **55**, 739e; 1961).
511. T. Yamada, *Okayama Igakkai Zasshi* **71**:7551–7557 (1959) (CA **55**, 739e; 1961).
512. T. F. Ponomareva and K. A. Drel', *Biokhimiya* **29**:185–190 (1964).
513. H. Gerber, *Schweiz. Arch. Tierheilk.* **106**:410–413 (1964).

514. M. R. Alioto, G. Napoleone, M. Ayala, and G. Rinaldi, *Boll. Soc. Ital. Sper.* **36**:1925–1927 (1960).
515. P. H. Khan and S. P. Karmarkar, *J. Biol. Sci. (Bombay)* **6**:19–27 (1963) (CA **60**, 7192g; 1964).
516. F. D'Abramo and E. Tomasos, *Ital. J. Biol.* **8**:173–180 (1959).
517. R. W. Albers, G. J. Koval, and W. B. Jakoby, *Exptl. Neurol.* **6**:85–89 (1962).
518. Z. N. Canellakis and P. P. Cohen, *J. Biol. Chem.* **222**:53–62 (1956).
519. O. Tangen, F. Fonnum, and R. Haavaldsen, *Biochim. Biophys. Acta* **96**:82–90 (1965).
520. C. Peraino and C. Pitot, *Biochim. Biophys. Acta* **73**:222–231 (1963).
521. S. R. Guha, H. S. Chakravarti, and J. J. Ghosh, *Ann. Biochem. Exptl. Med. (Calcutta)* **18**: 103–104 (1958) (CA **54**, 24924h; 1960).
522. M. Sugiura, *Japan J. Pharmacol.* **7**:1–5 (1957).
523. S. R. Mardashew, M. I. Lerman, and M. S. Benyumovich, *Vopr. Med. Khim.* **8**:547–549 (1962).
524. M. I. Lerman, *Vopr. Onkol.* **7**:40–44 (1961).
525. S. S. Guha and J. J. Ghosh, *Ann. Biochem. Exptl. Med.* **19**:33–36 (1959) (CA-**53**, 18136f; 1959).
526. F. Cacioppo and L. Pandolfo, *Boll. Soc. Ital. Biol. Sper.* **34**:929–932 (1958).
527. C. Nicotra, S. Macaione, and S. Campisi, *Boll. Soc. Ital. Biol. Sper.* **42**:1365–1368 (1966).
528. N. Canal and L. Frattola, *Med. Exptl.* **8**:129–134 (1963).
529. F. P. Kupiecki and M. J. Coon, *J. Biol. Chem.* **229**:743–754 (1957).
530. R. W. Albers and W. B. Jakoby, *in Inhibition in the Nervous System and γ-Aminobutyric Acid* (E. Roberts, ed.), p. 468, Pergamon Press, Oxford, London, and New York (1960).
531. E. Kowalski, A. Dancewicz, and Z. Szot, *Bull. Acad. Pol. Sci. Ser. Biol.* **5**:223–227 (1957).
532. R. Haavaldsen, *Nature* **196**:577–578 (1962).
533. H. George, R. Turner, and S. Gabay, *J. Neurochem.* **14**:841–845 (1967).
534. F. Fonnum and K. Larsen, *J. Neurochem.* **12**:589–598 (1965).
535. W. F. Bridgers, *J. Biol. Chem.* **240**:4591–4597 (1965).
536. D. A. Walsh and H. J. Sallach, *J. Biol. Chem.* **241**:4068–4076 (1966).
537. H. Hirsch and D. M. Greenberg, *J. Biol. Chem.* **242**:2283–2287 (1967).
538. S. P. Bessman, J. Rossen, and E. C. Layne, *J. Biol. Chem.* **201**:385–391 (1953).
539. F. Cacioppo, L. Pandolfo, and G. DiChiara, *Boll. Soc. Ital. Biol. Sper.* **35**:465–467 (1959).
540. C. F. Baxter and E. Roberts, *J. Biol. Chem.* **236**:3287–3294 (1961).
541. Y. Nishizawa, T. Kodama, and S. Konishi, *J. Vitaminol. (Osaka)* **5**:117–128 (1959).
542. A. Waksman and E. Roberts, *Biochemistry* **4**:2132–2139 (1965).
543. Z. W. Hall and E. A. Kravitz, *J. Neurochem.* **14**:45–54 (1967).
544. R. A. Salvador and R. W. Albers, *J. Biol. Chem.* **234**:922–925 (1959).
545. C. F. Baxter and E. Roberts, *J. Biol. Chem.* **233**:1135–1139 (1958).
546. Ke-Pang Chang and Hsiu-Fang Chen, *Sheng Li Hsueh Pao* **26**:275–281 (1963) (CA **62**, 13589c; 1965).
547. F. N. Pitts Jr., C. Quick, and E. Robins, *J. Neurochem.* **12**:93–101 (1965).
548. J. J. Sheridan, K. L. Sims, and F. N. Pitts, Jr., *J. Neurochem.* **14**:571–578 (1967).
549. L. Salganicoff and E. de Robertis, *Life Sci.* **2**:85–91 (1963).
550. E. J. Averinova, E. V. Bagdanova, I. A. Sytinskii, and G. M. Tokacheva, *Zh. Evol. Biokhim. Fiziol.* **2**:493–495 (1966) (CA **66**, 9200g; 1967).
551. B. Sisken, K. Sano, and E. Roberts, *J. Biol. Chem.* **236**:503–507 (1961).
552. G. Kh. Bunyatyan and R. R. Nersesyan, *Vopr. Biokhim. Mozga, Akad. Nauk. Arm. SSR, Inst. Biokhim.* **1**:5–13 (1964) (CA **63**, 15274f; 1965).
553. R. W. Albers, *in Inhibition in the Nervous System and Gamma-Aminobutyric Acid* (E. Roberts, ed.), pp. 196–201, Pergamon Press, Oxford, London, and New York (1960).
554. C. Long, *Biochem. J.* **50**:407–415 (1952).

555. M. Kerly and D. H. Leaback, *Biochem. J.* **67**:250–252 (1957).
556. H. M. Katzen and R. T. Schimke, *Proc. Natl. Acad. Sci.* **54**:1218–1225 (1965).
557. N. V. El'tsina and E. Heise, *Acta Biol. Med. Ger.* **14**:114–118 (1965).
558. M. Hite, *Dissertation Abstr.* **21**:747–748 (1960).
559. M. W. Slein, G. T. Cori, and C. F. Cori, *J. Biol. Chem.* **186**:763–780 (1950).
560. A. Sols and R. K. Crane, *J. Biol. Chem.* **206**:925–936 (1954).
561. F. Raggi and D. S. Kronfeld, *Nature* **209**:1353–1354 (1966).
562. L. Grossbard and R. T. Schimke, *J. Biol. Chem.* **241**:3546–3560 (1966).
563. D. Biesold and P. Teichgraeber, *Biochem. J.* **103**:13C–14C (1967).
564. J. E. Wilson, *Biochem. Biophys. Res. Commun.* **28**:123–127 (1967).
565. V. Jagannathan, *Indian J. Chem.* **1**:192–193 (1963).
566. G. P. Schwartz and R. E. Basford, *Biochemistry* **6**:1070–1078 (1967).
567. O. Meyerhof and J. R. Wilson, *Arch. Biochem.* **19**:502–508 (1948).
568. H. Weil-Malherbe and A. D. Bone, *Biochem. J.* **49**:339–347 (1951).
569. J. Stern, *Biochem. J.* **58**:536–542 (1954).
570. R. Nesbakken and L. Eldjarn, *Biochem. J.* **87**:526–532 (1963).
671. F. S. Rolleston and E. A. Newsholme, *Biochem. J.* **100**:64P–65P (1966).
572. E. A. Newsholme, F. S. Rolleston, and K. Taylor, *Biochem. J.* **104**:47P (1967).
573. M. Copley and H. J. Fromm, *Biochemistry* **6**:3503–3509 (1967).
574. Y. Takahashi and Y. Akabane, *Arch. Gen. Psychiat.* **3**:674–681 (1960).
575. E. L. Bennet, J. B. Drori, D. Krech, M. R. Rosenzweig, and S. Abraham, *J. Biol. Chem.* **237**:1758–1763 (1962).
576. O. H. Lowry, N. R. Roberts, K. Y. Leiner, M. L. Wu, A. L. Farr, and R. W. Albers, *J. Biol. Chem.* **207**:39–49 (1954).
577. A. V. Palladin and N. M. Polyakova, *Dokl. Akad. Nauk. SSSR* **91**:347–349 (1953).
578. R. K. Crane and A. Sols, *J. Biol. Chem.* **203**:273–292 (1953).
579. S. P. Damle, G. P. Talwar, and B. K. Anand, *Indian J. Med. Res.* **49**:852–856 (1961).
580. D. E. Smith, M. K. Jen, and E. Robins, *Res. Proc. Acad. Res. Nervous Mental Disease* **32**:305–310 (1953).
581. R. Tanaka and L. G. Abood, *J. Neurochem.* **10**:571–576 (1963).
582. D. G. Walker, in *Essays in Biochemistry* (P. N. Campbell and G. D. Greville, eds.) Vol. 2, pp. 33–67, Academic Press, London, (1966).
583. I. A. Rose and J. B. B. Warms, *Federation Proc.* **24**:297 (1965).
584. H. S. Bachelard, *Biochem. J.* **102**:21P (1967).
585. B. P. Setchell, *Biochem. J.* **72**:265–275 (1959).
586. F. Raggi, D. S. Kronfeld, J. Bartley, and J. R. Luick, *Nature*, **197**:190–191 (1963).
587. S. Abraham, B. Borrebaek, and I. L. Chaikoff, *J. Nutr.* **83**:273–288 (1964).
588. C. E. Cardini and L. F. Leloir, *Arch. Biochem. Biophys.* **45**:55–64 (1964).
589. L. Pandolfo and S. Macaione, *Boll. Soc. Ital. Biol. Sper.* **40**:750–752 (1964).
590. R. P. Harpur and J. H. Quastel, *Nature* **164**:693–694 (1949).
591. T. N. Pattabiraman and B. K. Bachhawat, *J. Sci. Ind. Res. Sect.* C **20C**:14–17 (1961, New Delhi).
592. T. N. Pattabiraman and B. K. Bachhawat, *J. Neurochem.* **11**:55–60 (1964).
593. T. N. Pattabiraman and B. K. Bachhawat, *Indian J. Exptl. Biol.* **1**:26–31 (1963).
594. J. A. Muntz, *Arch. Biochem. Biophys.* **42**:435–445 (1953).
595. O. H. Lowry and J. V. Passonneau, *J. Biol. Chem.* **241**:2268–2279 (1966).
596. S. M. Weissman, R. M. S. Smellie, and J. Paul, *Biochim. Biophys. Acta* **45**:101–110 (1960).
597. P. A. Kometiani, *Soobshcheniya Acad. Nauk. Gruzin. SSSR* **12**:531–538 (1951) (CA **48**, 12201e; 1954).
598. J. Wittenberg and A. Kornberg, *J. Biol. Chem.* **202**:431–444 (1953).
599. R. E. McCaman, *J. Biol. Chem.* **237**:672–676 (1962).

600. R. E. McCaman and K. Cook, *J. Biol. Chem.* **241**:3390–3394 (1966).
601. J. B. Solomon, *Biochem. J.* **66**:264–270 (1957).
602. M. Tsubosaka, *Bitamine* **35**:7–12 (1967).
603. D. B. McCormick and E. E. Snell, *J. Biol. Chem.* **236**:2085–2088 (1961).
604. D. B. McCormick, M. E. Gregory, and E. E. Snell, *J. Biol. Chem.* **236**:2076–2084 (1961).
605. Y. H. Loo and V. P. Whittaker, *J. Neurochem.* **14**:997–1011 (1967).
606. S. Garattini, *U.S. Dept. Com. Office Tech. Serv. A.D. 267/228*, 33 (1961) (CA **62**, 6911g; 1965).
607. M. Rabinowitz and F. Lipmann, *J. Biol. Chem.* **235**:1043–1050 (1960).
608. R. Rodnight and B. E. Lavin, *Biochem. J.* **93**:84–91 (1964).
609. H. Matsui, E. Orikabe, S. Ishikawa, and N. Shimazono, *J. Biochem. (Tokyo)* **57**:131–141 (1965).
610. L. E. Hokin, L. Yoda, and R. Sandhu, *Biochim. Biophys. Acta* **126**:100–116 (1966).
611. L. Décsi and R. Rodnight, *J. Neurochem.* **12**:791–796 (1965).
612. R. Rodnight and B. E. Lavin, *Biochem. J.* **101**:495–501 (1966).
613. H. Rodnight, D. A. Hems, and B. E. Lavin, *Biochem. J.* **101**:502–525 (1966).
614. D. A. Hems and R. Rodnight, *Biochem. J.* **101**:516–523 (1966).
615. R. v. Fellenberg, R. Richterich, and H. Aebi, *Enzymol. Biol. Clin.* **3**:240–250 (1963).
616. R. A. Pieringer and L. E. Hokin, *J. Biol. Chem.* **237**:653–658 (1962).
617. J. Jarnefelt, *Exptl. Cell Res.* **25**:211–213 (1961).
618. K. P. Strickland, *Can. J. Biochem. Physiol.* **40**:247–259 (1962).
619. L. E. Hokin and M. R. Hokin, *Biochim. Biophys. Acta* **31**:285–287 (1959).
620. M. R. Hokin and L. E. Hokin, *J. Biol. Chem.* **234**:1381–1386 (1959).
621. H. Brockerhoff and C. E. Ballou, *J. Biol. Chem.* **237**:49–52 (1962).
622. M. Colodzin and E. P. Kennedy, *J. Biol. Chem.* **240**:3771–3780 (1965).
623. M. Kai and J. N. Hawthorne, *Biochem. J.* **98**:23P–24P (1966).
624. H. Brockerhoff and C. E. Ballou, *J. Biol. Chem.* **237**:1764–1766 (1962).
625. J. Eichberg and R. M. C. Dawson, *Biochem. J.* **96**:644–650 (1965).
626. M. Kai, J. G. Salway, R. H. Michell, and J. N. Hawthorne, *Biochem. Biophys. Res. Commun.* **22**:370–375 (1966).
627. T. W. Rall and E. W. Sutherland, *J. Biol. Chem.* **232**:1065–1076 (1958).
628. E. W. Sutherland and T. W. Rall, *Pharmacol. Rev.* **12**:265–299 (1960).
629. E. W. Sutherland, T. W. Rall, and T. Menon, *J. Biol. Chem.* **237**:1220–1227 (1962).
630. L. M. Klainer, Y. M. Chi, S. L. Freidberg, T. W. Rall, and E. W. Sutherland, *J. Biol. Chem.* **237**:1239–1243 (1962).
631. T. W. Rall and E. W. Sutherland, *J. Biol. Chem.* **237**:1228–1232 (1962).
632. E. de Robertis, *Science* **156**:907–914 (1967).
633. E. de Robertis, G. Rodriguez de Lores Arnaiz, M. Alberici, R. W. Butcher, and E. W. Sutherland, *J. Biol. Chem.* **242**:3487–3493 (1967).
634. G. Domjan, *Enzymologia* **31**:1–8 (1966).
635. W. F. Bridgers, *J. Biol. Chem.* **242**:2080–2085 (1967).
636. C. Landriscina, V. Liso, M. N. Gadeleta, and A. Alifano, *Boll. Soc. Ital. Biol. Sper.* **42**:473–476 (1966).
637. D. R. Rao and P. Oesper, *Biochem. J.* **81**:405–411 (1961).
638. J. T. Oliver, *Biochem. J.* **61**:116–122 (1955).
639. A. Narayanaswami, *Biochem. J.* **52**:295–301 (1952).
640. A. Burger, R. Richterich, and H. Aebi, *Biochem. Z.* **339**:305–314 (1964).
641. H. Jakobs, H. W. Held, and M. Klingenberg, *Biochem. Biophys. Res. Commun.* **21**:346–353 (1965).
642. D. H. Deul and J. F. L. van Breemen, *Clin. Chim. Acta* **10**:276–283 (1964).
643. F. A. Graig and J. C. Smith, *Science* **156**:254–255 (1967).

644. T. Wood, *Biochem. J.* **87**:453–462 (1963).
645. K. Sjövall and A. Voigt, *Nature* **202**:701 (1964).
646. H. M. Eppenberger, D. M. Dawson, and N. O. Kaplan, *J. Biol. Chem.* **242**:204–209 (1967).
647. D. M. Dawson and I. H. Fine, *Arch. Neurol.* **16**:175–180 (1967).
648. D. M. Dawson, H. M. Eppenberger, and N. O. Kaplan, *Biochem. Biophys. Res. Commun.* **21**:346–353 (1965).
649. M. L. Tanzer and C. Gilvarg, *J. Biol. Chem.* **234**:3201–3204 (1959).
650. L. A. Tseitlin, *Biokhimiya* **18**:311–314 (1953).
651. T. Wood and P. D. Swanson, *J. Neurochem.* **11**:301–307 (1964).
652. T. Wood, *Biochem. J.* **89**:210–220 (1963).
653. E. K. Brownlow and D. B. Gammack, *Biochem. J.* **103**:47P–48P (1967).
654. T. O. Kleine, *Klin. Wschr.* **43**:504–510 (1965).
655. H. M. Eppenberger, M. Eppenberger, R. Richterich, and H. Aebi, *Develop. Biol.* **10**:1–16 (1964).
656. R. M. Smillie, *Arch. Biochem. Biophys.* **67**:213–224 (1957).
657. H. A. Krebs and R. Hems, *Biochem. J.* **61**:435–441 (1955).
658. R. P. Miech and R. E. Parks, Jr., *J. Biol. Chem.* **240**:351–357 (1965).
659. G. Giusti and M. Coltorti, *Rass. Med. Sper.* **2**:217–219 (1956).
660. P. E. Braun and G. I. Drummond, *J. Neurochem.* **13**:525–531 (1966).
661. D. A. Rappoport and R. R. Fritz, *Comp. Biochem. Physiol.* **4**:33–41 (1961).
662. L. G. Együd and W. J. Whelan, *Biochem. J.* **86**:11P–12P (1963).
663. A. V. Palladin, N. M. Polyakova, and V. K. Lishko, *J. Neurochem.* **10**:187–194 (1963).
664. N. M. Polyakova and N. A. Untina, *Vopr. Med. Khim.* **7**:524–527 (1961) (CA **56**, 15998f; 1962).
665. A. V. Palladin, *Acta Physiol. Acad. Sci. Hung.* **21**:105–111 (1962) (CA **58**, 1763a; 1963).
666. L. Mallucci and A. di Simone, *Boll. Soc. Ital. Biol. Sper.* **33**:1242–1243 (1957).
667. T. N. Pattabiraman and B. K. Bachhawat, *J. Sci. Industr. Res. (Biol.) (New Delhi) Sect. C*, **21**C:352–354 (1962) (CA **58**, 7203d; 1963).
668. V. W. Rodwell, J. C. Towne, and S. Grisolia, *J. Biol. Chem.* **228**:875–890 (1957).
669. M. Kurokawa, T. Kato, and H. Inamura, *Proc. Japan Acad.* **42**:1217–1222 (1966).
670. C. Deluca and N. O. Kaplan, *Biochim. Biophys. Acta* **30**:6–11 (1958).
671. M. Rohdewald, *in Hoppe-Seyler/Thierfelder, Handbuch der physiologisch und patholo-gisch-chemischen Analyse* (K. Lang and E. Lehnartz, eds.) Vol. 6B, pp. 528–710, Springer Verlag, Berlin, Göttingen, and Heidelberg, (1966).
672. A. S. Balasubramanian and N. K. Bachhawat, *Indian J. Exptl. Biol.* **1**:179–181 (1963).
673. N. Klein-Pete, and M. Wintzerith, *Compt. Rend. Soc. Biol.* **258**:5283–5286 (1964).
674. S. H. Barondes, *J. Neurochem.* **11**:663–669 (1964).
675. S. C. Bondy and H. Waelsch, *J. Neurochem.* **12**:751–756 (1965).
675. P. Mandel, A. R. Dravid, and N. Pete, *J. Neurochem.* **14**:301–306 (1967).
677. N. Pete, P. Chambon, and P. Mandel, *Prelimin. Abstr. of the VI. Int. Neurochem. Conf.*, p. 89, Oxford, (1965).
678. K. J. Isselbacher, *J. Biol. Chem.* **232**:429–444 (1958).
679. K. Kurahashi and E. P. Anderson, *Biochim. Biophys. Acta* **29**:498–501 (1958).
680. B. M. Breckenridge, S. Scott, J. L. Strominger, and E. J. Crawford, *J. Neurochem.* **7**:228–233 (1961).
681. D. K. Basu and B. K. Bachhawat, *J. Neurochem.* **7**:174–179 (1961).
682. D. Bertoli and St. Segal, *J. Biol. Chem.* **241**:4023–4029 (1966).
683. G. B. Ansell and T. Chojnacki, *Biochem. J.* **98**:303–310 (1966).
684. R. M. C. Dawson, *Essays in Biochemistry*, **2**:69–115 (1966).
685. L. F. Borkenhagen and E. P. Kennedy, *J. Biol. Chem.* **227**:951–962 (1957).
686. T. N. Pattabiraman and B. K. Bachhawat, *Biochim. Biophys. Acta* **50**, 129–134 (1961).

687. M. Shoyab, T. N. Pattabiraman, and B. K. Bachhawat, *J. Neurochem.* **11** :639–646 (1964).
688. M. Shoyab and B. K. Bachhawat, *Indian J. Biochem.* **2** :6–9 (1965).
689. G. Siebert, in *Hoppe-Seyler/Thierfelder, Handbuch der phsyiologisch- und pathologisch-chemischen Analyse* (K. Lang and E. Lehnartz, eds.), Vol. 6B, pp. 721–780, Springer Verlag, Berlin, Göttingen, Heidelberg, and New York (1966).
690. O. P. Chepinoga, E. B. Skvirskaya, L. P. Rukina, and T. P. Silich, *Ukrain. Biokhim. Zh.* **24**: 177–185 (1952) (CA **48**, 11599f; 1954).
691. M. R. V. Murthy and D. A. Rappoport, *Biochim. Biophys. Acta* **95**:132–145 (1965).
692. C. R. Jones and A. Janoff, *Comp. Biochem. Physiol.* **15**:77–80 (1965).
693. G. A. Nechaeva, *Biokhimiya* **30**:644–651 (1965).
694. G. A. Nechaeva, *Ukrain. Biokhim. Zh.* **36**:607 (1964) (*Federation Proc. Transl. Suppl.* 24, pp. 645–649 (1965).
695. R. K. Datta and J. J. Ghosh, *J. Neurochem.* **9**:463–464 (1962).
696. R. K. Datta, D. Bhattacharyya, and J. J. Ghosh, *J. Neurochem.* **11**:87–98 (1964).
697. R. K. Datta, *Brain Res.* **2**:301–322 (1966).
698. L. May and R. G. Grenell, *Proc. Soc. Exptl. Biol., N.Y.* **102**:235–239 (1959).
699. A. V. Palladin, in *Biochemistry of the Developing Nervous System* (H. Waelsch, ed.), pp. 177–184, Academic Press, New York (1955).
700. G. A. Nechaeva, *Dokl. Akad. Nauk. SSSR* **152**:225–227 (1963).
701. H. Koenig, *Trans. Am. Neurol. Assoc.* **88**:227–228 (1963).
702. H. Beaufay, A.-M. Berleur, and A. Doyen, *Biochem. J.* **66**:32P (1957).
703. R. Llamas and E. Coronas, *Anales Inst. Biol. (Univ. Nacl. Mex.)* **31**:3–11 (1960) (CA **55**, 21194d; 1961).
704. J. C. Houck, *J. Appl. Physiol.* **13**:273–277 (1958).
705. G. B. Ansell, T. Chojnacki, and R. F. Metcalfe, *J. Neurochem.* **12**:649–656 (1965).
706. R. J. Rossiter and K. P. Strickland, *Ann. N.Y. Acad. Sci.* **72**:790–802 (1959).
707. K. P. Strickland, *Can. J. Biochem. Physiol.* **40**:247–259 (1962).
708. K. P. Strickland, D. Subrahmanyam, E. T. Pritchard, W. Thompson, and R. J. Rossiter, *Biochem. J.* **87**:128–136 (1963).
709. M. Sribney and E. P. Kennedy, *J. Biol. Chem.* **233**:1315–1322 (1958).
710. W. Thompson, K. P. Strickland, and R. J. Rossiter, *Biochem. J.* **87**:136–142 (1963).
711. B. W. Agranoff, R. M. Bradley, and R. O. Brady, *J. Biol. Chem.* **233**:1077–1083 (1958).
712. R. J. Rossiter and F. B. Palmer, in *Lipoide, 16. Coll. Ges. für Physiol. Chem. in Mosbach* (E. Schütte, ed.), pp. 40–48, Springer Verlag, Berlin, Heidelberg, and New York (1966).
713. W. A. Himwich and J. P. Saunders, *Am. J. Physiol.* **153**:348–354 (1948).
714. A. Ruffo, H. Romano, and R. Giampaolo, *Boll. Soc. Ital. Biol. Sper.* **29**:46–48 (1953).
715. A. Koy and J. Frendo, *Acta Biochim. Polon.* **9**:373–379 (1962).
716. D. Reinwein, *Hoppe-Seyler's Z. Physiol. Chem.* **326**:94–101 (1961).
717. M. der Garabédian and P. Gonnard, *Compt. Rend. Soc. Biol.* **139**:311–312 (1945).
718. K. Lang, *Biochem. Z.* **259**:243–256 (1933).
719. L. Mallucci, *Boll. Soc. Ital. Biol. Sper.* **33**:1082–1083 (1957).
720. E. Castellá Bertrán, *Anales Fac. Vet. Univ. Madrid. Y Inst. Invest. Vet.* **4**:402–590 (1952), (CA **47**, 10667i; 1953).
721. A. S. Balasubramanian and B. K. Bachhawat, *J. Neurochem.* **11**:877–885 (1964).
722. G. M. McKhann, R. Levy, and W. Ho, *Biochem. Biophys. Res. Commun.* **20**:109–113 (1965).
723. A. S. Balasubramanian and B. K. Bachhawat, *Indian J. Biochem.* **2**:212–216 (1965).
724. G. M. McKhann and W. Ho, *J. Neurochem.* **14**:717–724 (1967).
725. O. Hoffmann-Ostenhof and R. Ehrenreich, in *Hoppe-Seyler/Thierfelder, Handbuch der physiologisch- und pathologisch-chemischen Analyse* (K. Lang and E. Lehnartz, eds.), Vol. 6B, pp. 877–920, Springer Verlag, Berlin, Göttingen, and Heidelberg (1966).

726. O. Eraenkoe, A. Kokko, and U. Soderholm, *Nature* **193**:778 (1962).
727. J. Bernsohn, L. Possley, and E. Liebert, *J. Neurochem.* **4**:191–201 (1959).
728. D. J. Ecobichon, *Can. J. Physiol. Pharmacol.* **44**:225–232 (1966).
729. J. Bernsohn, K. D. Barron, P. F. Doolin, A. R. Hess, and M. T. Hedrick, *J. Histochem. Cytochem.* **14**:455–472 (1966).
730. J. Bernsohn, K. D. Barron, and H. Norgello, *Biochem. J.* **91**:24c–26c (1964).
731. N. F. M. Lemkey, *Dissertation Abstr.* **23**:4520–4521 (1963).
732. A. Bignami, G. Palladini, and G. Venturini, *Brain Res.* **3**:207–209 (1966).
733. W. E. Page, *Dissertation Abstr.* **23**:4523 (1963).
734. O. Z. Sellinger and F. de Balbian Verster, *Anal. Biochem.* **3**:479–488 (1962).
735. J. Vandelli and F. Scaltriti, *Boll. Soc. Ital. Biol. Sper.* **18**:77–79 (1943).
736. W. N. Aldridge, *Biochem. J.* **93**:619–623 (1964).
737. J. R. Lagnado and M. Hardy, *Nature* **214**:1207–1210 (1967).
738. D. K. Myers, *Biochem. J.* **64**:740–747 (1956).
739. H. Blaschko and P. Holton, *Brit. J. Pharmacol.* **4**:181–184 (1949).
740. S. Edlbacher, E. Goldschmidt, and V. Schläppi, *Hoppe-Seyler's Z. Physiol. Chem.* **227**: 118–123 (1934).
741. M. Gozzano, *Boll. Soc. Ital. Biol. Sper.* **9**:167–169 (1934).
742. E. Bozzetti, *Boll. Soc. Ital. Biol. Sper.* **28**:1089–1090 (1952).
743. Y. Ishii, *Arch. Histol. Japan* **10**:551–563 (1956) (*Excerpta Med.* **11**, Sect. I, Abstr. No. 1121; 1957).
744. A. R. McNabb, *Can. J. Med. Sci.* **29**:208–215 (1951).
745. J. Cammermeyer, *J. Comp. Neurol.* **90**:121–149 (1949).
746. M. G. Ord and R. H. S. Thompson, *Biochem. J.* **51**:245–251 (1952).
747. J. C. George and P. T. Iype, *J. Animal. Morphol. Physiol.* **8**:42–47 (1961).
748. H. Oberlander and H. A. Schneiderman, *Nature* **212**:432–433 (1966).
749. E. D. Wills, *in Advances in Lipid Research*, Vol. 3 (R. Paoletti and D. Kritchevsky, eds.), pp. 197–240, Academic Press, New York, and London (1965).
750. J. Gallai-Hatchard, W. L. Magee, R. H. S. Thompson, and G. R. Webster, *J. Neurochem.* **9**:545–554 (1962).
751. G. R. Webster, *Biochem. J.* **98**:19P–20P (1966).
752. E. A. Marples and R. H. S. Thompson, *Biochem. J.* **74**:123–127 (1960).
753. E. L. Bennett, M. Rosenzweig, D. Krech, H. Karlsson, N. Dye, and A. Ohlander, *J. Neurochem.* **3**:144–152 (1958).
754. D. Krech, M. Rosenzweig, E. L. Bennett, H. K. Roderick, N. Dye, and A. Ohlander, *Am. J. Physiol.* **196**:31–32 (1959).
755. Th. H. Roderick, *Genetics* **45**:1123–1140 (1960).
756. R. L. Olson, *Dissertation Abstr.* **25**:6906 (1965).
757. K. H. Meinecke and H. Oettel, *Arch. Toxikol.* **22**:244–261 (1967).
758. C. N. Peiss, J. Field, and V. E. Hall, *Am. J. Physiol.* **155**:56–59 (1948).
759. D. Nachmansohn, *Nature* **140**:427 (1937).
760. E. A. Zeller, *Helv. Chim. Acta.* **32**:484–488 (1949).
761. A. Kaswin and A. Serfaty, *Bull. Muséum Natl. Hist. Nat. (Paris)* **18**:305–308 (1946) (CA **43**, 6324i; 1949).
762. M. H. Baslow and R. F. Nigrelli, *Zoologica* **49**:41–51 (1964).
763. D. Nachmansohn, *Nature* **145**:513–514 (1940).
764. D. Nachmansohn, *Compt. Rend. Soc. Biol.* **127**:894–896 (1938).
765. M. Rockstein, *J. Cellular. Comp. Physiol.* **35**:11–23 (1950).
766. E. Giacobini, *J. Neurochem.* **1**:234–244 (1957).
767. E. Giacobini, *Acta Physiol. Scand.* **45**:238–254 (1959).

768. E. Giacobini, *J. Histochem. Cytochem.* 8:419–424 (1960).
769. K. B. Augustinsson, *Acta Physiol. Scand.* Suppl. 52, 15 (1958).
770. M. G. Ord and R. H. S. Thompson, *Biochem. J.* 49:191–199 (1951).
771. O. H. Lowry, N. R. Roberts, Mei-Ling Wu, W. S. Hixon, and E. J. Crawford, *J. Biol. Chem.* 207:19–37 (1953).
772. H. C. Lawler, *Biochim. Biophys. Acta* 81:280–288 (1964).
773. R. L. Jackson and M. H. Aprison, *J. Neurochem.* 13:1351–1365 (1966).
774. S. S. Kaplay and V. Jagannathan, *Indian J. Biochem.* 3:54–55 (1966).
775. N. V. Kartasheva, A. N. Panyukov, and V. J. Rozengart, *Vopr. Med. Khim.* 13:104–105 (1967) (CA 66, 826034; 1967).
776. K. Got and J. B. Polya, *Nature* 198:884–885 (1963).
777. D. Nachmansohn and M. Rothenberg, *J. Biol. Chem.* 158:653–660 (1945).
778. K. N. Mehrotra and W. C. Dauterman, *J. Neurochem.* 10:119–126 (1963).
779. B. Holmstedt and G. Toschi, *Acta Physiol. Scand.* 47:280–283 (1959).
780. E. Moerman, *Arch. Intern. Pharmacodyn.* 143:287–297 (1963).
781. J. M. Little, *Am. J. Physiol.* 153:436–443 (1948).
782. O. Bodansky, *Ann. N.Y. Acad. Sci.* 47:521–547 (1946).
783. F. Bergmann and R. Segal, *Biochim. Biophys. Acta* 16:513–519 (1955).
784. M. B. Abou-Donia and D. B. Menzel, *Comp. Biochem. Physiol.* 21:99–108 (1967).
785. C. O. Hebb and K. Krnjevic, *in Neurochemistry* (K. H. C. Elliott, I. H. Page, and J. H. Quastel, eds.), 2nd ed. pp. 452–521, Charles C. Thomas, Springfield, Illinois (1962).
786. D. Nachmansohn, *in Neurochemistry* (K. H. C. Elliott, I. H. Page, and J. H. Quastel, eds.), 2nd ed. pp. 522–557, Charles C. Thomas, Springfield, Illinois (1962).
787. S. Okinaka, M. Yoshikawa, and J. Gotô, *Igaku to Seibutsugaku* (*Med. Biol.*) 18:38–40 (1951) (CA 45, 4759d; 1951).
788. E. L. Bennett, M. R. Rosenzweig, D. Krech. A. Ohlander, and H. Morimoto, *J. Neurochem.* 6:210–218 (1961).
789. B. Teokharov, *Compt. Rend. Acad. Bulgare Sci.* 15:325–328 (1962).
790. C. C. D. Shute and P. R. Lewis, *Nature* 199:1160–1164 (1963).
791. S. Mori, T. Maeda, and N. Shimizu, *Histochemie* 4:65–72 (1964).
792. R. L. Friede and L. M. Fleming, *J. Neurochem.* 11:1–7 (1964).
793. E. L. Bennett, M. C. Diamond, H. Morimoto, and M. Herbert, *J. Neurochem.* 13:563–572 (1966).
794. A. Pope, *J. Neurophysiol.* 15:115–130 (1952).
795. L. Austin and J. W. Phillis, *J. Neurochem.* 12:709–717 (1965).
796. P. de Giacomo, *Lavoro Neuropsichiat.* 27:3–14 (1960) (CA 55, 12579e; 1961).
797. M. H. Aprison and H. E. Himwich, *Am. J. Physiol.* 179:502–506 (1954).
798. B. Tigerman, J. Rebar, and C. D. Proctor, *Trans. Ill. State Acad. Sci.* 48:108–111 (1955) (CA 50, 16887f; 1956).
799. M. Papp and G. Bozsik, *J. Neurochem.* 13:697–703 (1966).
800. M. A. Gerebtzoff, *in Biochemistry of the Developing Nervous System* (H. Waelsch, ed.), pp. 315–326, Academic Press, New York (1955).
801. P. Kása, F. Joo, and B. Csillik, *J. Neurochem.* 12:31–35 (1965).
802. D. Glick, *Nature* 140:426–427 (1937).
803. D. Nachmansohn, *Compt. Rend. Soc. Biol.* 129:830–833 (1938).
804. L. Sperti, S. Sperti, and P. Zatti, *Rev. Arch. Ital. Biol.* 98:41–52 (1960).
805. A. A. Pokrovskii and L. G. Ponomareva, *Biokhimiya* 26:248–252 (1961/62).
806. A. A. Pokrovskii and L. G. Ponomareva, *Biokhimiya* 26:276–280 (1961/62).
807. E. Robins and D. E. Smith, *Res. Pubs. Assoc. Res. Nervous Mental Disease* 32:305–327 (1953).

808. D. Nachmansohn, *Compt. Rend. Soc. Biol.* **128**:24–26 (1938).
809. S. Okinaka, M. Yoshikawa, and J. Goto, *Igaku to Seibutsugaku* (*Med. Biol.*) **17**:82–85 (1950) (CA **45**, 1181c; 1951).
810. F. F. Foldes, E. K. Zsigmond, V. M. Foldes, and E. E. Erdös, *J. Neurochem.* **9**:559–572 (1962).
811. E. Appel, *Acad. Rep. Populare Romine, Studii Cercetari Neurol.* **8**:249–257 (1963) (CA **60**, 927d; 1964).
812. S. Okinaka, M. Yoshikawa, M. Uono, T. Muro, T. Mozai, A. Igata, H. Tanabe, S. Ueda, and M. Tomonaga, *Am. J. Phys. Med.* **40**:135–145 (1961).
813. P. de Giacomo and S. Fabri, *Biochim. Biol. Sper.* **1**:311–318 (1961/62).
814. M. Girgis, *J. Comp. Neurol.* **129**:85–95 (1967).
815. S. C. Shen, P. Greenfield, and E. J. Boell, *J. Comp. Neurol.* **102**:717–743 (1955).
816. G. Toschi, *Exptl. Cell Res.* **16**:232–255 (1959).
817. G. Rodriguez de Lores Arnaiz, *J. Histochem. Cytochem.* **12**:696–699 (1964).
818. H. F. Bradford, E. K. Brownlow, and D. B. Gammack, *J. Neurochem.* **13**:1283–1297 (1966).
819. B. Csillik and P. Kása, *Acta Neuroveg.* (*Vienna*) **29**:289–296 (1967).
820. P. Nathan and M. H. Aprison, *Federation Proc.* **14**:106 (1955).
821. M. G. Amadyan and E. I. Il'ina-Kakueva, *Zh. Vysshei Nervnoi Deyatel'nosti im I. P. Pavlova* **16**:514–518 (1966) (CA **65**, 9432d; 1966).
822. D. Richter and R. P. Hullin, *Biochem. J.* **48**:406–410 (1951).
823. N. L. Leibson, *Dokl. Akad. Nauk SSSR* **153**:1435–1438 (1963).
824. D. Nachmansohn, *Yale J. Biol. Med.* **12**:565–589 (1940).
825. C. J. Metzler and D. G. Humm, *Science* **113**:382–383 (1951).
826. J. Elkes and A. Todrick, in *Biochemistry of the Developing Nervous System* (H. Waelsch, ed.), pp. 309–314, Academic Press, New York (1955).
827. P. Cohn and D. Richter, *J. Neurochem.* **1**:166–172 (1956).
828. G. W. Atherton, *Histochemie* **3**:214–221 (1963).
829. G. J. Maletta and P. S. Timiras, *J. Neurochem.* **13**:75–84 (1966).
830. D. J. Rosca and C. Stanciu, *Stud. Univ. Babes-Bolyai Ser. Biol.* **2**:105–112 (1966) (CA **67**, 9254b; 1967).
831. B. Csillik, F. Joo, P. Kása, I. Tomity, and G. Kálmán, *Acta Biol. Acad. Sci. Hung.* **15**:11–17 (1964).
832. H. E. Himwich and M. H. Aprison, in *Biochemistry of the Developing Nervous System* (H. Waelsch, ed.), pp. 301–307, Academic Press, New York (1955).
833. F. Kavaler and V. M. Kimel, *J. Comp. Neurol.* **96**:113–119 (1952).
834. P. J. Parquet and J. Leonardelli, *Compt. Rend. Soc. Biol.* **160**:1229–1231 (1966).
835. A. V. Sakharova, *Tsitologia* **8**:54–59 (1966) (CA **64**, 18121f; 1966).
836. K. H. Sramstad, *Acta Physiol. Scand.* **36**:383–388 (1956).
837. D. Nachmansohn, *J. Neurophysiol.* **3**:396–402 (1940).
838. O. Svensmark, *Acta Physiol. Scand.* **52**:372–378 (1961).
839. J. Bernsohn, K. D. Barron, and A. R. Hess, *Nature* **195**:285–286 (1962).
840. J. Bernsohn, K. D. Barron, and M. T. Hedrick, *Biochem. Pharmacol.* **12**:761–763 (1963).
841. K. A. Youngstrom, *J. Neurophysiol.* **4**:473–477 (1941).
842. J. D. Utley, *Biochem. Pharmacol.* **15**:1–6 (1966).
843. W. A. Himwich, W. T. Sullivan, B. Kelly, H. B. W. Benaron, and B. E. Tucker, *J. Nervous Mental Disease* **122**:441–447 (1956).
844. D. Nachmansohn, *Compt. Rend. Soc. Biol.* **127**:670–673 (1938).
845. K. T. Rogers, L. A. L. de Vries, C. R. Kepler, and E. R. Speidel, *J. Exptl. Zool.* **144**:89–103 (1960).
846. L. Sperti, S. Sperti, and P. Zatti, *Rev. Arch. Ital. Biol.* **98**:53–59 (1960).

847. D. G. Shappirio, D. M. Eichenbaum, and B. R. Locke, *Biol. Bull.* **132**:108–125 (1967).
848. E. G. Lapetina, E. F. Soto, and E. de Robertis, *Biochim. Biophys. Acta* **135**:33–43 (1967).
849. J. D. Robinson and R. M. Bradley, *Nature* **197**:389–390 (1963).
850. M. G. Ord and R. H. S. Thompson, *Biochem. J.* **46**:346–352 (1950).
851. D. E. Woolley, *J. Neurochem.* **10**:447–452 (1963).
852. K. D. Barron, J. Bernsohn, and A. R. Hess, *J. Histochem. Cytochem.* **11**:139–156 (1963).
853. J. Bernsohn, K. D. Barron, and A. R. Hess, in *Progress in Brain Research*, Vol. 9 (W. A. Himwich and H. E. Himwich, eds.), pp. 161–164, Elsevier, Amsterdam, London, and New York (1964).
854. J. Suzuoki and T. Suzuoki, *J. Biochem. (Japan)* **40**:599–609 (1953).
855. J. Suzuoki and T. Suzuoki, *Nature* **173**:83–84 (1954).
856. R. Giuffre, G. Moricca, R. Cavaliere, and A. Massa, *Clin. Chim. Acta* **8**:54–57 (1963).
857. N. Seiler, L. Kamenikova, and G. Werner, *Hoppe-Seyler's Z. Physiol. Chem.* **348**:768–774 (1967).
858. F. Ogata, *Med. J. Shinshu Univ.* **9**:21–24 (1964) (CA **62**, 4435d; 1965).
859. R. C. Shope, *J. Biol. Chem.* **80**:127–132 (1928).
860. R. Clarenburg, A. B. Steinberg, J. H. Asling, and I. L. Chaikoff, *Biochemistry* **5**:2433–2440 (1966).
861. A. M. Bertolini, C. Guardamagna, and N. Massari, *Boll. Soc. Ital. Biol. Sper.* **36**:434–437 (1960).
862. E. T. Pritchard and N. E. Nichol, *Biochim. Biophys. Acta* **84**:781–782 (1964).
863. P. A. Srere, W. Seubert, and F. Lynen, *Biochim. Biophys. Acta* **33**:313–319 (1959).
864. E. E. Dekker, M. J. Schlesinger, and M. J. Coon, *J. Biol. Chem.* **233**:434–438 (1958).
865. G. Pfleiderer, in *Hoppe-Seyler/Thierfelder, Handbuch der physiologisch- und pathologisch-chemischen Analyse* (K. Lang and E. Lehnartz, eds.), Vol. 6C, pp. 677–681, Springer Verlag, Berlin, Göttingen, Heidelberg, and New York (1966).
866. H. J. Strecker, H. Sachs, and H. Waelsch, *J. Am. Chem. Soc.* **76**:3354–3355 (1954).
867. H. Sachs and H. Waelsch, *Arch. Biochem. Biophys.* **69**:422–434 (1957).
868. H. J. Strecker, P. Mela, and H. Waelsch, *J. Biol. Chem.* **212**:223–233 (1955).
869. O. Hoffmann-Ostenhof and R. Ehrenreich, in *Hoppe-Seyler/Thierfelder, Handbuch der physiologisch- und pathologisch-chemischen Analyse* (K. Lang and E. Lehnartz, eds.), Vol. 6B, pp. 963–1009, Springer Verlag, Berlin, Heidelberg, and New York (1966).
870. P. M. Charegaonkar and T. H. Rindani, *Proc. Indian. Acad. Sci.* **54**:113–116 (1961).
871. H. H. Fleischhacker, *J. Mental Sci.* **84**:947–959 (1938).
872. K. J. Kotkova, *Biochem. J. (Ukraine)* **13**:19–31 (1939) (CA **33**, 8635[4]; 1939).
873. J. J. Gordon, *Biochem. J.* **46**:96–99 (1950).
874. V. R. Cunningham and E. J. Field, *J. Neurochem.* **11**:281–285 (1964).
875. K. V. Giri and N. C. Datta, *Biochem. J.* **30**:1089–1096 (1936).
876. S. Saraswathi and B. K. Bachhawat, *J. Neurochem.* **13**:237–246 (1966).
877. R. K. Datta and J. J. Ghosh, *J. Neurochem.* **11**:779–786 (1964).
878. K. P. Strickland, R. H. S. Thompson, and R. G. Webster, *J. Neurol. Neurosurg. Psychiat.* **19**:12–16 (1956).
879. G. T. Adunts, *Izv. Akad. Nauk. Arm. SSR., Biol. Nauki* **16**:25–30 (1963).
880. D. Domyan, *Vopr. Fiziol. i. Patol. Nervn. Sistemy, Inst. Fiziol. Akad. Nauk. SSSR* **1958**:53–55 (CA **54**, 13325h; 1960).
881. D. Domyan, *Dokl. Akad. Nauk. SSSR* **126**:442–445 (1959) (CA **54**, 2532f; 1960).
882. H. B. Tewari and G. H. Bourne, *J. Anat.* **97**:65–72 (1963).
883. K. Felgenhauer, *Z. Zellforsch.* **60**:518–531 (1963).
884. H. Firket, C. Heusghem, M. A. Gerebtzoff, and G. Ninane, *Compt. Rend. Soc. Biol.* **143**:1407–1408 (1949).

885. K.-P. Cheng, S.-K. Chang, and H.-L. Ch'en, *Sheng Li Hsueh Pao* **25**: 42–48 (1962) (CA **59**, 13173f; 1963).
886. J. J. Gordon, *Biochem. J.* **55**:812–817 (1953).
887. G. Caradante, *Arch. Sci. Biol.* (*Italy*) **28**:13–21 (1942) (CA **37**, 54249; 1943).
888. R. L. Friede, *J. Neurochem.* **13**:197–203 (1966).
889. R. L. Holmes, *J. Endocrinol.* **23**:63 (1961).
890. N. Robinson and B. M. Phillips, *Clin. Chim. Acta* **10**:414–419 (1964).
891. E. Albert, *Hoppe-Seyler's Z. Physiol. Chem.* **302**:129–141 (1955).
892. K. T. Rogers, *Exptl. Cell Res.* **34**:100–110 (1964).
893. B. Smiechowska, *Folia Morphol.* **15**:227–256 (1964) (CA **62**, 15155d; 1965).
894. J. B. Flexner, C. L. Greenblatt, S. R. Copperband, and L. B. Flexner, *Am. J. Anat.* **98**:129–138 (1956).
895. E. V. Chirkovskaya, *Izv. Akad. Nauk. SSSR. Ser. Biol.* **1956**(6):19–25 (CA **51**, 5146i; 1957).
896. W. Albrecht, *Monatsschr. Kinderheilk.* **103**:412–414 (1955) (CA **50**, 1950a; 1955).
897. G. T. Adunts, *Vopr. Biokhim. Akad. Nauk. Arm. SSR* **2**:139–152 (1961) (CA **59**, 5549b; 1963).
898. K. T. Rogers, *J. Exptl. Zool.* **153**:15–20 (1963).
899. S. Loevtrup, *J. Exptl. Zool.* **147**:227–232 (1961).
900. R. L. Friede, *Topographic Brain Chemistry*, Academic Press, New York, London (1966).
901. F. Cedrangolo, *Boll. Soc. Ital. Biol. Sper.* **10**:374–376 (1935).
902. H. Josephy, *Arch. Neurol. Psychiat.* **61**:164–169 (1949).
903. P. J. Anderson, S. K. Song, and N. Christoff, *Intern. Congr. Neuropathol. Proc. 4, Munich 1961* **1**:75–79 (1962) (CA **59**, 15592a; 1963).
904. B. W. Moore and P. U. Angeletti, *Ann. N.Y. Acad. Sci.* **94**:659–667 (1961).
905. F. Cedrangolo and A. Ruffo, *Arch. Sci. Biol.* (*Italy*) **24**:59–69 (1938). (CA **32**, 9221⁵; 1938).
906. J. Osinchak, *J. Cell. Biol.* **21**:35–47 (1964).
907. T. Hiroshige, T. Nagatsugawa, T. Imazeki, and S. Itoh, *Japan J. Physiol.* **16**:103–112 (1966).
908. S. Goldfischer, *J. Neuropath. Exptl. Neurol.* **23**:36–45 (1964).
909. R. L. Friede and M. Knoller, *J. Neurochem.* **12**:441–450 (1965).
910. J. Mordoh, *J. Neurochem.* **12**:505–514 (1965).
911. K. Ogawa, Y. Shinonaga, and T. Suzuki, *J. Electronmicroscopy* (*Tokyo*) **11**:111–115 (1962) (CA **61**, 3489f; 1964).
912. O. Z. Sellinger and G. Rucker, *Life Sci.* **3**:1097–1102 (1964).
913. V. P. Whittaker, *Biochem. Pharmacol.* **1**:351 (1958/1959).
914. K. D. Barron and T. O. Tuncbay, *Am. J. Pathol.* **40**:637–652 (1962).
915. V. G. Avdeev and O. V. Palladin, *Ukr. Biokhim. Zh.* **39**:18–24 (1967) (CA **67**, 734a; 1967).
916. H. Koenig, *Nature* **195**:782–784 (1962).
917. H. Koenig, *J. Histochem. Cytochem.* **11**:120–121 (1963).
918. H. Koenig, *J. Histochem. Cytochem.* **11**:556–557 (1963).
919. N. F. M. Lemkey, *Dissertation Abstr.* **23**:4520–4521 (1963).
920. J. B. Flexner and L. B. Flexner, *J. Cellular. Comp. Physiol.* **31**:311–320 (1948).
921. A. M. Lassek and W. L. Hard, *Science* **102**:123–124 (1945).
922. K. Ahmed and J. D. Judah, *Biochim. Biophys. Acta* **93**:603–613 (1964).
923. R. W. Albers, G. Rodriguez de Lores Arnaiz, and E. de Robertis, *Proc. Natl. Acad. Sci. U.S.* **53**:557–564 (1966).
924. G. D. Hunter and G. C. Millson, *J. Neurochem.* **13**:375–383 (1966).
925. R. Coleman and G. Hübscher, *Biochim. Biophys. Acta* **56**:479–490 (1962).

926. B. W. Agranoff, *J. Lipid Res.* 3:190–196 (1962).
927. R. E. McCaman, M. Smith, and K. Cook, *J. Biol. Chem.* 240:3513–3517 (1965).
928. J. G. Salway, M. Kai, and J. N. Hawthorne, *J. Neurochem.* 14:1013–1024 (1967).
929. L. F. Pomazanskaya, *Zh. Evolyutsonnoi Biokhim. Fiziol.* 1:320–324 (1965) (CA **64**, 2467b; 1966).
930. L. F. Pomazanskaya, *Dokl. Akad. Nauk. SSSR.* 155:208–211 (1964) (CA **60**, 13619c; 1964).
931. P. L. Ipata, *Nature* 214:618 (1967).
932. T. G. Scott, *J. Histochem. Cytochem.* 13:657–667 (1965).
933. J. L. Reis, *Biochem. J.* 48:548–551 (1951).
934. W. K. Jordan and R. March, in *Biochemistry of the Developing Nervous System* (H. Waelsch, ed.), pp. 327–334, Academic Press, New York (1955).
935. D. Naidoo, *J. Histochem. Cytochem.* 10:421–434 (1962).
936. K. Nandy and G. H. Bourne, *Histochemie* 4:488–493 (1965).
937. T. G. Scott, *J. Comp. Neurol.* 129:97–113 (1967).
938. N. M. Polyakova and M. K. Malisheva, *Ukrain. Biokhim. Zh.* 33:713–730 (1961) (CA **56**, 7841d; 1962).
939. J. P. Greenstein and H. W. Chalkley, *Arch. Biochem.* 7:451–455 (1945).
940. J. P. Greenstein, C. E. Carter, H. W. Chalkley, and F. M. Leuthardt, *J. Natl. Cancer Inst.* 7:9–27 (1946).
941. G. Weber, in *Fructose-1.6-diphosphatase. Its role in Gluconeogenesis*, Symp. Univ. Virginia, 1961, pp. 57–84 (1964).
942. N. C. Ghosh, N. C. Kar, and I. Chatterjee, *Nature* 197:596–597 (1963).
943. H. B. Tewari and G. H. Bourne, *Exptl. Cell. Res.* 28:444–449 (1962).
944. H. B. Tewari and G. H. Bourne, *J. Histochem. Cytochem.* 11:121–122 (1963).
945. H. Kraut and F. Borkowsky, *Hoppe-Seyler' Z. Physiol. Chem.* 220:192–198 (1933).
946. A. A. Simonjan, *Izv. Akad. Nauk. Arm. SSR Biol. Nauki* 15:77–80 (1962) (CA **58**, 790e; 1963).
947. R. P. Samoilova, *Visnik. Kiiv. Univ. Ser. Biol.* 1958(1):163–166 (CA **54**, 16591g; 1960).
948. S. P. R. Rose and P. J. Heald, *Biochem. J.* 81:339–347 (1961).
949. S. Ando, *Gifu Ika Diagaku Kiyo* 11:345–353 (1963) (CA **61**, 11090f; 1964).
950. S. P. R. Rose, *Biochem. J.* 83, 614–622 (1962).
951. C.-N. Li and Y.-S. Ho, *Sheng Li Hua Hsüeh Pao* 1:87–95 (1958) (CA **54**, 21372f; 1960).
952. J. P. Greenstein and F. M. Leuthardt, *J. Natl. Cancer Inst.* 7:1–3 (1946).
953. M. L. Kornguth and E. A. Stubbs, *Arch. Biochem. Biophys.* 109:104–109 (1965).
954. S. Saraswathi and B. K. Bachhawat, *J. Neurochem.* 10:127–133 (1963).
955. K. H. Kiessling, *Acta Chem. Scand.* 14:1669–1670 (1960).
956. W. Thompson and R. M. C. Dawson, *Biochem. J.* 91:233 (1964).
957. G. H. Sloane-Stanley, *Biochem. J.* 53:613–619 (1953).
958. R. Rodnight, *Biochem. J.* 63:223–231 (1956).
959. R. M. C. Dawson and W. Thompson, *Biochem. J.* 91:244–250 (1964).
960. M. Chang and C. E. Ballou, *Biochem. Biophys. Res. Commun.* 26:199–205 (1967).
961. R. Joseph and B. K. Bachhawat, *J. Neurochem.* 11:517–526 (1964).
962. J. W. Healy, D. Stollar, M. J. Simon, and L. Levine, *Arch. Biochem. Biophys.* 103:461–468 (1963).
963. J. W. Healy, D. Stollar, and L. Levine, *Proc. Nucleic Acid Res.* 1966:188–191 (CA **65**, 10868d; 1966).
964. R. K. Datta and J. J. Ghosh, *J. Neurochem.* 10:285–286 (1963).
965. G. R. Webster, E. A. Marples, and R. H. S. Thompson, *Biochem. J.* 65:374–377 (1957).
966. K. V. Druzhinina and M. G. Kritsman, *Biokhimiya* 17:77–81 (1952).
967. A. Roitman and S. Gatt, *Israel J. Chem.* 1:190 (1963).

968. Y. Barenholz, A. Roitman, and S. Gatt, *J. Biol. Chem.* **241**:3731–3737 (1966).
969. W. Thompson, *Can. J. Biochem. Physiol.* **45**:853–861 (1967).
970. W. Thompson and R. M. C. Dawson, *Biochem. J.* **91**:237–243 (1964).
971. T. P. Silich and E. B. Skvirskaya, *Ukr. Biokhim. Zh.* **27**:41–46 (1955) (CA **49**, 10395e; 1955).
972. V. Desreux, R. Hacha, and E. Fredericq, *J. Gen. Physiol.*, Suppl., **45**:93–102 (1962).
973. G. Portier de Courcy and T. Terroine, *Arch. Sci. Physiol.* **20**:459–490 (1966).
974. K. Tempel and W. Rössner, *Arzneimittelforsch.* **15**:25–27 (1965).
975. R. Gautier and A. Leonard, *Exptl. Cell Res.* **28**:335–341 (1962).
976. G. I. Drummond and S. Perrot-Yee, *J. Biol. Chem.* **236**:1126–1129 (1961).
977. G. I. Drummond, N. T. Iyer, and J. Keith, *J. Biol. Chem.* **237**:3535–3539 (1962).
978. R. W. Butcher and E. W. Sutherland, *J. Biol. Chem.* **237**:1244–1250 (1962).
979. W.-Y. Cheung, *Biochemistry* **6**:1079–1087 (1967).
980. T. R. Shanta, W. D. Woods, M. B. Waitzman, and G. H. Bourne, *Histochemie* **7**:177–190 (1966).
981. W.-Y. Cheung and L. Salganicoff, *Nature* **214**:90–91 (1967).
982. K. B. Augustinsson and G. Heimbürger, *Acta Chem. Scand.* **8**:753–761 (1954).
983. K. S. Dogson, B. Spencer, and J. Thomas, *Biochem. J.* **53**:452–457 (1953).
984. C. Neuberg and E. Simon, *Biochem. Z.* **156**:365–373 (1925).
985. K. S. Dogson, B. Spencer, and C. H. Wynn, *Biochem. J.* **62**:500–507 (1956).
986. J. Austin, B. McAfee, D. Armstrong, M. O'Rourke, L. Shearer, and B. Bachhawat, *Biochem. J.* **93**:15C–17C (1964).
987. W. Bleszynski, *Enzymologia* **32**:169–181 (1967).
988. A. S. Balasubramanian and B. K. Bachhawat, *J. Neurochem.* **10**:201–211 (1963).
989. M. Kozig and A. Wenclewski, *Acta Histochem.* **21**:135–142 (1965).
990. L. M. Dzialoszynski and A. Wenclewski, *Clin. Chim. Acta* **8**:565–567 (1963).
991. A. Wenclewski and L. Dzialoszynski, *Acta Physiol. Polon.* **16**:635–640 (1965) (CA **64**, 14641d; 1966).
992. A. Wenclewski and I. Wysocka-Wenclewska, *Acta Physiol. Polon.* **16**:245–249 (1965) (CA **64**, 5507c; 1966).
993. J. Kuczinski, T. Pydzik, and A. Wenclewski, *Ginekol. Polska* **35**:373–378 (1964) (CA **61**, 13687e; 1964).
994. S. Okamoto, *J. Biochem.* (*Japan*) **37**:269–274 (1950).
995. O. Ya Rashba, *Ukr. Biokhim. Zh.* **20**:34–54 (1948) (CA **46**, 2147i; 1952).
996. B. Slovtzov, *Russ. Physiol. J.* **3**:(1–5) (1923).
997. M. L. Petrunkin, *Russ. Physiol. J.* **5**:1–2 (1922) (CA **18**, 2345[7]; 1924).
998. A. V. Palladin and O. Ya Rashba, *Ukr. Biokhim. Zh.* **20**:151–164 (1948) (CA **46**, 10221c; 1952).
999. I. A. Popova, A. J. Shubina-Vinitskaya, and E. L. Rozenfeld, *Biokhimiya* **29**:312–317 (1964) (CA **61**, 2120b; 1964).
1000. E. L. Rozenfeld and I. S. Lukomskaya, *Biokhimiya* **21**:412–416 (1956).
1001. F. Pasolini, *Acta Neurol.* (*Naples*) **8**:99–117 (1953) (CA **48**:7735f; 1954).
1002. R. Carubelli, R. E. Trucco, and R. Caputto, *Biochim. Biophys. Acta* **60**:196–197 (1962).
1003. I. V. Tsvetkova, *Biokhimiya* **30**:407–414 (1965).
1004. E. H. Morgan and C. B. Laurell, *Nature* **197**:921–922 (1963).
1005. G. L. Ada, *Biochim. Biophys. Acta* **73**:276–284 (1963).
1006. R. B. Cohen, S. H. Rutenburg, K.-C. Tsou, M. A. Woodbury, and A. M. Seligman, *J. Biol. Chem.* **195**:607–614 (1952).
1007. S. Gatt and M. M. Rapport, *Biochim. Biophys. Acta* **113**:567–576 (1966).
1008. S. Gatt, *Biochem. J.* **101**:687–691 (1966).
1009. S. Gatt and M. M. Rapport, *Biochem. J.* **101**:680–686 (1966).

1010. A. J. Furth and D. Robinson, *Biochem. J.* 97:59–66 (1965).
1011. G. Kh. Bunyatyan and E. E. Mkheyan, *Vopr. Biokhim. Akad. Nauk. Arm. SSR*, 3:5–9 (1963) (CA **62**, 2051b; 1965).
1012. J. Conchie, J. Findlay and G. A. Levvy, *Biochem. J.* 71:318–325 (1959).
1013. A. K. Hajra, D. M. Bowen, Y. Kishimoto, and N. S. Radin, *J. Lipid. Res.* 7:379–386 (1966).
1014. R. O. Brady, A. E. Gal, R. M. Bradley, and E. Martensson, *J. Biol. Chem.* 242:1021–1026 (1967).
1015. S. R. Korey and A. Stein, *in Brain Lipids, Lipoproteins, Leucodystrophies*, Proc. Neurochem. Symp. Rome 1961, pp. 71–82 (1963) (CA **59**, 13188g; 1963).
1016. G. Tettamani, L. Bertona, E. Bombardelle, and V. Zambotti, *Boll. Soc. Ital. Biol. Sper.* **39**:1974–1977 (1963).
1017. G. Kh. Bunyatyan and E. E. Mkheyan, *Vopr. Biokhim. Mozga, Akad. Nauk. Arm. SSR, Inst. Biokhim.* 2:130–135 (1966) (CA **66**, 62126s; 1967).
1018. E. Robins, H. K. Fisher, and I. P. Lowe, *J. Neurochem.* 8:96–104 (1961).
1019. K. Watanabe, *J. Biochem. (Tokyo)* 24:297–303 (1936).
1020. Y. Z. Frohwein and S. Gatt, *Biochemistry* 6:2775–2782 (1967).
1021. M. A. Verity, J. K. Gambell, and W. J. Brown, *Arch. Biochem. Biophys.* 118:310–316 (1967).
1022. Y. Z. Frohwein and S. Gatt, *Biochemistry* 6:2783–2787 (1967).
1023. O. Waltimo and S. Talanti, *Nature* 204:499–500 (1965).
1024. M. Panattoni, *Atti. Acad. Fisiocrit. Siena, Sez. Med. Fis.* 5:926–935 (1958) (CA **54**, 1693e; 1960).
1025. G. C. Millson, *J. Neurochem.* 12:461–468 (1965).
1026. J. H. Quastel and L. J. Zatman, *Biochim. Biophys. Acta* 10:256–267 (1953).
1027. K. B. Jacobson and N. O. Kaplan, *J. Biophys. Biochem. Cytol.* 3:31–43 (1957).
1028. A. Brunnemann und H. Coper, *Naunyn-Schmiedeberg's Arch. Exptl. Pathol. Pharmacol.* **250**:469–478 (1965).
1029. H. McIlwain and R. Rodnight, *Biochem. J.* 44:470–477 (1949).
1030. H. McIlwain, *Biochem. J.* 46:612–619 (1950).
1031. L. J. Zatman, N. O. Kaplan, and S. P. Colowick, *J. Biol. Chem.* 200:197–212 (1953).
1032. N. O. Kaplan and M. M. Ciotti, *J. Am. Chem. Soc.* 76:1713–1714 (1954).
1033. H. W. Dickerman and F. E. Stolzenbach, *Federation Proc.* 16:172 (1957).
1034. N. I. Swislocki and N. O. Kaplan, *J. Biol. Chem.* 242:1083–1088 (1967).
1035. M. E. Spaulding and W. D. Graham, *J. Biol. Chem.* 170:711–718 (1947).
1036. H. Cooper, *Naunyn-Schmiedeberg's Arch. Exptl. Pathol. Pharmacol.* 241:335–345 (1961).
1037. N. I. Swislocki, M. I. Kalish, F. I. Chasalow, and N. O. Kaplan, *J. Biol. Chem.* 242:1089–1094 (1967).
1038. R. M. Burton, *J. Neurochem.* 2:15–20 (1957).
1039. A. A. Abdel-Latif and L. G. Abood, *J. Neurochem.* 11:9–15 (1964).
1040. A. M. Nemeth and H. Dickerman, *J. Biol. Chem.* 235:1761–1764 (1960).
1041. S. Mahadevan and A. L. Tappel, *J. Biol. Chem.* 242:4568–4576 (1967).
1042. G. B. Ansell and S. Spanner, *Biochem. J.* 94:252–258 (1965).
1043. A. S. Brecher, *J. Neurochem.* 10:1–6 (1963).
1044. Y. Arai and T. Kusama, *Proc. Japan Acad.* 41:734–736 (1965).
1045. C. W. M. Adams and G. G. Glenner, *J. Neurochem.* 9:233–239 (1962).
1046. E. Deutsch and K. Irsliger, *New Istanbul Contrib. Clin. Sci.* 7:89–105 (1964) (CA **62**, 5649b; 1965).
1047. N. Marks, *Biochem. J.* 103:40P–41P (1967).
1048. M. W. Kies and S. Schwimmer, *J. Biol. Chem.* 145:685–691 (1942).
1049. L. L. Uzman, S. van den Noort, and M. K. Rumley, *J. Neurochem.* 9:241–252 (1962).

1050. A. S. Brecher, B. B. Oliphant, and R. E. Sobel, *Arch. Intern. Physiol. Biochim.* **74**: 429–434 (1966).

1051. K. Ichikawa, *Med. J. Shinshu Univ.* **4**:345–363 (1959) (CA **56**, 3943i; 1962).

1052. S. van den Noort and L. L. Uzman, *Proc. Soc. Exptl. Biol. Med.* **108**:32–34 (1961).

1053. L. L. Uzman, M. K. Rumley, and S. van den Noort, *J. Neurochem.* **10**:795–804 (1963).

1054. N. Koga, *Nichidai Igaku Zasshi* **18**:308–317 (1959) (CA **61**, 8696d; 1964).

1055. W. Grassmann, O. v. Schoenebeck, und G. Auerbach, *Hoppe-Seyler's Z. Physiol. Chem.* **210**:1–14 (1932).

1056. V. E. Price, A. Meister, J. B. Gilbert, and J. P. Greenstein, *J. Biol. Chem.* **181**:535–547 (1949).

1057. G. B. Ansell and D. Richter, *Biochim. Biophys. Acta* **13**:87–91 (1954).

1058. L. L. Uzman, M. K. Rumley, and S. van den Noort, *J. Neurochem.* **6**:299–310 (1961).

1059. H. v. Euler and H. Hasselquist, *Arkiv Kemi* **25**:129–133 (1966).

1060. E. Blum, A. I. Yakovchuk, and A. I. Yarmoshkevich, *Bull. Biol. Med. Exptl. URSS* **1**:17–18 (1936) (CA **31**, 7072⁶; 1937).

1061. H. v. Euler, H. Hasselquist, and K. Kyyroe, *Arkiv Kemi* **25**:151–157 (1966).

1062. H. v. Euler, H. Hasselquist, and K. Kyyroe, *Arkiv Kemi* **25**:257–262 (1966).

1063. H. v. Euler, K. Kyyroe, and H. Hasselquist, *Arkiv Kemi* **25**:97–107 (1966).

1064. A. Pope and C. B. Anfinsen, *J. Biol. Chem.* **173**:305–311 (1948).

1065. A. Pope, *J. Neurochem.* **4**:31–41 (1959).

1066. J. M. W. Bouma and M. Gruber, *Biochim. Biophys. Acta* **89**:545–547 (1964).

1067. G. B. Ansell and D. Richter, *Congr. Intern. Biochem. Résumés Commun., 2ᵉ Congr., Paris* **1952**:269 (CA **50**, 6632i; 1956).

1068. N. Marks and A. Lajtha, *Biochem. J.* **97**:74–83 (1965).

1069. N. M. Polyakova and V. K. Lishko, *Ukrain. Biochim. Zh.* **34**:208–216 (1962).

1070. K. F. Firfarova, A. D. Morozkin, and V. N. Orekhovich, *Biokhimiya* **29**:673–679 (1964).

1071. N. Marks and A. Lajtha, *Biochem. J.* **89**:438–447 (1963).

1072. A. Lajtha, *in Regional Neurochemistry* (S. S. Kety and J. Elkes, eds.), pp.25–36, Pergamon Press, Oxford, London, New York and Paris (1961).

1073. Y. V. Belik and V. I. Tyulenev, *Zh. Evolutsionnoi Biokhim. Fiziol.* **2**:333–338 (1966) (CA **65**, 19074d; 1966).

1074. G. B. Ansell and D. Richter, *Biochim. Biophys. Acta* **13**:92–97 (1954).

1075. N. W. Penn, *Biochim. Biophys. Acta* **37**:55–63 (1960).

1076. G. Guroff, *J. Biol. Chem.* **239**:149–155 (1964).

1077. E. E. Klein, E. G. Kurtskhaliya, and N. V. Gvaliya, *Soobshch. Akad. Nauk. Gruz. SSR* **44**:331–336 (1966) (CA **66**, 35862X; 1967).

1078. J. F. Woods and G. Nichols, Jr., *Nature* **208**:1325–1326 (1965).

1079. V. K. Lishko, *Ukr. Biokhim. Zh.* **36**:565–573 (1964) (CA **61**, 14962h; 1964).

1080. V. K. Lishko, *Ukr. Biokhim. Zh.* **37**:163–168 (1965) (CA **63**, 4591h; 1965).

1081. S. Ellis, *Biochem. Biophys. Res. Commun.* **12**:452–456 (1963).

1082. Z. Albert, J. Orlowska, M. Orlowski, and A. Szewczuk, *Acta Histochem.* **18**:78–89 (1964).

1083. Z. Albert, M. Orlowski, Z. Rzucidlo, and J. Orlowska, *Acta Histochem.* **25**:312–320 (1966).

1084. K. C. Hooper, *Biochem. J.* **90**:584–587 (1964).

1085. C. C. Toh, *J. Physiol. (London)* **173**:420–430 (1964).

1086. T. Turský and E. Valovičová, *J. Neurochem.* **11**: 99–108 (1964).

1087. H. A. Krebs, *Biochem. J.* **47**:605–614 (1950).

1088. D. Hiwatashi, *Tohoku J. Exptl. Med.* **41**:310–316 (1941) (CA **42**, 8231h; 1948).

1089. H. A. Krebs, *Biochem. J.* **29**:1951–1969 (1935).

1090. M. Errera and J. P. Greenstein, *J. Biol. Chem.* **178**:495–502 (1949).

1091. K. Horiki, *Naika Hokan* **4**:1044–1047 (1957) (CA **53**, 9306d; 1959).

1092. N. L. Blumson, *Biochem. J.* **65**:138–143 (1957).

1093. A. V. Palladin, N. M. Polyakova, and M. K. Malysheva, *Dokl. Akad. Nauk. SSSR* **134**:1236–1239 (1960).

1094. J. D. Utley, *Biochem. Pharmacol.* **13**:1383–1392 (1964).

1095. S. R. Guha and J. J. Ghosh, *Ann. Biochem. Exptl. Med.* (*Calcutta*) **19**:163–168 (1959) (CA **54**, 13340h; 1960).

1096. S. Oeriu, G. Costescu, and O. Teodorescu, *Ukr. Biokhim. Zh.* **35**:163–165 (1963) (CA **59**, 6783c; 1963).

1097. G. Porcellati, *Boll. Soc. Ital. Biol. Sper.* **31**:141–144 (1955).

1098. J. P. Greenstein, *Adv. Enzymol.* **8**:117–169 (1948).

1099. N. Allen, *Exptl. Neurol.* **1**:155–165 (1959).

1100. A. Mori, *Folia Psychiat. Neurol.* (*Japan*) **15**:4–9 (1961) (CA **59**, 7958a; 1963).

1101. M. H. Baslow and J. F. Lenney, *Can. J. Biochem.* **45**:337–340 (1967).

1102. S. Gatt, *J. Biol. Chem.* **241**:3724–3730 (1966).

1103. S. Gatt, in *Inborn Disorders of Sphingolipid Metabolism* (S. M. Aronson and B. W. Volk, eds.), pp. 261–266, Pergamon Press, Oxford, London, and New York (1967).

1104. E. Koch, F. Heiss, and R. Schneider. *Med. Wschr.* **1952**:1184–1186.

1105. E. Koch, H. Bohn, F. Heiss, und R. Schneider, *Naunyn-Schmiedeberg's Arch. Exptl. Pathol. Pharmakol.* **220**:157–183 (1953).

1106. E. Effenberger and H. P. R. Seeliger, *Z. Hyg.* **146**:260–266 (1960).

1107. D. M. Greenberg, in *Hoppe-Seyler/Thierfelder, Handbuch der physiologisch- und patho-logisch-chemischen Analyse* (K. Lang and E. Lehnartz, eds.), Vol. 6C, pp. 354–368, Springer Verlag, Berlin, Heidelberg, and New York (1966).

1108. M. B. Sporn, W. Dingman, A. J. Defalco, and R. K. Davies, *J. Neurochem.* **5**:62–67 (1959).

1109. S. Ratner, H. Morell, and E. Carvalho, *Arch. Biochem. Biophys.* **91**:280–289 (1960).

1110. H. C. Buniatian and M. A. Davtian, *J. Neurochem.* **13**:743–753 (1966).

1111. S. Tomlinson and R. G. Westall, *Nature* **188**:235–236 (1960).

1112. G. P. Talwar, B. K. Goel, M. Mansoor, and N. C. Panda, *J. Neurochem.* **8**:310–311 (1961).

1113. V. P. Korotkoruchko and E. G. Goncharenko, *Ukr. Biokhim. Zh.* **34**:720–726 (1962) (CA **58**, 3788b; 1963).

1114. R. Levine, T. C. Hall, and C. A. Harris, *Cancer* **16**:269–272 (1963).

1115. Y. Wakabayasi, *J. Biochem.* (*Tokyo*) **28**:185–197 (1938).

1116. M. Mansoor, G. D. Kalyankar, and G. P. Talwar, *Biochim. Biophys. Acta* **77**:307–317 (1963).

1117. S. Kumar, K. K. Tewari, and P. S. Krishnan, *J. Neurochem.* **13**:1550–1552 (1966).

1118. S. Kumar, V. Josan, K. C. S. Sanger, K. K. Tewari, and P. S. Krishnan, *Biochem. J.* **102**:691–704 (1967).

1119. W. K. Jordan, R. March, O. B. Houchin, and E. Popp, *J. Neurochem.* **4**:170–174 (1959).

1120. W. K. Jordan, R. March, and R. A. Messing, in *Progress in Neurobiology: 1*, "*Neuro-chemistry*," (S. R. Korey and J. I. Nurnberger, eds.), pp. 101–113, P. B. Hoeber, New York (1956).

1121. E. Hirschberg, M. R. Murray, E. R. Peterson, J. Kream, R. Schafranek, and J. L. Pool, *Cancer Res.* **13**:153–157 (1953).

1122. T. G. Brady and C. I. O'Donovan, *Comp. Biochem. Physiol.* **14**:101–120 (1965).

1123. M. K. Malysheva, N. M. Polyakova, and T. I. Emchuk, *Ukr. Biokhim. Zh.* **36**:323–333 (1964) (CA **61**, 10934g; 1964).

1124. E. J. Conway and R. Cooke, *Biochem. J.* **33**:479–492 (1939).

1125. K. C. Smith, *J. Neurochem.* **9**:277–280 (1962).

1126. J. A. Muntz, *J. Biol. Chem.* **201**:221–233 (1953).
1127. H. Weil-Malherbe and R. H. Green, *Biochem. J.* **61**:218–224 (1955).
1128. M. K. Malysheva, *Ukr. Biokhim. Zh.* **37**:370–378 (1965) (CA **63**, 15165d; 1965).
1129. B. Setlow and J. M. Lowenstein, *J. Biol. Chem.* **242**:607–615 (1967).
1130. P. L. Ipata, G. Ronca, and C. A. Rossi, *Biochem. J.* **97**:18P (1965).
1131. J. Mendicino and J. A. Muntz, *J. Biol. Chem.* **233**:178–183 (1958).
1132. B. Setlow, R. Burger and J. M. Lowenstein, *J. Biol. Chem.* **241**:1244–1245 (1966).
1133. M. K. Malysheva and N. M. Polyakova, *Ukr. Biokhim. Zh.* **37**:360–369 (1965) (CA **63**, 12063b; 1965).
1134. L. G. Abood and L. Romanchek, *Exptl. Cell Res.* **8**:459–465 (1955).
1135. G. K. Bunyatyan and S. G. Movsesyan, *Vopr. Biokhim. Mozga Akad. Nauk Arm. SSR, Inst. Biokhim.* **2**:5–22 (1966) (CA **66**, 61972c; 1967).
1136. O. Hoffmann-Ostenhof, *in Hoppe-Seyler/Thierfelder, Handbuch der physiologisch- und pathologisch-chemischen Analyse* (K. Lang and E. Lehnartz, eds.), Vol. 6C, pp. 399–412, Springer Verlag, Berlin, Heidelberg, and New York (1966).
1137. B. Naganna, *Current Sci.* (*India*) **20**:231–232 (1951) (CA **46**, 8230b; 1952).
1138. U. S. Seal and F. Binkley, *J. Biol. Chem.* **228**:193–199 (1957).
1139. J. J. Gordon, *Nature* **164**:579–580 (1949).
1140. O. M. Fedorov, *Akad. Nauk. Ukr. SSR* **56**:70 (1964) (CA **62**, 10914c; 1965).
1141. O. M. Fedorov, *Ukr. Biokhim. Zh.* **38**:128–131 (1966) (CA **65**, 4355e; 1966).
1142. S. Lovtrup and L. Svennerholm, *Exptl. Cell Res.* **29**:298–313 (1963).
1143. R. J. A. Hosie, *Biochem. J.* **96**:404–412 (1965).
1144. J. Somogyi, A. Fonyo and I. Vincze, *Acta Physiol. Acad. Sci. Hung.* **21**:295–300 (1962).
1145. H. M. Pappius and K. A. C. Elliott, *Can. J. Biochem. Physiol.* **32**:471–483 (1954).
1146. D. Naidoo and O. E. Pratt, *Biochem. J.* **62**:465–469 (1956).
1147. M. B. R. Gore, *Biochem. J.* **50**:18–24 (1951).
1148. A. V. Palladin and T. M. Shtutman, *Ukr. Biokhim. Zh.* **20**:311–320 (1948) (CA **48**, 5895d; 1954).
1149. J. Rebar, Jr., N. Tigerman, and C. P. Proctor, *Trans. Illinois State Acad. Sci.* **48**:104–107 (1955) (CA **50**, 17047a; 1956).
1150. C. Lenti and N. Bargoni, *Med. Sper.* **28**:48–54 (1956) (CA **50**, 16921d; 1956).
1151. J. Somogyi and I. Vincze, *Acta Physiol. Acad. Sci. Hung.* **20**:325–337 (1961) (CA **57**, 2697b; 1961).
1152. R. Tanaka and L. G. Abood, *J. Biol. Chem.* **237**:2999–3004 (1962).
1153. J. R. Lagnado, R. Balázs and D. Richter, *Biochem. J.* **73**:18P (1959).
1154. D. H. Deul and H. McIlwain, *J. Neurochem.* **8**:246–256 (1956).
1155. J. B. Chappell and G. B. Greville, *Nature* **174**:930–931 (1954).
1156. S. E. Epelbaum, G. S. Sheves, and A. A. Kobylin, *Biokhimiya* **14**:107–112 (1949) (CA **43**, 6263d; 1949).
1157. O. V. Kirsenko, *Tretya Vses. Konf. Biokhim. Nervnoi Sistemy Akad. Nauk Arm. SSR, Inst. Biokhim. Sb. Dokl. Erevan* **1962**:55–65 (1963) (CA **60**, 13506a; 1964).
1158. Y. Nakamaru, M. Kosakai, and K. Konishi, *Arch. Biochem. Biophys.* **120**:15–21 (1967).
1159. G. Rendina, *J. Neurochem.* **13**:683–695 (1966).
1160. H. H. Hess and A. Pope, *J. Neurochem.* **3**:287–299 (1959).
1161. O. V. Kirsenko, A. V. Palladin, O. M. Rozhmanova, and S. S. Eismont, *Ukr. Biokhim. Zh.* **35**:807–815 (1963) (CA **60**, 12462e; 1964).
1162. L. C. Mokrasch, *Am. J. Physiol.* **199**:950–954 (1960).
1163. H. H. Hess and A. Pope, *J. Neurochem.* **8**:299–309 (1961).
1164. R. Tanaka and L. G. Abood, *Arch. Biochem. Biophys.* **105**:554–562 (1964).
1165. D. Biesold and V. Bigl, *Wiss. Z. Karl-Marx Univ. Leipzig, Math. Naturw. Reihe* **14**:687–689 (1965) (CA **65**, 4133b; 1966).

1166. Z. P. Kometiani and A. A. Kalandarishvili, *Soobshch. Akad. Nauk. Gruz. SSR.* **43**: 375–381 (1966) (CA **65**, 17455c; 1966).
1167. M. Germain and P. Proulx, *Biochem. Pharmacol.* **14**:1815–1819 (1965).
1168. M. Ruscak, D. Ruscakova, and E. Macejova, *Physiol. Bohemoslov.* **14**:261–265 (1965) (CA **63**, 5964d; 1965).
1169. H. Sugawara and S. Uchida, *Sci. Papers Coll. Gen. Educ. Univ. Tokyo* **13**:73–82 (1963) (CA **59**, 11976d; 1963).
1170. M. Hayashi, J. V. Auditore, and R. Uchida, *Biochim. Biophys. Acta* **81**:624–626 (1964).
1171. I. Krzhechek and F. Shtepanek, *Physiol. Bohemoslov.* **2**:297–302 (1953) (CA **52**, 8321d; 1958).
1172. V. N. Nikitin and R. I. Golubitskaya, *Zh. Khim. Biol. Khim.* **1955**:1271 (CA **50**, 5056b; 1956).
1173. G. K. Bunyatyan and A. A. Simonyan, *Dokl. Akad. Nauk. Arm. SSR* **41**:97–103 (1965) (CA **64**, 931f; 1966).
1174. T. M. Shtutman, *Ukr. Biokhim. Zh.* **21**:73–75 (1949) (CA **48**, 4667h; 1954).
1175. M. Chiga, in *Hoppe-Seyler/Thierfelder, Handbuch der physiologisch- und pathologisch-chemischen Analyse* (K. Lang and E. Lehnartz, eds.), Vol. 6C, pp. 424–438, Springer Verlag, Berlin, Heidelberg, and New York (1966).
1176. G. Siebert, in *Hoppe-Seyler/Thierfelder, Handbuch der physiologisch- und pathologisch-chemischen Analyse* (K. Lang and E. Lehnartz, eds.), Vol. 6C, pp. 439–460, Springer Verlag, Berlin, Heidelberg, and New York (1966).
1177. J. Jarnefelt, *Biochim. Biophys. Acta* **48**:104–110 (1961).
1178. J. C. Skou, *Biochim. Biophys. Acta* **58**:314–325 (1962).
1179. J. Somogyi, *Biochim. Biophys. Acta* **92**:615–617 (1964).
1180. R. F. Squires, *Biochem. Biophys. Res. Commun.* **19**:27–32 (1965).
1181. M. Hayashi and J. V. Auditore, *J. Neurochem.* **11**:671–677 (1964).
1182. R. Gibbs, P. M. Roddy, and E. Titus, *J. Biol. Chem.* **240**:2181–2187 (1965).
1183. J. R. Cooper and H. McIlwain, *Biochem. J.* **102**:675–783 (1967).
1184. J. Jarnefelt, *Biochim. Biophys. Acta* **59**:643–654 (1962).
1185. J. Jarnefelt, *Biochem. Biophys. Res. Commun.* **17**:330–334 (1964).
1186. H. H. Hess, *J. Neurochem.* **9**:613–621 (1962).
1187. J. Somogyi and I. Vincze, *Acta Physiol. Acad. Sci. Hung.* **21**:29–41 (1962).
1188. J. Somogyi and M. Budai, *Hoppe-Seyler's Z. Physiol. Chem.* **336**:264–270 (1964).
1189. N. Gruener and Y. Avi-Dor, *Biochem. J.* **100**:762–767 (1966).
1190. L. J. Fenster and J. H. Copenhaver, Jr., *Biochim. Biophys. Acta* **137**:406–408 (1967).
1191. H. McIlwain and S. Balakrishnan, *Biochem. J.* **79**:1P–2P (1961).
1192. H. Yoshida and H. Fujisawa, *Biochim. Biophys. Acta* **60**:443–444 (1962).
1193. S. P. R. Rose, *Nature* **199**:375–377 (1963).
1194. F. Medzihradsky, M. H. Kline, and L. E. Hokin, *Arch. Biochem. Biophys.* **121**:311–316 (1967).
1195. P. D. Swanson and W. L. Stahl, *Biochem. J.* **99**:396–403 (1966).
1196. K. Nagano, T. Kanazawa, N. Mizuno, Y. Tashima, T. Nakao, and M. Nakao, *Biochem. Biophys. Res. Commun.* **19**:759–764 (1965).
1197. R. Whittam and D. M. Blond, *Biochem. J.* **92**:147–158 (1964).
1198. J. Cummins and H. Hydén, *Biochim. Biophys. Acta* **60**:271–283 (1962).
1199. K. Nagano, N. Mizuno, M. Fujita, Y. Tashima, T. Nakao, and M. Nakao, *Biochim. Biophys. Acta* **143**:239–248 (1967).
1200. C. J. Skou and C. Hilberg, *Biochim. Biophys. Acta* **110**:359–369 (1965).
1201. F. J. Samaha, *J. Neurochem.* **14**:333–341 (1967).
1202. S. L. Bonting, L. L. Caravaggio, and N. M. Hawkins, *Arch. Biochem. Biophys.* **98**:413–419 (1962).

1203. J. C. Skou, *Physiol. Rev.* **45**:596–617 (1965).
1204. J. Jarnefelt, *Biochim. Biophys. Acta* **59**:655–662 (1962).
1205. E. Lewin and H. H. Hess, *J. Neurochem.* **11**:473–481 (1964).
1206. W. N. Aldridge, *Biochem. J.* **83**:527–533 (1962).
1207. O. Z. Sellinger, F. de Balbian Verster, R. J. Sullivant, and C. Lamar, *J. Neurochem.* **13**:501–513 (1966).
1208. D. Voth, *Brain Res.* **4**:60–80 (1967).
1209. F. Turba, H. Buescher, H. Fasold, J. Finke, and J. Gerlach, *Z. gesamte Exptl. Med.* **141**:343–349 (1966) (CA **66**, 63318t; 1967).
1210. V. P. Whittaker, *Ann. N.Y. Acad. Sci.* **137**:982–998 (1966).
1211. L. J. Côté, *Life Sci.* **3**:899–901 (1964).
1212. F. E. Samson, Jr., and D. J. Quinn, *J. Neurochem.* **14**:421–427 (1967).
1213. G. Palladini and G. Venturini, *Atti Acad. Nazl. Lincei, Rend. Classe Sci. Fis. Mat. Nat.* **41**:122–125 (1966) (CA **66**, 62891n; 1967).
1214. L. G. Abood, R. W. Gerard, J. Banks, and R. D. Tschirgi, *Am. J. Physiol.* **168**:728–738 (1952).
1215. M. F. Utter, *J. Biol. Chem.* **185**:499–517 (1950).
1216. W. K. Jordan and R. March, *J. Histochem. Cytochem.* **4**:301–307 (1956).
1217. L. Malucci, *Boll. Soc. Ital. Biol. Sper.* **33**:1080–1081 (1957).
1218. F. Moog, *J. Exptl. Zool.* **105**:209–220 (1947).
1219. J. Jarnefelt, *Biochim. Biophys. Acta* **48**:111–116 (1961).
1220. P. D. Swanson, *J. Neurochem.* **13**:229–236 (1966).
1221. M. Nakao, K. Nagano, T. Nakao, N. Mizuno, Y. Tashima, M. Fujita, H. Maeda, and H. Matsudaira, *Biochem. Biophys. Res. Commun.* **29**:588–592 (1967).
1222. M. Fujita, K. Nagano, N. Mizuno, Y. Tashima, T. Nakao, and M. Nakao, *Biochem. J.* **106**:113–121 (1968).
1223. K. Bowler and C. J. Duncan, *Comp. Biochem. Physiol.* **24**:223–227 (1968).
1224. Y. Israel and E. Titus, *Biochim. Biophys. Acta* **139**:450–459 (1967).
1225. W. Bleszynski and A. Leźnicki, *Enzymologia* **33**:373–389 (1967).
1226. L. Raijman, S. Grisolia, and H. Edelhoch, *J. Biol. Chem.* **235**:2340–2342 (1960).
1227. A. Guerritore, A. Zanobini, and G. Ramponi, *Boll. Soc. Ital. Biol. Sper.* **35**:2163–2166 (1959).
1228. F. Melani, G. Ramponi and A. Guerritore, *Boll. Soc. Ital. Biol. Sper.* **37**:1268–1270 (1961).
1229. G. K. Bunyatyan and G. T. Adunts, *Vopr. Biokhim. Akad. Nauk. Arm. SSR* **1**:149–158 (1960) (CA **55**, 23724c; 1961).
1230. G. T. Adunts and I. G. Aslanyan, *Biol. Nauki* **18**:22–28 (1965) (CA **63**, 16837b; 1965).
1231. T. N. Pattabiraman, T. N. Sekhara Varma, and B. K. Bachhawat, *Biochim. Biophys. Acta* **83**:74–83 (1964).
1232. D. Naidoo and O. E. Pratt, *J. Neurol. Neurosurg. Physchiat.* **15**:164–168 (1952).
1233. D. Naidoo, *J. Histochem. Cytochem.* **10**:580–591 (1962).
1234. S. S. Lazarus and J. Wallace, *J. Histochem. Cytochem.* **12**:729–736 (1964).
1235. L. A. Tseitlin, *Biokhimiya* **17**:208–213 (1952) (CA **46**, 7603g; 1952).
1236. J. Meyer and J. P. Weinmann, *J. Histochem. Cytochem.* **5**:354–397 (1957).
1237. J. A. Sinden and E. Scharrer, *Proc. Soc. Exptl. Biol. Med.* **72**:60–62 (1949).
1238. P. M. Dreyfus and G. Hauser, *Biochim. Biophys. Acta* **104**:78–84 (1965).
1239. R. A. Felicioli, F. Gabrielli, and C. A. Rossi, *Experientia* **22**:728–729 (1966).
1240. M. Suga, *Okayama Igakkai Zasshi* **71**:2113–2117 (1959) (CA **54**, 16595i; 1960).
1241. M. Suga, *Okayama Igakkai Zasshi* **71**:2119–2122 (1959) (CA **54**, 16595i; 1960).
1242. J. L. Hedrick and H. J. Sallach, *Arch. Biochem. Biophys.* **105**:261–269 (1964).

1243. W. E. Davies, A. K. Hajra, S. S. Parmar, N. S. Radin, and J. F. Mead, *J. Lipid Res.* **7**:270–276 (1966).
1244. J. F. Mead, *Metab. Lipids Related Atherosclerosis, Symp. Urbana,* **1963**:198–206 (1965) (CA **65**, 1154d; 1966).
1245. G. M. Levis, *Biochim. Biophys. Acta* **99**:194–197 (1965).
1246. G. M. Levis and J. F. Mead, *J. Biol. Chem.* **239**:77–80 (1964).
1247. R. C. MacDonald, (CA **66**:43959z; 1967).
1248. S. Nakagawa and W. R. Ruegamer, *Biochemistry* **6**:1249–1261 (1967).
1249. B. Weissmann, G. Rowin, J. Marshall, and D. Friederici, *Biochemistry* **6**:207–214 (1967).
1250. J. D. Robinson, *Biochemistry* **6**:3250–3258 (1967).
1251. E. Roberts and S. Frankel, *J. Biol. Chem.* **188**:789–795 (1951).
1252. I. P. Lowe, E. Robins, and G. S. Eyerman, *J. Neurochem.* **3**:8–18 (1958).
1253. J. Awapara, A. J. Landua, R. Fuerst, and B. Seale, *J. Biol. Chem.* **187**:35–39 (1950).
1254. H. Langemann and H. Ackermann, *Helv. Physiol. Pharmacol. Acta* **19**:399–406 (1961).
1255. N. Frontali, *Nature* **191**:178–179 (1961).
1256. J. P. Szusz, B. Haber, and E. Roberts, *Biochemistry* **5**:2870–2877 (1966).
1257. S. O. Petrova, A. J. Komkova, and I. A. Sytinskii, *Vestn. Leningr. Univ. Ser. Biol.* **19**(3): 141–144 (1964) (CA **62**, 5507h; 1965).
1258. W. J. Wingo and J. Awapara, *J. Biol. Chem.* **187**:267–271 (1950).
1259. E. Roberts and D. G. Simonsen, *Biochem. Pharmacol.* **12**:113–134 (1963).
1260. C. F. Baxter, E. Roberts, and E. Eidelberg, *Federation Proc.* **18**:187 (1959).
1261. E. Roberts, P. J. Harman, and S. Frankel, *Proc. Soc. Exptl. Biol. Med.* **78**:799–803 (1951).
1262. R. W. Albers and R. O. Brady, *J. Biol. Chem.* **234**:926–928 (1959).
1263. R. W. Albers, *in The Neurochemistry of Nucleotides and Amino Acids* (R. O. Brady and D. B. Tower, eds.), pp. 146–158, John Wiley, New York, London (1960).
1264. P. B. Mueller and H. Langemann, *J. Neurochem.* **9**:399–401 (1962).
1265. I. A. Sytinskii, V. A. Bernshtam, and T. N. Priyatkina, *Nervnaya Sistema, Leningr. Gos. Univ. Fiziol. Inst.* (6)19–26 (1965) (CA **65**, 19072f; 1966).
1266. B. Sisken, E. Roberts, and C. F. Baxter, *in Inhibition in the Nervous System and γ-Aminobutyric Acid"* (E. Roberts, ed.), pp. 219–225, Pergamon Press, Oxford, London, New York, Paris (1960).
1267. S. Lovtrup, *J. Neurochem.* **8**:243–245 (1961).
1268. N. F. Shatunova and I. A. Sytinskii, *J. Neurochem.* **11**:701–708 (1964).
1269. H. Weinstein, E. Roberts, and T. Kakefuda, *Biochem. Pharmacol.* **12**:503–509 (1963).
1270. E. A. Kravitz, *J. Neurochem.* **9**:363–370 (1962).
1271. W. A. Himwich, J. C. Petersen, and J. P. Graves, *Recent Advances Biol. Psychiat.* **3**:218–226 (1961).
1272. C. F. Baxter, J. P. Schadé, and E. Roberts, *in Inhibition in the Nervous System and γ-Aminobutyric Acid* (E. Roberts, ed.), pp. 214–218, Pergamon Press, Oxford, London, New York, and Paris (1960).
1273. S. Hisada and T. Nakashima, *Bitamin (Kyoto)* **21**:81–85 (1960) (CA **61**, 16358f; 1964).
1274. L. P. Bouthillier and Y. Binette, *Can. J. Biochem. Physiol.* **39**:1930–1933 (1961).
1275. R. W. Schayer, R. L. Smiley, and K. J. Davis, *Am. J. Physiol.* **187**:63–65 (1956).
1276. P. Holtz and E. Westermann, *Naunyn-Schmiedeberg's Arch. Exptl. Pathol. Pharmacol.* **227**:538–546 (1956).
1277. T. Naito and K. Kuriaki, *Naunyn-Schmiedeberg's Arch. Exptl. Pathol. Pharmacol.* **232**:481–486 (1958).
1278. G. P. Gulidova, *Zh. Evol. Biokhim. Fiziol.* **3**:300–303 (1967).
1279. H. S. Bachelard, *Nature* **215**:959–960 (1967).

1280. A. Bertler and E. Rosengren, *Acta Physiol. Scand.* **47**:350–361 (1959).
1281. V. E. Davis and J. Awapara, *J. Biol. Chem.* **235**:124–127 (1960).
1282. L. S. Dietrich, *J. Biol. Chem.* **204**:587–591 (1953).
1283. R. S. de Ropp and A. Furst, *Brain Res.* **2**:323–332 (1966).
1284. A. Bertler and E. Rosengren, *Experientia* **15**:382–384 (1959).
1285. N. E. Andén, *Acta Physiol. Scand.* **64**:197–203 (1965).
1286. R. Kuntzman, P. A. Shore, D. Bogdanski, and B. B. Brodie, *J. Neurochem.* **6**:226–232 (1961).
1287. P. Holtz and E. Westermann, *Naturwiss.* **42**:647–648 (1955).
1288. G. R. Pscheidt and B. Haber, *J. Neurochem.* **12**:613–618 (1965).
1289. S. E. Smith, R. S. Stacey, and I. M. Young, *Biochem. Pharmacol.* **8**:32 (1961).
1290. G. B. West, *J. Pharmacy, Pharmacol. Suppl.* **10**:92T–97T (1958).
1291. S. Schanberg and N. J. Giarman, *Biochim. Biophys. Acta* **41**:556–558 (1960).
1292. H. Blaschko and T. L. Chrusciel, *J. Physiol. (London)* **151**:272–284 (1960).
1293. W. Lovenberg, H. Weissbach, and S. Udenfriend, *J. Biol. Chem.* **237**:89–93 (1962).
1294. W. Sacks, *J. Appl. Physiol.* **16**:1050–1054 (1961).
1295. K. Freter, H. Weissbach, S. Udenfriend, and N. Witkop, *Proc. Soc. Exptl. Biol. Med.* **94**:725–728 (1957).
1296. S. Udenfriend, D. F. Bogdanski, and H. Weissbach, *in Metabolism of the Nervous System* (D. Richter, ed.), pp. 566–577, Pergamon Press, London, New York, Paris, and Los Angeles (1957).
1297. M. K. Paasonen, P. D. McLean, and N. J. Giarman, *J. Neurochem.* **1**:326–333 (1957).
1298. G. Rodriguez de Lores Arnaiz and E. de Robertis, *J. Neurochem.* **11**:213–218 (1964).
1299. S. E. Smith, R. S. Stacey, and I. M. Young, *Proc. Intern. Pharm. Meeting, Stockholm, 1961*, **8**:101–105 (1962).
1300. A. Tissari, *Biochem. Pharmacol.* Suppl. 39, **12**: (1963).
1301. G. Pepeu and N. J. Giarman, *J. Gen. Physiol.* **45**:575–583 (1962).
1302. F. Chatagner and B. Bergeret, *Bull. Soc. Chim. Biol.* **38**:1159–1163 (1956).
1303. J. G. Jacobsen, L. L. Thomas, and L. H. Smith, Jr., *Biochim. Biophys. Acta* **85**:103–116 (1964).
1304. R. S. Piha and H. Saukkonen, *Suomen Kemistilehti* **B39**:112–114 (1966).
1305. J. D. Wilson, K. D. Gibson, and S. Udenfriend, *J. Biol. Chem.* **235**:3539–3543 (1960).
1306. A. A. Abdel-Latif and L. G. Abood, *Neurochem.* **13**:1189–1196 (1966).
1307. R. S. de Ropp, *J. Neurochem.* **14**:693–694 (1967).
1308. J. Dancis, J. Hutzler, and M. Levitz, *Biochim. Biophys. Acta* **52**:60–64 (1961).
1309. Y. Kasé, M. Kataoka, and T. Miyata, *Life Sci.* **6**:2427–2431 (1967).
1310. F. Schapira, *Bull. Soc. Chim. Biol.* **43**:1357–1365 (1961).
1311. P. Christen, U. Rensing, A. Schmidt, and F. Leuthardt, *Helv. Chim. Acta* **49**:1872–1875 (1966).
1312. C. J. Foxwell, E. J. Cran, and D. N. Baron, *Biochem. J.* **100**:44P–45P (1966).
1313. A. V. Palladin and N. M. Polyakova, *Ukr. Biokhim. Zh.* **21**:341–349 (1949).
1314. M. K. Johnson, *Biochem. J.* **79**:24P (1961).
1315. U. Rensing, A. Schmidt, and F. Leuthardt, *Hoppe-Seyler's Z. Physiol. Chem.* **348**:921–928 (1967).
1316. D. E. Comb and S. Roseman, *J. Biol. Chem.* **235**:2529–2537 (1960).
1317. P. Brunetti and S. Roseman, *J. Biol. Chem.* **237**:2447–2453 (1962).
1318. S. Ochoa, *Adv. Enzymol.* **15**:183–270 (1954).
1319. P. A. Srere, *J. Biol. Chem.* **234**:2544–2547 (1959).
1320. M. Wollemann, *Acta Physiol. Acad. Sci. Hung.* **10**:171–183 (1956).
1321. S. Tuček, *J. Neurochem.* **14**:519–529 (1967).
1322. S. Tuček, *Biochem. J.* **104**:749–756 (1967).
1323. H. van Goor, *Enzymologia* **8**:113–128 (1940).

1324. W. Ashby and D. V. Chan, *J. Biol. Chem.* **151**:521–527 (1943).
1325. W. Ashby, *Proc. Soc. Exptl. Biol. Med.* **67**:17–22 (1948).
1326. N. A. Verzhbinskaya, *Izvest. Akad. Nauk SSSR, Ser. Biol.* **1949**:598–607 (CA **44**, 2129e; 1950).
1327. E. Yu. Chenykaeva, *Fiziol. Zh. SSSR* **40**:70–75 (1954) (CA **48**, 6534d; 1954).
1328. M. R. Klee and M. Liefländer, *Hoppe-Seyler's Z. Physiol. Chem.* **341**:143–145 (1965).
1329. W. Ashby and D. V. Chan, *J. Biol. Chem.* **156**:323–329 (1944).
1330. W. Ashby and D. V. Chan, *J. Biol. Chem.* **156**:331–341 (1944).
1331. W. Ashby and G. D. Weickhardt, *J. Nervous Mental Disease* **106**:540–549 (1947).
1332. T. Nishimura, H. Tanimukai, and K. Nishinuma, *J. Neurochem.* **10**:257–261 (1963).
1333. R. Karler and D. M. Woodbury, *Biochem. J.* **75**:538–543 (1960).
1334. E. Giacobini, *J. Neurochem.* **9**:169–177 (1962).
1335. W. D. Gray, C. E. Rauh, and R. W. Shanahan, *Biochem. Pharmacol.* **8**:307–316 (1961).
1336. W. Ashby and E. M. Schuster, *J. Biol. Chem.* **184**:109 (1950).
1337. G. Millichap, *Proc. Soc. Exptl. Biol. Med.* **96**:125–129 (1957).
1338. G. Gerov, T. Radev, and R. Georgleva, *Izv. Inst. Biol. Patol. Razmnozhavaneto Selskostopansk. Zhivotni, Akad. Selskostopansk, Nauki Bulgar.* **4**:193–201 (1963) (CA **62**, 13583b; 1965).
1339. Z. D. Pigareva, *Dokl. Akad. Nauk. SSSR.* **60**:185–189 (1948) (CA **42**, 6388g; 1948).
1340. D. Tăutu and N. Voiculet, *Comm. Acad. Rep. Populare Romine* **8**:233–239 (1958) (CA **53**, 2408a; 1959).
1341. D. A. Fisher, *Proc. Soc. Exptl. Biol. Med.* **107**:359–363 (1961).
1342. J. A. Shepherd and G. Kalnitsky, *J. Biol. Chem.* **207**:605–611 (1954).
1343. R. V. Coxon, *Bull. Soc. Chim. Belges* **65**:68–75 (1956).
1344. T. Wood, *Biochem. J.* **91**:453–560 (1964).
1345. D. B. Hope, *Federation Proc.* **18**:249 (1959).
1346. H. Shimizu, Y. Kakimoto, and I. Sano, *J. Neurochem.* **13**:65–73 (1966).
1347. R. W. Schayer, *Am. J. Physiol.* **202**:66–72 (1962).
1348. S. Linberg, S. E. Lindell, and H. Westling, *Acta Obst. Gynec. Scandinav.* Suppl. 1, **42**: 49–54 (1963).
1349. J. R. Stern, A. del Campillo, and I. Raw, *J. Biol. Chem.* **218**:971–983 (1946).
1350. J. Brüggemann, K. Schlossmann, M. Merkenschlager, and M. Waldschmidt, *Biochem. Z.* **335**:392–399 (1962).
1351. H. Gibian, *Hoppe-Seyler's Z. Physiol. Chem.* **291**:6–13 (1952).
1352. J. R. Tata, *Proc. Soc. Exptl. Biol. Med.* **95**:362–364 (1957).
1353. S. Lissitzky, *Bull. Soc. Chim. Biol.* **42**:1187–1206 (1960).
1354. S. Lissitzky and R. Michel, *in Hoppe-Seyler/Thierfelder Handbuch der physiologisch- und pathologisch-chemischen Analyse* (K. Lang and E. Lehnartz, eds.), Vol. 6C, pp. 700–706, Springer Verlag, Berlin, Heidelberg, and New York (1966).
1355. G. Pfleiderer, *in Hoppe-Seyler/Thierfelder, Handbuch der physiologisch- und pathologisch-chemischen Analyse* (K. Lang and E. Lehnartz, eds.), Vol. 6C, pp. 676–681, Springer Verlag, Berlin, Heidelberg, and New York (1966).
1356. A. S. Keston, *J. Biol. Chem.* **239**:3241–3251 (1964).
1357. J. M. Bailey, P. G. Pentchev, and J. Woo, *Biochim. Biophys. Acta* **94**:124–129 (1965).
1358. P. Oesper and O. Meyerhof, *Arch. Biochem. Biophys.* **27**:223–233 (1950).
1359. P. Faulkner and J. H. Quastel, *Nature* **177**:1216-1218 (1956).
1360. T. N. Pattabiraman and B. K. Bachhawat, *Biochim. Biophys. Acta* **54**: 273–283 (1961).
1361. L. F. Leloir and C. E. Cardini, *Biochim. Biophys. Acta* **20**:33–42 (1956).
1362. H. L. Krueskemper, E. Forchielli, and H. J. Ringold, *Steroids* **3**:295–309 (1964).
1363. F. Lipmann, *in Metabolism of the Nervous System* (D. Richter, ed.), pp. 329–340, Pergamon Press, London, New York, Paris, and Los Angeles (1957).

1364. M. Satake, Y. Takahashi, K. Mase, and K. Ogata, *J. Biochem.* (*Japan*) **57**:184–191 (1965).
1365. A. L. Rubin and K. H. Stenzel, *Proc. Natl. Acad. Sci. U.S.* **53**:963–968 (1965).
1366. Y. Takahashi and S. Abe, *Experientia* **19**:186–187 (1963).
1367. R. S. Piha, R. K. Äiras, and L. I. Ääri, *Suomen Kemistilehti* **B39**:204–208 (1966).
1368. M. R. V. Murthy and D. A. Rappoport, *Biochim. Biophys. Acta* **95**:121–131 (1965).
1369. M. Wender and M. Hierowski, in *Progress in Brain Research* (D. Purpura and J. P. Schadé, eds.), Vol. 4, pp. 273–280, Elsevier, Amsterdam, London, and New York (1964).
1370. D. H. Clouet and D. Richter, *J. Neurochem.* **3**:219–229 (1959).
1371. R. Rendi and T. Hultin, *Exptl. Cell Res.* **19**:253–266 (1960).
1372. M. Satake, K. Mase, Y. Takahashi, and K. Ogata, *Biochim. Biophys. Acta* **41**:366–367 (1960).
1373. G. Acs, A. Neidle, and H. Waelsch, *Biochim. Biophys. Acta* **50**:403–404 (1961).
1374. C. E. Zomzely, S. Roberts, and D. Rapaport, *J. Neurochem.* **11**:567–582 (1964).
1375. S. Roberts and B. S. Morelos, *J. Neurochem.* **12**:373–387 (1965).
1376. S. Roberts and C. E. Zomzely, in *Protides of the Biological Fluids, Proc. 13th Coll.* (H. Peeters, ed.), pp. 91–102, Elsevier, Amsterdam, London, and New York (1965).
1377. D. H. Clouet, M. Ratner, and N. Williams, *Biochim. Biophys. Acta* **123**:142–150 (1966).
1378. A. T. Campagnoni and H. R. Mahler, *Biochemistry* **6**:956–967 (1967).
1379. M. K. Campbell, *Dissertation Abstr.* **B27**:4274 (1967).
1380. A. von der Decken, in *Techniques in Protein Biosynthesis* (P. N. Campbell and J. R. Sargent, eds.), Vol. 1, pp. 64–131, Academic Press, New York, and London (1967).
1381. I. D. Herriman and G. D. Hunter, *J. Neurochem.* **12**:937–947 (1965).
1382. M. Satake, Y. Takahashi, K. Mase, and K. Ogata, *J. Biochem.* (*Japan*) **56**:504–511 (1964).
1383. C. S. Bondy and S. V. Perry, *J. Neurochem.* **10**:603–609 (1963).
1384. K. H. Stenzel, R. F. Aronson, and A. L. Rubin, *Biochemistry* **5**:930–936 (1966).
1385. K. Susuki, S. R. Korey, and R. D. Terry, *J. Neurochem.* **11**:403–412 (1964).
1386. S. Furst, A. Lajtha, and H. Waelsch, *J. Neurochem.* **2**:216–225 (1958).
1387. M. R. V. Murthy and D. A. Rappoport, *Biochim. Biophys. Acta* **95**:132–145 (1965).
1388. S. Yamagami, R. R. Fritz, and D. A. Rappoport, *Biochim. Biophys. Acta* **129**:532–547 (1966).
1389. D. H. Adams and L. Lim, *Biochem. J.* **99**:261–265 (1966).
1390. L. Lim and D. H. Adams, *Biochem. J.* **104**:229–238 (1967).
1391. G. Feuer and M. Wollemann, *Acta Physiol. Acad. Sci. Hung.* **5**:253–255 (1954).
1392. G. Feuer and M. Wollemann, *Acta Physiol. Acad Sci. Hung.* **7**:343–359 (1955).
1393. G. A. Rao, I. A. Hansen, and B. K. Bachhawat, *J. Sci. Ind. Res.* (*India*) **20C**:284 (1961).
1394. E. M. Gál, *Biochem. Biophys. Res. Commun.* **9**:563–568 (1962).
1395. J. Schuberth, *Biochim. Biophys. Acta* **98**:1–7 (1965).
1396. D. S. Beattie and R. E. Basford, *J. Biol. Chem.* **241**:1412–1418 (1966).
1397. D. S. Beattie and R. E. Basford, *J. Biol. Chem.* **241**:1419–1423 (1966).
1398. E. A. Hosein, M. Emblem, and J. N. Smoly, *Arch. Biochem. Biophys.* **119**:29–35 (1967).
1399. M. Wollemann, *Acta Physiol. Acad. Sci. Hung.* **16**:153–154 (1959).
1400. J. F. Mead, *Proc. Robert A. Welch Found. Conf. Chem. Res.* **5**:359–362 (1961) (CA **63**, 4772h; 1965).
1401. J. D. Robinson, R. M. Bradley and R. O. Brady, *J. Biol. Chem.* **238**:528–532 (1963).
1402. E. Grossi, P. Paoletti, and R. Paoletti, *Arch. Intern. Physiol. Biochim.* **66**:564–572 (1958).
1403. A. Lajtha, in *Hoppe-Seyler/Thierfelder, Handbuch der physiologisch- und pathologisch-chemischen Analyse* (K. Lang and E. Lehnartz, eds.), Vol. 6C, pp. 740–744, Springer Verlag, Berlin, Heidelberg, and New York (1966).

1404. Ch. Wu, *Comp. Biochem. Physiol.* **8**:335–351 (1963).
1405. M. Schon, N. Grossowicz, A. Lajtha, and H. Waelsch, *Nature* **167**:818–819 (1951).
1406. D. Rudnick, P. Mela, and H. Waelsch, *Nature* **172**:253–254 (1953).
1407. A. Lajtha, P. Mela, and H. Waelsch, *J. Biol. Chem.* **205**:553–564 (1953).
1408. W. H. Elliott, in *Methods in Enzymology* (S. P. Colowick and N. O. Kaplan, eds.), Vol. 2, pp. 337–342, Academic Press, New York (1955).
1409. V. Pamiljans, P. R. Krishnaswamy, G. Dumville, and A. Meister, *Biochemistry* **1**:153–158 (1962).
1410. K. Schnackerz and L. Jaenicke, *Hoppe-Seyler's Z. Physiol. Chem.* **347**:127–144 (1966).
1411. V. G. Kolpakov, *Genetika* **1966**:94–97 (CA **65**, 19070d; 1966).
1412. V. P. Wellner and A. Meister, *Biochemistry* **5**:872–879 (1966).
1413. O. Sellinger, *Biochim. Biophys. Acta* **132**:514–516 (1967).
1414. J. Greenberg and N. Lichtenstein, *J. Biol. Chem.* **234**:2337–2339 (1959).
1415. H. M. Kagan and A. Meister, *Biochemistry* **5**:725–732 (1966).
1416. F. Leuthardt and M. Staehelin, *Helv. Physiol. Pharmacol. Acta* **11**:C61–62 (1953).
1417. O. Z. Sellinger and F. R. Domer, *Experientia* **20**:686–687 (1964).
1418. S. Berl, *Biochemistry* **5**:916–922 (1966).
1419. H. Waelsch, in *Biochemistry of the Central Nervous System* (F. Brücke, ed.), pp. 37–45, Pergamon Press, London, New York, Paris, and Los Angeles (1959).
1420. Ch. Wu. *Arch. Biochem. Biophys.* **106**:394–401 (1964).
1421. V. A. Kheruvimova, *Dokl. Akad. Nauk SSSR* **136**:968–970 (1961) (CA **55**, 15681e; (1961).
1422. D. Rudnick and H. Waelsch, in *Biochemistry of the Developing Nervous System* (H. Waelsch, ed.), pp. 335–338, Academic Press, New York (1955).
1423. R. A. Felicioli, *Boll. Soc. Ital. Biol. Sper.* **41**:168–171 (1965).
1424. S. Berl, G. Takagaki, D. D. Clarke, and H. Waelsch, *J. Biol. Chem.* **237**:2570–2573 (1962).
1425. C. A. Rossi, S. Berl, D. D. Clarke, D. P. Purpura, and H. Waelsch, *Life Sci.* **1**:533–539 (1962).
1426. H. Waelsch, S. Berl, C. A. Rossi, D. D. Clarke, and D. P. Purpura, *J. Neurochem.* **11**:717–728 (1964).
1427. R. A. Felicioli, F. Gabrielli, and C. A. Rossi, *Life Sci.* **6**:133–143 (1967).
1428. C. Elison and H. W. Elliott, *Biochem. Pharmacol.* **12**:1363–1366 (1963).
1429. M. Knecht, *Naturwiss.* **53**:85 (1966).
1430. O. H. Lowry, *J. Histochem. Cytochem.* **1**:420–428 (1953).
1431. H. Hydén, *Nature* **184**:433–435 (1959).
1432. B. I. Roots and B. V. Johnston, *J. Ultrstruct. Res.* **10**:350–361 (1964).
1433. M. Satake and S. Abe, *J. Biochem. (Tokyo)* **59**:72–75 (1966).
1434. M. B. Sporn, T. Wanko, and W. Dingman, *J. Cell. Biol.* **15**:109–120 (1962).
1435. C. F. Emanuel and I. L. Chaikoff, *J. Neurochem.* **5**:236–244 (1960).
1436. T. Kato and M. Kurokawa, *J. Cell. Biol.* **32**:649–662 (1967).
1437. J. Altman and S. L. Chorover, *J. Physiol. (London)* **169**:770–780 (1963).
1438. T. M. Brody and S. A. Bain, *J. Biol. Chem.* **195**:685–689 (1952).
1439. V. P. Whittaker, in *Regional Neurochemistry* (S. S. Kety and J. Elkes, eds.), pp. 259–263, Pergamon Press, Oxford, London, and New York (1961).
1440. E. G. Gray and V. P. Whittaker, *J. Anat. (London)* **96**:79–88 (1962).
1441. E. de Robertis, A. Pellegrino de Iraldi, G. Rodriguez de Lores Arnaiz, and C. J. Gomez, *J. Biophys. Biochem. Cytol.* **9**:229–235 (1961).
1442. E. De Robertis and H. S. Bennett, *J. Biophys. Biochem. Cytol.* **1**:47–58 (1955).
1443. V. P. Whittaker, *Progr. Biophys. Mol. Biol.* **15**:39–96 (1965).

1444. H. Fernández-Morán, in *Metabolism of the Nervous System* (D. Richter, ed.), pp. 1–34, Pergamon Press, London (1957).
1445. H. McIlwain, *Biochemistry and the Central Nervous System*, 3rd ed., p. 244, J. A. Churchill, London (1966).
1446. L. Autilio, W. T. Norton, and R. D. Terry, *J. Neurochem.* **11**:17–27 (1964).
1447. J. Eichberg, V. P. Whittaker, and R. M. C. Dawson, *Biochem. J.* **92**:91–100 (1964).
1448. C. W. M. Adams, A. N. Davison, and N. A. Gregson, *J. Neurochem.* **10**:383–395 (1963).
1449. R. Balázs, K. Magyar, and D. Richter, in *Comparative Neurochemistry* (D. Richter, ed.), pp. 225–248, Pergamon Press, Oxford, London, and New York (1964).
1450. Y. Machiyama, R. Balázs, and F. Julian, *Biochem. J.* **96**:68P–69P (1965).
1451. G. M. McKhann, R. W. Albers, L. Sokoloff, O. Mickelsen, and D. B. Tower, in *Inhibition in the Nervous System and γ-Aminobutyric Acid* (E. Roberts, ed.), pp. 169–181, Pergamon Press, Oxford, London, and New York (1960).
1452. G. M. McKhann and D. B. Tower, *J. Neurochem.* **7**:26–32 (1961).
1453. K. A. C. Elliott, *Brit. Med. Bull.* **21**:70–75 (1965).
1454. K. Krnjevic and S. Schwartz, *Nature* **211**:1372–1374 (1966).
1455. L. L. Iversen, E. A. Kravitz, and M. Otsuoka, *J. Physiol. (London)* **188**:21P–22P (1967).
1456. E. De Robertis, G. Rodriguez de Lores Arnaiz, and O. Z. Sellinger, *Nature* **212**:537–538 (1966).
1457. W. Feldberg and M. Vogt, *J. Physiol. (London)* **107**:372–381 (1948).
1458. C. W. M. Adams, *Neurohistochemistry*, pp. 290–297, Elsevier, Amsterdam, London, and New York (1965).
1459. B. B. Brodie and P. A. Shore, *Ann. N.Y. Acad. Sci.* **66**:631–642 (1957).
1460. N.-E. Andén, A. Dahlström, K. Fuxe, K. Larsson, L. Olson, and U. Ungerstedt, *Acta Physiol. Scand.* **67**:313–326 (1966).
1461. J. Glowinski and R. J. Baldessarini, *Pharmacol. Rev.* **18**:1201–1238 (1966).
1462. E. Rosengren, *Acta Physiol. Scand.* **49**:364–369 (1960).
1463. R. M. Marchbanks, *J. Neurochem.* **13**:1481–1493 (1966).
1464. A. Pletscher, K. F. Gey, and P. Zeller, in *Progress in Drug Research* (E. Jucker, ed.), Vol. 2, pp. 417–590, Birkhäuser Verlag, Basel, Stuttgart (1960).
1465. E. Costa and B. B. Brodie, in *Progress in Brain Research* (H. E. Himwich and W. A. Himwich, eds.), Vol. 8, pp. 168–185, Elsevier, Amsterdam, London, and New York (1964).
1466. J. Axelrod, in *Progress in Brain Research* (H. E. Himwich and W. A. Himwich, eds.), Vol. 8, pp. 81–89, Elsevier, Amsterdam, London, and New York (1964).
1467. I. J. Kopin and E. K. Gordon, *J. Pharmacol. Exptl. Therap.* **140**:207–216 (1963).
1468. J. Glowinski, L. L. Iversen, and J. Axelrod, *J. Pharmacol.* **151**:385–399 (1966).
1469. J. Axelrod, *Pharmacol. Rev.* **18**:95–113 (1966).
1470. A. Philippu, R. Pfeiffer, and H. J. Schümann, *Naunyn-Schmiedeberg's Arch. Pharmakol. Exptl. Pathol.* **257**:321 (1967).
1471. N. H. Becker, S. Goldfischer, W. Y. Shin, and A. B. Novikoff, *J. Biophys. Biochem. Cytol.* **8**:649–663 (1960).
1472. K. D. Barron and S. Sklar, *Neurology* **11**:866–875 (1961).
1473. P. J. Anderson and K. S. Song, *J. Neuropath. Exptl. Neurol.* **21**:263–283 (1962).
1474. E. Robins, in *Regional Neurochemistry* (S. S. Kety and J. Elkes, eds.), p. 213–219, Pergamon Press, Oxford, London, and New York (1961).
1475. A. Vorbrodt, *J. Histochem. Cytochem.* **9**:647–655 (1961).
1476. R. M. Torack and R. J. Barrnett, *Exptl. Neurol.* **6**:224–244 (1962).
1477. H. Koenig, in *Progress in Brain Research* (H. E. Himwich and W. A. Himwich, eds.), Vol. 8, pp. 137–141, Elsevier, Amsterdam, London, and New York (1964).
1478. L. Gordis, *J. Pediat.* **68**:638–649 (1966).

1479. O. Z. Sellinger and G. D. Rucker, *Life Sci.* **5**:163–167 (1966).
1480. A. Lavelle, *in Progress in Brain Research* (W. A. Himwich and H. E. Himwich, eds.), Vol. 9, pp. 93–96, Elsevier, Amsterdam, London, and New York (1964).
1481. L. B. Flexner, *in Biochemistry of the Developing Nervous System* (H. Waelsch, ed.), pp. 281–300, Academic Press, New York (1955).
1482. H. E. Himwich, *Brain Metabolism and Cerebral Disorders*, p. 125, Williams and Wilkons, Baltimore (1951).
1483. D. B. Tyler and A. van Harreveld, *Am. J. Physiol.* **136**:600–603 (1942).
1484. H. E. Himwich and J. F. Fazekas, *Am. J. Physiol.* **132**:454–459 (1941).
1485. F. E. Samson, W. H. Balfour, and R. J. Jacobs, *Am. J. Physiol.* **199**:693–696 (1960).
1486. R. L. Friede, *J. Neurochem.* **4**:101–111 (1959).
1487. R. L. Friede, *Arch. Psychiat. Z. Neurol.* **196**:196–204 (1957).
1488. R. L. Friede, *J. Neurochem.* **4**:111–123 (1959).
1489. R. L. Friede, *J. Neurochem.* **4**:290–303 (1959).
1490. R. L. Friede, *J. Neurochem.* **5**:156–171 (1960).
1491. R. L. Friede, *J. Neurochem.* **6**:190–199 (1960).
1492. J. Knolle, *Z. Zellforsch.* **50**:183–231 (1959).
1493. M. Gordon, P. Hawley, and M. K. Gaitonde, *Neurology*, Suppl. 1, **8**:75–76 (1958).
1494. F. Dickens and G. E. Glock, *Biochem. J.* **50**:81–95 (1951).
1495. D. Di Pietro and S. Weinhouse, *Arch. Biochem. Biophys.* **80**:268–282 (1959).
1496. D. B. Tower, *J. Neurochem.* **3**:185–205 (1959).
1497. A. Liuzzi and P. U. Angeletti, *Experientia* **20**:512–514 (1964).
1498. A. Geiger and J. Magnes, *Biochem. J.* **33**:866–876 (1939).
1499. D. B. Tyler, *Proc. Soc. Exptl. Biol. Med.* **49**:537–539 (1942).
1500. H. E. Himwich, A. O. Bernstein, H. Herrlich, A. Chesler, and J. F. Fazekas, *Am. J. Physiol*, **135**:387–391 (1941/1942).
1501. P. Mandel, R. Bieth, and J. D. Weill, *in Metabolism of the Nervous System* (D. Richter, ed.), pp. 291–296, Pergamon Press, London, New York, and Paris (1957).
1502. A. Chesler and H. E. Himwich, *Am. J. Physiol.* **142**:544–549 (1944).
1503. A. Chesler and H. E. Himwich, *Arch. Biochem. Biophys.* **2**:173–181 (1943).
1504. R. L. Friede, *Hoppe-Seyler's Z. Physiol. Chem.* **310**:4–7 (1958).
1505. E. Penhoet, T. Rajkumar, and W. J. Rutter, *Proc. Natl. Acad. Sci. U.S.* **56**:1275–1282 (1966).
1506. E. Roberts and K. Kuriyama, *Brain Res.* **8**:1–35 (1968).
1507. I. B. Fritz, *Advan. Lipid Res.* **1**:285–334 (1963).
1508. C. Bode and M. Klingenberg, *Biochem. Z.* **341**:271–299 (1965).
1509. K. P. DuBois and V. R. Potter, *J. Biol. Chem.* **150**:185–195 (1943).
1510. M. Crevier, *Can. J. Biochem. Physiol.* **36**:275–288 (1958).
1511. Y. Ishii, *Arch. Histol. Japan* **12**:612–637 (1957).
1512. S. I. Zacks, *Anat. Rec.***118**:509–537 (1954).
1513. A. Bonichon, *Ann. Histochim.* **3**:85–93 (1958).
1514. G. Filogamo, *Compt. Rend. Ass. Anat.* **46**:251–255 (1960).
1515. B. S. Wenger, *Federation Proc.* **10**:268–269 (1951).
1516. A. G. Karczmar, *in Heffter-Heubner Handbook of Experimental Pharmacology* (O. Eichler, A. Farah, and G. B. Koelle, eds.), Vol. 15, p. 129, Springer Verlag, Berlin, Heidelberg, and New York (1963).
1517. M. H. Aprison, *Recent Adv. Biol. Psychiat.* **4**:133–146 (1961).
1518. R. Kato, *J. Neurochem.* **5**:202 (1960).
1519. W. Liese, *Z. Mikroskop. Anat. Forsch.* **70**:48–61 (1963).
1520. D. G. McKay, E. C. Adams, A. T. Hertig, and S. Danziger, *Anat. Rec.* **122**:125–151 (1955).

1521. F. Rossi and E. Reale, *Acta Anat.* **30**:656–681 (1957).
1522. F. Moog, *Biol. Bull.* **86**:51–80 (1944).
1523. P. J. Anderson and S. K. Song, *J. Neuropath. Exptl. Neurol.* **21**:263–283 (1962).
1524. P. Meyer, *Acta Neurol. Scand.* **39**:123–138 (1963).
1525. K. D. Barron and R. Boshes, *J. Histochem. Cytochem.* **9**:455–457 (1961).
1526. H. B. Tewari and G. H. Bourne, *Acta Histochem.* **17**:197–207 (1964).
1527. D. P. Purpura, R. J. Shofer, E. M. Housepian, and C. R. Noback, *in Progress in Brain Research* Vol. 4, pp. 187–221 (D. P. Purpura and J. P. Schadé, eds.), Elsevier, Amsterdam, London, and New York (1964).
1528. S. Berl and D. P. Purpura, *J. Neurochem.* **13**:293–304 (1966).
1529. S. C. Bondy and H. Waelsch, *Life Sci.* **3**:633–669 (1964).
1530. C. W. M. Adams and A. N. Davison, *J. Neurochem.* **4**:282–289 (1959).
1531. C. W. M. Adams and A. N. Davison, *J. Neurochem.* **5**:293–296 (1960).
1532. D. Richter, *in Regional Development of the Brain in Early Life*, (A. Minkowski, ed.), pp. 137–156, Blackwell Scientific Publications, Oxford, Edinburgh (1967).

APPENDIX

Compiled by W. Himwich

TABLE I

Body and Brain Weights and Moisture Content of the Brain

	Rat				Human		
Age	Body weight (g)	Brain weight (g)	Moisture (%)		Age (years)	Brain weight (g)	Moisture (%)
− 17 day	0.89	0.08	90.7		21–30[a]	1394	76.8
− 18 day	1.81	0.08	91.4		31–40	1375	76.8
− 19 day	2.69	0.13	88.9		41–50	1358	77.1
− 20 day	2.38	0.14	89.1		51–60	1345	77.2
− 21 day	3.76	0.17	90.5		61–70	1310	77.5
Newborn	6.20	0.27	89.5		71–74[a]	1284	78.3
1 day	6.01	0.23	89.5		75–80	1259	78.3
2 day	6.72	0.27	90.1		81–84	1254	78.0
3 day	7.54	0.32	89.9		85–90	1232	78.0
4 day	7.99	0.40	89.8				
5 day	9.65	0.40	90.2				
6 day	12.0	0.52	89.8				
7 day	13.0	0.60	90.1				
8 day	13.0	0.60	90.4				
9 day	18.9	0.86	90.4				
10 day	18.7	0.84	89.7				
12 day	21.5	1.04	89.3				
25 day	38.6	1.29	86.5				
Adult	312	1.69	83.2				

[a] F. W. Appel and E. M. Appel, Intracranial variation in the weight of the human brain, *Human Biol.* **14**:48–68 (1942). M. Bürger, Die chemische Biomorphose des menschlichen Gehirns, *Abhandl. Saechs. Akad. Wiss. Leipzig, Math.-Naturw. Kl.* **45**:5–62 (1957). The reference to Appel and Appel concerns data for ages 71–80. The reference to Bürger concerns data for ages 21–69 and 81–90 and also for all the human moisture values.

TABLE II

Body and Brain Weights and Moisture Content of the Brain[a]

Age	Cat			Dog			Guinea pig			Mouse			Rabbit		
	Body weight (g)	Brain weight (g)	Moisture (%)	Body weight (g)	Brain weight (g)	Moisture (%)	Body weight (g)	Brain weight (g)	Moisture (%)	Body weight (g)	Brain weight (g)	Moisture (%)	Body weight (g)	Brain weight (g)	Moisture (%)
Newborn	106	4.87	91.8	338	8.3	92.1	91.0	2.67		1.27	0.095	88.4	43.9	1.52	95.0
1 day	142	6.45	91.1				73.0	2.53	83.5	1.61	0.098	88.4	44.4	1.50	88.4
7 day	140	5.85	90.7	410	13.2	91.2	141	2.99		3.88	0.25	88.7	97.2	3.01	87.2
14 day	287	11.2	89.4	710	23.2	90.9	165	3.18	82.6[b]	6.89	0.38	85.2	166	4.34	84.3
21 day	399	15.1	88.3	851	29.3	90.2	257	3.29	82.2[c]	13.7	0.39	82.9	321	5.33	82.8
30 day	362	16.5	88.1	1330	36.6	90.2	341	3.59	81.7	21.8	0.37	82.3	527	5.93	81.1
60 day	490	18.8	85.6	2422	57.4	86.8				24.9	0.43		1215	7.04	79.6
90 day	1264	22.2	83.5	6356	65.4	85.8				34.5	0.40		1254	7.16	
120 day	1351	22.5	82.9	6430	71.8	84.9				—	0.43		2559	7.38	
Adult	2611	25.5	82.3	10,139	70.1	84.6	1143	4.22	79.5	41.7	0.44	80.3	2610	7.04	77.0

[a] References: J. Graves and H. E. Himwich, Age and the water content of rabbit brain parts, *Am. J. Physiol.* **180**:205 (1955). H. C. Agrawal, J. M. Davis, and W. A. Himwich, Water content of dog brain parts in relation to maturation of the brain, *Am. J. Physiol.* **215**:846 (1968).
[b] This value is for 10 day.
[c] This value is for 20 day.

SUBJECT INDEX